Unless Recalled Earlier
**DATE DUE**

# Progress in Mathematics
Volume 90

*Series Editors*
Hyman Bass
Joseph Oesterlé
Alan Weinstein

V. Srinivas

# Algebraic K-Theory
*Second Edition*

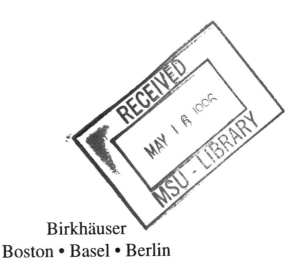

Birkhäuser
Boston • Basel • Berlin

V. Srinivas
School of Mathematics
Tata Institute of Fundamental Research
Bombay, India

QA
612.33
.S67
1996

**Library of Congress Cataloging-in-Publication Data**

Srinivas, V.
 Algebraic K-theory / V. Srinivas. -- 2nd ed.
   p.  cm. -- (Progress in mathematics ; v. 90)
  Includes bibliographical references.
  ISBN 0-8176-3702-8
  1. K-theory. I. Title.  II. Series: Progress in mathematics
QA612.33.S67   1993                    93-9417
512'.55--dc20                          CIP

Printed on acid-free paper
© Birkhäuser Boston 1996                 *Birkhäuser*

Copyright is not claimed for works of U.S. Government employees.
All rights reserved. No part of this publication may be reproduced, stored in a retrieval system, or transmitted, in any form or by any means, electronic, mechanical, photocopying, recording, or otherwise, without prior permission of the copyright owner.

Permission to photocopy for internal or personal use of specific clients is granted by Birkhäuser Boston for libraries and other users registered with the Copyright Clearance Center (CCC), provided that the base fee of $6.00 per copy, plus $0.20 per page is paid directly to CCC, 222 Rosewood Drive, Danvers, MA 01923, U.S.A. Special requests should be addressed directly to Birkhäuser Boston, 675 Massachusetts Avenue, Cambridge, MA 02139, U.S.A.

ISBN 0-8176-3702-8
ISBN 3-7643-3702-8
Layout and typesetting by Martin Stock, Cambridge, MA
Printed and bound by Quinn-Woodbine, Woodbine, NJ
Printed in the United States of America

9 8 7 6 5 4 3 2 1

*Dedicated to my parents.*

# Contents

Preface to the First Edition .............................. xi

Preface to the Second Edition .......................... xvii

1. "Classical" $K$-Theory ........................................... 1

   Review of parts of Milnor's book: definitions of $K_0$, $K_1$, $K_2$ of rings; computation of $K_1$ of a noncommutative local ring; definition of symbols; statement of Matsumoto's theorem; examples of symbols (norm residue symbol, Galois symbol, differential symbol); presentation for $K_2$ of a commutative local ring.

2. The Plus Construction ....................................... 18

   The plus construction; computation that $\pi_2(BGL(R)^+) \cong K_2(R)$; $H$-space structure of $BGL(R)^+$ and products in $K$-theory (following Loday); statement of Quillen's theorem on $K_i$ of a finite field.

3. The Classifying Space of a Small Category .................. 31

   Simplicial sets; geometric realization; classifying space of a small category; elementary theorems about classifying spaces (compatibility with products, natural transformations give homotopies, adjoint functors give homotopy inverses, filtering categories are contractible); example of the classifying space of a discrete group as the classifying space of the category with one object, whose endomorphisms equal the group.

4. Exact Categories and Quillen's $Q$-Construction ............. 38

   Exact categories; admissible mono- and epimorphisms; definition of $Q\mathcal{C}$ for a small exact category $\mathcal{C}$; definition of $K_i(\mathcal{C})$ for a small exact category $\mathcal{C}$; statements of theorems about $K_i$ ($K_0$ agrees with that defined "classically", theorem on exact sequences of functors, resolution theorem, dévissage theorem, localization theorem); "bare hands" construction of a homomorphism $K_0(\mathcal{C}) \to \pi_1 BQ\mathcal{C}$.

5. **The $K$-Theory of Rings and Schemes** .................... 46

   Statement of the theorem comparing the definitions of $K_i$ of a ring using the plus and $Q$ constructions; definition of $G_i(A)$ as $K_i$ of finitely generated $A$-modules, for Noetherian rings $A$; computations of $G_i(A[t])$, $G_i(A[t, t^{-1}])$ for Noetherian $A$, and hence $K_i(A[t])$, $K_i(A[t, t^{-1}])$ for Noetherian regular $A$; definition of $K_i(X)$, $G_i(X)$ for schemes, using vector bundles and coherent sheaves, respectively; construction of direct image and inverse image maps for $K_i$ and $G_i$ of Noetherian schemes for morphisms satisfying appropriate conditions; action of $K_0$ on $K_i$, $G_i$ and projection formulas; $K_i$, $G_i$ commute with filtered direct limits; localization for $G_i$ of a closed subscheme and the open complement; Mayer–Vietoris for $G_i$; $G_i$ of affine and projective space bundles; filtration by codimension of support and the $BGQ$ spectral sequence; Gersten's conjecture for power series rings, and semilocal rings of finite sets of points on a smooth variety over an infinite field; Bloch's formula; $K_i$ of projective bundles, of $\mathbf{P}^1$ over a noncommutative ring, and of Severi–Brauer schemes.

6. **Proofs of the Theorems of Chapter 4** ........................ 89

   Proofs of the following theorems: $\pi_1 BQ\mathcal{C}) \cong K_0(\mathcal{C})$; Theorems A and B of Quillen; the theorem on exact sequences of functors; the resolution theorem; the dévissage theorem; the localization theorem.

7. **Comparison of the Plus and $Q$-Constructions** ............... 126

   Monoidal categories; localization of the action of a monoidal category on a small category; computation of the homology of the classifying space of a localized category; the $\mathcal{S}^{-1}\mathcal{S}$ construction, viewed as a "functorial" version of the plus construction; construction of the homotopy equivalence $\mathcal{S}^{-1}\mathcal{S} \to \Omega BQ\mathcal{C}$ for any exact category $\mathcal{C}$ in which all exact sequences are split, where $\mathcal{S}$ is the category of isomorphisms in $\mathcal{C}$; corollary that the plus and $Q$-constructions yield the same $K$-groups for projective modules over a ring.

8. **The Merkurjev–Suslin Theorem** ........................... 145

   The Galois symbol; statement of the Merkurjev–Suslin theorem; Hilbert's Theorem 90 for $K_2$; proof of the Merkurjev–Suslin theorem; torsion in $K_2$; torsion in $CH^2$.

9. **Localization for Singular Varieties** ........................ 194

   Quillen's localization theorem for the complement of an effective Cartier divisor in a quasi-projective scheme with affine complement; discussion of naturality of this sequence (after Swan); proof of the "Fundamental Theorem" on $K_i$ of polynomial and Laurent polynomial rings; Levine's localization theorem; computation of $K_0$ of the category of modules of finite length and finite projective dimension over the local ring of a normal surface singularity, in terms of $H^1(\mathcal{K}_2)$ of the resolution; computation of this $K_0$ for quotient singularities; Chow groups of surfaces with quotient singularities.

Contents                                                          ix

Appendix A. Results from Topology .......................... 230

(A.1) Compactly generated spaces; (A.2)–(A.6) Homotopy groups, Hurewicz theorems; (A.7) Products; (A.8)–(A.12) CW-complexes, Whitehead theorem, Milnor's theorem on the homotopy type of mapping spaces, comparison of singular and cellular homology and cohomology; (A.13)–(A.15) Local coefficients, homology and cohomology with local coefficients for CW-complexes via cellular chains; (A.16) Obstruction theory for maps and homotopies between CW-complexes (which may not be simply connected); (A.17)–(A.22) Fibrations, the homotopy lifting property, long exact homotopy sequence, fiber homotopy equivalence, fibrations over a contractible base are fiber homotopy equivalent to a product, local coefficient systems of the homology and cohomology groups of the fibers of a fibration; (A.23)–(A.26) Leray–Serre spectral sequence for homology and cohomology of a fibration over a CW-complex; (A.27) Homotopy fibers; (A.28) Spectral sequences for the homology and cohomology of a covering space; (A.29)–(A.35) Quasi fibrations (some results of Dold and Thom); (A.36)–(A.42) NDR-pairs and cofibrations (following Steenrod); (A.43)–(A.47) $H$-spaces; (A.48)–(A.50) Covering spaces of simplicial sets; (A.51)–(A.54) Hurewicz and Whitehead theorems for non-simply connected $H$-spaces; (A.55) Milnor's theorem on the geometric realization of a product of simplicial sets.

Appendix B. Results from Category Theory .................... 276

Small categories; equivalences; Abelian categories; construction of the quotient of a small Abelian category by a Serre subcategory; examples of quotients; adjoint functors; filtering categories and direct limits.

Appendix C. Exact Couples ................................. 287

The spectral sequence of an exact couple; bigraded couples; elementary discussion of convergence; the BGQ spectral sequence; the spectral sequence of a filtered complex.

Appendix D. Results from Algebraic Geometry ................. 295

(D.1)–(D.14) Sheaves; (D.15)–(D.20) Schemes; (D.21)–(D.41) Some properties of schemes; (D.42)–(D.59) Coherent and quasi-coherent sheaves; (D.60)–(D.66) Cohomology and direct images of quasi-coherent and coherent sheaves; (D.67)–(D.70) Some miscellaneous topics.

Bibliography ............................................... 339

# Preface to the First Edition

These notes are based on a course of lectures I gave at the Tata Institute during 1986–87. The aim of the course was to give an introduction to higher $K$-theory, and in particular, to expose in detail the results of Quillen, contained in the following basic papers:

1. D. Quillen: *Higher Algebraic K-Theory I*, Lecture Notes in Math. No. 341, Springer-Verlag, New York (1973).
2. D. Grayson: *Higher Algebraic K-Theory II (after Daniel Quillen)*, Lecture Notes in Math. No. 551, Springer-Verlag, New York (1976).

The audience consisted of colleagues and some graduate students who were mainly algebraists and algebraic geometers, and were interested in learning about $K$-theory because of its applications to these fields. Most members of the audience had a limited background in topology. As such, one of my aims during the course was to give proofs of the topological results needed, assuming the minimum possible. The two applications (beyond Quillen's results) which are discussed also reflect the tastes of the audience (and the lecturer). In a few places, I chose not to prove results in the maximum possible generality, when I felt that the ideas behind the proofs might be obscured by technical details.

Algebraic $K$-theory is an active area of research, which has connections with algebra, algebraic geometry, topology, and number theory. Some recent interesting results in algebra and related fields proved using $K$-theoretic methods are the following:

(i) Merkurjev and Suslin's theorem on Brauer groups of fields, its generalizations due to Merkurjev and Suslin, Levine, and the work of Rost and others on Milnor's conjecture relating $K_2$ and Witt groups; these results have interesting consequences for the Chow groups of algebraic cycles modulo rational equivalence on a smooth algebraic variety.

(ii) Serre's conjecture on the vanishing of intersection multiplicities for modules over a regular local ring, proved by Gillet and Soulé (this was independently proved by Paul Roberts, using intersection theory, as developed in Fulton's book).

(iii) Levine's computation in terms of $K_1$, for the Grothendieck group of modules of finite length and finite projective dimension over the local ring of an isolated Cohen–Macaulay singular point of a variety, which leads to a new proof of the results of Dutta, Hochster, and MacLaughlin on modules of finite projective dimension with negative intersection multiplicity. Levine's results also have applications to Chow groups of singular varieties. A generalization of Levine's results has been announced by Thomason and Trobaugh, which should have similar applications.

Chapters 8 and 9 touch on the first and third "algebraic" applications mentioned above.

Lack of knowledge prevents me from giving a detailed list of results in topology and number theory, but I will mention Waldhausen's algebraic $K$-theory of spaces, which is a key ingredient in the recently announced proof by Hsiang and Madsen of the Novikov conjecture on the surgery obstruction map; the higher dimensional generalization of class field theory due to Bloch, Kato, Saito, and others; work of Bloch, Colliot-Thelene, Sansuc, and others on the torsion and cotorsion of Chow groups of varieties over number fields and local fields; and finally, results of Bloch, Beilinson, and others relating the ranks of $K$-groups and Chow groups of varieties over number fields to the orders of vanishing of $L$-functions, leading to the celebrated Beilinson conjectures. For the algebraic geometers, I should also mention Beilinson's generalization of the Hodge conjectures, which relate certain groups of transcendental cohomology classes to $K$-theory.

The work of Quillen, cited above, provides the foundation for much of this work, and forms the core of these notes.

The more algebraically minded reader may prefer, at a first reading, to read Chapters 1, 3, 4, and 5, skip the more topological Chapters 2, 6, and 7, and go on to applications of interest. The somewhat long Appendix A should, I hope, help such a reader to eventually work through the "topological" chapters.

The detailed contents of the notes are as follows (this is for the benefit of readers who may already have some acquaintance with $K$-theory, say at the level of $K_0$, and to state where we have omitted topics from Quillen's papers).

Chapter 1 contains a quick review of "classical" $K$-theory (i.e., $K_0$, $K_1$, $K_2$), mainly based on Milnor's book. In Chapter 2, Quillen's *plus* construction is given, leading to the definition $K_i(R) = \pi_i(BGL(R)^+)$ for $i \geq 1$, for the higher $K$-groups of a ring. We construct products $K_i(R) \otimes K_j(R) \to K_{i+j}(R)$, following Loday. We also show that $K_1$, $K_2$ agree with the "classical" $K_1$, $K_2$ of Chapter 1.

Chapters 3–7 contain our exposition of "Higher Algebraic $K$-Theory I", and the comparison of the *plus* and $Q$ constructions for $K_i$. Chapter 3 introduces the language of simplicial sets, and leads to the basic notion of the classifying space $B\mathcal{C}$ of a small category $\mathcal{C}$. This leads to a "dictionary" between category theory and topology, under which we have the following correspondences:

$$\begin{aligned}
\text{small category} &\longrightarrow \text{topological space (CW complex)} \\
\text{functor} &\longrightarrow \text{cellular map} \\
\text{natural transformation} &\longrightarrow \text{homotopy} \\
\text{adjoint functors} &\longrightarrow \text{homotopy inverse pair of maps} \\
\text{category with an initial} &\longrightarrow \text{contractible space} \\
\text{or final object} &
\end{aligned}$$

In Chapter 4, we give Quillen's $Q$ construction, using the $K$-groups of a small exact category $\mathcal{C}$ in terms of the homotopy groups of the classifying space of its associated Quillen category $Q\mathcal{C}$,

$$K_i(\mathcal{C}) = \pi_{i+1}(BQ\mathcal{C}).$$

We then state a number of purely "$K$-theoretic" results, contained in the first part of "Algebraic $K$-Theory I" (the computation of $K_0$, and the theorems on *exact sequences of functors, resolution, dévissage,* and *localization*). Given that $K_0$ is the same as that defined "classically", these theorems express standard facts about $K_0$. However, the extensions to the higher $K_i$ involve a lot of new machinery. As such, we postpone the proofs of these results to Chapter 6.

In Chapter 5, we apply the results of Chapter 4 to study the $K$-theory of rings and schemes (the second part of "Higher Algebraic $K$-Theory I"). If $R$ is a ring (or a Noetherian ring), let $K_i(R)$ (or $G_i(R)$, respectively) be the $K$-groups of the category of finitely generated projective $R$-modules (or of all finitely generated $R$-modules, respectively). We first prove the formulas $G_i(R[t]) \cong G_i(R)$, $G_i(R[t,t^{-1}]) \cong G_i(R) \oplus G_{i-1}(R)$. (Quillen deduces these formulas from more general results about filtered rings, which we omit; we also omit the applications to computing $K_i$ for certain division rings.)

For a (Noetherian separated) scheme $X$, define $K_i(X)$ (or $G_i(X)$, respectively) using the category of vector bundles (or coherent sheaves, respectively). We first study the groups $G_i(X)$, and in particular construct the $BGQ$-spectral sequence

$$E_1^{p,q} = \oplus_{\text{codim } x = p} K_{-p-q}(k(x)) \Longrightarrow G_{-p-q}(X)$$

using the filtration of the category of coherent sheaves given by "codimension of support." We give Quillen's proof of Gersten's conjecture for a power series ring, and for the semi-local ring obtained from a finite set of smooth points on a variety over an infinite field (Quillen proves it, more generally, assuming only that the variety is regular over a field). This is used to obtain Bloch's formula

$$CH^p(X) = H^p(X, \mathcal{K}_{p,X})$$

where $CH^p(X)$ is the Chow group of algebraic cycles of codimension $p$ on $X$ modulo rational equivalence, and $\mathcal{K}_{p,X}$ is a sheaf for the Zariski topology with stalks $K_p(\mathcal{O}_{x,X})$. This generalizes the familiar formula $Pic(X) = H^1(X, \mathcal{O}_X^*)$ for the Picard group of invertible sheaves on $X$. Chapter 5 ends with the computation of $K_i$ for a projective bundle, and for a Severi–Brauer scheme.

Chapter 6 contains the proofs of the results stated in Chapter 4. We begin by computing $K_0$ using simplicial coverings. We then prove Quillen's Theorem A, which gives a criterion for a functor to induce a homotopy equivalence on classifying spaces, and Theorem B, which identifies the homotopy fiber of such a map on classifying spaces, in certain cases. These proofs make use of various topological results proved (or discussed) in Appendix A. Theorems A and B are then used to give "algebraic" proofs of the remaining results of Chapter 4 (exactness, resolution, dévissage, and localization).

Chapter 7 is devoted to the proof that "$+ = Q$", using the notions of monoidal categories, and actions of these on other categories. A monoidal category is, roughly speaking, a (small) category $\mathcal{S}$, together with a functor $+ : \mathcal{S} \times \mathcal{S} \to \mathcal{S}$ which is "associative" and has an "identity". Thus the classifying space $B\mathcal{S}$ becomes an $H$-space ("homotopy monoid"). The basic example for us is $\mathcal{S} = Iso\mathcal{P}(R)$, the category whose objects are finitely generated projective modules, and whose arrows are all isomorphisms of projective modules; the operation $+$ is the direct sum. The set $\pi_0(B\mathcal{S})$ of path components is just the monoid of isomorphism classes of projective $R$-modules. The construction of $\mathcal{S}^{-1}\mathcal{S}$ yields an $H$-group ("group up to homotopy") with an $H$-map $B\mathcal{S} \to B\mathcal{S}^{-1}\mathcal{S}$, which is "universal up to homotopy" among such maps; on $\pi_0$, this just yields the Grothendieck group $K_0(R)$. One first shows that there is a homotopy equivalence $K_0(R) \times BGL(R)^+ \cong B\mathcal{S}^{-1}\mathcal{S}$ by computing the homology of $B\mathcal{S}^{-1}\mathcal{S}$. Then, using Quillen's extension construction, one shows that $B\mathcal{S}^{-1}\mathcal{S}$ is homotopy-equivalent to the loop space of $BQ\mathcal{P}(R)$. This yields the isomorphisms $\pi_i(BGL(R)^+) \cong \pi_{i+1}(BQ\mathcal{P}(R))$, relating the two definitions of $K_i(R)$, and in particular identifying $K_1(R)$, $K_2(R)$ with the groups of Chapter 1.

Chapter 8 gives the proof of the theorem of Merkurjev and Suslin, relating $K_2$ and the Brauer group of a field. Let $F$ be a field containing a primitive $n$-th root of unity. Then the theorem states that the natural

map (the "Galois Symbol" or "Norm Residue homomorphism") gives an isomorphism:
$$K_2(F) \otimes \mathbf{Z}/n\mathbf{Z} \cong {}_n\mathrm{Br}(F).$$
We give the proof of this theorem in detail, based on an expository article by Suslin. We omit only one step in the proof—the argument (see Prop. (8.7)(c)) using Gillet's Riemann–Roch theorem for the vanishing of certain differentials in the $BGQ$ spectral sequence, up to torsion; the proof of the Riemann–Roch theorem involves tools from $K$-theory and topology beyond the scope of these notes (e.g., the homotopy theory of simplicial sheaves). We then prove the relevant easy case of the results of Bloch and Ogus, and deduce the result that if $X$ is a smooth variety over an algebraically closed field, then the $n$-torsion subgroup ${}_nCH^2(X)$ of the Chow group of codimension-2 cycles on X is finite. We also prove Roitman's theorems on torsion-zero cycles.

Chapter 9 begins with Quillen's localization theorem for singular varieties, contained in the latter half of "Higher Algebraic $K$-Theory II", leading to the Fundamental Theorem (computation of $K_i(R[t])$, $K_i(R[t, t^{-1}])$). Next, we give a generalization of Quillen's localization theorem due to Levine, and use it to obtain a presentation for $K_0(\mathcal{C}_R)$, where $\mathcal{C}_R$ is the category of $R$-modules of finite length and finite projective dimension over the local ring $R$ of a normal surface singularity. This is used to show that quotient singularities do not contribute to the Chow group of zero cycles on a normal surface.

Appendix A discusses the topological results needed in the main text. We sketch proofs for standard results when these are not too long, and give references in other cases. We also give more detailed proofs for the results of Dold and Thom on quasi fibrations, and of some results on $H$-spaces, and on simplicial sets, which are perhaps less standard. Appendix B discusses category theory, and in particular contains the construction of the quotient of a small Abelian category by a Serre subcategory. Appendix C deals with spectral sequences from the point of view of exact couples. Though this is standard in topology, it seems to be less familiar to algebraists. We give an *ad hoc* treatment of convergence that suffices for our purposes.

A word about sources — Chapters 4–7, and the first half of Chapter 9, are based on the work of Quillen cited above. For the other chapters, we list a few main sources at the beginning of the chapter, and give other references in the course of the text. The absence of a specific reference, however, does not imply any claim to originality on my part; in fact, all the material covered in these notes (with the exception of parts of Chapter 9 on singular surfaces) is based on other sources.

## Acknowledgements

I must thank the people who attended the original course of lectures, and made many comments that clarified my ideas, primarily the members of the "algebra school" at the Tata Institute. I began learning algebraic geometry and $K$-theory when I was a graduate student at Chicago, from Bloch, Murthy, and Swan; at the Tata Institute, I have also learned a lot from Mohan Kumar and Madhav Nori; to all these people, I owe a considerable debt. I received generous help from several people on topics connected with these notes: Balwant Singh, Bhatwadekar, Coombes, Dalawat, Esnault, Lemaire, Levine, Parimala, Paranjape, Pati, Raghunathan, Ramanan, Roy, Simha, Soule, Sridharan, Stienstra, Suslin, and Vaserstein. Thanks are also due to M.K. Priyan for the major effort of typing the first version of the notes, and to K.P. Shivaraman for help with the final version.

# Preface to the Second Edition

Apart from the improved typescript, now rendered in TEX, there are two main changes in the second edition. Several people with more of a background in topology, but who were unfamiliar with algebraic geometry, commented to me that the first edition of the book was difficult for them to follow. In response to these comments, I have added Appendix D, entitled "Results from Algebraic Geometry," which contains all the definitions and results (with sketches of proofs, and/or suitable references) needed to understand Chapter 5. I have also added suitable cross-references to this appendix throughout Chapter 5. The reader who masters this material ought to be able to also read Chapters 8 and 9, consulting the references given there.

Secondly, I have rewritten major parts of Chapter 8, based on a later proof of the Merkurjev-Suslin theorem given by Merkurjev in *Contemp. Math.* 55, Part II, A.M.S. (1986). This proof is more elementary than the one given in the first edition of this book, in that higher $K$-theory and $K$-cohomology are needed only to prove Hilbert's Theorem 90 for $K_2$. The treatment of the Merkurjev-Suslin theorem now given is thus self-contained (at least with respect to $K$-theory), as we do not need to invoke the theory of Chern classes or the Riemann-Roch theorem for higher $K$-groups.

A reader interested in the connections with Cyclic Homology (or "additive $K$-theory") can consult the book

J.-L. Loday, *Cyclic Homology*, Grundlehren Math. 301, Springer-Verlag (1992).

The reader would also do well to consult the book

J. Rosenberg, *Algebraic K-Theory and Its Applications*, Graduate Texts No. 147, Springer-Verlag (1994).

His treatment of $K$-theory assumes less background of the reader than mine, but covers different ground. The book also gives some historical background, discusses the connections with number theory, $C^*$-algebras, etc. and contains a more complete list of references.

Thanks are due to Birkhäuser-Boston, and in particular Ann Kostant, for the job of rendering the manuscript into TEX, to obtain the much-improved result now before the reader.

V. Srinivas<br>Bombay, 1995

# 1. "Classical" $K$-Theory

The main reference used here is Milnor's book: *Introduction to Algebraic K-Theory*, Annals of Mathematical Studies, No. 72, Princeton University Press (1971).

Let $R$ be an associative ring (with 1), and let $\mathcal{P}(R)$ denote the category of finitely generated projective $R$-modules. We define the Grothendieck group $K_0(R)$ to be the quotient

$$K_0(R) = \mathcal{F}/\mathcal{R},$$

$\mathcal{F}$ = free Abelian group on the isomorphism classes of projective modules in $\mathcal{P}(R)$,

$\mathcal{R}$ = subgroup generated by elements

$$[P \oplus Q] - [P] - [Q], \text{ for all } P, Q \in \mathcal{P}(R).$$

Thus, for any $P, Q \in \mathcal{P}(R)$, $[P] = [Q]$ in $K_0(R) \iff P \oplus P' \cong Q \oplus P'$ for some $P' \in \mathcal{P}(R) \iff P \oplus R^n \cong Q \oplus R^n$ for some $n \geq 0$. Indeed, if $[P] = [Q]$ in $K_0(R)$, then we have a relation in $\mathcal{F}$ of the form

$$[P] - [Q] = \sum_{i=1}^{r}([P_i \oplus Q_i] - [P_i] - [Q_i]) - \sum_{j=1}^{s}([P'_j \oplus Q'_j] - [P'_j] - [Q'_j]).$$

Hence

$$[Q] + \sum_{i=1}^{r}[P_i \oplus Q_i] + \sum_{j=1}^{s}([P'_j] + [Q'_j]) = [P] + \sum_{i=1}^{r}([P_i] + [Q_i]) + \sum_{j=1}^{s}[P'_j \oplus Q'_j]$$

in $\mathcal{F}$, the free Abelian group on isomorphism classes in $\mathcal{P}(R)$. Hence the terms on the right must be a permutation of the set of terms on the left. In particular, if we let

$$P' = \left(\bigoplus_{i=1}^{r}(P_i \oplus Q_i)\right) \oplus \left(\bigoplus_{j=1}^{s}(P'_j \oplus Q'_j)\right)$$

then $P \oplus P' \cong Q \oplus P'$. Thus, we have shown that $[P] = [Q]$ in $K_0(R) \implies P \oplus P' \cong Q \oplus P'$ for some $P' \in \mathcal{P}(R)$. The converse is obvious. Further, we can find $Q' \in \mathcal{P}(R)$ such that $P' \oplus Q' \cong R^n$ for some $n$, since $P'$ is a quotient of some $R^n$ ($P'$ is finitely generated) and $P'$ is projective. Hence $P \oplus P' \cong Q \oplus P' \implies P \oplus R^n \cong Q \oplus R^n$.

If $f : R \to S$ is a homomorphism of rings, $f$ induces a functor $\mathcal{P}(R) \to \mathcal{P}(S)$ given by $P \mapsto S \otimes_R P$. This preserves direct sums, and hence induces a homomorphism $f_* : K_0(R) \to K_0(S)$.

**Example (1.1).** *Let $(R, \mathcal{M})$ be a local ring, i.e., $R$ is a possibly non-commutative ring (with 1), $\mathcal{M} \subset R$ is a 2-sided maximal ideal, and $R - \mathcal{M} = R^*$, the group of units. Then $K_0(R) = \mathbb{Z}$, with a generator given by the class of the free $R$-module of rank 1.*

Indeed, there is a natural homomorphism

$$\rho : K_0(R) \longrightarrow K_0(R/\mathcal{M}) = \mathbb{Z},$$

since $R/\mathcal{M}$ is a division ring, and a finitely generated projective (left) $R/\mathcal{M}$-module is a left vector space, i.e., a free $R/\mathcal{M}$-module of rank equal to the dimension of the vector space. Thus $\rho$ is surjective, since $\rho([R^n])$ is the class of a vector space of dimension $n$; we prove that in fact every projective $R$-module is free, so that $\rho$ is an isomorphism.

Let $P \in \mathcal{P}(R)$, and let $\dim_{R/\mathcal{M}} P/\mathcal{M}P = n$. Choose $x_1, \ldots, x_n \in P$ whose images $\bar{x}_1, \ldots, \bar{x}_n \in P/\mathcal{M}P$ give a basis for the $(R/\mathcal{M}R)$-vector space. We claim that $x_1, \ldots, x_n$ give a basis for $P$ as a free $R$-module.

Let $Q \in \mathcal{P}(R)$ such that $P \oplus Q \cong R^{n+m}$ is a free module, where we must have $m = \dim_{R/\mathcal{M}} Q/\mathcal{M}Q$. Let $x_{n+1}, \ldots, x_{m+n} \in Q$ map to a basis of $Q/\mathcal{M}Q$ over $R/\mathcal{M}R$, so that $x_1, \ldots, x_{m+n} \in R^{m+n}$ give a basis of $(R/\mathcal{M}R)^{m+n}$. We claim then that $x_1, \ldots, x_{m+n}$ give a basis for the free $R$-module $R^{m+n}$, which immediately implies that $x_1, \ldots, x_n \in P$, $x_{n+1}, \ldots, x_{m+n} \in Q$ give bases, so that $P, Q$ are free $R$-modules of ranks $n, m$, respectively.

Let $x_i = (a_{i1}, \ldots, a_{i,m+n}) \in R^{m+n}$. It suffices to prove that $A = [a_{ij}] \in GL_{m+n}(R)$, i.e., $A$ has a 2-sided inverse. If $a_{ij} \mapsto \bar{a}_{ij} \in R/\mathcal{M}$, and $\bar{A} = [\bar{a}_{ij}]$, then $\bar{A} \in GL_{m+n}(R/\mathcal{M})$, so there exists $B \in M_{m+n}(R)$, with $B \mapsto \bar{B} \in GL_{m+n}(R/\mathcal{M})$, and $\bar{A}\bar{B} = \bar{B}\bar{A} = \bar{I}_{m+n}$, where $I_{m+n}$ is the identity matrix. Then $AB \equiv BA \equiv I_{m+n} \pmod{\mathcal{M}}$, and so $AB = [c_{ij}]$ with $c_{ii} \in R^*$, $c_{ij} \in \mathcal{M}$ for $i \neq j$. Hence, there exists an elementary matrix $E$ with $ABE = \text{diag}(c_1, \ldots, c_{m+n})$ for suitable $c_1, \ldots, c_{m+n} \in R^*$, i.e., $AB$ can be diagonalized by right column operations, involving adding a right multiple of a column to another column—we may first add a suitable multiple of the first column to each of the other columns to make all off-diagonal entries on the first row vanish; this does not alter the condition that other off diagonal entries lie in $\mathcal{M}$, and diagonal entries are units, since $R$ is local; now perform a similar reduction of the second row, etc.

# 1. "Classical" K-Theory

Since $\text{diag}(c_1, \ldots, c_{m+n})$ is invertible, $A$ has a right inverse. By a similar argument using row operations on $BA$, we see that $A$ has a left inverse. Since matrix multiplication is associative, $A$ is invertible.

**Example (1.2).** *Let $R$ be a Dedekind domain, i.e., $R$ is a commutative Noetherian integrally closed domain such that every non-zero prime ideal of $R$ is maximal. Then $K_0(R) \cong \mathbb{Z} \oplus C\ell(R)$ where $C\ell(R)$ is the ideal class group of $R$, defined to be the group of isomorphism classes of invertible ideals (with tensor product as the group operation—see Milnor's book for details).*

**Definition of $K_1$.** Let $R$ be an associative ring (with 1), $GL_n(R)$ the group of invertible matrices of size $n$ over $R$; let $E_n(R)$ be the subgroup of *elementary matrices*, defined to be the group generated by the matrices $e_{ij}^{(n)}(\lambda)$, $1 \leq i \neq j \leq n$, $\lambda \in R$, where $e_{ij}^{(n)}(\lambda)$ is the unipotent matrix whose only non-trivial off-diagonal entry is $\lambda$ in the $(i,j)$th position. Thus, if $i < j$, then $e_{ij}^{(n)}(\lambda)$ has the form

$$e_{ij}^{(n)}(\lambda) = \begin{bmatrix} 1 & 0 & \cdots\cdots & 0 \\ 0 & 1 & & \vdots \\ \vdots & & & \\ 0 & 0 & \cdots 1 \cdots & \lambda \\ \vdots & & & 1 & \cdots \\ 0 & 0 & \cdots\cdots & 0 & 1 \end{bmatrix} \quad \longleftarrow i\text{th row.}$$

with $j$th column indicated.

Let $GL_n(R) \hookrightarrow GL_{n+1}(R)$ by

$$A \longmapsto \begin{bmatrix} A & 0 \\ 0 & 1 \end{bmatrix}$$

and let $GL(R) = \varinjlim GL_n(R)$. Similarly, let $E(R) = \varinjlim E_n(R)$. Since $e_{ij}^{(n)}(\lambda) \longmapsto e_{ij}^{(n+1)}(\lambda)$ under $E_n(R) \hookrightarrow E_{n+1}(R)$, we obtain matrices $e_{ij}(\lambda) \in E(R)$ as the common image of all the $e_{ij}^{(n)}(\lambda)$ for $n \geq i,j$, and $E(R)$ is the subgroup of $GL(R)$ generated by the $e_{ij}(\lambda)$. The $e_{ij}(\lambda)$ satisfy the following identities:

(1.3) a) $e_{ij}(\lambda) \cdot e_{ij}(\mu) = e_{ij}(\lambda + \mu)$, $\forall \lambda, \mu \in R$
b) $[e_{ij}(\lambda), e_{k\ell}(\mu)] = 1$ for $j \neq k$, $i \neq \ell$
c) $[e_{ij}(\lambda), e_{jk}(\mu)] = e_{ik}(\lambda\mu)$ for $i \neq k$, $\forall \lambda, \mu \in R$.

These identities are deduced from similar identities for the $e_{ij}^{(n)}(\lambda)$. We deduce immediately that $E_n(R)$ is perfect for $n \geq 3$ (i.e., $[E_n(R), E_n(R)] = E_n(R)$) and $E(R)$ is perfect.

**Lemma (1.4)** (Whitehead). *For any $A \in GL_n(R)$,*
$$\begin{bmatrix} A & 0 \\ 0 & A^{-1} \end{bmatrix} \in E_{2n}(R).$$

**Proof.** For any $B \in M_n(R)$, it is easy to see that $\begin{bmatrix} I_n & B \\ 0 & I_n \end{bmatrix}, \begin{bmatrix} I_n & 0 \\ B & I_n \end{bmatrix}$ lie in $E_{2n}(R)$. Now use the identity
$$\begin{bmatrix} A & 0 \\ 0 & A^{-1} \end{bmatrix} = \begin{bmatrix} I_n & 0 \\ A^{-1} - I_n & I_n \end{bmatrix} \begin{bmatrix} I_n & I_n \\ 0 & I_n \end{bmatrix} \begin{bmatrix} I_n & 0 \\ A - I_n & I_n \end{bmatrix} \begin{bmatrix} I_n & -A^{-1} \\ 0 & I_n \end{bmatrix}.$$

**Proposition (1.5).** $E(R) = [E(R), E(R)] = [GL(R), GL(R)]$.

**Proof.** We have already noted that $E(R)$ is perfect, so it suffices to note that for any $A, B \in GL_n(R)$,
$$\begin{bmatrix} ABA^{-1}B^{-1} & 0 \\ 0 & I_n \end{bmatrix} = \begin{bmatrix} AB & 0 \\ 0 & (AB)^{-1} \end{bmatrix} \begin{bmatrix} A^{-1} & 0 \\ 0 & A \end{bmatrix} \begin{bmatrix} B^{-1} & 0 \\ 0 & B \end{bmatrix}$$
and the right side lies in $E_{2n}(R)$ by Lemma (1.4).

**Definition.**
$$K_1(R) = GL(R)/[GL(R), GL(R)]$$
$$= GL(R)/E(R) = H_1(GL(R), \mathbb{Z})$$

(the definitions are equivalent by Proposition (1.5), and the isomorphism $H_1(G, \mathbb{Z}) \cong G^{ab} = G/[G, G]$ for any $G$).

**Example (1.6).** *Let $(R, \mathcal{M})$ be a (possibly non-commutative) local ring. Then the natural map $GL_1(R) \to K_1(R)$ induces an isomorphism*
$$R^*/[R^*, R^*] \cong K_1(R).$$

Indeed, let
$$\bar{R}^* = (R^*)^{ab}.$$
If $A = [a_{ij}] \in GL_n(R)$, then by adding another column to the first one if necessary, we may assume $a_{11} \in R^*$ (any row or column must have an entry in $R^*$, since $\bar{A} = [\bar{a}_{ij}] \in GL_n(R/\mathcal{M})$ has a non-zero entry in every row and column). Adding multiples of the first column to the others, we

can make $a_{1i} = 0$ for all $i > 1$. Similarly, we work on the second row to make $a_{22} \in R^*$, $a_{2i} = 0$ for $i \neq 2$, etc. Thus, after column operations, given by right multiplication by an element of $E_n(R)$, we can make $A$ diagonal without changing its image in $K_1(R)$. Now by the Whitehead lemma (1.4), we deduce that the image of $A$ in $K_1(R)$ lies in the image of $R^* = GL_1(R)$. Hence $R^* \twoheadrightarrow K_1(R)$. Since $K_1(R)$ is Abelian, we have an induced surjection

$$\bar{R}^* = R^*/[R^*, R^*] \twoheadrightarrow K_1(R).$$

Next, if $x, y \in R$ such that $1 - xy \in R^*$, we claim that $1 - yx \in R^*$ also, and $(1 - xy)(1 - yx)^{-1} \in [R^*, R^*]$. If $1 - xy$, $y \in R^*$ then

$$(1 - xy) = (y^{-1} - x)y = y(y^{-1} - x)(y^{-1} - x)^{-1}y^{-1}(y^{-1} - x)y = (1 - yx)y_1$$

with $y_1 \in [R^*, R^*]$ proving our claim in this case. A similar argument works if $x \in R^*$. So we may assume $x, y \in \mathcal{M}$, in which case clearly $1 - yx$, $1 - xy \in R^*$. If $x_1 = 1 + x - xy$, then $x_1$, $1 - yx_1$, $1 - x_1 y \in R^*$, and we compute that

$$(1 - xy)(1 - y) = 1 - y - xy + xy^2 = 1 - x_1 y, \text{ and}$$
$$(1 - yx)(1 - y) = 1 - y - yx + yxy = 1 - yx_1.$$

Hence

$$(1 - xy)(1 - yx)^{-1} = (1 - x_1 y)(1 - yx_1)^{-1} \in [R^*, R^*].$$

as claimed. (I learned this argument from L. Vaserstein, who has used analogous arguments to compute $K_1$ for most semi-local rings.)

We now show that there is a well-defined determinant homomorphism $\det : GL(R) \longrightarrow \bar{R}^*$, satisfying

(i) $\det(AB) = \det A \cdot \det B$

(ii) $\det A = 1$ for all $A \in E(R)$

(iii) the composite $R^* = GL_1(R) \longrightarrow GL(R) \xrightarrow{\det} \bar{R}^*$ is the natural quotient map.

Our construction of 'det' follows the treatment given in Artin's book *Geometric Algebra* of the determinant over division rings. It suffices to construct a compatible family of maps $\det_n : GL_n(R) \to \bar{R}^*$ such that

1)$_n$ if $A \in GL_n(R)$, and $A'$ is obtained from $A$ by multiplying a column on the right by $\mu \in R^*$, then $\det_n A' = \bar{\mu} \cdot \det_n A$, where $\mu \mapsto \bar{\mu} \in \bar{R}^*$

2)$_n$ if $A \in GL_n(R)$ and $A'$ is obtained from $A$ by adding a right multiple of a column to another column, then $\det_n A' = \det_n A$

3)$_n$ if $I_n \in GL_n(R)$ is the identity matrix, $\det_n I_n = \bar{1}$, the image of $1 \in R^*$ in $\bar{R}^*$.

We prove by induction on $n$ that $\det_n$ exists, and observe at once that $\det_n$, if it exists, is characterized by the above properties, since any $A \in GL_n(R)$ can be transformed by operations as in 2)$_n$ to a matrix

$$\begin{bmatrix} a & & & & 0 \\ & 1 & & & \\ & & 1 & & \\ & & & \ddots & \\ 0 & & & & 1 \end{bmatrix}, \quad a \in R^*,$$

so that $\det_n A = \bar{a} \in \bar{R}^*$, by 1)$_n$ and 3)$_n$. Further, if $\det_n$ exists, it must satisfy:

a) $\det_n\big|_{GL_{n-1}(R)} = \det_{n-1}$
b) $\det_n(AB) = \det_n(A) \cdot \det_n(B)$
c) if $A'$ is obtained from $A$ by interchanging 2 columns, $\det_n A' = (\overline{-1}) \cdot \det_n A$.

We establish b), and leave a), c) to the reader (see Artin's book for details): write $B = \mathrm{diag}(b, 1, \ldots, 1)$. $B'$ with $b \in R^*$, $B' \in E_n(R)$. Then $\det_n B = \bar{b}$ by 2)$_n$. Similarly

$$\begin{aligned} \det_n(AB) &= \det_n(A \cdot \mathrm{diag}(b, 1, \ldots, 1)) \quad \text{(by 2)}_n) \\ &= \bar{b} \cdot \det_n A \quad \text{(by 1)}_n) \\ &= \det_n A \cdot \det_n B, \text{ since } \bar{R}^* \text{ is commutative.} \end{aligned}$$

**Proof That $\det_n$ Exists.** Clearly $\det_1$ exists and is the natural quotient $R^* \to \bar{R}^*$. Assume $n > 1$, and by induction, $\det_{n-1}$ exists, and hence is well determined and satisfies a), b), c) above. Let $A \in GL_n(R)$ have columns $A_1, \ldots, A_n$, so that $A = [A_1, \ldots, A_n]$. Then the columns $A_i$ give a basis for the free right $R$-module $R^n$ of columns of length $n$, so that there is a unique linear combination $\sum A_i \lambda_i = \begin{pmatrix} 1 \\ 0 \\ \vdots \\ 0 \end{pmatrix}$, the first standard basis vector. Clearly at least one $\lambda_i \in R^*$. Write $A_i = \begin{bmatrix} a_{1i} \\ B_i \end{bmatrix}$, $B_i \in R^{n-1}$, $a_{1i} \in R$, so that $\sum a_{1i} \lambda_i = 1$ and $\sum B_i \lambda_i = 0$. For any $i$, let $C_i = [B_1, \ldots, B_{i-1}, B_{1+i}, \ldots]$ be the matrix with columns $B_1, \ldots, B_{i-1}, B_{i+1}, \ldots, B_n$. If $\lambda_i \in R^*$, then clearly $A_1, \ldots, A_{i-1}, A_{i+1}, \ldots, A_n, \begin{pmatrix} 1 \\ 0 \\ \vdots \\ 0 \end{pmatrix}$ form another basis for the free right module of columns $R^n$, so that $B_1, \ldots, B_{i-1}, B_{i+1}, \ldots, B_n$ form a basis for the free right module of columns $R^{n-1}$. Hence $\lambda_i \in R^* \Longrightarrow C_i \in GL_{n-1}(R)$.

If $\lambda_i, \lambda_j \in R^*$ with $i < j$, then by $2)_{n-1}$ we get

$$\det_{n-1}[B_1, \ldots, B_{i-1}, B_i\lambda_i, \ldots, B_{j-1}, B_{j+1}, \ldots, B_n]$$
$$= \det_{n-1}[B_1, \ldots, B_{i-1}, -B_j\lambda_j, B_{i+1}, \ldots, B_{j-1}, B_{j+1}, \ldots, B_n],$$

where both matrices lie in $GL_{n-1}(R)$ since $\lambda_i, \lambda_j \in R^*$, and $C_i, C_j \in GL_{n-1}(R)$. Thus we have (the overbar denotes the image in $\bar{R}^*$)

$$\det_{n-1} C_j = \overline{(-1)}^{j-i-1} \bar{\lambda}_i^{-1} \overline{(-\lambda_j)} \det_{n-1}(C_i)$$

i.e.,

$$\overline{(-1)}^{i+1} \bar{\lambda}_i^{-1} \det_{n-1}(C_i) = \overline{(-1)}^{j+1} \bar{\lambda}_j^{-1} \det_{n-1}(C_j),$$

if $\lambda_i \lambda_j \in R^*$. Define $\det_n A = \overline{(-1)}^{i+1} \bar{\lambda}_i^{-1} \det_{n-1}(C_i)$ for any $i$ such that $\lambda_i \in R^*$; this is well defined by the above remarks. We have to verify $1)_n$, $2)_n$, $3)_n$, of which $3)_n$ is obvious.

$1)_n$: Suppose $A_i$ is replaced by $A_i\mu$, for some $\mu \in R^*$. Then $\lambda_i$ is replaced by $\mu^{-1}\lambda_i$, and the remaining $\lambda_j$ are unchanged. If $\lambda_i \in R^*$, $C_i$ is unchanged, and $\mu^{-1}\lambda_i \in R^*$, so that

$$\det_n A' = \overline{(-1)}^{i+1} (\overline{\mu^{-1}\lambda_i})^{-1} \det_{n-1}(C_i)$$
$$= \bar{\mu} \det_n A.$$

If $\lambda_j \in R^*$ with $j \neq i$, a column of $C_j$ is multiplied by $\mu$ to get $C'_j$, so that

$$\det_n A' = \overline{(-1)}^{i+1} \bar{\lambda}_j^{-1} \cdot \det_{n-1}(C'_j)$$
$$= \overline{(-1)}^{i+1} \bar{\lambda}_j^{-1} (\bar{\mu} \cdot \det_{n-1}(C_j)) = \bar{\mu} \cdot \det_n A.$$

$2)_n$: Suppose $A_i$ is replaced by $A_i + A_j\mu$ for some $\mu$ in $R$. Then $\lambda_j$ is replaced by $\lambda_j - \mu\lambda_i$, while $\lambda_k$ is unchanged for $k \neq j$, since

$$(A_i + A_j\mu)\lambda_i + A_j(\lambda_j - \mu\lambda_i) = A_i\lambda_i + A_j\lambda_j.$$

If some $\lambda_k \in R^*$, $k \neq i, j$ then $C'_k$ is obtained from $C_k$ by a column operation, and so $\det_n A' = \overline{(-1)}^{k+1} \bar{\lambda}_k^{-1} \det_n C'_k = \det_n A$. If $\lambda_i \in R^*$, $C'_i = C_i$ and $\lambda'_i = \lambda_i$, so that $\det_n A' = \det_n A$. Finally, if $\lambda_j \in R^*$ and $\lambda_k \in M$ for $k \neq j$, then we compute that if $i < j$ (the computations when $i > j$ are similar),

$$C'_j = [B_1, \ldots, B_{i-1}, B_i + B_j\mu, B_{i+1}, \ldots, B_{j-1}, B_{j+1}, \ldots]$$
$$= [B_1, \ldots, B_{i-1}, B_i(1 - \lambda_i\lambda_j^{-1}\mu)$$
$$\quad - \sum_{k \neq i,j} B_k\lambda_k\lambda_j^{-1}\mu, B_{i+1}, \ldots, B_{j-1}, B_{j+1}, \ldots]$$
$$\longrightarrow [B_1, \ldots, B_{i-1}, B_i(1 - \lambda_i j^{-1}\mu), B_{i+1}, \ldots, B_{j-1}, B_{j+1}, \ldots],$$

where $\longrightarrow$ denotes the result of a column operation. This last matrix is obtained from $C_j$ by multiplying the $i$th column of the right by $1-\lambda_i\lambda_j^{-1}\mu \in R^*$ (since $\lambda_i \in \mathcal{M}$). Hence

$$\det_n A' = (-1)^{j+1}(\lambda_j - \mu\lambda_i)^{-1}(1 - \lambda_i\lambda_j^{-1}\mu) \cdot \det_n C_j$$
$$= \overline{(\lambda_j - \mu\lambda_i)^{-1}} \cdot \overline{(1 - \lambda_i\lambda_j^{-1}\mu)} \cdot \bar{\lambda}_j \cdot \det_n A.$$

So we must show that

$$\overline{\lambda_j - \mu\lambda_i} = \bar{\lambda} = \bar{\lambda}_j\overline{(1 - \lambda_i\lambda_j^{-1})} \text{ in } \bar{R}^*$$
$$\iff \overline{(1 - \mu\lambda_i\lambda_j^{-1})} = \overline{(1 - \lambda_i\lambda_j^{-1}\mu)} \text{ since } \overline{\phantom{x}} \text{ is a homomorphism}$$
$$\iff \overline{(1 - \mu \cdot (\lambda_i\lambda_j^{-1}))} = \overline{(1 - (\lambda_i\lambda_j^{-1})\mu)}$$

which has the form $\overline{(1 - xy)} = \overline{(1 - yx)}$, with $x = \mu$, $y = \lambda_i\lambda_j^{-1} \in \mathcal{M}$. But we have seen that this relation holds in $\bar{R}^*$. This completes the proof that $\det_n$ has all the required properties.

**Corollary (1.7)** (Dieudonné). *If $D$ is a division ring, then $K_1(D) \cong D^*/[D^*, D^*]$, induced by the Dieudonné determinant $GL(D) \to (D^*)^{ab}$.*

**Example (1.8).** *Let $R$ be a commutative ring. The determinant gives a surjection $GL(R) \to R^*$ split by $GL_1(R) \hookrightarrow GL(R)$. Let $SL(R) \subset GL(R)$ be the group of matrices with determinant 1. Then $[SL(R), SL(R)] = E(R)$. Let $SK_1(R) = SL(R)/[SL(R), SL(R)] = SL(R)/E(R)$. Then we have a natural split exact sequence*

$$0 \longrightarrow SK_1(R) \longrightarrow K_1(R) \longrightarrow R^* \longrightarrow 0.$$

**Example (1.9)** (Mennicke symbol). *Let $R$ be a commutative ring, $a, b \in R$ such that $aR + bR = R$. Choose $c, d \in R$ so that $ad - bc = 1$, and define the Mennicke symbol*

$$[a, b] \in SK_1(R)$$

*to be the class of $\begin{bmatrix} a & b \\ c & d \end{bmatrix}$ in $SK_1(R)$. Then*

(i) *$[a, b]$ is well defined.*
(ii) *$[a, b] = [b, a]$; if $a \in R^*$, $[a, b] = 1$ for all $b \in R$.*
(iii) *$[a_1 a_2, b] = [a_1, b][a_2, b]$ if $a_1 a_2 R + bR = R$.*
(iv) *$[a, b] = [a + \lambda b, b]$ for all $\lambda \in R$.*
(v) *If $R$ is Noetherian of Krull dimension $\leq 1$, then the Mennicke symbols generate $SK_1(R)$ (see Bass: Algebraic K-Theory).*

(vi) *If $R$ is Noetherian of Krull dimension $\leq 1$ with finite residue fields at all maximal ideals, then $SK_1(R)$ is torsion.*

(vii) *If $R$ is a Euclidean domain (e.g., $\mathbb{Z}$, $\mathbb{Z}[i]$, $k[t]$ where $k$ is a field) then $SK_1(R) = 0$.*

(viii) *If $R$ is the ring of algebraic integers in an algebraic number field (a finite algebraic extension of $\mathbb{Q}$) then $SK_1(R) = 0$ (see Milnor's book, Ch. 16).*

**Definition of $K_2$.** Let $R$ be a ring. The $n$th Steinberg group $St_n(R)$ is defined to be the quotient of the free group on symbols $x_{ij}^{(n)}(\lambda)$ for $1 \leq i, j \leq n$, $i \neq j$, and for all $\lambda \in R$, modulo the normal subgroup generated by the words:

(i) $x_{ij}^{(n)}(\lambda) \cdot x_{ij}^{(n)}(\mu) \cdot x_{ij}^{(n)}(\lambda + \mu)^{-1}$ for all $i, j$, for all $\lambda, \mu \in R$

(ii) $[x_{ij}^{(n)}(\lambda), x_{k\ell}^{(n)}(\mu)]$ for $i \neq \ell$, $k \neq j$, for all $\lambda, \mu \in R$

(iii) $[x_{ij}^{(n)}(\lambda), x_{jk}^{(n)}(\mu)] \cdot x_{ik}^{(n)}(\lambda\mu)^{-1}$ for $i \neq k$, for all $\lambda, \mu \in R$.

From (1.3) we have a natural surjection $\phi_n : St_n(R) \to E_n(R)$, given by $\phi_n(x_{ij}^{(n)}(\lambda)) = e_{ij}^{(n)}(\lambda)$. We also have natural homomorphisms $St_n(R) \to St_{n+1}(R)$ (which need not be injective), and so we obtain the infinite Steinberg group

$$St(R) = \varinjlim St_n(R),$$

and the surjection $\phi : St(R) \longrightarrow E(R)$.

**Definition.** $K_2(R) = \ker \phi$.
  Let $x_{ij}(\lambda) \in St(R)$ be the common image of $x_{ij}^{(n)}(\lambda)$, for any $n \geq i, j$, where $i \neq j$, and $\lambda \in R$.

**Lemma (1.9).** *$K_2(R)$ is the center of $St(R)$.*

**Proof.** It is easy to see that the center of $E(R)$ is trivial, so that the center of $St(R)$ lies in $K_2(R)$. Hence it suffices to prove that $K_2(R)$ is central in $St(R)$.

Suppose $\alpha \in K_2(R) \cap \text{image}(St_{n-1}(R) \to St(R))$, i.e., $\alpha$ can be expressed as a word in the $x_{ij}(\lambda)$ with $i, j < n$. Let $P_n$ be the subgroup of $St(R)$ generated by all the $x_{in}(\mu)$, $1 \leq i \leq n-1$, $\mu \in R$. Then from the relation (ii) in $St_m(R)$, $m \geq n$, we deduce that $P_n$ is commutative. From (i), $x_{ij}(0) = 1$, so that each element of $P_n$ has a unique expression in the form

$$x_{1n}(\mu_1) \ldots x_{n-1,n}(\mu_{n-1}), \mu_1, \ldots, \mu_{n-1} \in R,$$

and $\phi\big|_{P_n}$ is an isomorphism onto the group of matrices in $SL_n(R) \subset GL(R)$ of the form
$$\begin{bmatrix} 1 & 0 & \cdots & 0 & \mu_1 \\ 0 & 1 & \cdots & 0 & \mu_2 \\ 0 & & \cdots & 1 & \mu_{n-1} \\ 0 & & \cdots & 0 & 1 \end{bmatrix}.$$

Now we compute that if $i, j < n$ then
$$x_{ij}(\lambda) x_{kn}(\mu) x_{ij}(-\lambda) = \begin{cases} x_{kn}(\mu) & \text{if } j \neq k \\ x_{in}(\lambda\mu) \cdot x_{kn}(\mu) & \text{if } j = k \end{cases}$$
so that the given class $\alpha \in K_2(R)$ normalizes $P_n$. Since $\phi\big|_{P_n}$ is injective, while $\phi(\alpha) = 1$, $\alpha$ centralizes $P_n$, i.e., $[\alpha, x_{in}(\lambda)] = 0$ for all $1 \leq i \leq n-1$, $\lambda \in R$. By a similar argument we see that $[\alpha, x_{nj}(\lambda)] = 0$ for all $1 \leq j \leq n-1$, $\lambda \in R$. Hence $\alpha$ commutes with $[x_{jn}(\lambda), x_{ni}(\mu)] = x_{ji}(\lambda\mu)$ for all $\lambda, \mu \in R$, $1 \leq j \neq i \leq n-1$. Since $n$ can be taken to be arbitrarily large, we are done. Thus
$$0 \longrightarrow K_2(R) \longrightarrow St(R) \longrightarrow E(R) \longrightarrow 0$$
is a central extension of $E(R)$.

**Definition.** If $G$ is a group, a central extension
$$(E) \ldots \qquad 0 \longrightarrow K \longrightarrow H \longrightarrow G \longrightarrow 0$$
is called a *universal central extension* of $G$ if for any other central extension
$$(E') \ldots \qquad 0 \longrightarrow K' \longrightarrow H' \longrightarrow G \longrightarrow 0,$$
there is a unique homomorphism $f : H \longrightarrow H'$ over $G$.

**Remark.** In the above situation, $f\big|_K : K \longrightarrow K'$ in fact determines $(E')$ up to isomorphism, in the following sense. If $K'$ is any Abelian group, $g : K \longrightarrow K'$ any homomorphism, then the pushout of
$$\begin{array}{ccc} K & \longrightarrow & H \\ {\scriptstyle g}\downarrow & & \\ K' & & \end{array}$$
yields a central extension
$$(E'') \ldots \qquad 0 \longrightarrow K' \longrightarrow H \times_K K' \longrightarrow G \longrightarrow 0.$$
If $g = f\big|_K$ where $f : H \longrightarrow H'$ is as above, then by the universal property of the pushout, there is a map $H \times_K K' \longrightarrow H'$ giving a diagram
$$\begin{array}{ccccccccc} 0 & \longrightarrow & K' & \longrightarrow & H \times_K K' & \longrightarrow & G & \longrightarrow & 0 \\ & & \| & & \downarrow & & \| & & \\ 0 & \longrightarrow & K' & \longrightarrow & H' & \longrightarrow & G & \longrightarrow & 0 \end{array}$$

Hence $(E'')$ and $(E')$ are isomorphic central extensions of $G$. From homological algebra, it is standard that central extensions of $G$ by an Abelian group $K'$, up to isomorphism, are classified by elements of $H^2(G, K')$, where $K'$ is regarded as a $G$-module with trivial action. From the above remarks, if $(E)$ is a universal central extension with kernel $K$, then we have an isomorphism $H^2(G, K') \cong \operatorname{Hom}(K, K')$ of functors on the category of Abelian groups. From the proposition below, we also have $H_1(G, \mathbb{Z}) = 0$, since $G$ has a universal central extension; hence by the universal coefficient theorem, $H^2(G, K') \cong \operatorname{Hom}(H_2(G, \mathbb{Z}), K')$ is an isomorphism of functors. Thus, $K \cong H_2(G, \mathbb{Z})$.

**Proposition (1.10).** (a) *A central extension*

$(E)\ldots \qquad 0 \longrightarrow K \longrightarrow H \longrightarrow G \longrightarrow 0$

*is universal* $\iff$ $H$ *is perfect (i.e., $H_1(H, \mathbb{Z}) = H/[H, H] = 0$) and every central extension of $H$ is split (i.e., $H_2(H, \mathbb{Z}) = 0$, from the above remarks).*

(b) $G$ *has a universal central extension* $\iff$ $G$ *is perfect*.

**Outline of Proof.** (a) ($\Leftarrow$) let $(E')$ be an arbitrary central extension of $G$:

$(E')\ldots \qquad 0 \longrightarrow K' \longrightarrow H' \longrightarrow G \longrightarrow 0.$

The pull-back of $(E')$ along $H \longrightarrow G$ is a central extension of $H$, which must be trivial. Hence the projection

$$p_2 : H' \times_G H \longrightarrow H$$

has a section $s : H \longrightarrow H' \times_G H$, and $f = p_1 \circ s : H \longrightarrow H'$ is a map over $G$. To prove that $f$ is unique, we see that if $f'$ is another map $H \longrightarrow H'$ over $G$, and $x, y \in H$ are arbitrary, then $f'(x) = af(x)$, $f'(y) = bf(y)$ with $a, b \in K' \cap \operatorname{center}(H')$. Thus $f'([x, y]) = [f'(x), f'(y)] = [f(x), f(y)] = f([x, y])$ since $f, f'$ are homomorphisms. Since $H = [H, H]$, $f' = f$.
($\Longrightarrow$) If

$(E)\ldots \qquad 0 \longrightarrow K \longrightarrow H \xrightarrow{\theta} G \longrightarrow 0$

is a central extension such that $H$ is not perfect, then there is a non-zero map $\psi : H \longrightarrow A$ for some Abelian group $A$. If $H' = A \times G$ is the split central extension of $G$ by $A$, then $f = (0, \theta)$ and $f' = (\psi, \theta)$ are 2 distinct homomorphisms $H \longrightarrow H'$ over $G$. Hence $(E)$ cannot be universal, unless $H$ is perfect.

If $(E)$ is universal, and if

$$0 \longrightarrow K' \longrightarrow H' \xrightarrow{\psi} H \longrightarrow 0$$

is a central extension, then one shows that the induced surjection $H' \longrightarrow G$ has a central kernel; hence there is a (unique) map $H \longrightarrow H'$ over $G$, which must be a section of $H' \longrightarrow H$ (one first shows that $[H', H']$ is perfect, and $\psi([H', H']) = H$; if $x_0 \in \psi^{-1}(K)$, then $x \mapsto x_0 \, x \, x_0^{-1}$ gives a map $[H', H'] \longrightarrow [H', H']$ over $H$).

(b) If $G$ has a universal central extension

$$(E)\ldots \qquad 0 \longrightarrow K \longrightarrow H \longrightarrow G \longrightarrow 0$$

then $H = [H, H]$ by (a) above, so $G = [G, G]$. Conversely, if $G = [G, G]$, and $F \longrightarrow G$ is a surjection from a free group to $G$, giving a presentation

$$0 \longrightarrow R \longrightarrow F \longrightarrow G \longrightarrow 0,$$

then

$$0 \longrightarrow R/[F, R] \longrightarrow F/[F, R] \longrightarrow G \longrightarrow 0$$

is a central extension, and

$$0 \longrightarrow \frac{R \cap [F, F]}{[F, R]} \longrightarrow [F, F]/[F, R] \longrightarrow G \longrightarrow 0$$

is a central extension with $[F, F]/[F, R]$ as a perfect group. One directly verifies that this is a universal central extension: given any central extension

$$(E)\ldots \qquad 0 \longrightarrow K \longrightarrow H \longrightarrow G \longrightarrow 0,$$

there is a map $F \longrightarrow H$ over $G$ (as $F$ is free), which kills $[F, R]$ since $(E)$ is central; as in (a), the restriction of this map to $[F, F]$ is independent of the choice of $F \longrightarrow H$ over $G$.

**Proposition (1.11).** (a) $St(R)$ and $St_n(R)$, $n \geq 3$ are perfect.
(b) $St(R)$ and $St_n(R)$, $n \geq 5$ have no non-split central extensions.

**Corollary (1.12).** *The extension*

$$0 \longrightarrow K_2(R) \longrightarrow St(R) \longrightarrow E(R) \longrightarrow 0$$

*is a universal central extension of $E(R)$. In particular,*

$$K_2(R) = H_2(E(R), \mathbb{Z}).$$

**Proof of (1.11).** (a) is clear from the definitions of $St_n(R)$, $St(R)$.
(b) (Outline) Let

$$(E)\ldots \qquad 0 \longrightarrow K \longrightarrow H \xrightarrow{\psi} St(R) \longrightarrow 0$$

be a central extension. For each $i \neq j$ and each $x_{ij}(\lambda)$, choose $h \neq i, j$, and let $y_1 \in \psi^{-1}(x_{ih}(1))$, $y_2 \in \psi^{-1}(x_{hj}(\lambda))$. Then $y_{ij}(\lambda) = [y_1, y_2] \in$

$\psi^{-1}(x_{ij}(\lambda))$ is independent of the choices of $y_1, y_2$, since $(E)$ is a central extension. One checks that

(i) $y_{ij}(\lambda)$ does not depend on the choice of $h \neq i, j$

(ii) $\{y_{ij}(\lambda)\}$ satisfy the Steinberg identities (1.3)(a), (b), (c).

Thus, $x_{ij}(\lambda) \longmapsto y_{ij}(\lambda)$ gives a section of $\psi$. A similar argument works for $St_n(R)$, $n \geq 5$.

**Products.** Suppose $R$ is a commutative ring. Then the tensor product $(P, Q) \longrightarrow P \otimes_R Q$ on projective modules induces a pairing

$$K_0(R) \otimes_{\mathbb{Z}} K_0(R) \longrightarrow K_0(R)$$

making $K_0(R)$ into a commutative, associative ring, with identity element given by the class $[R]$ of the free module of rank 1.

We claim that there are natural pairings

$$K_0(R) \otimes_{\mathbb{Z}} K_i(R) \longrightarrow K_i(R), \quad i = 1, 2$$

making $K_i(R)$, $i = 1, 2$ into $K_0(R)$-modules. Given a projective module $P$ and a matrix $A \in GL_n(R)$, choose a projective module $Q$ such that $P \oplus Q \cong R^m$ is free, and fix such an isomorphism. Then we have an automorphism

$$(A \otimes 1_P) \oplus (1_{R^n} \otimes 1_Q) \in \text{Aut}(R^n \otimes_R (P \oplus Q)) \cong GL_{mn}(R).$$

If $h_P : GL_n(R) \longrightarrow GL(R)$ is the resulting map, then it is well defined up to an inner conjugation on $GL(R)$. If $\tilde{h}_P : GL_n(R) \longrightarrow K_1(R)$ is induced by composing with the natural quotient map, then $\tilde{h}_{P_1 \oplus P_2} = \tilde{h}_{P_1} + \tilde{h}_{P_2}$, so $\tilde{h}_P$ depends only on the class $[P] \in K_0(R)$. Thus one gets a well defined product $K_0(R) \otimes_{\mathbb{Z}} K_1(R) \longrightarrow K_1(R)$.

Next, we note that $h_P(E_n(R))$ lies in the commutator subgroup $E(R) \subset GL(R)$, and one checks[†] that the induced map $E_n(R) \longrightarrow E(R)$ is well defined up to an inner conjugation of $E(R)$. Hence $h_P$ induces a map $(\overline{h_P})_* : H_2(E_n(R), \mathbb{Z}) \longrightarrow H_2(E(R), \mathbb{Z})$. Further, one can check $(\overline{h_{P \oplus Q}})_* = (\overline{h_P})_* + (\overline{h_Q})_*$. Hence the induced map

$$(h_P)_* : H_2(E(R), \mathbb{Z}) = \varinjlim H_2(E_n(R), \mathbb{Z}) \longrightarrow H_2(E(R), \mathbb{Z})$$

depends only on the class $[P] \in K_0(R)$. This gives a product $K_0(R) \otimes_{\mathbb{Z}} K_2(R) \longrightarrow K_2(R)$.

---

[†] To verify that $h_P : E_n(R) \longrightarrow E(R)$ is well defined up to an inner conjugation, we note that it factors through $E_{mn}(R)$; now conjugation by $B \in GL_{mn}(R)$ agrees with conjugation by $\begin{bmatrix} B & 0 \\ 0 & B^{-1} \end{bmatrix} \in E_{2mn}(R) \subset E(R)$ on the subgroup $E_{mn}(R)$.

**Example (1.13).** *Let $R$ be a commutative ring, $R[t,t^{-1}]$ the Laurent polynomials over $R$. Change of rings yields a homomorphism $\alpha : K_0(R) \longrightarrow K_0(R[t,t^{-1}])$. Since $t \in R[t,t^{-1}]^* \subset GL(R)$, $t$ gives a class in $K_1(R[t,t^{-1}])$. Composing $\alpha$ with multiplication by $[t] \in K_1(R[t,t^{-1}])$ yields a map $\psi : K_0(R) \longrightarrow K_1(R[t,t^{-1}])$. It turns out that $\psi$ is always injective, and its image is (functorially) a direct summand in $K_1(R[t,t^{-1}])$.*

Next, for commutative rings $R$, we construct a pairing
$$K_1(R) \otimes_{K_0(R)} K_1(R) \longrightarrow K_2(R).$$
Suppose $\alpha, \beta \in E(R)$ commute; we define a 'product' $\alpha \star \beta \in K_2(R)$ by $\alpha \star \beta = [\tilde{\alpha}, \tilde{\beta}]$, where $\tilde{\alpha}, \tilde{\beta} \in St(R)$ are inverse images in $St(R)$ of $\alpha, \beta$ respectively. Then $\alpha \star \beta$ does not depend on the specific lifts $\tilde{\alpha}, \tilde{\beta}$, and has the following properties:

(a) if $\alpha_1$ and $\alpha_2$ both commute with $\beta$, then
$$(\alpha_1 \alpha_2) \star \beta = (\alpha_1 \star \beta) \cdot (\alpha_2 \star \beta)$$
(b) for any $\nu \in E(R)$, $\nu \alpha \nu^{-1} \star \nu \beta \nu^{-1} = \alpha \star \beta$
(c) $\alpha \star \beta = -(\beta \star \alpha)$.

Now let $A \in GL_n(R)$, $B \in GL_m(R)$, and let $I_n \in GL_n(R)$, $I_m \in GL_m(R)$ be the respective identity matrices. Then
$$\alpha = (A \otimes I_m, A^{-1} \otimes I_m, I_n \otimes I_m),$$
$$\beta = (I_n \otimes B, I_n \otimes I_m, I_n \otimes B^{-1})$$
give commuting elements of $E_{3mn}(R)$. We define
$$\{A, B\} = \alpha \star \beta.$$
One checks that this determines a well-defined skew-symmetric pairing
$$K_1(R) \otimes K_1(R) \longrightarrow K_2(R)$$
which is $K_0(R)$-bilinear (see Milnor's book, Ch. 8 for details).

In particular, if $m = n = 1$, we have a product
$$R^* \otimes_{\mathbb{Z}} R^* \longrightarrow K_2(R)$$
called the *Steinberg symbol*, given by
$$(a \otimes b) \longmapsto \{a, b\} = \begin{bmatrix} a & 0 & 0 \\ 0 & a^{-1} & 0 \\ 0 & 0 & 1 \end{bmatrix} \star \begin{bmatrix} b & 0 & 0 \\ 0 & 1 & 0 \\ 0 & 0 & b^{-1} \end{bmatrix}.$$
One knows (Lemma (9.8) of Milnor's book) that the Steinberg symbol has the property that if $a \in R^*$ such that $1 - a \in R^*$, then the *Steinberg relation* $\{a, 1-a\} = 1$ is valid in $K_2(R)$; also, for any $a \in R^*$, $\{a, -a\} = 1$.

**Theorem (1.14)** (Matsumoto). *If $F$ is a (commutative) field, then $K_2(F)$ has a presentation as the free Abelian group on the symbols $\{a,b\}$ with $a, b \in F^*$ subject to the relations:*

(i) $\{a_1 a_2, b\} = \{a_1, b\} \{a_2, b\}$ *for all* $a_1, a_2, b \in F^*$

(ii) $\{a, b\} = \{b, a\}^{-1}$ *for all* $a, b \in F^*$

(iii) $\{a, 1-a\} = 1$ *for all* $a \in F^* - \{1\}$.

**Symbols on a Field.** Let $F$ be a field. A *symbol* on $F$ with values in an Abelian group $G$ is a map

$$(\,,\,) : F^* \times F^* \longrightarrow G$$

satisfying

(i) $(a_1 a_2, b) = (a_1, b)(a_2, b)$ for all $a_1, a_2, b \in F^*$

(ii) $(a, b) = (b, a)^{-1}$ for all $a, b \in F^*$

(iii) $(a, 1-a) = 1$ for all $a \in F^* - \{1\}$.

From Matsumoto's theorem, there is a bijection between symbols on $F$ with values in $G$, and homomorphisms $K_2(F) \longrightarrow G$. There are many interesting examples of symbols arising "in nature". We list some below.

**Example (1.15)** (Tame Symbol). *Let $F$ be a field with a discrete valuation $v$, whose residue field is $k(v)$. Let $U \subset F^*$ be the kernel of $v : F^* \longrightarrow \mathbb{Z}$, and let $\psi : U \longrightarrow k(v)^*$ be the quotient map. Define the tame symbol*

$$T_v : F^* \times F^* \longrightarrow k(v)^*$$

*by the formula*

$$T_v(a, b) = \psi((-1)^{v(a)v(b)} a^{v(b)} / b^{v(a)}).$$

(We also denote the resulting map $K_2(F) \longrightarrow k(v)^*$ by $T_v$ and call it the tame symbol; we follow this practice in the other examples too.)

**Example (1.16)** (Galois symbol). *Let $F$ be a field, $n$ a positive integer such that $\frac{1}{n} \in F$. Assume that $F$ contains the nth roots of unity, and let $\zeta \in F^*$ be a primitive nth root of unity. The nth order Galois symbol is a map*

$$F^* \times F^* \longrightarrow {}_n Br(F),$$

*where ${}_n Br(F)$ is the n-torsion subgroup of the Brauer group of similarity classes of central simple algebras over $F$. The map is given by*

$$(a, b) \longrightarrow [D_\zeta(a, b)] \in Br(F)$$

where $D_\zeta(a,b)$ is the cyclic algebra of dimension $n^2$ over $F$ with generators $X, Y$ satisfying $X^n = a$, $Y^n = b$, $XY = \zeta YX$.

**Example (1.17)** (Differential symbol). *Let $F$ be a field, $\Omega^1_F = \Omega^1_{F/\mathbb{Z}}$ the $F$-vector space of absolute Kähler differentials of $F$; thus $\Omega^1_F$ is spanned by symbols $da$, for $a \in F$, subject to the relations $d(a_1 a_2) = a_1 \cdot da_2 + a_2 \cdot da_1$ for all $a_1, a_2 \in F$. Let $\Omega^2_F = \wedge^2_F \Omega^1_F$. The differential symbol is the map*

$$F^* \times F^* \longrightarrow \Omega^2_F$$

*given by $(a,b) \longrightarrow \frac{da}{a} \wedge \frac{db}{b}$, for all $a,b \in F^*$.*

**Example (1.18)** (Norm-Residue symbols). Let $k$ be a local field with $\frac{1}{m} \in k$, and containing the $m$th roots of unity. Let $k'$ be the extension field obtained by adjoining the $m$th roots of all elements of $k$. By Kummer theory, we have an isomorphism

$$k^* \otimes_\mathbb{Z} \mathbb{Z}/m\mathbb{Z} \cong \mathrm{Hom}(\mathrm{Gal}(k'/k), \mu_m),$$

and by local class-field theory, we have an isomorphism

$$\mathrm{Gal}(k'/k) \cong k^*/N_{k'/k}(k')^*$$

where $N_{k'/k}$ is the norm (note that $k'/k$ is a finite extension, since $k^* \otimes_\mathbb{Z} \mathbb{Z}/m\mathbb{Z}$ is finite). Thus we have a pairing

$$k^* \otimes_\mathbb{Z} k^* \xrightarrow{\langle,\rangle} \mu_m$$

called the *local $m$th-power norm-residue symbol*; one can check that it does indeed satisfy the conditions (i)–(iii) in the definition of a symbol. In fact, if $k \neq \mathbb{C}$, then the $m$-torsion subgroup of $Br(k)$ is known to be isomorphic to $\mathbb{Z}/m\mathbb{Z} \cong \mu_m$; for a suitable choice of this isomorphism, one can identify the local norm residue symbol with the Galois Symbol (Ex. (1.16)).

Now let $K$ be a number field. For any place $v$ of $K$ which is either real or non-Archimedean, if $m(v)$ is the number of roots of unity in the local field $K_v$, let $\langle\ ,\ \rangle_v$ denote the composition of $K_2(K) \longrightarrow K_2(K_v)$ with the local $m(v)$th power norm-residue symbol, taking values in the group $\mu_v \subset K_v^*$ of roots of unity. If $m$ is the order of the group $\mu_K$ of roots of unity in $K$, we have surjections $\mu_v \longrightarrow \mu_K$ given by raising to the $\left(\frac{m(v)}{m}\right)$-th power.

**Theorem** (Moore Reciprocity Law). *The sequence*

$$K_2(K) \xrightarrow{\alpha} \bigoplus_v \mu_v \xrightarrow{\beta} \mu_K \longrightarrow 0$$

is exact, where $\alpha$ is the sum of the maps $\langle\ ,\ \rangle_v$ (and in particular, almost all $\langle\ ,\ \rangle_v$ are trivial on a given class in $K_2(K)$), and $\beta$ is the sum of the surjections $\mu_v \longrightarrow \mu_K$ above.

**Example (1.19).** *Let $F$ be a finite field. Then $K_2(F) = 0$.*

This is easily proved using Matsumoto's theorem (1.14), and the fact that the multiplicative group $F^*$ of non-zero elements of $F$ is a cyclic group. Indeed, let $u \in F^*$ be a generator. It suffices to prove that $\{u, u\}$ is trivial in $K_2(F)$. If $F$ has characteristic 2, then $\{u, u\} = \{u, -u\}$ is trivial. So we may assume $F$ has odd characteristic. If $F$ has cardinality $q$, then $u^{(q-1)/2} = -1$, so $\{u, u^{(q+1)/2}\} = \{u, -u\}$ is trivial. Hence $\{u, u\}$ has order dividing $(q+1)/2$, as well as $q-1$ (as $u$ has order $q-1$ in $F^*$), i.e., $\{u, u\}$ has order at most 2.

Now $F^* - \{1\}$ is invariant under the bijection $v \mapsto 1 - v$. Since $F^* - \{1\}$ has $(q-1)/2$ non-squares, and $(q-3)/2$ squares, there must be a non-square $v$ such that $1 - v$ is also a non-square. Then $v = u^i$, $1 - v = u^j$ with $i, j$ odd, so that the triviality of $\{v, 1 - v\}$ implies that $\{u, u\}$ has odd order. Hence $\{u, u\}$ is trivial.

**Remark.** M. Stein has shown that if $R$ is a commutative semi-local ring, which is generated by $R^*$ as an additive group, then $K_2(R)$ is generated by Steinberg symbols (see "Surjective stability in dimension 0 for $K_2$ and related functors", *Trans. A.M.S.* 178 (1973), 165–191). W. van der Kallen has further shown that the relations (i), (ii), (iii), in Theorem (1.14) determine a presentation for $K_2(R)$, provided all residue fields have at least 5 elements (see "The $K_2$ of rings with many units", *Ann. Sci. Ecole Norm. Sup.* 10 (1977), 473–515).

## 2. The Plus Construction

In this section, all spaces, pairs, etc. have the homotopy type of a $CW$-complex. The main reference for this chapter is: J.-L. Loday, $K$-théorie algébrique et représentations de groupes, *Ann. Sci. Ecole Norm. Sup.* 9 (1976). A general reference for topology is: G.W. Whitehead, *Elements of Homotopy Theory*, Grad. Texts in Math., No. 61, Springer-Verlag. These are cited below as [L] and [W], respectively.

For any associative ring $R$, we regard $GL(R)$ as a topological group with the discrete topology, and let $BGL(R)$ denote the 'classifying space' of $GL(R)$. For our purposes, it is only important to know that $BGL(R)$ is an Eilenberg–MacLane space $K(GL(R), 1)$, i.e., $BGL(R)$ is a connected space with $\pi_1(BGL(R)) \cong GL(R)$, $\pi_i(BGL(R)) = 0$ for $i \geq 2$, and that these properties characterize $BGL(R)$ up to homotopy equivalence (since we are assuming that all spaces considered here have the homotopy type of a $CW$-complex). We give a construction of the classifying space of a discrete group in the next chapter (Example (3.10)).

Since $\pi_1(BGL(R)) \cong GL(R)$, $\pi_1(BGL(R))$ has a perfect normal subgroup $E(R)$. We will construct below a space $BGL(R)^+$ by applying the plus construction of Quillen to the pair $(BGL(R), E(R))$. There is an inclusion $i : BGL(R) \longrightarrow BGL(R)^+$ such that

(i) $i_* : \pi_1(BGL(R)) \longrightarrow \pi_1(BGL(R)^+)$ is the natural quotient map $GL(R) \longrightarrow GL(R)/E(R)$.

(ii) for any local coefficient system $L$ on $BGL(R)^+$,
$$i_* : H_n(BGL(R), i^*L) \longrightarrow H_n(BGL(R)^+, L)$$
is an isomorphism for all $n \geq 0$.

These properties will characterize $BGL(R)^+$ up to homotopy equivalence. Quillen's first definition of higher $K$-groups is: $K_i(R) = \pi_i(BGL(R)^+)$, for $i \geq 1$.

The plus construction is described in the following result.

**Theorem (2.1)** (Quillen). *Let $(X, x)$ be a path connected space, $N \triangleleft \pi_1(X, x)$ a perfect normal subgroup. Then there exists a continuous map of pairs $f : (X, x) \longrightarrow (X^+, x^+)$ such that*

(a) *there is an exact sequence*

$$0 \longrightarrow N \longrightarrow \pi_1(X,x) \xrightarrow{f_*} \pi_1(X^+, x^+) \longrightarrow 0$$

(b) *for any local coefficient system $L$ on $X^+$,*

$$f_* : H_n(X, f^*L) \longrightarrow H_n(X^+, L)$$

*is an isomorphism for any $n \geq 0$*

(c) *if $g : (X,x) \longrightarrow (Y,y)$ is a continuous map such that*

$$N \subset \ker(g_* : \pi_1(X,x) \longrightarrow \pi_1(Y,y)),$$

*then there exists a continuous map $h : (X^+, x^+) \longrightarrow (Y,y)$, unique up to homotopy, making the diagram*

$$\begin{array}{ccc} (X,x) & \xrightarrow{f} & (X^+, x^+) \\ {\scriptstyle g} \searrow & & \swarrow {\scriptstyle h} \\ & (Y,y) & \end{array}$$

*commute.*

**Proof.** We construct $X^+$ by attaching 2-cells and 3-cells to $X$, so that $(X^+, X)$ is a *CW*-pair of (relative) dimension 3, and take $x^+ = x$. We construct $X^+$ in a number of steps:

**Step (i).** First choose classes $e_\alpha \in \pi_1(X,x)$, for $\alpha$ running over a suitable index set $\mathcal{A}$, such that $e_\alpha$ generate $N$ as a normal subgroup. For each $\alpha \in \mathcal{A}$, choose a loop $\gamma_\alpha$ representing $e_\alpha$, and attach a 2-cell $a_\alpha$ to $X$ using $\gamma_\alpha$ on the boundary. Let $X_1$ be the resulting space. Then by van Kampen's theorem, $\pi_1(X_1, x)$ is the quotient of $\pi_1(X,x)$ by the normal subgroup generated by the $e_\alpha$, i.e., $\pi_1(X_1, x) = \pi_1(X,x)/N$.

**Step (ii).** Let $\tilde{X}_1 \longrightarrow X_1$ be the universal covering, and let $\hat{X} \longrightarrow X$ be the induced covering, so that we have a pullback diagram

$$\begin{array}{ccc} \hat{X} & \longrightarrow & \tilde{X}_1 \\ \downarrow & & \downarrow{\scriptstyle \pi} \\ X & \longrightarrow & X_1 \end{array}$$

Then $\tilde{X}_1$ is obtained from $\hat{X}$ by attaching 2-cells; since $\tilde{X}_1$ is connected, so is $\hat{X}$. Thus $\hat{X} \longrightarrow X$ is the covering space corresponding to the subgroup $N \subset \pi_1(X,x)$, and is Galois with group $\pi_1(X,x)/N$. For each 2-cell $a_\alpha$ of $(X_1, X)$, $\pi_1(X,x)$ acts transitively on the 2-cells in $\pi^{-1}(a_\alpha)$, with the isotropy of any of these 2-cells being $N$. Hence $H_2(\tilde{X}_1, \hat{X}; \mathbb{Z})$ is a free $\mathbb{Z}[\pi_1(X,x)/N]$-module on generators $[\tilde{a}_\alpha]$, where $\tilde{a}_\alpha$ is a 2-cell in $\pi^{-1}(a_\alpha)$.

**Step (iii).** We have the following diagram, whose vertical maps are Hurewicz maps,

$$\begin{array}{ccccccc} \pi_2(\hat{X}) & \longrightarrow & \pi_2(\tilde{X}_1) & \longrightarrow & \pi_2(\tilde{X}_1, \hat{X}) & \longrightarrow & \pi_1(\hat{X}) \\ \downarrow & & \downarrow & & \downarrow & & \downarrow \\ H_2(\hat{X}, \mathbb{Z}) & \longrightarrow & H_2(\tilde{X}_1, \mathbb{Z}) & \xrightarrow{j} & H_2(\tilde{X}_1, \hat{X}; \mathbb{Z}) & \longrightarrow & H_1(\hat{X}, \mathbb{Z}). \end{array}$$

Now $H_1(\hat{X}, \mathbb{Z}) = \pi_1(\hat{X})^{ab} = N^{ab} = 0$ since $N$ is perfect. Next, since $\pi_1(\tilde{X}_1) = 0$, $\pi_2(\tilde{X}_1) \cong H_2(\tilde{X}_1, \mathbb{Z})$ by the Hurewicz theorem. Since $j$ is onto, we can thus find maps $\tilde{f}_\alpha : S^2 \longrightarrow \tilde{X}_1$ such that the homotopy class $[\tilde{f}_\alpha] \in \pi_2(\tilde{X}_1)$ maps to $[\tilde{a}_\alpha] \in H_2(\tilde{X}_1, \hat{X}; \mathbb{Z})$ under the composite

$$\pi_2(\tilde{X}_1) \longrightarrow H_2(\tilde{X}_1, \mathbb{Z}) \xrightarrow{j} H_2(\tilde{X}_1, \hat{X}; \mathbb{Z}).$$

Let $f_\alpha : S^2 \longrightarrow X_1$ be the composite $\pi \circ \tilde{f}_\alpha$, and let $X^+$ be the space obtained by attaching 3-cells $b_\alpha$ to $X_1$ using $f_\alpha$ along the boundary, for each $\alpha \in \mathcal{A}$.

By construction, $(X^+, X_1)$ is a relative $CW$-complex with only 3-cells, so that $\pi_1(X_1, x) \cong \pi_1(X^+, x) \cong \pi_1(X, x)/N$. This proves (a) of the theorem for $(X, x) \longrightarrow (X^+, x^+)$ where $x^+ = x$.

Let $\tilde{X}^+$ be the universal cover of $X^+$; then we can extend our earlier diagram as follows:

$$\begin{array}{ccccc} \hat{X} & \longrightarrow & \tilde{X}_1 & \longrightarrow & \tilde{X}^+ \\ \downarrow & & \downarrow & & \downarrow \\ X & \longrightarrow & X_1 & \longrightarrow & X^+ \end{array}$$

The 2 squares are Cartesian (i.e., are pullbacks).

The relative cellular chain complex (see (A.14)) for $(\tilde{X}^+, X)$ looks like:

$$\cdots 0 \longrightarrow C_3(\tilde{X}^+, \hat{X}) \xrightarrow{d} C_2(\tilde{X}^+, \hat{X}) \longrightarrow 0 \cdots$$

with non-zero terms only in dimensions 2 and 3. We also have isomorphisms (from the definition of the relative cellular chain complex)

$$C_3(\tilde{X}^+, \hat{X}) \cong C_3(\tilde{X}^+, \tilde{X}_1) \cong H_3(\tilde{X}^+, \tilde{X}_1; \mathbb{Z}),$$
$$C_2(\tilde{X}^+, \hat{X}) \cong C_2(\tilde{X}_1, \hat{X}) \cong H_2(\tilde{X}_1, \hat{X}; \mathbb{Z}),$$

such that $d$ is identified with the boundary map in the homology sequence of the triple $(\tilde{X}^+, \tilde{X}_1, \hat{X})$:

$$\cdots \longrightarrow H_3(\tilde{X}_1, \hat{X}; \mathbb{Z}) \longrightarrow H_3(\tilde{X}^+, \hat{X}; \mathbb{Z}) \longrightarrow H_3(\tilde{X}^+, \tilde{X}_1; \mathbb{Z})$$
$$\xrightarrow{d} H_2(\tilde{X}_1, \hat{X}; \mathbb{Z}) \longrightarrow \cdots .$$

From the construction of the homology sequence of a triple, $d = j \circ \partial$ where $\partial$ is the boundary map

$$\cdots H_3(\tilde{X}^+, \mathbb{Z}) \longrightarrow H_3(\tilde{X}^+, \tilde{X}_1; \mathbb{Z}) \xrightarrow{\partial} H_2(\tilde{X}_1, \mathbb{Z}) \longrightarrow \cdots$$

## 2. The Plus Construction

in the homology sequence of the pair $(\tilde{X}^+, \tilde{X}_1)$, and $j$ is the natural map $H_2(\tilde{X}_1, \mathbb{Z}) \longrightarrow H_2(\tilde{X}_1, \hat{X}; \mathbb{Z})$ (used in (ii) above to construct $X^+$).

Let $\tilde{b}_\alpha$ be the 3-cell of $\tilde{X}^+$ with attaching map $\tilde{f}_\alpha$. We have a diagram with vertical Hurewicz maps

$$\begin{array}{ccccc}
\longrightarrow & \pi_3(\tilde{X}^+) & \longrightarrow & \pi_3(\tilde{X}^+, \tilde{X}_1) & \longrightarrow & \pi_2(\tilde{X}_1) \\
& \downarrow & & \downarrow & & \downarrow \\
\longrightarrow & H_3(\tilde{X}^+, \mathbb{Z}) & \longrightarrow & H_3(\tilde{X}^+, \tilde{X}_1, \mathbb{Z}) & \xrightarrow{\partial} & H_2(\tilde{X}_1, \mathbb{Z}).
\end{array}$$

Then $\tilde{b}_\alpha$ also determine elements of $\pi_3(\tilde{X}^+, \tilde{X}_1)$, which we also denote by $[\tilde{b}_\alpha]$, whose images under the Hurewicz map are the class $[\tilde{b}_\alpha] \in H_3(\tilde{X}^+, \tilde{X}_1; \mathbb{Z})$ corresponding to the 3-cells $\tilde{b}_\alpha$.

If $[\tilde{f}_\alpha] \in \pi_2(\tilde{X}_1)$ is the class determined by $\tilde{f}_\alpha$, the attaching map of $\tilde{b}_\alpha$, then $[\tilde{b}_\alpha]$ maps to $[\tilde{f}_\alpha]$ under the boundary map $\pi_3(\tilde{X}^+, \tilde{X}_1) \longrightarrow \pi_2(\tilde{X}_1)$, from the definition of this boundary map. By construction, $[\tilde{f}_\alpha] \in \pi_2(\tilde{X}_1)$ maps to $[\tilde{a}_\alpha] \in H_2(\tilde{X}_1, \hat{X}; \mathbb{Z})$ under the composite

$$\pi_2(\tilde{X}_1) \longrightarrow H_2(\tilde{X}_1, \mathbb{Z}) \xrightarrow{j} H_2(\tilde{X}_1, \hat{X}; \mathbb{Z}).$$

We conclude that $d : H_3(\tilde{X}^+, \tilde{X}_1; \mathbb{Z}) \longrightarrow H_2(\tilde{X}_1, \hat{X}; \mathbb{Z})$ maps $[\tilde{b}_\alpha]$ to $[\tilde{a}_\alpha]$.

As in (ii) above, $H_3(\tilde{X}^+, \tilde{X}_1; \mathbb{Z})$ is a free $[\pi_1(X, x)/N]$-module on the generators $[\tilde{b}_\alpha]$. Hence, $C_3(\tilde{X}^+, \hat{X})$ and $C_2(\tilde{X}^+, \hat{X})$ are both free $(\pi_1(X, x)/N)$-modules on generators indexed by the same set $\mathcal{A}$, and $d$ puts the two sets of generators in bijection. Hence $d$ is an isomorphism of $(\pi_1(X, x)/N)$-modules.

Now let $L$ be a local coefficient system on $X^+$, i.e., $L$ is an Abelian group on which $\pi_1(X^+, x^+)$ acts. Equivalently, $L$ is a $(\pi_1(X, x)/N)$-module. The relative homology groups $H_n(X^+, X; L)$ are computed as the homology groups of the tensor product complex

$$L \otimes_{\mathbb{Z}[\pi_1(X,x)/N]} C_*(\tilde{X}^+, \hat{X}).$$

But this complex is acyclic, so $H_n(X^+, X; L) = 0 \; \forall n$. From the exact homology sequence for $(X^+, X)$ with local coefficients in $L$, we see that $H_n(X, i^*L) \cong H_n(X^+, L) \; \forall n$, where $i : X \to X^+$. This proves (b).

The proof of (c) is by obstruction theory; we omit the details (however, see (A.16)).

**Example (2.2).** Let $X = BS_\infty$, where $S_\infty = \bigcup_{n \geq 1} S_n$ is the *infinite permutation group*. Let $A_\infty \subset S_\infty$ be the infinite alternating group $\bigcup_{n \geq 1} A_n$. Then $A_\infty \triangleleft S_\infty = \pi_1(BS_\infty)$ is a perfect normal subgroup, so we can form the space $BS_\infty^+$, with $\pi_1(BS_\infty^+) = \mathbb{Z}/2\mathbb{Z}$. In fact, by the theorem of Barrat, Priddy and Quillen, one has a homotopy equivalence $\mathbb{Z} \times BS_\infty^+ \cong \Omega^\infty \Sigma^\infty(S^0)$ where $S^0$ is the 0-sphere (discrete 2 point space) (see S. Priddy, On $\Omega^\infty S^\infty$ and the infinite symmetric group, *Proc. Symp. Math.* 22, A. M. S. (1971), pp. 217–220). We note that for any $(X,x)$, if $\Sigma X$ is the reduced suspension and $\Omega X$ the loop space of loops based at $x$, with natural base points, then there is a natural continuous map $X \longrightarrow \Omega \Sigma X$; by applying this to $\Sigma^n X$ and applying the functor $\Omega^n$, we obtain a direct system

$$X \longrightarrow \Omega \Sigma X \longrightarrow \Omega^2 \Sigma^2 X \longrightarrow \cdots \longrightarrow \Omega^n \Sigma^n X \longrightarrow \cdots$$

whose direct limit is called $\Omega^\infty \Sigma^\infty X$. Then we have

$$\pi_i(\Omega^n \Sigma^n X) \cong \pi_{i+n}(\Sigma^n X),$$

so that we have a direct system of groups

$$\pi_i(X) \longrightarrow \pi_{i+1}(\Sigma X) \longrightarrow \pi_{i+2}(\Sigma^2 X) \longrightarrow \cdots$$

whose direct limit is the $i$th *stable homotopy group* of $X$.

Let $\pi_i$ be the $i$th stable homotopy group of $S^0$,

$$\pi_i = \varinjlim \pi_{i+k}(S^k) \cong \pi_i(BS_\infty^+), \quad i \geq 1.$$

Now $S_\infty \subset GL(\mathbb{Z})$ as the group of permutation matrices, and $A_\infty \subset E(\mathbb{Z})$. Hence, by the functoriality of the plus construction, there is a map $BS_\infty^+ \to BGL(\mathbb{Z})^+$. This induces maps on homotopy groups

$$\pi_i \cong \pi_i(BS_\infty^+) \longrightarrow \pi_i(BGL(\mathbb{Z})^+) = K_i(\mathbb{Z}).$$

For $i = 1$, the image of the generator of $\pi_1 = \mathbb{Z}/2\mathbb{Z}$ is $(-1) \in K_1(\mathbb{Z})$.

**Proposition (2.3).** *Let* $(\hat{X}, \hat{x}) \longrightarrow (X, x)$ *be the covering of* $X$ *corresponding to the subgroup* $N \triangleleft \pi_1(X, x)$, *and let* $(\tilde{X}^+, \tilde{x}^+)$ *be the universal covering of* $(X^+, x^+)$. *Then* $(\tilde{X}^+, \tilde{x}^+)$ *is the result of applying the plus construction to* $(\hat{X}, \hat{x})$.

**Proof.** This was built into our construction of $X^+$.

**Proposition (2.4).** *Let* $f_i : (X_i, x_i) \longrightarrow (X_i^+, x_i^+)$, $i = 1, 2$ *be obtained from the plus construction, for given perfect normal subgroups* $N_i \triangleleft \pi_1(X_i, x_i)$. *Then* $(f_1, f_2) : (X_1 \times X_2, (x_1, x_2)) \longrightarrow (X_1^+ \times X_2^+, (x_1^+, x_2^+))$ *is homotopy equivalent to the result of applying the plus construction to* $N_1 \times N_2 \triangleleft \pi_1(X_1 \times X_2, (x_1, x_2))$.

**Proof.** The properties (2.1) (a), (b) are easily verified. By (c), this characterizes the homotopy type of the resulting space.

We now return to the special case when $X = BGL(R)$, $N = E(R)$.

## 2. The Plus Construction

Let $F(R)$ be the homotopy fiber of $BGL(R) \longrightarrow BGL(R)^+$; since $BGL(R)^+$ is connected, the homotopy type of $F(R)$ is independent of the choices of the base points (see (A.27)).

**Proposition (2.5).**
(a) $F(R)$ is acyclic, i.e., $\tilde{H}_n(F(R), \mathbb{Z}) = 0$ for all $n \geq 0$.
(b) $\pi_1(F(R)) \cong St(R)$, the Steinberg group.
(c) $\pi_1(F(R))$ acts trivially on $\pi_i(F(R))$, $i \geq 2$ i.e., $F(R)$ is simple in dimensions $\geq 2$.

**Proof.** (a) If we replace $BGL(R)^+$ by its universal cover, and $BGL(R)$ by the induced covering, this does not change the homotopy type of $F(f)$ (see (A.27)); we again use $F(R)$ to denote the homotopy fiber of

$$\widetilde{BGL(R)} \longrightarrow \widetilde{BGL(R)^+},$$

where $\widetilde{BGL(R)^+}$ is the universal cover of $BGL(R)^+$, and $\widetilde{BGL(R)}$ is the induced covering of $BGL(R)$, which is just the covering space associated to the subgroup $E(R) \subset GL(R) = \pi_1(BGL(R))$ (thus $\widetilde{BGL(R)}$ has the homotopy type of $BE(R)$, the Eilenberg–MacLane space with $\pi_1(BE(R)) \cong E(R)$, $\pi_i(BE(R)) = 0$, $i \neq 1$).

We have a spectral sequence (see (A.27))

$$E^2_{p,q} = H_p(\widetilde{BGL(R)^+}, H_q(F(R), \mathbb{Z})) \Longrightarrow H_{p+q}(\widetilde{BGL(R)}, \mathbb{Z})$$

(where the $E^2$-term is the usual homology group with coefficients in $H_q(F(R), \mathbb{Z})$, since the local coefficient system associated to $H_q(F(R), \mathbb{Z})$ is trivial on the simply connected space $\widetilde{BGL(R)^+}$). Further, from Proposition (2.3) and Theorem (2.1)(b), the edge homomorphisms

$$H_n(\widetilde{BGL(R)}, \mathbb{Z}) \twoheadrightarrow E^\infty_{n,0} \longrightarrow E^2_{n,0} = H_n(\widetilde{BGL(R)^+}, \mathbb{Z})$$

are isomorphisms, i.e., $E^2_{n,0} = E^\infty_{n,0}$ and $E^\infty_{p,q} = 0$ for $q \neq 0$.

Now suppose $F(R)$ is not acyclic; since $\widetilde{BGL(R)^+}$ is simply connected and $\widetilde{BGL(R)}$ is connected, $F(R)$ is path connected. Thus if $q$ is the smallest integer such that $\tilde{H}_q(F(R), \mathbb{Z}) \neq 0$, then $q > 0$. Then $E^2_{p,q'} = E^r_{p,q'} = 0$ for all $p$, and all $q'$ with $0 < q' < q$. Since $d_r : E^r_{p,q} \longrightarrow E^r_{p-r,q+r-1}$, we see that $E^2_{0,q} \cong E^{q+1}_{0,q}$, $E^{q+2}_{0,q} \cong E^\infty_{0,q}$, and $E^{q+2}_{0,q}$ is the cokernel of $d_{q+1} : E^{q+1}_{q+1,0} \longrightarrow E^{q+1}_{0,q}$. But $E^2_{n,0} = E^\infty_{n,0}$, so that $d_{q+1} = 0$ also. Hence $E^2_{0,q} = E^\infty_{0,q} = 0$ as seen above. But on the other hand,

$$E^2_{0,q} = H_0(BGL(R)^+, H_q(F(R), \mathbb{Z}) \neq 0$$

as we assumed $H_q(F(R), \mathbb{Z}) = \tilde{H}_q(F(R), \mathbb{Z}) \neq 0$. This contradiction proves that $F(R)$ is acyclic.

(b), (c): Let $G = \pi_1(F(R))$, and consider the spectral sequence (see (A.28)) for the universal covering $\tilde{F}(R) \longrightarrow F(R)$. This has the form

$$E^2_{p,q} = H_p(G, H_q(\tilde{F}(R), \mathbb{Z})) \Longrightarrow H_{p+q}(F(R), \mathbb{Z})$$

where by (a) we have $\tilde{H}_n(F(R), \mathbb{Z}) = 0 \,\forall\, n$. Also, $\tilde{F}(R)$ is simply connected, so $H_1(\tilde{F}(R), \mathbb{Z}) = 0$. This forces

$$E^2_{1,0} = E^\infty_{1,0} = 0, \quad E^2_{2,0} = E^\infty_{2,0} = 0;$$

further $E^2_{3,0} \cong E^3_{3,0} \xrightarrow{d_3} E^3_{0,2} \cong E^2_{0,2}$ must be an isomorphism, since the kernel and cokernel are respectively $E^\infty_{3,0} = 0$ and $E^\infty_{0,2} = 0$. Thus

$$H_1(G, \mathbb{Z}) = H_2(G, \mathbb{Z}) = 0, \text{ and } H_3(G, \mathbb{Z}) \cong H_0(G, H_2(\tilde{F}(R))).$$

The homotopy exact sequence (see (A.18)) for

$$F(R) \longrightarrow \widehat{BGL}(R) \longrightarrow \widehat{BGL}(R)^+$$

together with

$$\pi_i(\widehat{BGL}(R)^+) \cong \pi_i(BGL(R)^+), \quad i \geq 2,$$
$$\pi_i(\widehat{BGL}(R)) = 0, \quad i \geq 2,$$

and $\pi_1(BGL(R)) = E(R)$ yields isomorphisms

$$\pi_{i+1}(BGL(R)^+) \cong \pi_i(F(R)), \quad i \geq 2,$$

and an exact sequence

(*)
$$0 \longrightarrow \pi_2(BGL(R)^+) \longrightarrow \pi_1(F(R)) \longrightarrow E(R) \longrightarrow 0.$$
$$\phantom{0 \longrightarrow \pi_2(BGL(R)^+) \longrightarrow } \| $$
$$\phantom{0 \longrightarrow \pi_2(BGL(R)^+) \longrightarrow } G$$

By Lemma (A.26), the conjugation action of $G$ on itself is trivial on $\ker(G \longrightarrow E(R))$, and $G$ acts trivially on $\pi_i(F(R))$, $i \geq 2$ (i.e., $F(R)$ is simple in dimensions $i \geq 2$). Thus the above exact sequence (*) is a central extension of $E(R)$, such that $H_1(G, \mathbb{Z}) = H_2(G, \mathbb{Z}) = 0$, i.e., by Proposition (1.10) is isomorphic to the universal central extension

$$0 \longrightarrow K_2(R) \longrightarrow St(R) \longrightarrow E(R) \longrightarrow 0.$$

In particular $G \cong St(R)$.

**Corollary (2.6).** $\pi_i(BGL(R)^+) \cong K_i(R)$, $i = 1, 2$, and

$$\pi_3(BGL(R)^+) \cong H_3(St(R), \mathbb{Z}).$$

**Proof.** The only thing left to prove is the formula for $\pi_3(BGL(R)^+)$. But $\pi_3(BGL(R)^+) \cong \pi_2(F(R)) \cong \pi_2(\tilde{F}(R)) \cong H_2(\tilde{F}(R), \mathbb{Z})$ (by Hurewicz); we also had an isomorphism

$$H_3(G, \mathbb{Z}) \cong H_0(G, H_2(\tilde{F}(R))) \cong H_0(G, \pi_2(F(R))) \cong \pi_2(F(R)),$$

since $F(R)$ is simple in dimension 2.

This corollary motivates Quillen's first definition of higher $K$-theory (for projective modules over a ring).

**Definition.** $K_i(R) = \pi_i(BGL(R)^+)$, $i \geq 1$.

As a consequence, we obtain

**Corollary (2.7).** $K_3(R) = H_3(St(R), \mathbb{Z})$.

**Remark.** We note that there are Hurewicz homomorphisms

$$K_i(R) = \pi_i(BGL(R)^+) \longrightarrow H_i(BGL(R)^+, \mathbb{Z}) \cong H_i(BGL(R), \mathbb{Z}).$$

Following Loday, we describe below a natural $H$-space structure on $BGL(R)^+$, which we then use to construct products

$$K_i(R) \otimes K_j(R) \longrightarrow K_{i+j}(R).$$

One other consequence of the $H$-space structure is that by a theorem of Milnor and Moore (see J. Milnor, J.C. Moore, On the Structure of Hopf Algebras, *Ann. Math.* 81 (1965), 211–264), the Hurewicz maps above are injective up to torsion, and the $\mathbb{Q}$-subspace

$$K_i(R) \otimes_{\mathbb{Z}} \mathbb{Q} \subset H_i(BGL(R), \mathbb{Q}) = H_i(GL(R), \mathbb{Q})$$

is identified with the subspace of *primitive elements* for the comultiplication of the natural Hopf algebra structure on $H_*(GL(R), \mathbb{Q})$ (for any connected space $X$, there is a comultiplication

$$\Delta_* : H_*(X, \mathbb{Q}) \longrightarrow H_*(X, \mathbb{Q}) \times_{\mathbb{Q}} H_*(X, \mathbb{Q})$$

induced by the diagonal $\Delta : X \longrightarrow X \times X$, and the Kunneth isomorphism $H_*(X \times X, \mathbb{Q}) \cong H_*(X, \mathbb{Q}) \otimes_{\mathbb{Q}} H_*(X, \mathbb{Q})$; the primitive elements $H_n(X, \mathbb{Q})$ are those elements $x$ satisfying $\Delta(x) = x \otimes 1 + 1 \otimes x$). This result has been used by A. Borel (*Stable Real Cohomology of Arithmetic Groups*, Ann. Sci. E.N.S. 7 (1974), 235–272) to compute the ranks of the higher $K$-groups of the ring of algebraic integers in a number field.

The following discussion is based closely on [L]. Let $R$ be a ring; for $\alpha, \beta \in GL(R)$ let $\alpha \oplus \beta$ be defined by

$$(\alpha \oplus \beta)_{ij} = \begin{cases} \alpha_{k\ell} & \text{if } i = 2k-1, j = 2\ell - 1 \\ \beta_{k\ell} & \text{if } i = 2k, j = 2\ell \\ 0 & \text{otherwise.} \end{cases}$$

Schematically, if

$$\alpha = \begin{bmatrix} \star & \star & \star & \cdots \\ \star & \star & \star & \cdots \\ \cdot & \cdot & \cdot & \cdots \\ \cdot & \cdot & \cdot & \cdots \\ \cdot & \cdot & \cdot & \cdots \end{bmatrix}, \quad \beta = \begin{bmatrix} x & x & x & \cdots \\ x & x & x & \cdots \\ \cdot & \cdot & \cdot & \cdots \\ \cdot & \cdot & \cdot & \cdots \\ \cdot & \cdot & \cdot & \cdots \end{bmatrix}$$

then

$$\alpha \oplus \beta = \begin{bmatrix} \star & 0 & \star & 0 & \cdots \\ 0 & x & 0 & x & \cdots \\ \star & 0 & \star & 0 & \cdots \\ 0 & x & 0 & x & \cdots \\ \vdots & \vdots & \vdots & \vdots & \end{bmatrix} \in GL(R).$$

**Lemma (2.8).** $\oplus : GL(R) \times GL(R) \longrightarrow GL(R)$ is a homomorphism.

**Proof.** Left as an easy exercise.

**Remark.** Let $\alpha \in GL_n(R)$, $\beta \in GL_m(R)$,

$$\gamma = \begin{bmatrix} \alpha & 0 \\ 0 & \beta \end{bmatrix} \in GL_{m+n}(R),$$

and let $\alpha^s$, $\beta^s$, $\gamma^s$ denote their images in $GL(R)$. Then $\alpha^s \oplus \beta^s \neq \gamma^s$, but the two homomorphisms $GL_n(R) \times GL_m(R) \longrightarrow GL(R)$, given by $(\alpha, \beta) \mapsto \alpha^s \oplus \beta^s$, and $(\alpha, \beta) \mapsto \gamma^s$, are conjugate by an element of $GL(R)$ (even by one of $E(R)$, since the two maps factor through $GL_d(R)$, $d = 2\max(m,n)$, on which conjugation by $A \in GL_d(R)$ equals conjugation by $\begin{bmatrix} A & 0 \\ 0 & A^{-1} \end{bmatrix} \in E_{2d}(R)$). In particular, for any $\alpha \in GL(R)$, $\alpha \oplus \alpha^{-1} \in E(R)$.

We define a product on $BGL(R)$, as follows. By Proposition (2.4), the natural map

$$k : B(GL(R) \times GL(R))^+ \longrightarrow BGL(R)^+ \times BGL(R)^+$$

is a homotopy equivalence. Choose a homotopy inverse $k^{-1}$, and let $+$ be the composite

$$+ : BGL(R)^+ \times BGL(R)^+ \xrightarrow{k^{-1}} B(GL(R) \times GL(R))^+ \xrightarrow{\oplus^+} BGL(R)^+,$$

where we note that (by Lemma (2.8)) there is a map $B(GL(R) \times GL(R)) \to BGL(R)$, such that the induced map on fundamental groups carries $E(R) \times E(R)$ into the commutator subgroup $E(R) \subset GL(R)$, and hence induces a map $\oplus^+$ between the plus constructions.

**Proposition (2.9).** $(BGL(R)^+, +)$ *is a homotopy commutative and associative, connected H-space, hence a commutative H-group.*

The proof will depend on a few simple lemmas, which we prove first.

Let $u : \mathbb{N} \to \mathbb{N}$ be an injective self-map of the set of positive integers. Define $u_\bullet : GL(R) \longrightarrow GL(R)$ by

$$u_\bullet(\alpha)_{ij} = \begin{cases} \alpha_{k\ell} & \text{if } (i,j) = (u(k), u(\ell)) \\ \delta_{ij} & \text{(Kronecker delta) otherwise.} \end{cases}$$

We call $u_\bullet$ a 'pseudo-conjugation' of $GL(R)$.

**Lemma (2.10).** *For each pseudo-conjugation $u_\bullet$, there is an induced map $u^+ : BGL(R)^+ \longrightarrow BGL(R)^+$ which is a homotopy equivalence.*

**Proof.** Since $u_\bullet$ is a homomorphism, by an easy computation, it induces a map $BGL(R)^+ \longrightarrow BGL(R)^+$; it also induces a $GL(R)/E(R)$-equivariant self-map on the universal covering space $B\widetilde{GL}(R)^+ \cong BE(R)^+$ (by Prop. (2.3), and the fact that $BE(R) \longrightarrow BGL(R)$ is precisely the covering associated to $E(R) \subset GL(R) = \pi_1(BGL(R))$). Let $\bar{u} : BE(R)^+ \longrightarrow BE(R)^+$ be this induced map; since $BE(R)^+$ is simply connected, if we show that it induces an isomorphism on integral homology groups, then it is a homotopy equivalence (see (A.10)).

Let $x \in H_n(BE(R), \mathbb{Z})$ be the class of the cycle $\Sigma n_i(g_1^{(i)}, \ldots, g_n^{(i)})$ (in the standard complex for $E(R)$, say). Then $\bar{u}_*(x)$ is the class of the cycle $\Sigma n_i(u_\bullet(g_1^{(i)}), \ldots, u_\bullet(g_n^{(i)}))$. Now the $g_j^{(i)}$ range over a finite set in $E(R)$, contained in $E_m(R)$, say. The map $u_\bullet|_{E_m(R)} : E_m(R) \longrightarrow E(R)$ is equal to conjugation by some $C \in E(R)$ (we can take $C$ to be an even permutation matrix), so that $\bar{u}_*(x)$ is also represented by $C(\Sigma n_i(g_1^{(i)}, \ldots, g_n^{(i)}))C^{-1}$. But inner conjugation induces the identity map on group homology, so that $\bar{u}_*(x) = x$.

**Lemma (2.11).** *Let $M$ be the monoid (under composition) of injective self-maps of the set $\mathbb{N}$ of natural numbers. Then the Grothendieck group $K_0(M) = 0$ (i.e., any monoid homomorphism from $M$ to a group is trivial).*

**Proof.** Suppose $u \in M$ has infinitely many fixed points; let $i \in M$ be given by $i(n) = n$th fixed point of $u$. Then $u \circ i = i$, so that the class $[u] \in K_0(M)$ is trivial. In general, for any $v \in M$, we claim there exists $u \in M$ such that $u, vu$ both have infinitely many fixed points. The proof of the claim is left as an exercise to the reader.

**Corollary (2.12).** *For any $u \in M$, $u^+ : BGL(R)^+ \longrightarrow BGL(R)^+$ is homotopic to the identity* (through maps preserving the base point).

**Proof.** The assignment $u \longrightarrow [u^+]$ gives a monoid homomorphism from $M$ to the group of homotopy classes of base point preserving self-homotopy equivalences of $BGL(R)^+$.

**Remark.** The fact that $\bar{u} : BE(R)^+ \longrightarrow BE(R)^+$ induces the identity on homology groups does not suffice to conclude that it is homotopic to the identity. In fact, there exist spaces with self-maps not homotopic to the identity, but which induce the identity on homology and homotopy (where we assume the map fixes a base point, to make sense of the statement about homotopy groups). M. Lemaire has constructed such an example.

**Definition.** Let $G$ be a group. Two homomorphisms $f, g : G \longrightarrow GL(R)$ are *pseudo-conjugate* if there exists $u \in M$ such that either $u_\bullet \circ f$ and $g$, or $u_\bullet \circ g$ and $f$, are conjugate by an element of $GL(R)$.

**Corollary (2.13).** *If $f, g : G \longrightarrow GL(R)$ are pseudo-conjugate, then the induced maps $f^+, g^+ : BG \longrightarrow BGL(R)^+$ are homotopic as maps preserving the respective base points.*

**Proof.** For any map $f : G \longrightarrow GL(R)$, let $f \oplus 1$ denote the map given by $x \mapsto f(x) \oplus 1$, for any $x \in G$. Then $f^+$ is homotopic to $(f \oplus 1)^+$, preserving the base point, because $f \oplus 1 = (u_0)_\bullet \circ f$, where $u_0 \in M$ is defined by $u_0(i) = 2i - 1$.

Now suppose the given maps $f, g$ satisfy $g = (u_\bullet \circ f)^\alpha$ for $\alpha \in GL(R)$ (where $(u_\bullet \circ f)^\alpha(x) = \alpha(u_\bullet \circ f(x))\alpha^{-1}$ for all $x \in G$). If $\beta = \alpha \oplus \alpha^{-1} \in E(R)$, then $g \oplus 1 = ((u_\bullet \circ f) \oplus 1)^\beta$. Thus, the induced maps $BG \longrightarrow BGL(R)$ are freely homotopic, such that under a homotopy, the image of the base point of $BG$ is a loop homotopic to $[\beta] \in \pi_1(BGL(R)) \cong GL(R)$ (Lemma (A.50)). The induced maps $BG \longrightarrow BGL(R)^+$ are homotopic preserving the base points, since $[\beta] \mapsto 0$ in $\pi_1(BGL(R)^+)$ (see (A.41)).

**Proof of Proposition (2.9).** Let $u_0, v_0 \in M$ be given by $u_0(i) = 2i - 1$, $v_0(i) = 2i$, for all $i \in \mathbb{N}$. Then $(u_0)_\bullet(\alpha) = \alpha \oplus 1$, and $(v_0)_\bullet(\alpha) = 1 \oplus \alpha$ for any $\alpha \in GL(R)$. By construction, the map $+ : BGL(R)^+ \times BGL(R)^+ \longrightarrow BGL(R)^+$, when restricted to $BGL(R)^+ \times \{\star\}$, is homotopic to $(u_0)^+$, and when restricted to $\{\star\} \times BGL(R)^+$, is homotopic to $(v_0)^+$. Hence the base point $\star \in BGL(R)^+$ is a 2-sided identity, up to homotopy, for the operation $+$. By definition, this makes $(BGL(R)^+, +)$ a $H$-space.

Next, there exist $w, w' \in M$ such that for any $x, y, z \in GL(R)$, $x \oplus y = w_\bullet \circ (y \oplus x)$, and $(x \oplus y) \oplus z = w'_\bullet \circ (x \oplus (y \oplus z))$. Hence $BGL(R)^+$ is homotopy commutative and associative. Since it is connected, it is an $H$-group (Lemma (A.47)).

**Loday's Product in $K$-Theory.** Following [L], we construct natural products
$$K_i(R) \otimes K_j(R) \longrightarrow K_{i+j}(R), \quad i,j \geq 1$$
which generalize the product for $i = j = 1$ considered in §1.

First, fix a choice of a bijection $\phi : \mathbb{N} \times \mathbb{N} \longrightarrow \mathbb{N}$. Given rings $R$, $R'$ and elements $\alpha \in GL(R)$, $\alpha' \in GL(R')$, $\phi$ enables us to convert $\alpha \otimes \alpha'$ into an element of $GL(R \otimes_\mathbb{Z} R')$. Thus we obtain a homomorphism $GL(R) \times GL(R') \longrightarrow GL(R \otimes_\mathbb{Z} R')$, which clearly maps $E(R) \times E(R')$ into the commutator subgroup $E(R \otimes_\mathbb{Z} R')$, and hence induces a continuous map
$$g : BGL(R)^+ \times BGL(R')^+ \longrightarrow BGL(R \otimes_\mathbb{Z} R')^+.$$
Let $x_0, x_0'$ be the respective base points of $BGL(R)^+$, $BGL(R')^+$ and define a new map
$$\tilde{g} : BGL(R)^+ \times BGL(R')^+ \longrightarrow BGL(R \otimes_\mathbb{Z} R')^+$$
by $\tilde{g}(x, x') = g(x, x') - g(x_0, x') - g(x, x_0') + g(x_0, x_0')$ where the operations on the right side are performed using the $H$-group structure on the space $BGL(R \otimes_\mathbb{Z} R')^+$ given by Proposition (2.9). Then by construction $\tilde{g}$ is null homotopic when restricted to $BGL(R)^+ \vee BGL(R')^+ = BGL(R)^+ \times \{x_0'\} \cup \{x_0\} \times BGL(R')^+$. Hence there exists a mapping on the smash product
$$h : BGL(R)^+ \wedge BGL(R')^+ \longrightarrow BGL(R \otimes_\mathbb{Z} R')^+,$$
unique up to homotopy, making the diagram below commute up to homotopy:

$$\begin{array}{ccc} BGL(R)^+ \times BGL(R')^+ & \longrightarrow & BGL(R)^+ \wedge BGL(R')^+ \\ & \searrow \quad \swarrow & \\ & BGL(R \otimes_\mathbb{Z} R')^+ & \end{array}$$

This induces bilinear pairings on homotopy groups (see (A.7))
$$h_* : K_i(R) \otimes_\mathbb{Z} K_j(R) \longrightarrow K_{i+j}(R \otimes_\mathbb{Z} R'),$$
from the homeomorphisms $S^i \wedge S^j \xrightarrow{\cong} S^{i+j}$. Further, the "switch map" $S^i \wedge S^j \xrightarrow{\cong} S^j \wedge S^i$ has degree $(-1)^{ij}$ (compute it on homology) so that if $R$ is commutative, the homomorphism $R \otimes_\mathbb{Z} R \longrightarrow R$ given by multiplication induces bilinear products
$$\star : K_i(R) \otimes_\mathbb{Z} K_j(R) \longrightarrow K_{i+j}(R), \quad i,j \geq 1$$
satisfying $x \star y = (-1)^{ij} y \star x$ for $x \in K_i(R)$, $y \in K_j(R)$. Loday computes that for $i = j = 1$, this product is the *negative* of the product constructed in §1 following Milnor (see [L], Proposition (2.2.3)).

Finally, we remark that Quillen has computed the $K$-groups of a finite field in the following paper: D. Quillen, On the Cohomology and $K$-Theory of the General Linear Group over a Finite Field, *Ann. Math.* 96 (1972), 552–586. In fact, the computations in this paper motivated the definition of the plus construction. The result is as follows: If $\mathbb{F}_q$ denotes the field with $q$ elements, then

$$K_0(\mathbb{F}_q) = \mathbb{Z};$$
$$K_{2i}(\mathbb{F}_q) = 0, \quad \text{for } i > 0;$$
$$K_{2i-1}(\mathbb{F}_q) \cong \mathbb{Z}/(q^i - 1)\mathbb{Z}, \quad \text{for } i > 0.$$

# 3. The Classifying Space of a Small Category

The references for simplicial sets are: J. Milnor, The geometric realization of a semi-simplicial complex, *Ann. Math.* 65 (1957), 357–362, and J.P. May, *Simplicial Objects in Algebraic Topology*, Midway reprints (1982). Apart from Quillen's paper, *Higher Algebraic K-Theory – I*, a reference for the classifying space of a category is: G. Segal, Classifying spaces and spectral sequences, *Publ. Math. IHES* 34 (1968), 105–112.

**(Semi-)Simplicial Sets.** Let $\Delta$ be the following category: for each non-negative integer $n$, let $\underline{n} = \{0 < 1 < \cdots < n\}$ be the ordered set consisting of $0, 1, \ldots, n$; the *objects* of $\Delta$ are the ordered sets $\underline{n}$, and *morphisms* are monotonic maps (i.e., maps $f : \underline{m} \longrightarrow \underline{n}$ such that $f(i) \leq f(j)$ for $i < j$; in particular equality $f(i) = f(j)$ is permitted).

For each positive $n$, we have $n+1$ maps in $\Delta$ $\partial_i^n : \underline{n-1} \longrightarrow \underline{n}$ which are injective, given by

$$\partial_i^n(j) = \begin{cases} j & \text{if } j < i \\ j+1 & \text{if } j \geq i. \end{cases}$$

These are called the *face maps*. Next, we have $n$ maps $s_i^{n-1} : \underline{n} \longrightarrow \underline{n-1}$ which are surjective, given by

$$s_i^{n-1}(j) = \begin{cases} j & \text{if } j \leq i \\ j-1 & \text{if } j > i. \end{cases}$$

These are called the *degeneracy maps*. The face and degeneracy maps satisfy certain obvious identities (see May's book, pg. 1) and any morphism in $\Delta$ can be written as a composition of face and degeneracy maps (if $f : \underline{m} \longrightarrow \underline{n}$, we can uniquely write $f = g \circ h$ where $h : \underline{m} \longrightarrow \underline{p}$ and $g : \underline{p} \rightarrowtail \underline{n}$ are respectively surjective and injective arrows in $\Delta$; now $h$ is a composition of degeneracies and $g$ a composition of faces).

**Definition.** A *simplicial object* in a category $\mathcal{C}$ is a contravariant functor $\Delta \longrightarrow \mathcal{C}$ i.e., a functor $\Delta^{\text{op}} \longrightarrow \mathcal{C}$, where $\Delta^{\text{op}}$ is the opposite category. A morphism of simplicial objects in $\mathcal{C}$ is a natural transformation.

Thus, a *simplicial set* is a functor $\Delta^{\text{op}} \longrightarrow \underline{\text{Set}}$, where $\underline{\text{Set}}$ denotes the category of sets; similarly a *simplicial space* is a functor $\Delta^{\text{op}} \longrightarrow \underline{\text{Top}}$ where

Top denotes the category of topological spaces. In the older terminology, a simplicial set was called a semi-simplicial set.

Suppose $F : \Delta^{\mathrm{op}} \longrightarrow \underline{\mathrm{Set}}$ is a simplicial set. Then for each non-negative integer $n$, $F(\underline{n})$ is a set, called the *set of n-simplices* of $F$. The maps $\partial_i^n$ give rise to $n+1$ maps of sets $F(\underline{n}) \longrightarrow F(\underline{n-1})$, called the face maps, which associate to each $n$-simplex in $F(\underline{n})$ a collection of $n+1$ $(n-1)$-simplices in $F(\underline{n-1})$, called its *faces*. Similarly the $n$ maps $s_i^{n-1}$ give maps $F(\underline{n-1}) \longrightarrow F(\underline{n})$, associating to each $(n-1)$-simplex a collection of $n$ "degenerate" $n$-simplices; these maps $F(\underline{n-1}) \longrightarrow F(\underline{n})$ are called degeneracies. For $\delta \in F(\underline{n})$, we call $F(\partial_i^n)(\delta)$ the $i$th face of $\delta$, and $F(s_i^n)(\delta) \in F(\underline{n+1})$ the $i$th degenerate simplex of $\delta$.

**Example (3.1).** Let $X$ be a topological space. Let $S(X)$ denote the *total singular complex* of $X$, so that $S_n(X)$, the set of $n$-simplices of $S(X)$, is just the set of singular $n$-simplices in $X$, i.e., $S_n(X)$ is the set of all continuous maps $\Delta_n \longrightarrow X$, where $\Delta_n$ is the standard $n$-simplex

$$\Delta_n = \{(t_0,\ldots,t_n) \in \mathbb{R}^{n+1} \mid t_i \geq 0, \Sigma t_i = 1\}.$$

To make $S(X) = \{S_n(X)\}_{n\geq 0}$ into a simplicial set, we must show that to any morphism $\underline{m} \longrightarrow \underline{n}$ in $\Delta$, it is possible to assign a map of sets $S_n(X) \longrightarrow S_m(X)$, compatible with compositions. If $f : \underline{m} \longrightarrow \underline{n}$ is a morphism in $\Delta$, i.e., a monotonic map of ordered sets, we first describe a continuous map $\tilde{f} : \Delta_m \longrightarrow \Delta_n$, such that $\underline{m} \longmapsto \Delta_m$, $f \longmapsto \tilde{f}$ is a functor $\Delta \longrightarrow \underline{\mathrm{Top}}$. Since $S_n(X) = \mathrm{Hom}_{\mathrm{Top}}(\Delta_n, X)$, and $Y \longrightarrow \mathrm{Hom}_{\mathrm{Top}}(Y, X)$ is a functor $(\underline{\mathrm{Top}})^{\mathrm{op}} \longrightarrow \underline{\mathrm{Set}}$, we will have shown that $S(X)$ is a simplicial set. If

$$\Delta_m = \{(s_0,\ldots,s_m) \in \mathbb{R}^{m+1} \mid s_j \geq 0, \Sigma s_j = 1\},$$

let $\tilde{f} : \Delta_m \longrightarrow \Delta_n$ be the map $\tilde{f}((s_0,\ldots,s_m)) = (t_0,\ldots,t_n)$ where

$$t_i = \sum_{f(j)=i} s_j,$$

where we define $t_j = 0$ if $\{j \mid f(j) = i\}$ is empty. One easily verifies that this gives a functor $\Delta \longrightarrow \underline{\mathrm{Top}}$.

**Example (3.2)** ("The $n$-sphere"). Let $(S^n)$ be the simplicial set whose set of $m$-simplices is given by

$$(S^n)_m = \begin{cases} \{e_m\} & \text{if } 0 \leq m < n \\ \{e_n, f_n\} & \text{if } m = n \\ \{e_m\} \cup \{\text{all degenerate} \\ \quad m\text{-simplices associated to } f_n\}, & \text{if } m > n. \end{cases}$$

The faces of $e_m$ all equal $e_{m-1}$ (for $m \geq 1$), and the degenerate simplices of $e_m$ all equal $e_{m+1}$ (for $m \geq 0$). The faces of $f_n$ all equal $e_{n-1}$, if $n > 0$ (if

$n = 0$, there are no face maps to be defined). We leave it to the reader to check that these conditions do uniquely define a simplicial set; the reason for the name ("The $n$-sphere") is explained below (Ex. (3.4)).

**(3.3) Geometric Realization.** To each simplicial set $F : \Delta^{\text{op}} \longrightarrow \underline{\text{Set}}$, we can associate a topological space $|F|$ called the *geometric realization* of $F$. This is defined to be the quotient space

$$\left(\coprod_{n \geq 0} F(\underline{n}) \times \Delta_n\right) / \sim$$

where for each $n \geq 0$, $F(\underline{n})$ is regarded as a discrete space. The equivalence relation $\sim$ is defined as follows: given $f : \underline{m} \longrightarrow \underline{n}$ in $\Delta$, let $\tilde{f} : \Delta_m \longrightarrow \Delta_n$ be the continuous map defined in Ex. (3.1) above. Then for any $\delta \in F(\underline{n})$, we set

$$(\delta, \tilde{f}(y)) \sim (F(f)\delta, y)$$

for all $y \in \Delta_m$, where $(\delta, \tilde{f}(y)) \in F(\underline{n}) \times \Delta_n$, and $(F(f)\delta, y) \in F(\underline{m}) \times \Delta_m$. Let $\sim$ be the equivalence relation so generated, and let $|F|$ be the quotient space, with the quotient topology (an open subset is a set whose inverse image in $\coprod_n F(\underline{n}) \times \Delta_n$ is open).

Clearly the construction of the geometric realization is functorial, i.e., if $F \longrightarrow G$ is a morphism of simplicial sets (i.e., a natural transformation of functors $\Delta^{\text{op}} \longrightarrow \underline{\text{Set}}$) we get a continuous map $|F| \longrightarrow |G|$. For any simplicial set $F$, a simplex $\delta \in F(\underline{n})$ is called *non-degenerate* if it is not the degenerate simplex assigned to any $(n-1)$-simplex by one of the degeneracies. Then $|F|$ is homeomorphic to a $CW$-complex, which has one $n$-cell corresponding to each non-degenerate $n$-simplex of $F$. If $F, G$ are simplicial sets, let $F \times G$ denote the simplicial set whose $n$-simplices are $F(\underline{n}) \times G(\underline{n})$, with obvious maps. Then by a theorem of Milnor (see the references cited earlier) the natural continuous map $|F \times G| \longrightarrow |F| \times |G|$ (induced by the maps of simplicial sets $F \times G \longrightarrow F$, $F \times G \longrightarrow G$) is a continuous bijection, and is a homeomorphism if $|F| \times |G|$ is given the compactly generated topology associated to the product topology (i.e., the product is formed in the category of compactly generated spaces). In particular, if $|F|$ or $|G|$ is locally compact (e.g., if $F$ has only a finite number of non-degenerate simplices, so that $|F|$ is compact) then $|F \times G|$ is homeomorphic to $|F| \times |G|$. Let $\Delta(n) = \text{Hom}_\Delta(-, \underline{n})$ be the simplicial set naturally associated to $\underline{n} \in \text{Ob}\,\Delta$; then one sees easily that $|\Delta(n)| = \Delta_n$. Milnor's theorem follows from the special case $F = \Delta(n)$, $G = \Delta(m)$, which we prove in the Appendix (see (A.55)) (see also Lemma (6.8)).

Finally, we note that the homology of $|F|$ can be computed as follows. Let $C_n(F)$ denote the free Abelian group on $F(\underline{n})$, and let $\tilde{\partial}_i^n$ :

$C_n(F) \longrightarrow C_{n-1}(F)$ be the map induced by $F(\partial_i^n)$. If $d_n = \Sigma(-1)^i \tilde{\partial}_i^n$, then $C(F) = (C_n(F), d_n)_{n \geq 1}$ is a chain complex. Then, for any Abelian group $A$, $H_*(|F|, A) \cong H_*(C(F) \otimes_{\mathbb{Z}} A)$, and $H^*(|F|, A) \cong H^*(\mathrm{Hom}_{\mathbb{Z}}(C(F), A))$ (see May's book, Prop. (16.2) and Corollary (22.3)).

Let $X$ be a topological space, and let $S(X)$ denote the total singular complex (Ex. (3.1)). Then there is a continuous surjective map $f : |S(X)| \longrightarrow X$. If $x_0 \in X$ is a base point, then $S(x_0) \subset S(X)$ is a subcomplex (i.e., $S(x_0)(\underline{n}) \subset S(X)(\underline{n})$ for all $n$) such that $|S(x_0)|$ is a point whose image under $f$ is $x_0$. We have:

**Theorem.** $f : (|S(X)|, |S(x_0)|) \longrightarrow (X, x_0)$ *induces isomorphisms on homotopy groups. Hence if $X$ is a CW-complex, it is a homotopy equivalence* (see May's book, Theorem (16.6)).

**Example (3.4).** Let $(S^n)$ be the simplicial set described in Ex. (3.2). Then the geometric realization $|(S^n)|$ is naturally homeomorphic to the $n$-sphere $S^n$. Indeed, the only 2 non-degenerate simplices of $(S^n)$ are $e_0 \in (S^n)(\underline{0})$ and $f_n \in (S^n)(\underline{n})$, so $|(S^n)|$ is a $CW$ complex with exactly 2 cells. If $n > 0$, all faces of $f_n$ are degenerate, so that $|(S^n)|$ is obtained by attaching an $n$-cell to the 0-cell $|e_0|$ with a constant attaching map. This is the standard description of $S^n$ as a $CW$-complex.

**(3.5) The Classifying Space of a Category.** Let $\mathcal{C}$ be a small category, i.e., a category whose objects form a set. The *nerve* of $\mathcal{C}$, denoted $N\mathcal{C}$ (or $N(\mathcal{C})$) is defined to be the following simplicial set: an $n$-simplex of $N\mathcal{C}$ is a diagram

$$A_0 \xrightarrow{f_1} A_1 \xrightarrow{f_2} A_2 \xrightarrow{f_3} \cdots \xrightarrow{f_n} A_n$$

with $A_i \in \mathrm{Ob}\,\mathcal{C}$, $f_i \in \mathrm{Mor}\,\mathcal{C}$. Given a map $f : \underline{m} \longrightarrow \underline{n}$ in $\Delta$, the corresponding map $N\mathcal{C}(\underline{n}) \longrightarrow N\mathcal{C}(\underline{m})$ maps the above $n$-simplex to the $m$-simplex

$$B_0 \xrightarrow{g_1} B_1 \xrightarrow{g_2} B_2 \longrightarrow \cdots \xrightarrow{g_m} B_m$$

where $B_j = A_{f(j)}$, and $B_j \longrightarrow B_{j+1}$ is the composite map $A_{f(j)} \longrightarrow A_{f(j+1)}$, where if $f(j) = f(j+1)$, let $A_{f(j)} \longrightarrow A_{f(j+1)}$ be the identity map. In particular the $i$th face of the above $n$-simplex is the $(n-1)$-simplex

$$A_0 \xrightarrow{f_1} A_1 \longrightarrow \cdots \longrightarrow A_{i-1} \xrightarrow{f_{i+1} \circ f_i} A_{i+1} \longrightarrow \cdots \longrightarrow A_n,$$

while the $i$th degenerate simplex is the $(n+1)$-simplex

$$A_0 \xrightarrow{f_1} A_1 \longrightarrow \cdots \longrightarrow A_i \xrightarrow{1} A_i \xrightarrow{f_{i+1}} A_{i+1} \longrightarrow \cdots \longrightarrow A_n.$$

## 3. The Classifying Space of a Small Category

The *classifying space* of $\mathcal{C}$ is defined to be the geometric realization of $N\mathcal{C}$ and is denoted by $B\mathcal{C}$ (or $B(\mathcal{C})$); thus

$$B\mathcal{C} = |N\mathcal{C}|.$$

Clearly, if $F : \mathcal{C} \longrightarrow \mathcal{D}$ is a functor between small categories $\mathcal{C}, \mathcal{D}$ then there is an induced map of simplicial sets $N\mathcal{C} \longrightarrow N\mathcal{D}$, and hence an induced continuous map $BF : B\mathcal{C} \longrightarrow B\mathcal{D}$.

A simple and useful example is given by the category $\{0 < 1\}$, consisting of two objects $0, 1$ and a unique non-identity morphism $0 \longrightarrow 1$. Then one checks easily that $B\{0 < 1\} = I$, the unit interval, since the only non-degenerate simplices in $N\{0 < 1\}$ are $\{0\}, \{1\}, \{0 \longrightarrow 1\}$ (the first two are 0-simplices, and the third is a 1-simplex).

Let $\mathcal{C}, \mathcal{C}'$ be small categories, $\mathcal{C} \times \mathcal{C}'$ their product, so that

$$\mathrm{Ob}(\mathcal{C} \times \mathcal{C}') = (\mathrm{Ob}\,\mathcal{C}) \times (\mathrm{Ob}\,\mathcal{C}'), \text{ and}$$
$$\mathrm{Hom}_{\mathcal{C}\times\mathcal{C}'}((A,B),(C,D)) = \mathrm{Hom}_{\mathcal{C}}(A,C) \times \mathrm{Hom}_{\mathcal{C}'}(B,D).$$

Then clearly $N(\mathcal{C} \times \mathcal{C}')$ is the product simplicial set $N\mathcal{C} \times N\mathcal{C}'$. Hence $B(\mathcal{C}\times\mathcal{C}') \longrightarrow B\mathcal{C}\times B\mathcal{C}'$ is a continuous bijection, which is a homeomorphism if the product on the right is given the compactly generated topology; in particular it is a homeomorphism if either $B\mathcal{C}$ or $B\mathcal{C}'$ is locally compact. Thus $B(\mathcal{C} \times \{0 < 1\}) \cong B\mathcal{C} \times I$. A functor $H : \mathcal{C} \times \{0 < 1\} \longrightarrow \mathcal{D}$ is just a pair of functors $F, G : \mathcal{C} \longrightarrow \mathcal{D}$ (given by $F(A) = H((A,0))$, $G(A) = H((A,1))$) together with a natural transformation $\eta : F \longrightarrow G$ (given by $\eta(A) = H((1_A, 0 \longrightarrow 1)))$. Hence we deduce the following lemma.

**Lemma (3.6).** *Let $F, G : \mathcal{C} \longrightarrow \mathcal{D}$ be functors between small categories, such that there is a natural transformation $F \longrightarrow G$. Then*

$$BF, BG : B\mathcal{C} \longrightarrow B\mathcal{D}$$

*are homotopic maps.*

**Corollary (3.7).** *Let $F : \mathcal{C} \longrightarrow \mathcal{D}$ be a functor between small categories. Suppose $F$ has either a left or a right adjoint. Then $BF$ is a homotopy equivalence. In particular, if $\mathcal{C}$ is a small category with either an initial or a final object, then $B\mathcal{C}$ is contractible.* (See Appendix B for the definitions of adjoint functors.)

**Lemma (3.8).** *Let $I$ be a small filtering category, and let $\{\mathcal{C}_i\}_{i\in I}$ be a family of small categories indexed by $I$. Let $\mathcal{C} = \varinjlim_{I} \mathcal{C}_i$ be the direct limit category. Let $X_i \in \mathcal{C}_i$ be a family of objects such that $X_i \longmapsto X_j$ under* the transition functors $\mathcal{C}_i \longrightarrow \mathcal{C}_j$ of the given family (corresponding to morphisms $i \to j$ in $I$), and let $X \in \mathcal{C}$ be the common image of the $X_i$. Then

$$\pi_n(B\mathcal{C}, \{X\}) = \varinjlim_I \pi_n(B\mathcal{C}_i, \{X_i\}) \text{ for all } n \geq 0.$$

(See Appendix B for the definitions of filtering category, etc.)

**Proof.** Any finite diagram in $\mathcal{C}$ is the image of a similar diagram in some $\mathcal{C}_i$. Thus $N\mathcal{C} = \varinjlim N\mathcal{C}_i$ (where $(\varinjlim N\mathcal{C}_i)(\underline{n}) = \varinjlim N\mathcal{C}_i(\underline{n})$), and any finite subcomplex of $N\mathcal{C}$ (a simplicial subset of $N\mathcal{C}$ with a finite number of non-degenerate simplices) is the isomorphic image of a subcomplex of some $N\mathcal{C}_i$. Since $\pi_n(B\mathcal{C}, \{X\})$ is the direct limit of $\pi_n$ of all finite subcomplexes of $N\mathcal{C}$ containing $X$ (regarded as an element of $N\mathcal{C}(\underline{0})$), and a similar claim holds for each $B\mathcal{C}_i$, the result follows easily.

By abuse of terminology, a category is called *contractible* if its classifying space is; similarly a functor $F$ is called a *homotopy equivalence* if $BF$ is one, etc.

**Corollary (3.9).** *Any small filtering category is contractible.*

**Proof.** Let $I$ be a filtering category. For each $i \in I$, let $I/i$ denote the category of objects over $i$, consisting of pairs $(j, j \xrightarrow{u} i)$ with morphisms $(j_1, j_1 \xrightarrow{u_1} i_1) \longrightarrow (j_2, j_2 \xrightarrow{u_2} i_2)$ being commutative diagrams:

$$\begin{array}{ccc} j_1 & \longrightarrow & j_2 \\ {}_{u_1}\searrow & & \swarrow_{u_2} \\ & i & \end{array}$$

In fact, since $I$ is filtering, one verifies easily that for this direct system $i \longrightarrow I/i$, the naturally defined functor $\varinjlim I/i \longrightarrow I$ is an equivalence. But $(i, i \xrightarrow{1} i)$ is a final object of $I/i$, so that $I/i$ is contractible. Hence $BI$ is weakly contractible (has vanishing homotopy groups). Since it is a $CW$-complex, it is contractible, by Whitehead's theorem (see (A.9)).

**Remark.** The classifying space functor is not full. An interesting example is given by the natural homeomorphism $B\mathcal{C} \cong B\mathcal{C}^{\mathrm{op}}$, where $\mathcal{C}^{\mathrm{op}}$ is the opposite category.

**Example (3.10).** Let $G$ be a discrete group. Let $\underline{G}$ be the category with one object $\star$ such that the monoid $\mathrm{Hom}_{\underline{G}}(\star, \star)$ is the group $G$. We claim that $B\underline{G}$ is the classifying space of $G$. In fact, let $\tilde{G}$ denote the category whose objects are in bijection with the elements of $G$, and the following arrows—if $[g]$ is the object of $\tilde{G}$ corresponding to $g \in G$, then $\text{Hom}_{\tilde{G}}([g],[h])$ consists of a unique arrow $\delta(g,h)$. Then the $\delta(g,h)$ are forced to satisfy the composition rules $\delta(g_2,g_3) \circ \delta(g_1,g_2) = \delta(g_1,g_3)$; one checks that these rules do define a category. There is a functor $\tilde{G} \longrightarrow \underline{G}$ given by $[g] \longrightarrow \star$, $\delta(g,h) \longrightarrow hg^{-1} \in G = \text{Hom}_{\underline{G}}(\star,\star)$. The group $G$ acts on $\tilde{G}$ by $g([h]) = [hg^{-1}]$, $g(\delta(g_1,g_2)) = \delta(g_1 g^{-1}, g_2 g^{-1})$. Since the isotropy group (for the $G$-action) of any object of $\tilde{G}$ is trivial, the isotropy group of any $n$-simplex in $N(\tilde{G})$ is trivial. Hence $G$ acts freely on the classifying space $B\tilde{G}$. The functor $\tilde{G} \longrightarrow \underline{G}$ is $G$-equivariant for the trivial $G$-action on $\underline{G}$, so that $B\tilde{G} \longrightarrow B\underline{G}$ is also $G$-equivariant for the trivial $G$-action on $B\underline{G}$. From the criterion of (A.49), we see that $B\tilde{G} \longrightarrow B\underline{G}$ is a locally trivial covering space with (discrete) fiber isomorphic to $G$, considered as the 0-skeleton of $B\tilde{G}$, and the group $G$ acts transitively on this fiber. Thus $B\underline{G} \cong B\tilde{G}/G$, and $B\tilde{G} \longrightarrow B\underline{G}$ is a Galois covering space with group $G$. Since $\tilde{G}$ has an initial object (any object is an initial object) $B\tilde{G}$ is contractible. Thus $G \cong \pi_1(B\underline{G})$, and $\pi_i(B\underline{G}) = 0$ for $i \neq 0$. One also sees that $B\tilde{G} \longrightarrow B\underline{G}$ is a principal $G$-bundle with contractible total space, which is the usual defining property of the classifying space, which characterizes it up to homotopy equivalence.

# 4. Exact Categories and Quillen's Q-Construction

For our purposes, an *exact category* $\mathcal{C}$ is an additive category $\mathcal{C}$ embedded as a full (additive) subcategory of an Abelian category $\mathcal{A}$, such that if $0 \longrightarrow M' \longrightarrow M \longrightarrow M'' \longrightarrow 0$ is an exact sequence in $\mathcal{A}$ with $M', M'' \in \mathcal{C}$, then $M$ is isomorphic to an object of $\mathcal{C}$. An *exact sequence* in $\mathcal{C}$ is then defined to be an exact sequence in $\mathcal{A}$ whose terms lie in $\mathcal{C}$. Let $\mathcal{E}$ be the class of exact sequences in $\mathcal{C}$. One can give an intrinsic definition of an exact category $\mathcal{C}$ in terms of a class $\mathcal{E}$ of diagrams in the additive category $\mathcal{C}$, satisfying suitable axioms (see Quillen's paper for details). In all cases relevant to us, the category embeds naturally in some Abelian category $\mathcal{A}$, such that $\mathcal{C}$ is closed under extensions in $\mathcal{A}$.

An *exact functor* $F: \mathcal{C} \longrightarrow \mathcal{D}$ between exact categories $\mathcal{C}, \mathcal{D}$ is an additive functor such that if $0 \longrightarrow M' \longrightarrow M \longrightarrow M'' \longrightarrow 0$ is exact in $\mathcal{C}$, then
$$0 \longrightarrow F(M') \longrightarrow F(M) \longrightarrow F(M'') \longrightarrow 0$$
is exact in $\mathcal{D}$.

If $\mathcal{C}$ is a small exact category, then we define the Grothendieck group $K_0(\mathcal{C})$ in the usual way: $K_0(\mathcal{C}) = \mathcal{F}/\mathcal{R}$, where $\mathcal{F}$ is the free Abelian group on the objects of $\mathcal{C}$, and $\mathcal{R}$ is the subgroup generated by classes $[M] - [M'] - [M'']$ for each exact sequence
$$0 \longrightarrow M' \longrightarrow M \longrightarrow M'' \longrightarrow 0$$
in $\mathcal{C}$ (the more standard definition involves taking the free Abelian group on isomorphism classes of objects of $\mathcal{C}$, but is easily seen to be isomorphic to the above group, using sequences with $M'' = 0$). If $\mathcal{C}$ is an exact category, we form a new category $Q\mathcal{C}$ defined as follows: $Q\mathcal{C}$ has the same objects as $\mathcal{C}$, but a morphism $X \longrightarrow Y$ is given by an isomorphism class of diagrams

(*) $$X \xleftarrow{q} Z \xrightarrowtail{i} Y$$

where $i$ is an *admissible monomorphism* and $q$ an *admissible epimorphism*

## 4. Exact Categories and Quillen's Q-Construction

in $\mathcal{C}$; by definition, this means that there are exact sequences

$$0 \longrightarrow Z \xrightarrow{i} Y \longrightarrow Y' \longrightarrow 0$$
$$0 \longrightarrow X' \longrightarrow Z \xrightarrow{q} X \longrightarrow 0$$

in $\mathcal{C}$. Another diagram

$(*)'$  $\qquad\qquad\qquad X \xleftarrow{q'} Z' \xrightarrow{i'} Y$

gives the same morphism (in which case we say the two diagrams $(*)$, $(*)'$, are *isomorphic*) if there is an isomorphism $Z \longrightarrow Z'$ making the diagram

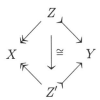

commute. Composition of arrows in $Q\mathcal{C}$ is defined as follows: given $X \leftarrowtail Z \rightarrowtail Y$ and $Y \leftarrowtail V \rightarrowtail T$, form the diagram (in the ambient category $\mathcal{A}$)

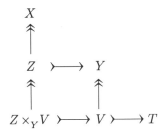

Since $\mathcal{C}$ is closed under extensions, and $\ker(Z \times_Y V \to Z) \cong \ker(V \twoheadrightarrow Y) \in \mathcal{C}$, $(Z \times_Y V) \in \mathcal{C}$ and $Z \times_Y V \twoheadrightarrow X$, $Z \times_Y V \rightarrowtail Y$ are respectively admissible epi and mono. Hence the diagram $X \leftarrowtail Z \times_Y V \rightarrowtail T$ defines an arrow in $Q\mathcal{C}$ from $X$ to $T$. One checks that the isomorphism class of this diagram depends only on the isomorphism classes of $X \leftarrowtail Z \rightarrowtail Y$ and $Y \leftarrowtail V \rightarrowtail T$, so that we have a well-defined composition rule for morphisms; next, one verifies that composition is associative. Hence if $\mathcal{C}$ is a small category, then $Q\mathcal{C}$ is well defined, and is a small category.

Let $0 \in \mathrm{Ob}\,\mathcal{C} = \mathrm{Ob}\,Q\mathcal{C}$ be a zero object, so that $\{0\}$ is a point of $BQ\mathcal{C}$. Quillen's first result in this context is:

**Theorem (4.0).** *There is a natural isomorphism $\pi_1(BQ\mathcal{C}, \{0\}) \cong K_0(\mathcal{C})$ (where $K_0(\mathcal{C})$ is as defined above).*

This motivates Quillen's second definition of higher $K$-theory.

**Definition.** $K_i(\mathcal{C}) = \pi_{i+1}(BQ\mathcal{C}, \{0\})$, $i \geq 0$.

We note that this does not depend on the choice of a zero object, since given another zero object $0'$, there is a canonical isomorphism $0 \xrightarrow{\cong} 0'$ in $\mathcal{C}$, and hence in $Q\mathcal{C}$, giving a canonical choice of a path joining $\{0\}$ and $\{0'\}$ in $BQ\mathcal{C}$. The isomorphism of Theorem (4.0) can be described explicitly as follows: given an object $M \in \mathcal{C}$, there are 2 canonically defined maps $0 \to M$ in $Q\mathcal{C}$, given by the diagrams $0 \leftarrow 0 \rightarrowtail M$ and $0 \leftarrow M \xrightarrow{1} M$ (where 1 denotes the identity map on $M$). These give rise to a loop $\{0\} \leftrightarrows \{M\}$ in $BQ\mathcal{C}$. The map $K_0(\mathcal{C}) \longrightarrow \pi_1(BQ\mathcal{C}, \{0\})$ is given by mapping $[M] \in K_0(\mathcal{C})$ to the class of the above loop in $\pi_1$. We defer the proof of this theorem to chapter 6 (however, see Ex. (4.10) below).

At this point, we merely list certain properties of the $K$-groups defined above, whose proofs will be given in chapter 6. To illustrate the theory, and to provide motivation to work through the details of the proofs (if it is required!), we will use these properties in chapter 5 to get interesting results on the $K$-theory of rings and schemes; a high point will be the proof of the formula $H^p(X, \mathcal{K}_{p,X}) \cong CH^p(X)$ for the Chow groups of a smooth variety. These properties of $K_i$ are generalizations of familiar properties of the Grothendieck group functor $K_0$.

Let $\mathcal{C}$ be a small exact category, $\mathcal{E}$ the category of (short) exact sequences in $\mathcal{C}$. If $\mathcal{C} \subset \mathcal{A}$ is an embedding as a full exact subcategory of an Abelian category, then $\mathcal{E}$ can be regarded as a full additive subcategory of the Abelian category of complexes in $\mathcal{A}$ of the form

$$0 \longrightarrow M' \longrightarrow M \longrightarrow M'' \longrightarrow 0,$$

where a morphism of complexes is a (commutative) diagram in $\mathcal{A}$

$$\begin{array}{ccccccccc} 0 & \longrightarrow & M' & \longrightarrow & M & \longrightarrow & M'' & \longrightarrow & 0 \\ & & \downarrow & & \downarrow & & \downarrow & & \\ 0 & \longrightarrow & N' & \longrightarrow & N & \longrightarrow & N'' & \longrightarrow & 0. \end{array}$$

Hence $\mathcal{E}$ is an exact category in a natural way, such that if $E', E, E'' \in \mathcal{E}$, then $0 \longrightarrow E' \longrightarrow E \longrightarrow E'' \longrightarrow 0$ is exact precisely when the corresponding diagram in $\mathcal{C}$ has exact rows and columns. There are exact functors $s, t, q : \mathcal{E} \longrightarrow \mathcal{C}$ such that if $E \in \mathcal{E}$, then

$$0 \longrightarrow sE \longrightarrow tE \longrightarrow qE \longrightarrow 0$$

is the corresponding exact sequence in $\mathcal{C}$. We can now state the first property of the functors $K_i$.

**Theorem (4.1).** *$(s, q) : Q\mathcal{E} \longrightarrow Q\mathcal{C} \times Q\mathcal{C}$ is a homotopy equivalence.*

## 4. Exact Categories and Quillen's $Q$-Construction

**Corollary (4.2).** $t_* = s_* + q_* : K_i(\mathcal{E}) \longrightarrow K_i(\mathcal{C})$ *for all $i$.*

**Corollary (4.3).** *Let $0 \longrightarrow F' \longrightarrow F \longrightarrow F'' \longrightarrow 0$ be an exact sequence of exact functors $\mathcal{C} \longrightarrow \mathcal{D}$ between small exact categories (i.e., there are natural transformations $F' \longrightarrow F$ and $F \longrightarrow F''$ such that $0 \longrightarrow F'(M) \longrightarrow F(M) \longrightarrow F''(M) \longrightarrow 0$ is exact in $\mathcal{D}$ for any $M \in \mathcal{C}$). Then $F_* = F'_* + F''_* : K_i(\mathcal{C}) \longrightarrow K_i(\mathcal{D})$ for all $i$.*

If $F : \mathcal{C} \longrightarrow \mathcal{D}$ is an additive functor between small exact categories, an admissible filtration $0 = F_0 \subset F_1 \subset \cdots \subset F_{n-1} \subset F_n = F$ by subfunctors is a sequence of additive subfunctors $F_i : \mathcal{C} \longrightarrow \mathcal{D}$ such that $F_{p-1}(M) \longrightarrow F_p(M)$ is an admissible monomorphism in $\mathcal{D}$ for each $M \in \mathcal{C}$. Hence the quotients $F_p/F_{p-1}$, $p \geq 1$, make sense as additive functors $\mathcal{C} \longrightarrow \mathcal{D}$.

**Corollary (4.4).** *If $\{F_i\}_{0 \leq i \leq n}$ is an admissible filtration of an additive functor $F : \mathcal{C} \longrightarrow \mathcal{D}$ such that $F_p/F_{p-1}$ is an exact functor for each $p$, then $F$ is exact, and*

$$F_* = \sum_{p=1}^{n} (F_p/F_{p-1})_* : K_i(\mathcal{C}) \longrightarrow K_i(\mathcal{D}) \quad \text{for all } i.$$

**Corollary (4.5).** *If*

$$0 \longrightarrow F_0 \longrightarrow F_1 \longrightarrow \cdots \longrightarrow F_n \longrightarrow 0$$

*is an admissible long exact sequence of exact functors $\mathcal{C} \longrightarrow \mathcal{D}$ (in the obvious sense), then for all $i \geq 0$,*

$$\sum_{j=0}^{n} (-1)^j (F_j)_* = 0 : K_i(\mathcal{C}) \longrightarrow K_i(\mathcal{D}).$$

Next, we give the resolution and dévissage theorems which allow us to replace one exact category by another, in certain situations, without changing the $K$-groups. Let $\mathcal{M}$ be an exact category, $\mathcal{P} \subset \mathcal{M}$ a full additive subcategory, closed under extensions in $\mathcal{M}$; then $\mathcal{P}$ is an exact category in a natural way such that the inclusion $\mathcal{P} \longrightarrow \mathcal{M}$ is an exact functor.

**Theorem (4.6)** (Resolution theorem). *Assume that $\mathcal{P} \subset \mathcal{M}$ are as above, and assume further that*

(a) *if $0 \longrightarrow M' \longrightarrow M \longrightarrow M'' \longrightarrow 0$ is exact in $\mathcal{M}$ and $M, M'' \in \mathcal{P}$, then $M' \in \mathcal{P}$*

(b) *for any object $M \in \mathcal{M}$, there is a finite resolution*

$$0 \longrightarrow P_n \longrightarrow P_{n-1} \longrightarrow \cdots \longrightarrow P_0 \longrightarrow M \longrightarrow 0,$$

*with $P_i \in \mathcal{P}$.*

Then $BQ\mathcal{P} \longrightarrow BQ\mathcal{M}$ is a homotopy equivalence, and hence $K_i(\mathcal{P}) \cong K_i(\mathcal{M})$ for all $i$.

**Corollary (4.7).** *Let $\{T_i\}_{i \geq 1}$ be a $\delta$-functor from the exact category $\mathcal{M}$ to the Abelian category $\mathcal{A}$ (i.e., $T_i : \mathcal{M} \longrightarrow \mathcal{A}$ are additive functors such that if $0 \longrightarrow M' \longrightarrow M \longrightarrow M'' \longrightarrow 0$ is an exact sequence in $\mathcal{M}$, then there are functorial boundary maps $T_{i+1}(M'') \longrightarrow T_i(M')$, $i > 1$, giving a long exact sequence*

$$\longrightarrow T_{i+1}(M'') \longrightarrow T_i(M') \longrightarrow T_i(M) \longrightarrow T_i(M'')$$
$$\longrightarrow T_{i-1}(M') \longrightarrow \cdots \longrightarrow T_1(M'')).$$

*Let $\mathcal{P} \subset \mathcal{M}$ be the full exact subcategory of $T$-acyclic objects (i.e., objects $P$ with $T_i(P) = 0 \ \forall i$). Assume that for every object $M \in \mathcal{M}$, there is an admissible epimorphism $P \longrightarrow M$ with $P \in \mathcal{P}$, and that $T_n(M) = 0$ for all sufficiently large $n$ (depending on $M$). Then $BQ\mathcal{P} \longrightarrow BQ\mathcal{M}$ is a homotopy equivalence.*

**Theorem (4.8)** (Dévissage theorem). *Let $\mathcal{A}$ be an Abelian category, $\mathcal{B}$ a full Abelian subcategory which is closed under taking subobjects, quotients and finite products in $\mathcal{A}$. Suppose each object $M \in \mathcal{A}$ has a finite filtration in $\mathcal{A}$*

$$0 = M_0 \subset M_1 \subset \cdots \subset M_n = M$$

*with $M_i/M_{i-1} \in \mathcal{B}$ for all $i \geq 1$. Then $BQ\mathcal{B} \longrightarrow BQ\mathcal{A}$ is a homotopy equivalence, so that $K_i(\mathcal{B}) \cong K_i(\mathcal{A})$.*

Finally, we state the localization theorem, which under certain circumstances gives us long exact sequences of $K$-groups. We recall that if $\mathcal{A}$ is an Abelian category, $\mathcal{B} \subset \mathcal{A}$ a full Abelian subcategory closed under taking subobjects, quotients and extensions in $\mathcal{A}$, then $\mathcal{B}$ is an Abelian category, and is called a *Serre subcategory* of $\mathcal{A}$. Under these conditions one can construct a quotient Abelian category $\mathcal{A}/\mathcal{B}$ (see Appendix B), which, in various concrete cases of interest to us, is naturally equivalent to a suitable "localization" of the category $\mathcal{A}$; indeed the construction of $\mathcal{A}/\mathcal{B}$ is a generalization of the construction of the localization of a ring, and of modules over the ring, with respect to a central multiplicative set.

**Theorem (4.9)** (Localization). *Let $\mathcal{B}$ be a Serre subcategory of the Abelian category $\mathcal{A}$, and let $\mathcal{C}$ be the quotient Abelian category $\mathcal{A}/\mathcal{B}$. Let $s : \mathcal{B} \longrightarrow \mathcal{A}$, $p : \mathcal{A} \longrightarrow \mathcal{C}$ be the natural exact functors. Then*

$$BQ\mathcal{B} \xrightarrow{BQs} BQ\mathcal{A} \xrightarrow{BQp} BQ\mathcal{C}$$

4. Exact Categories and Quillen's $Q$-Construction        43

is a *homotopy fibration* (i.e., the natural map $BQ\mathcal{B} \longrightarrow F(BQp)$ is a homotopy equivalence, where $F(BQp)$ is the homotopy fiber (see (A.27) of $BQp$)). Hence there is a long exact sequence

$$\cdots \longrightarrow K_{i+1}(\mathcal{C}) \longrightarrow K_i(\mathcal{B}) \longrightarrow K_i(\mathcal{A}) \longrightarrow K_i(\mathcal{C}) \longrightarrow \cdots \longrightarrow K_0(\mathcal{A})$$
$$\longrightarrow K_0(\mathcal{C}) \longrightarrow 0.$$

It is possible to give a "naive" construction of the map $K_0(\mathcal{C}) \longrightarrow \pi_1(BQ\mathcal{C}, \{0\})$ of Theorem (4.0); we discuss this below. This may give the reader a little practice in thinking about the $Q$-construction.

**Example (4.10).** As noted above, the isomorphism

$$\psi : K_0(\mathcal{C}) \xrightarrow{\cong} \pi_1(BQ\mathcal{C}, \{0\})$$

is explicitly given by associating to each $M \in \mathcal{C}$ a certain loop $r_M$ based at 0, such that $\psi([M]) = [r_M] \in \pi_1(BQ\mathcal{C}, \{0\})$. There is a canonical admissible mono $i_M : 0 \rightarrowtail M$ and an admissible epi $q_M : M \twoheadrightarrow 0$, associated to any object $M \in \mathcal{C}$. Given any admissible mono $i : M_1 \rightarrowtail M_2$, there is an arrow $i_! : M_1 \longrightarrow M_2$ in $Q\mathcal{C}$, corresponding to the (class of the) diagram $M_1 \xleftarrow{1} M_1 \xrightarrow{i} M_2$. Similarly, given any admissible epi $q : M_1 \twoheadrightarrow M_2$, there is an arrow $q^! : M_2 \longrightarrow M_1$ in $Q\mathcal{C}$, corresponding to the diagram $M_2 \xleftarrow{q} M_1 \xrightarrow{1} M_1$. One sees at once that if $M \xleftarrow{q} M' \xrightarrow{i} N$ is a diagram representing an arrow $u : M \longrightarrow N$ in $Q\mathcal{C}$, then from the definition of composition of morphisms, $u = i_! \circ q^!$.

The two arrows $0 \longrightarrow M$ in $Q\mathcal{C}$ given by $i_{M!}$ and $q_M^!$ give 2 paths $\{0\} \longrightarrow \{M\}$ in $BQ\mathcal{C}$, denoted $(i_{M!})$ and $(q_M^!)$. Let $r_M = (i_{M!}) \circ (q_M^!)^{-1}$ where the inverse and composition are the usual operations on paths; thus $r_M$ is the oriented loop obtained by first following $(i_{M!})$ and then following $(q_M^!)$ in reverse. To see that $[M] \mapsto [r_M]$ defines a homomorphism $K_0(\mathcal{C}) \longrightarrow \pi_1(BQ\mathcal{C}, \{0\})$, we must show that if

$(E) \ldots \qquad 0 \longrightarrow M' \xrightarrow{i} M \xrightarrow{q} M'' \longrightarrow 0$

is an exact sequence in $\mathcal{C}$, then $[r_M] = [r_{M'}] \cdot [r_{M''}]$ in $\pi_1(BQ\mathcal{C})$ where $\cdot$ denotes the group operation in $\pi_1$. From the split exact sequences

$$0 \longrightarrow M' \longrightarrow M' \oplus M'' \longrightarrow M'' \longrightarrow 0,$$
$$0 \longrightarrow M'' \longrightarrow M' \oplus M'' \longrightarrow M' \longrightarrow 0$$

one sees immediately that the classes $[r_{M'}]$, $[r_{M''}]$ commute, so the homomorphism $\psi : K_0(\mathcal{C}) \longrightarrow \pi_1(BQ\mathcal{C}, \{0\})$ will be well defined.

We note that from the sequence $(E)$ above, and the facts that $i_M = i \circ i_{M'}$, $q_M = q_{M''} \circ q$ in $\mathcal{C}$, giving $i_{M!} = i_! \circ i_{M'!}$ and $q_M^! = q^! \circ q_{M''}^!$ in $Q\mathcal{C}$, we have a diagram where the shaded triangles commute:

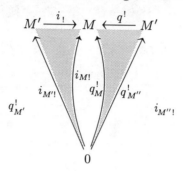

The shaded triangles give 2-simplices in $BQC$. From the diagram, we also notice 2 more arrows $0 \longrightarrow M$ in $QC$, namely $i_! \circ q^!_{M'}$, and $q^! \circ i_{M'''!}$. We claim that in fact $i_! \circ q^!_{M'} = q^! \circ i_{M'''!}$. By definition the composite $i_! \circ q^!_{M'}$, corresponding to the diagram $0 \xleftarrow{q_{M'}} M' \xrightarrow{i} M$ in $C$, represents an arrow $u : 0 \longrightarrow M$ in $QC$. On the other hand, the composition law gives that $q^! \circ i_{M'''!}$ is represented by

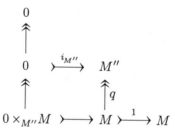

But there is an isomorphism $0 \times_{M''} M \xrightarrow{\cong} M'$ such that

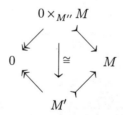

commutes, since $0 \longrightarrow M' \xrightarrow{i} M \xrightarrow{q} M'' \longrightarrow 0$ is exact. Thus $u = i_! \circ q^!_{M'}$ and $q^! \circ i_{M'''!}$ are represented by isomorphic diagrams, and so give equal arrows in $QC$. Thus, in the earlier diagram with 2 shaded triangles, we can add on a third arrow $u : 0 \longrightarrow M$, and add 2 shaded triangles

## 4. Exact Categories and Quillen's Q-Construction

(corresponding to 2-simplices in $BQ\mathcal{C}$), from the diagram

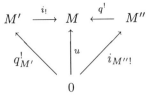

We do not draw the resulting diagram with 4 shaded triangles, but the reader can imagine it as yielding a $CW$-complex homeomorphic to a 2-sphere with 3 holes, such that the boundary circles have a common point $\{0\}$, and (properly oriented) are just the 3 loops $r_M, r_{M'}, r_{M''}$. Taking the orientations into account one checks that $r_M$ is homotopic to $r_{M'} \cdot r_{M''}$.

# 5. The $K$-Theory of Rings and Schemes

If $R$ is a ring, let $\mathcal{P}(R)$ denote the category of finitely generated projective (left) $R$-modules. This is a full subcategory of the Abelian category of left $R$-modules, so that $\mathcal{P}(R)$ is an exact category where all exact sequences are split. We will prove the following result, comparing the plus and $Q$ constructions, in Chapter 7.

**Theorem (5.1).** *There is a homotopy equivalence*

$$BGL(R)^+ \longrightarrow (\Omega BQ\mathcal{P}(R))^0 \ \textit{which is natural up to homotopy}.$$

(where $\Omega$ denotes the loop space, and the superscript 0 denotes the connected component of the trivial loop at $\{0\} \in BQ\mathcal{P}(R)$). Hence there are natural isomorphisms $K_i(\mathcal{P}(R)) \cong \pi_i(BGL(R)^+)$, $i \geq 1$.

From Theorem (5.1), we see that Quillen's new definition $K_i(R) = K_i(\mathcal{P}(R))$ agrees, for $i \geq 1$, with the definition given by the plus construction; in particular, $K_0$, $K_1$ and $K_2$ agree with the earlier definitions. This fact will be needed in a couple of places for certain computations.

We will assume in the following discussion that *all rings are left Noetherian*, unless specified otherwise. If $A$ is a Noetherian ring, let $\mathcal{M}(A)$ denote the category of finitely generated (left) $A$-modules. Then $\mathcal{M}(A)$ is equivalent to a small full subcategory, and any two such full subcategories are naturally equivalent to their union. Hence we can define $K_i(\mathcal{M}(A))$. A similar convention is used for all other exact categories which we will deal with, and has already been tacitly used in the definition of $K_i(\mathcal{P}(R))$ above.

We define $G_i(A) = K_i(\mathcal{M}(A))$; this is also sometimes denoted by $K'_i(A)$. The inclusion $\mathcal{P}(A) \subset \mathcal{M}(A)$ induces a natural map $K_i(A) \longrightarrow G_i(A)$. By the resolution theorem (4.6), if $A$ is (left) regular, then $K_i(A) \cong G_i(A)$ (recall that a Noetherian ring $A$ is left regular if every finitely generated $A$-module has a finite resolution by finitely generated projective $A$-modules).

5. The K-Theory of Rings and Schemes

Our first goal is to prove:

**Theorem (5.2).** *Let $A$ be Noetherian. Then there are natural isomorphisms for all $i \geq 0$*

(i) $G_i(A) \cong G_i(A[t])$, *induced by change of rings;*

(ii) $G_i(A[t, t^{-1}]) \cong G_i(A) \oplus G_{i-1}(A)$.

*(in (ii), for $i = 0$ we define $G_{-1}(A) = 0$).*

**Proof (ii).** We first prove (ii), assuming (i). Let $\mathcal{B} \subset \mathcal{M}(A[t])$ be the Serre subcategory consisting of modules annihilated by a power of $t$ (strictly, we must first replace $\mathcal{M}(A[t])$ by an equivalent small full subcategory, and let $\mathcal{B}$ be the Serre subcategory of this small Abelian category consisting of modules annihilated by a power of $t$—we will in future suppress such points, leaving it to the careful reader to make the necessary modifications). The quotient Abelian category $\mathcal{M}(A[t])/\mathcal{B}$ is naturally equivalent to $\mathcal{M}(A[t, t^{-1}])$. Hence the localization theorem (4.9) yields an exact sequence

$$\cdots \longrightarrow G_{i+1}(A[t, t^{-1}]) \longrightarrow K_i(\mathcal{B}) \longrightarrow G_i(A[t]) \longrightarrow G_i(A[t, t^{-1}]) \longrightarrow \cdots.$$

Now $\mathcal{M}(A) \subset \mathcal{B}$ as the full subcategory of modules annihilated by $t$. Hence by dévissage (Theorem (4.8)), $K_i(\mathcal{B}) \cong G_i(A)$. Hence the localization sequence can be rewritten as

$$\cdots \longrightarrow G_i(A) \longrightarrow G_i(A[t]) \xrightarrow{f} G_i(A[t, t^{-1}]) \longrightarrow G_{i-1}(A) \longrightarrow \cdots.$$

From (i), the flat change of rings $A \to A[t]$ gives an exact functor $\mathcal{M}(A) \to \mathcal{M}(A[t])$, $M \mapsto A[t] \otimes_A M$, which gives an isomorphism $G_i(A) \to G_i(A[t])$. Since $A \to A[t, t^{-1}]$ also gives an exact functor $i : \mathcal{M}(A) \to \mathcal{M}(A[t, t^{-1}])$ we have a commutative diagram

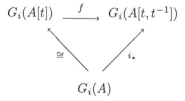

It suffices to prove that $i_*$ is a split inclusion; then the same holds for $f$, and (ii) follows immediately.

Let $\mathcal{M}_1(A[t, t^{-1}]) \subset \mathcal{M}(A[t, t^{-1}])$ be the full subcategory of modules $M$ satisfying $\mathrm{Tor}_1^{A[t,t^{-1}]}(M, A) = 0$, where we let $A$ be the $A[t, t^{-1}]$-module $A[t, t^{-1}]/(t-1)$. Then by the resolution theorem (or rather, Cor. (4.7)), $K_i(\mathcal{M}_1(A[t, t^{-1}])) \cong G_i(A[t, t^{-1}])$. The functor $i : \mathcal{M}(A) \longrightarrow \mathcal{M}(A[t, t^{-1}])$

clearly factors through $\mathcal{M}_1(A[t,t^{-1}])$. Also, there is an *exact* functor $j : \mathcal{M}_1(A[t,t^{-1}]) \longrightarrow \mathcal{M}(A)$ given by $M \longrightarrow A \otimes_{A[t,t^{-1}]} M$ (where as above, $A$ is the cyclic $A[t,t^{-1}]$-module with annihilator $(t-1)$). Clearly $j \circ i$ is isomorphic to the identity functor. Hence $j_* \circ i_* : G_i(A) \longrightarrow G_i([t,t^{-1}]) \longrightarrow G_i(A)$ is the identity.

**Proof (i).** We begin by proving a lemma. Let $B = A[t_1,\ldots,t_n]$ be the polynomial ring in $n$ variables over $A$, with its usual grading ($\deg t_i = 1$ for all $i$, $\deg a = 0$ for all $a \in A$). Let $\mathcal{M}gr B$ denote the category of positively graded, finite $B$-modules. Thus each $N \in \mathcal{M}gr B$ can be written as $N = \bigoplus_{p \geq 0} N_p$ where the $N_p$ are finite $A$-modules. Then $N$ has a finite increasing filtration $\{F_p N\}_{p \geq 0}$, where $F_p N$ is the $B$-submodule generated by $N_0 \oplus \cdots \oplus N_p$. Regard $A$ as a graded $B$-module annihilated by $t_1,\ldots,t_n$ and concentrated in degree 0. Then $A \otimes_B N$ is a graded $B$-module with

$$(A \otimes_B N)_p \cong N_p/B_1 N_{p-1} + \cdots + B_p N_0$$

as $A$-modules, where $B_i \subset B$ is the $A$-submodule of homogeneous polynomials of degree $i$. Thus

$$(*) \qquad (A \otimes_B F_p N)_m = \begin{cases} 0 & \text{if } p < m \\ (A \otimes_B N)_m & \text{if } p \geq m. \end{cases}$$

For any $N \in \mathcal{M}gr B$, and any $p > 0$, let $N(-p) \in \mathcal{M}gr B$ be the graded module with underlying $B$-module $N$ and grading

$$N(-p)_m = \begin{cases} N_{m-p} & \text{if } m \geq p \\ 0 & \text{if } m < p. \end{cases}$$

**Lemma (5.3).** *Let $N \in \mathcal{M}gr B$ such that $\operatorname{Tor}_1^B(A,N) = 0$. Then for each $p \geq 0$ the natural map of graded $B$-modules*

$$\phi_p : B(-p) \otimes_A (A \otimes_B N)_p \longrightarrow F_p N/F_{p-1} N$$

*is an isomorphism.*

**Proof.** Since $(A \otimes_B N)_p \cong (F_p N/F_{p-1} N)_p$, and $F_p N/F_{p-1} N$ is generated over $B$ by its $A$-submodule of elements of degree $p$, there is a natural map $\phi_p$ as in the statement of the lemma, which is surjective. We prove that under the condition $\operatorname{Tor}_1^B(A,N) = 0$, $\phi_p$ is an injection for each $p$. We work by descending induction on $p$; for large $p$, $N = F_p N = F_{p-1} N$ so that both sides vanish. We also have $\operatorname{Tor}_1^B(A, F_{p-1} N) = 0$.

Assume that $\phi_{p+1}$ is an isomorphism, and that $\operatorname{Tor}_1^B(A, F_p N) = 0$; we will deduce that $\phi_p$ is an isomorphism, and that $\operatorname{Tor}_1^B(A, F_{p-1} N) = 0$.

From the sequence $0 \longrightarrow F_{p-1}N \longrightarrow F_pN \longrightarrow F_pN/F_{p-1}N \longrightarrow 0$ we have an exact sequence

$$\operatorname{Tor}_1^B(A, F_pN) \to \operatorname{Tor}_1^B(A, F_pN/F_{p-1}N) \to A \otimes_B F_{p-1}N \xrightarrow{i} A \otimes_B F_pN \to$$
$$\shortparallel \qquad\qquad\qquad\qquad\qquad\qquad\qquad \to A \otimes_B (F_pN/F_{p-1}N) \to 0.$$
$$0$$

From (∗), $A \otimes_B F_{p-1}N \longrightarrow A \otimes_B F_p N$ is an isomorphism in degrees $\leq p-1$, and $A \otimes_B F_{p-1}N$ vanishes in all higher degrees. Hence the map $i$ is injective, and so $\operatorname{Tor}_1^B(A, F_pN/F_{p-1}N) = 0$.

Now consider the exact sequence (which defines $M$)

$$0 \longrightarrow M \longrightarrow B(-p) \otimes_A (A \otimes_B N)_p \xrightarrow{\phi_p} F_pN/F_{p-1}N \longrightarrow 0.$$

Tensoring with the $B$-module $A$, we obtain

$$0 \to \operatorname{Tor}_1^B(A, F_pN/F_{p-1}N) \to A \otimes_B M \to (A \otimes_B N)_p \to$$
$$\shortparallel \qquad\qquad\qquad\qquad\qquad\qquad \to (A \otimes F_pN/F_{p-1}N)_p \to 0.$$
$$0$$

Since $\phi_p$ is an isomorphism in degree $p$, we obtain $A \otimes_B M = 0$. Since $M$ is a finite graded $B$-module, this forces $M = 0$. Hence $\phi$ is an isomorphism. Since $B$ is a polynomial ring, $\operatorname{Tor}_i^B(A, B(-p) \otimes_A K) = 0$ for all $i > 0$, for any $A$-module $K$. Hence, using the isomorphism $\phi_p$,

$$\operatorname{Tor}_i^B(A, F_pN/F_{p-1}N) = 0 \text{ for all } i > 0.$$

Hence $\operatorname{Tor}_1^B(A, F_pN) = 0 \implies \operatorname{Tor}_1^B(A, F_{p-1}N) = 0$. This completes the inductive step, and the proof of the lemma.

The operation $N \longrightarrow N(-1)$ gives an exact functor from $\mathcal{M}gr B$ to itself. Hence there is a natural $\mathbb{Z}[t]$-module structure on $K_i(\mathcal{M}gr B)$, where $t$ acts by shifting the grading by $-1$. The change of rings map gives a homomorphism $G_i(A) \longrightarrow K_i(\mathcal{M}gr B)$, and hence a homomorphism $\psi : G_i(A) \otimes_{\mathbb{Z}} \mathbb{Z}[t] \longrightarrow K_i(\mathcal{M}gr B)$, where $\psi(x \otimes t^n) = \psi_n(x)$, and $\psi_n$ is induced by the functor $\mathcal{M}(A) \longrightarrow \mathcal{M}gr B$, $P \mapsto B(-n) \otimes_A P$.

**Proposition (5.4).** $\psi$ gives an isomorphism $G_i(A) \otimes_{\mathbb{Z}} \mathbb{Z}[t] \cong K_i(\mathcal{M}gr B)$.

**Proof.** Let $\mathcal{N} \subset \mathcal{M}gr B$ denote the full subcategory consisting of all modules $N$ such that $\operatorname{Tor}_j^B(A, N) = 0$ for all $j > 0$. Since the functors $T_j = \operatorname{Tor}_j^B(A, -)$ vanish for all $j > n$ (the number of polynomial variables), Cor. (4.7) to the resolution theorem implies that $K_i(\mathcal{N}) \cong K_i(\mathcal{M}gr B)$.

Let $\mathcal{N}_p \subset \mathcal{N}$ be the full subcategory of modules $N$ satisfying $F_pN = N$ i.e., $N$ is generated by elements of degree $\leq p$. There are exact functors

$$\mathcal{M}(A)^{p+1} \xrightarrow{a} \mathcal{N}_p \xrightarrow{b} \mathcal{M}(A)^{p+1}$$

given by
$$a(M_0, \ldots, M_p) = \bigoplus_{j=0}^{p} B(-j) \otimes_A M_j$$
and
$$b(N) = ((A \otimes_B N)_0, \ldots, (A \otimes_B N)_p).$$
Clearly $b \circ a : \mathcal{M}(A)^{p+1} \longrightarrow \mathcal{M}(A)^{p+1}$ is the identity. Now the identity functor $I_p : \mathcal{N}_p \longrightarrow \mathcal{N}_p$ has an admissible filtration by additive subfunctors $F_q : N \mapsto F_q N$, $0 \leq q \leq p$. If $0 \longrightarrow N' \longrightarrow N \longrightarrow N'' \longrightarrow 0$ is an exact sequence in $\mathcal{N}$, then
$$0 \longrightarrow A \otimes_B N' \longrightarrow A \otimes_B N \longrightarrow A \otimes_B N'' \longrightarrow 0$$
is exact. Hence, from Lemma (5.3), the sequences
$$0 \longrightarrow F_q N'/F_{q-1} N' \longrightarrow F_q N/F_{q-1} N \longrightarrow F_q N''/F_{q-1} N'' \longrightarrow 0$$
are exact, for $0 < q \leq p$. Hence by Cor. (4.4),
$$\sum_{q=1}^{p} (F_q/F_{q-1})_* = (I_p)_* : K_i(\mathcal{N}_p) \longrightarrow K_i(\mathcal{N}_p).$$
But by Lemma (5.3), the map $\sum (F_q/F_{q-1})_*$ is precisely the map on $K_i(\mathcal{N}_p)$ induced by the composite functor $a \circ b$. Hence the functor $a$ induces an isomorphism
$$a_* : G_i(A)^{p+1} \longrightarrow K_i(\mathcal{N}_p).$$
Taking the direct limit over $p$ (Lemma (3.8)), the proposition follows.

**Proof of Theorem (5.2)** (i). Let $B = A[t, u]$ be the polynomial ring, $C = A[t, u, u^{-1}]$ the localization with respect to powers of $u$. As above let $\mathcal{M}grB$ denote the category of finite positively graded $B$-modules. Let $\mathcal{M}grC$ denote the category of finite $\mathbb{Z}$-graded $C$-modules. Then $\mathcal{M}grC$ is naturally equivalent to the quotient of $\mathcal{M}grB$ by the Serre subcategory $\mathcal{A}$ of modules annihilated by a power of $u$. Clearly $\mathcal{M}gr(A[t]) \subset \mathcal{A}$ as the full subcategory of modules annihilated by $u$; by dévissage,
$$K_i(\mathcal{M}gr(A[t])) \cong K_i(\mathcal{A}).$$
Also, we have a natural equivalence of categories $\mathcal{M}grC \xrightarrow{\cong} \mathcal{M}(A[x])$, where $x = \frac{t}{u} \in C$, given by $N \mapsto N_0$ (the inverse functor is given by change of rings). Hence from the localization theorem we have an exact sequence
$$\longrightarrow K_i(\mathcal{M}gr(A[t])) \xrightarrow{\phi} K_i(\mathcal{M}grB) \longrightarrow G_i(A[x]) \longrightarrow K_{i-1}(\mathcal{M}gr(A[t])) \longrightarrow.$$

Now
$$K_i(\mathcal{M}\mathit{gr}(A[t])) \cong G_i(A) \otimes_{\mathbb{Z}} \mathbb{Z}[y]$$
$$K_i(\mathcal{M}\mathit{gr}\,B) \cong G_i(A) \otimes_{\mathbb{Z}} \mathbb{Z}[y]$$
where multiplication by $y$ corresponds to a shift in grading by $-1$ in both cases. Hence $\phi$ is $\mathbb{Z}[y]$-linear, so to compute it, it suffices to compute $\phi|_{G_i(A)}$.

For any $A$-module $M$ we have an exact sequence of graded $B$-modules
$$0 \longrightarrow B(-1) \otimes_A M \xrightarrow{u} B \otimes_A M \longrightarrow A[t] \otimes_A M \longrightarrow 0$$
where $A[t] = B/uB$. Thus the composite functor
$$i : \mathcal{M}(A) \longrightarrow \mathcal{M}\mathit{gr}(A[t]) \longrightarrow \mathcal{M}\mathit{gr}\,B$$
$$M \mapsto A[t] \otimes_A M$$
fits into an exact sequence of functors $\mathcal{M}(A) \longrightarrow \mathcal{M}\mathit{gr}\,B$
$$0 \longrightarrow j(-1) \longrightarrow j \longrightarrow i \longrightarrow 0,$$
where $j(-n) : \mathcal{M}(A) \longrightarrow \mathcal{M}\mathit{gr}\,B$ is $j(-n)(M) = B(-n) \otimes_A M$. Hence $i_* = (1-y) \cdot j_* : G_i(A) \longrightarrow K_i(\mathcal{M}\mathit{gr}\,B)$. Thus $\phi$ is identified with multiplication by $1 - y$. In particular $\phi$ is injective with cokernel isomorphic to $G_i(A)$, where the isomorphism is given by
$$j_* : G_i(A) \longrightarrow K_i(\mathcal{M}\mathit{gr}\,B) \longrightarrow K_i(\mathcal{M}\mathit{gr}\,B)/(1-y) \cdot K_i(\mathcal{M}\mathit{gr}\,B).$$
Hence $G_i(A) \longrightarrow G_i(A[x])$, given by change of rings, is an isomorphism.

**Corollary (5.5).** *Let $A$ be a regular ring. Then*

(i) $K_i(A) \cong K_i(A[t])$

(ii) $K_i(A[t, t^{-1}]) \cong K_i(A) \oplus K_{i-1}(A)$

(where if $i = 0$, $K_{-1}(A)$ is defined to be 0).

**(5.6) $K$-Theory of Schemes.** For terminology from algebraic geometry, we refer to Appendix D. If $X$ is an arbitrary scheme, let $\mathcal{P}(X)$ denote the category of locally free sheaves of finite rank, which is a full subcategory of the Abelian category of quasi-coherent sheaves of $\mathcal{O}_X$-modules on $X$ (see (D.10), (D.17), (D.43)). Define $K_i(X) = K_i(\mathcal{P}(X))$.

If $X$ is a Noetherian scheme (D.24), let $\mathcal{M}(X)$ be the Abelian category of coherent sheaves (see (D.44)) on $X$; define $G_i(X) = K_i(\mathcal{M}(X))$ (sometimes this is also denoted by $K'_i(X)$). Since we will mainly study $G_i$ at this point, we assume until further notice that all schemes under consideration are Noetherian and separated (D.33), unless explicitly mentioned otherwise.

There is a natural homomorphism $K_i(X) \longrightarrow G_i(X)$ induced by $\mathcal{P}(X) \subset \mathcal{M}(X)$. If $X$ is regular (D.25), then (since $X$ is Noetherian, hence quasi-compact (D.23)) every coherent sheaf on $X$ is a quotient of a locally free sheaf of finite rank (see (D.54), (D.55)), and hence has a finite resolution by locally free sheaves of finite rank. Hence by the resolution theorem, $K_i(X) \cong G_i(X)$.

If $\mathcal{E}$ is a locally free sheaf on $X$, then the assignment (see (D.8) for tensor products)
$$\mathcal{F} \mapsto \mathcal{E} \otimes_{\mathcal{O}_X} \mathcal{F}, \quad \mathcal{F} \in \mathcal{M}(X)$$
gives an exact functor $[\mathcal{E}]_\bullet : \mathcal{M}(X) \longrightarrow \mathcal{M}(X)$, and hence a map $[\mathcal{E}]_* : G_i(X) \longrightarrow G_i(X)$ for each $i$. If $0 \longrightarrow \mathcal{E}' \longrightarrow \mathcal{E} \longrightarrow \mathcal{E}'' \longrightarrow 0$ is an exact sequence in $\mathcal{P}(X)$, then $[\mathcal{E}]_* = [\mathcal{E}']_* + [\mathcal{E}'']_*$, since we have an exact sequence of functors $\mathcal{M}(X) \longrightarrow \mathcal{M}(X)$
$$0 \longrightarrow [\mathcal{E}']_\bullet \longrightarrow [\mathcal{E}]_\bullet \longrightarrow [\mathcal{E}'']_\bullet \longrightarrow 0$$
corresponding to the exact sequences in $\mathcal{M}(X)$
$$0 \longrightarrow \mathcal{E}' \otimes \mathcal{F} \longrightarrow \mathcal{E} \otimes \mathcal{F} \longrightarrow \mathcal{E}'' \otimes \mathcal{F} \longrightarrow 0.$$

Thus, $\mathcal{E} \mapsto [\mathcal{E}]_*$ extends to a homomorphism $K_0(X) \longrightarrow \text{End}(G_i(X))$, i.e., a pairing $K_0(X) \otimes_{\mathbb{Z}} G_i(X) \longrightarrow G_i(X)$. Since $K_0(X)$ is a ring with the multiplication induced by tensor products of locally free sheaves, the above pairing in fact makes $G_i(X)$ a module over the commutative ring $K_0(X)$. Similarly one can define $K_0(X)$-module structures on all the $K_i(X)$, such that $K_i(X) \longrightarrow G_i(X)$ is a $K_0(X)$-module homomorphism.

If $X = \text{Spec } A$ is affine (see (D.15)), then we have natural equivalences of categories $\mathcal{P}(X) \xrightarrow{\cong} \mathcal{P}(A)$, $\mathcal{M}(X) \xrightarrow{\cong} \mathcal{M}(A)$. Hence $K_i(X) \cong K_i(A)$, $G_i(X) \cong G_i(A)$, and we identify these groups under these natural isomorphisms.

**Remark (5.7).** From Theorem (5.1), $K_i(A)$ defined using the $Q$-construction agrees with the earlier definition in Chapter 3 as $\pi_i(BGL(A)^+)$, for $i \geq 1$. The latter groups have the product $K_i(A) \otimes_{\mathbb{Z}} K_j(A) \longrightarrow K_{i+j}(A)$ defined in Chapter 3 for $i, j \geq 1$; one can show that this is $K_0(A)$-bilinear, making $\bigoplus_{i \geq 0} K_i(A)$ into a graded anti-commutative ring.

More generally, F. Waldhausen (Algebraic K-Theory of Generalized Free Products, *Ann. Math.* 108 (1978), 135–256) has shown that if $\mathcal{A}, \mathcal{B}, \mathcal{C}$ are exact categories, and $\mathcal{A} \times \mathcal{B} \xrightarrow{F} \mathcal{C}$ is bi-exact (i.e., $F(M, -)$ and $F(-, N)$ are exact functors for each $M \in \mathcal{A}$, $N \in \mathcal{B}$) then there is a product (for each $i, j \geq 0$)
$$K_i(\mathcal{A}) \otimes K_j(\mathcal{B}) \longrightarrow K_{i+j}(\mathcal{C})$$

satisfying certain naturality properties, and which agrees with Loday's product if $\mathcal{A} = \mathcal{B} = \mathcal{C} = \mathcal{P}(A)$. This new product will make $\bigoplus_{i \geq 0} K_i(X)$ into a graded anti-commutative ring, and $\bigoplus_{i \geq 0} G_i(X)$ into a graded $\bigoplus_{i \geq 0} K_i(X)$-module (if $X$ is Noetherian) such that

$$\bigoplus_{i \geq 0} K_i(X) \longrightarrow \bigoplus_{i \geq 0} G_i(X)$$

is a module homomorphism.

**(5.8) Functoriality Properties.** If $f : X \longrightarrow Y$ is a morphism (see (D.16), (D.17)) of schemes, the exact functor $f^* : \mathcal{P}(Y) \longrightarrow \mathcal{P}(X)$ (see (D.41) for $f^*$)gives homomorphisms $f^* : K_i(Y) \longrightarrow K_i(X)$. Clearly $K_i$ becomes a contravariant functor from schemes to Abelian groups.

If $f : X \longrightarrow Y$ is a flat morphism (D.40) of Noetherian schemes, we have an exact functor $f^* : \mathcal{M}(Y) \longrightarrow \mathcal{M}(X)$, and hence a map on $K$-groups $G_i(Y) \longrightarrow G_i(X)$. Thus $G_i$ is a contravariant functor on the category of Noetherian schemes and flat morphisms.

**Lemma (5.9).** (i) *Let $i \mapsto X_i$ be a filtered inverse system of schemes such that the transition morphisms $X_i \longrightarrow X_j$ are affine (D.28), and let $X = \underset{\longrightarrow}{\lim} X_i$. Then $K_q(X) = \underset{\longrightarrow}{\lim} K_q(X_i)$.*

(ii) *If in (i), $X, X_i$ are all Noetherian, and the transition morphisms are also flat, then $G_q(X) = \underset{\longrightarrow}{\lim} G_q(X_i)$.*

**Proof.** (i) We wish to apply Lemma (3.8), since $\mathcal{P}(X)$ is essentially the direct limit of the $\mathcal{P}(X_i)$. However $i \mapsto \mathcal{P}(X_i)$ does not form a directed system of categories indexed by $I^{\mathrm{op}}$, since for morphisms $i \to j \to k$ in $I$,

$$(i \to k)^* : \mathcal{P}(X_k) \longrightarrow \mathcal{P}(X_i) \text{ and}$$
$$(i \to j)^* \circ (j \to k)^* : \mathcal{P}(X_k) \longrightarrow \mathcal{P}(X_i)$$

are not equal, but only isomorphic (the tensor product involved in defining the pullback of a sheaf is not strictly associative, but only associative up to isomorphism).

One gets around this problem by the following device. Let $I' = I \cup \{\emptyset\}$ where $\emptyset$ is an initial object, and let $X_\emptyset = X$. For $i \in I'$, let $I'/i$ be the full subcategory of $I'$ consisting of objects $j$ preceding $i$; then there is a unique morphism $j \longrightarrow i$ for $j \in I'/i$ since $I^{\mathrm{op}}$, hence $(I')^{\mathrm{op}}$, is directed. Let $\mathcal{P}_i$ denote the following category, for each $i \in I'$—an object in $\mathcal{P}_i$ is a family $\{(P_j, \theta_j) \mid j \in I'/i\}$ where $P_j \in \mathcal{P}(X_j)$, and $\theta_j : (j \to i)^* P_i \longrightarrow P_j$ is an isomorphism. An arrow $f : \{(P_j, \theta_j)\} \longrightarrow \{(P'_j, \theta'_j)\}$ is a collection of arrows $f_j : P_j \longrightarrow P'_j$ in $\mathcal{P}(X_j)$ such that the diagram below commutes:

$$\begin{array}{ccc}
(j \to i)^* P_i & \xrightarrow{\theta_j} & P_j \\
(j \to i)^* f_i \downarrow & & \downarrow f_j \\
(j \to i)^* P'_i & \xrightarrow{\theta'_j} & P'_j
\end{array}$$

Then $\mathcal{P}_i \longrightarrow \mathcal{P}(X_i)$ given by $\{(P_j, \theta_j)\} \longrightarrow P_i$ is an equivalence of categories for each $i \in I'$, and we claim (see the remark below) $i \longrightarrow \mathcal{P}_i$ is a functor from $(I')^{\mathrm{op}}$ to small categories. Further, since $(I')^{\mathrm{op}}$ has a final object $\emptyset$, the direct limit is just $\mathcal{P}_\emptyset$, which is equivalent to $\mathcal{P}(X)$. Hence the lemma follows from Lemma (3.8), and the fact that

$$Q(\varinjlim \mathcal{P}_i) = \varinjlim Q\mathcal{P}_i.$$

(ii) The proof is similar and is left to the reader.

**Remark.** The transition functor $\mathcal{P}_i \longrightarrow \mathcal{P}_j$ for a morphism $j \to i$ in $I'$ is given by $\{(P_\ell, \theta_\ell) \mid \ell \in I'/i\} \longrightarrow \{(P_\ell, \psi_\ell) \mid \ell \in (I'/j)\}$, where $I'/j \subset I'/i$, so it makes sense to consider the subfamily of $P_\ell$ with $\ell \in I'/j$ of the given family of $P_\ell$, $\ell \in I'/i$; the map $\psi_\ell$ is the unique isomorphism making the following diagram commute:

$$\begin{array}{ccc}
(\ell \to j)^* \circ ((j \to i)^* P_i) & \xrightarrow{(\ell \to j)^* \theta_j} & (\ell \to j)^* P_j \\
\eta_{j\ell} \downarrow & & \downarrow \psi_\ell \\
(\ell \to i)^* P_i & \xrightarrow{\theta_\ell} & P_\ell
\end{array}$$

where $\eta_{j\ell}$ is the "natural" isomorphism

$$(\ell \to j)^* \circ ((j \to i)^* P_i) \cong (\ell \to j)^* P_i,$$

given by

$$\mathcal{O}_{X_\ell} \otimes (\mathcal{O}_{X_j} \otimes P_i) \longrightarrow \mathcal{O}_{X_\ell} \otimes P_i,$$

which is locally given by $a \otimes (b \otimes p) \mapsto ab \otimes p$.

So one has to verify that with the above transition functors,

$$(\ell \longrightarrow i)^* = (\ell \longrightarrow j)^* \circ (j \longrightarrow i)^* : \mathcal{P}_i \longrightarrow \mathcal{P}_\ell,$$

for $\ell \longrightarrow j \longrightarrow i$ in $I'$. Both operations take a family $\{(P_m, \theta_m) \mid m \in I'/i\}$ to the same family of objects $\{P_m \mid m \in I'/\ell\}$, and possibly two sets of isomorphisms $\chi_m, \chi'_m : (m \to \ell)^* P_\ell \longrightarrow P_m$; we have to verify that $\chi_m = \chi'_m$ for every $m \in I'/\ell$. This is a local question, so we are reduced to the affine case.

Now the situation is as follows: we are given a direct system of rings $\{A_i\}_{i \in J}$, where $J = (I')^{\mathrm{op}}$ is a filtering category. Let $\mathcal{P}_j = \{(P_j, \theta_j) \mid P_j \in \mathcal{P}(A_j), \theta_j : A_j \otimes_{A_i} P_i \xrightarrow{\cong} P_j\}$. The transition functors $(i < j)^* : \mathcal{P}_i \longrightarrow \mathcal{P}_j$

are given by $(i < j)^* \{(P_\ell, \theta_\ell)\} = \{(P_\ell, \psi_\ell)\}$ where for $j < \ell$, $\psi_\ell$ is the unique isomorphism making the following diagram commute:

$$
\begin{array}{ccc}
A_\ell \otimes (A_j \otimes P_i) & \xrightarrow{1_\ell \otimes \theta_j} & A_\ell \otimes P_j \\
{\scriptstyle \eta_{\ell j}} \downarrow & & \downarrow {\scriptstyle \psi_\ell} \\
A_\ell \otimes P_i & \xrightarrow{\theta_\ell} & P_\ell
\end{array}
$$

i.e., $\psi_\ell = \theta_\ell \circ \eta_{\ell j} \circ (1_\ell \otimes \theta_j)^{-1}$. Here $\eta_{\ell j}$ is induced by $a \otimes (b \otimes p) \mapsto ab \otimes p$. Now suppose $i < j < \ell < m$; then

$$(j < \ell)^* \circ (i < j)^* \{(P_m, \theta_m)\} = \{(P_m, \chi_m)\}$$

where $\chi_m$ is defined by the commutativity of

$$
\begin{array}{ccc}
A_m \otimes (A_\ell \otimes P_j) & \xrightarrow{1_m \otimes \psi_\ell} & A_m \otimes P_\ell \\
{\scriptstyle \eta'_{m\ell}} \downarrow & & \downarrow {\scriptstyle \chi_m} \\
A_m \otimes P_j & \xrightarrow{\psi_m} & P_m
\end{array}
$$

(and $\eta'_{m\ell}(a \otimes (b \otimes p)) = ab \otimes p$). On the other hand

$$(i < \ell)^* \{(P_m, \theta_m)\} = \{(P_m, \chi'_m)\},$$

where $\chi'_m$ is the unique map making the following diagram commute:

$$
\begin{array}{ccc}
A \otimes (A \otimes P_i) & \xrightarrow{1_m \otimes \theta_\ell} & A_m \otimes P_i \\
{\scriptstyle \eta_{m\ell}} \downarrow & & \downarrow {\scriptstyle \chi'_m} \\
A_m \otimes P_i & \xrightarrow{\theta_m} & P_m.
\end{array}
$$

Since $\psi_m = \theta_m \circ \eta_{mj} \circ (1_m \otimes \theta_j)^{-1}$, the desired equality $\chi_m = \chi'_m$ follows from the commutativity of the diagram:

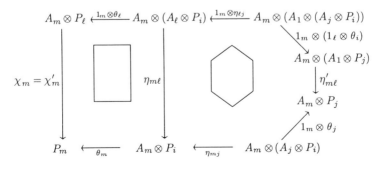

The commutativity of the outer border, with left vertical arrow $\chi_m$, is by the definition of $\chi_m$, while the commutativity of the left rectangle, with left vertical arrow $\chi'_m$, is by definition of $\chi'_m$. Hence, the commutativity of the hexagon proves $\chi_m = \chi'_m$.

To prove commutativity of the hexagon, let $a \in A_m$, $b \in A_\ell$, $c \in A_j$, $p \in P_i$; then $a \otimes (b \otimes (c \otimes p)) \in A_m \otimes (A_\ell \otimes (A_j \otimes P_i))$ yields the hexagon

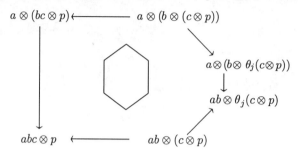

We note that the commutativity of the hexagon reflects a basic coherence property of the tensor product (see MacLane's book *Categories for the Working Mathematician*, Grad. Texts, No. 5, Springer-Verlag (1972)).

**(5.10)** Let $f : X \longrightarrow Y$ be a morphism (of Noetherian schemes) of *finite Tor-dimension*, i.e., there is an integer $N > 0$ such that the $\mathcal{O}_X$-modules $\text{Tor}_i^{\mathcal{O}_Y}(\mathcal{O}_X, \mathcal{F}) = 0$ for all $i \geq N$ for any $\mathcal{F} \in \mathcal{M}(Y)$. Let $\mathcal{M}(Y, f) \subset \mathcal{M}(Y)$ denote the full subcategory of sheaves $\mathcal{F}$ satisfying $\text{Tor}_i^{\mathcal{O}_Y}(\mathcal{O}_X, \mathcal{F}) = 0$ for all $i > 0$; then $\mathcal{P}(Y) \subset \mathcal{M}(Y, f)$. Hence, if every coherent sheaf on $Y$ is a quotient of a vector bundle, then by Cor. (4.7) of the resolution theorem, $K_i(\mathcal{M}(Y, f)) \cong G_i(Y)$ for all $i$. Also, $f^* : \mathcal{M}(Y, f) \longrightarrow \mathcal{M}(X)$, $\mathcal{F} \mapsto f^*\mathcal{F}$, is an exact functor in any case, yielding a map $f^* : K_i(\mathcal{M}(Y, f)) \longrightarrow G_i(X)$, so that in the above situation, we have a map $f^* : G_i(Y) \longrightarrow G_i(X)$. The condition that every coherent sheaf is a quotient of a vector bundle holds if either (i) $Y$ is regular (D.54), or (ii) if $Y$ supports an *ample* invertible sheaf (D.57) (e.g., $Y$ is quasi-projective over a Noetherian ring (D.52)). Finally the formula $(f \circ g)^* = g^* \circ f^*$ is valid for maps $f : X \longrightarrow Y$, $g : Y \longrightarrow Z$, both of which have finite Tor-dimension, such that coherent sheaves are quotients of vector bundles on $Y$ and $Z$.

**(5.11)** Let $f : X \longrightarrow Y$ be a proper morphism (see (D.34), (D.38)), so that for any $\mathcal{F} \in \mathcal{M}(X)$, $R^i f_* \mathcal{F} \in \mathcal{M}(Y)$ for each $i$, and the functors $R^i f_*$ vanish for sufficiently large $i$ on $\mathcal{M}(X)$ (see (D.12), (D.61), (D.62)). Let $\mathcal{M}(X, f) \subset \mathcal{M}(X)$ be the full subcategory of sheaves $\mathcal{F}$ satisfying $R^i f_* \mathcal{F} = 0$ for all $i > 0$. Then if every coherent sheaf on $X$ is a subsheaf of a sheaf in $\mathcal{M}(X, f)$, then $K_i(\mathcal{M}(X, f)) \cong K_i(\mathcal{M}(X, f)^{\text{op}}) \cong K_i(\mathcal{M}(X)^{\text{op}}) \cong$

$K_i(\mathcal{M}(X))$, where the middle isomorphism follows from Corollary (4.7) of the resolution theorem, and we have used the fact that $\mathcal{QC} \cong \mathcal{QC}^{\mathrm{op}}$ for any small exact category $\mathcal{C}$ (this is proved later, Cor. (6.3)). Hence, under this hypothesis on $\mathcal{M}(X)$, there is a direct image map $f_* : G_i(X) \longrightarrow G_i(Y)$; and for appropriate maps $f, g$ we have the formula $(f \circ g)_* = f_* \circ g_*$.

The hypothesis above on $\mathcal{M}(X)$ (namely, that any coherent sheaf is a subsheaf of an object of $\mathcal{M}(X, f)$) is satisfied if either (i) $f$ is finite, in which case $\mathcal{M}(X) = \mathcal{M}(X, f)$, or (ii) $X$ has an ample line bundle $\mathcal{L}$. In the latter case, $\mathcal{L}^{\otimes n}$ is generated by a finite number of global sections (as an $\mathcal{O}_X$-module) for all sufficiently large $n$. For any coherent sheaf $\mathcal{F}$ on $X$, let $\mathcal{F}(n)$ denote $\mathcal{F} \otimes \mathcal{L}^{\otimes n}$. Then for all sufficiently large $n$, $R^i f_* \mathcal{F}(n) = 0$ for all $i > 0$ i.e., $\mathcal{F}(n) \in \mathcal{M}(X, f)$ (this follows from (D.57), combined with (D.62)–(D.65)). If we fix such a value of $n$, which is large enough so that $\mathcal{O}_X(n)$ is also generated by global sections, then we can find $N$ global sections which generate $\mathcal{O}_X(n)$ for some $N > 0$, i.e., a surjection $\mathcal{O}_X^{\oplus N} \twoheadrightarrow \mathcal{O}_X(n)$ from a trivial bundle of rank $N$, whose kernel is a vector bundle on $X$. Applying the functor $\mathcal{H}om_{\mathcal{O}_X}(-, \mathcal{O}_X(n))$ (i.e., dualizing and tensoring with $\mathcal{O}_X(n)$) we obtain an exact sequence of vector bundles

$$0 \longrightarrow \mathcal{O}_X \longrightarrow \mathcal{O}_X(n)^{\oplus N} \longrightarrow \mathcal{E}_n \longrightarrow 0,$$

and hence on tensoring with $\mathcal{F}$ an exact sequence in $\mathcal{M}(X)$

$$0 \longrightarrow \mathcal{F} \longrightarrow \mathcal{F}(n)^{\oplus N} \longrightarrow \mathcal{E}_n \otimes \mathcal{F} \longrightarrow 0,$$

which gives an inclusion of $\mathcal{F}$ into a sheaf in $\mathcal{M}(X, f)$. Note that in this construction, if $\mathcal{F} \in \mathcal{P}(X)$, then $\mathcal{F}(n)^{\oplus N} \in \mathcal{P}(X, f) = \mathcal{P}(X) \cap \mathcal{M}(X, f)$. This is used below, in the proof of Prop. (5.12).

**Proposition (5.12)** (Projection formula). *Let $f : X \longrightarrow Y$ be a proper morphism of finite Tor-dimension between schemes supporting ample line bundles, so that $f^* : G_i(Y) \longrightarrow G_i(X)$, $f_* : G_i(X) \longrightarrow G_i(Y)$ are both defined. Then*

(i) *there is a well defined map $f_* : K_i(X) \longrightarrow K_i(Y)$ giving a commutative diagram*

$$\begin{array}{ccc} K_i(X) & \longrightarrow & G_i(X) \\ f_* \downarrow & & \downarrow f_* \\ K_i(Y) & \longrightarrow & G_i(Y). \end{array}$$

(ii) *for any $x \in K_0(X)$, $y \in G_i(Y)$, we have the formula $f_*(x) \cdot y = f_*(x \cdot f^* y)$*

(iii) *for any $x \in K_0(X)$, $y \in K_i(Y)$, we have the formula $f_*(x) \cdot y = f_*(x \cdot f^* y)$*

(iv) *for any $y \in K_0(Y)$, $x \in G_i(X)$, we have the formula $f_*(f^*(y) \cdot x) = y \cdot f_*(x)$.*

(in (ii), (iii), (iv) above, $\cdot$ denotes the $K_0$-module structure; in (iv), we do not need the hypothesis that $Y$ supports an ample line bundle).

**Proof.** The projection formula for sheaves (see (D.41))

$$f_*(f^*\mathcal{E} \otimes_{\mathcal{O}_X} \mathcal{F}) \cong \mathcal{E} \otimes_{\mathcal{O}_Y} f_*\mathcal{F},$$

for any $\mathcal{E} \in \mathcal{P}(Y)$, $\mathcal{F} \in \mathcal{M}(X)$, immediately yields (iv). We prove (i) and (ii) below; the proof of (iii) is similar and is left to the reader.

Let $\mathcal{P}(X, f) = \mathcal{P}(X) \cap \mathcal{M}(X, f)$ be the category of vector bundles $\mathcal{E}$ on $X$ satisfying $R^i f_*\mathcal{E} = 0$ for all $i > 0$. As noted at the end of (5.11), since $X$ supports an ample line bundle, for any $\mathcal{E} \in \mathcal{P}(X)$ there is an admissible monomorphism $\mathcal{E} \longrightarrow \mathcal{E}'$ with $\mathcal{E}' \in \mathcal{P}(X, f)$ (i.e., such that $0 \longrightarrow \mathcal{E} \longrightarrow \mathcal{E}' \longrightarrow \mathcal{E}'/\mathcal{E} \longrightarrow 0$ is an exact sequence *in* $\mathcal{P}(X)$). Hence $K_i(\mathcal{P}(X, f)) \cong K_i(\mathcal{P}(X)) = K_i(X)$, as in (5.11).

Let $\mathcal{E} \in \mathcal{P}(X, f)$; then we claim that $f_*\mathcal{E} \in \mathcal{H}(Y)$, the full subcategory of $\mathcal{M}(Y)$ consisting of sheaves of finite homological dimension (see (D.56)); since $Y$ supports an ample line bundle, these are precisely the sheaves which have finite resolutions by vector bundles (so that $K_i(Y) \cong K_i(\mathcal{H}(Y))$ by the resolution theorem). Since being of finite homological dimension is a local property, our claim (that $f_*\mathcal{E} \in \mathcal{H}(Y)$) is local on $Y$. Hence for proving the claim we may assume $Y = \operatorname{Spec} A$ is affine.

Now for any affine open subset $U = \operatorname{Spec} B$ in $X$, we know that $\operatorname{Tor}_i^A(B, -) = 0$ for all $i \geq N$ (for some $N$ independent of $U$), since $f$ has finite Tor-dimension. Let $U_i$ be an affine open cover of $X$ by a finite number of open sets, such that $\mathcal{E}|_{U_i}$ is trivial; since $f$ is separated, all the intersections $U_{i_1} \cap \cdots \cap U_{i_m}$ are affine (see (D.33)), and we can compute the cohomology groups $H^i(X, \mathcal{E})$ from the Čech complex associated to this affine cover (D.61). But $R^i f_*\mathcal{E}$ is just the sheaf on the affine scheme $Y = \operatorname{Spec} A$ associated to the $A$-module $H^i(X, \mathcal{E})$, for each $i \geq 0$ (see (D.43)(iv)). Since $\mathcal{E} \in \mathcal{P}(X, f)$, we have $H^i(X, \mathcal{E}) = 0$ for $i > 0$, and an exact sequence (with a finite number of non-zero terms)

$$0 \longrightarrow H^0(X, \mathcal{E}) \longrightarrow \coprod_i H^0(U_i, \mathcal{E}) \longrightarrow \coprod_{i,j} H^0(U_i \cap U_j, \mathcal{E}) \longrightarrow \cdots.$$

Each term in this sequence except the first is an $A$-module $M$ which is given to have the property $\operatorname{Tor}_i^A(M, -) = 0$ for $i \geq N$. Hence the functors $\operatorname{Tor}_i^A(H^0(X, \mathcal{E}), -)$ also vanish for $i \geq N$, from repeated applications of the long exact sequence of Tors. But $f_*\mathcal{E}$ is the coherent $\mathcal{O}_Y$-module associated to the finite $A$-module $H^0(X, \mathcal{E})$, which satisfies $\operatorname{Tor}_i^A(H^0(X, \mathcal{E}), -) = 0$ for

$i \geq N$. Hence $H^0(X, \mathcal{E})$ has projective dimension $< N$ over $A$, i.e., $f_*\mathcal{E}$ has finite homological dimension. This completes the proof of (i).

(ii) Let $\mathcal{E} \in \mathcal{P}(X, f)$ be fixed, and let

$$0 \longrightarrow \mathcal{E}_m \longrightarrow \mathcal{E}_{m-1} \longrightarrow \cdots \longrightarrow \mathcal{E}_0 \longrightarrow f_*\mathcal{E} \longrightarrow 0$$

be a finite resolution by vector bundles on $Y$ (which exists, as $f_*\mathcal{E} \in \mathcal{H}(Y)$). Then by definition, the action of $[f_*\mathcal{E}] \in K_0(Y)$ on $G_i(Y)$ is given by $y \mapsto \sum_{j=0}^{m}(-1)^j x_j \cdot y$, where $y \in G_i(Y)$, and $x_j = [\mathcal{E}_j]$.

Let $\mathcal{N} \subset \mathcal{M}(Y, f)$ be the full subcategory of sheaves satisfying

$$\mathrm{Tor}_i^{\mathcal{O}_Y}(f_*\mathcal{E}, \mathcal{F}) = 0$$

for all $i > 0$ (this Tor sheaf is an $\mathcal{O}_Y$-module). Then $K_i(\mathcal{N}) \cong G_i(Y)$ by the resolution theorem. For any $\mathcal{F} \in \mathcal{N}$, we have an exact sequence, natural in $\mathcal{F}$,

$$0 \longrightarrow \mathcal{E}_m \otimes \mathcal{F} \longrightarrow \mathcal{E}_{m-1} \otimes \mathcal{F} \longrightarrow \cdots \longrightarrow \mathcal{E}_0 \otimes \mathcal{F} \longrightarrow f_*\mathcal{E} \otimes \mathcal{F} \longrightarrow 0$$

which we interpret as an exact sequence of functors $\mathcal{N} \longrightarrow \mathcal{M}(Y)$. Hence the functor $\mathcal{N} \longrightarrow \mathcal{M}(Y)$ given by tensoring with $f_*\mathcal{E}$ represents the action of $[f_*\mathcal{E}] \in K_0(Y)$. Hence if $x = [\mathcal{E}] \in K_0(X)$, $y \mapsto f_*(x) \cdot y$ is represented by the functor $\mathcal{N} \longrightarrow \mathcal{M}(Y)$ given by tensoring with $f_*\mathcal{E}$.

Next, we want a similar representation of $y \mapsto f_*(x \cdot f^*(y))$ by a functor $\mathcal{N} \longrightarrow \mathcal{M}(Y)$. We claim that for any $\mathcal{F} \in \mathcal{N}$, we have (a) $R^i f_*(\mathcal{E} \otimes f^*\mathcal{F}) = 0$ for all $i > 0$, (b) there is a natural isomorphism $f_*(\mathcal{E} \otimes f^*\mathcal{F}) \cong (f_*\mathcal{E}) \otimes \mathcal{F}$. In view of the claimed naturality in (b), both statements are local on $Y$, so to prove them, we may assume that $Y = \mathrm{Spec}\, A$ is affine. Then $\mathcal{F} = \widetilde{M}$ is the sheaf associated to a finite $A$-module $M$. Since $\mathrm{Tor}_i^{\mathcal{O}_Y}(f_*\mathcal{E}, \mathcal{F}) = 0$ for $i > 0$, $\mathrm{Tor}_i^A(H^0(X, \mathcal{E}), M) = 0$ for $i > 0$ (as $f_*\mathcal{E}$ is just the sheaf associated to $H^0(X, \mathcal{E})$). Further, since $\mathcal{F} \in \mathcal{N} \subset \mathcal{M}(Y, f)$, if $U = \mathrm{Spec}\, B$ is any affine open subset of $X$, then $\mathrm{Tor}_i^A(B, M) = 0$ for all $i > 0$. Thus, if $\{U_i\}$ is an affine open cover of $X$ by a finite number of open sets such that $\mathcal{E}|_{U_i}$ is trivial, then the exact sequence (obtained from the Čech complex)

$$(*) \quad \cdots 0 \longrightarrow H^0(X, \mathcal{E}) \longrightarrow \prod_i H^0(U_i, \mathcal{E}) \longrightarrow \prod_{i,j} H^0(U_i \cap U_j, \mathcal{E}) \longrightarrow \cdots$$

remains exact on tensoring with the $A$-module $M$, since for each term in the sequence, $\mathrm{Tor}_i^A(-, M) = 0$ for all $i > 0$. For any affine open subset $U = \mathrm{Spec}\, B \subset X$, $f^*\mathcal{F}|_U$ is the sheaf associated to the finite $B$-module $B \otimes_A M$, and so $(\mathcal{E} \otimes f^*\mathcal{F})|_U$ is the sheaf associated to the $B$-module $H^0(U, \mathcal{E}) \otimes_B (B \otimes_A M) \cong H^0(U, \mathcal{E}) \otimes_A M$. Thus the sequence $(*)$, tensored with $M$ (over $A$), is the analogous complex obtained from the Čech complex for $\mathcal{E} \otimes f^*\mathcal{F}$ by adjoining $H^0(X, \mathcal{E}) \otimes_A M$ at the beginning. Since this complex is exact, we conclude that $H^i(X, \mathcal{E} \otimes f^*\mathcal{F}) = 0$ for $i > 0$, and

$H^0(X, \mathcal{E} \otimes f^*\mathcal{F}) \cong H^0(X, \mathcal{E}) \otimes_A M$ as $A$-modules. This means precisely that $R^i f_* \mathcal{E} \otimes f^*\mathcal{F} = 0$ for $i > 0$, and $f_*(\mathcal{E} \otimes f^*\mathcal{F}) \cong (f_*\mathcal{E}) \otimes \mathcal{F}$, proving the claims (a) and (b).

Now the exact functor $\mathcal{N} \longrightarrow \mathcal{M}(X, f)$ given by $\mathcal{F} \mapsto \mathcal{E} \otimes f^*\mathcal{F}$ represents $G_i(Y) \longrightarrow G_i(X)$, $y \mapsto x \cdot f^* y$. Hence the exact functor $\mathcal{N} \longrightarrow \mathcal{M}(Y)$ given by $\mathcal{F} \mapsto f_*(\mathcal{E} \otimes f^*\mathcal{F})$ represents $G_i(Y) \longrightarrow G_i(Y)$, $y \mapsto f_*(x \cdot f^* y)$. Since we have an isomorphism of functors $\mathcal{N} \longrightarrow \mathcal{M}(Y)$ given by the natural isomorphism $f_*(\mathcal{E} \otimes f^*\mathcal{F}) \cong (f_*\mathcal{E}) \otimes \mathcal{F}$, we obtain the formula $(f_*x) \cdot y = f_*(x \cdot f^* y)$ for any $y \in G_i(Y)$ and $x = [\mathcal{E}]$, with $\mathcal{E} \in \mathcal{P}(X, f)$. But such classes $x$ generate $K_0(X)$, and both sides are additive in $x$.

**Proposition (5.13).** *Let*

$$\begin{array}{ccc} X' & \xrightarrow{g'} & X \\ f' \downarrow & & \downarrow f \\ Y' & \xrightarrow{g} & Y \end{array}$$

*be a fiber product diagram of schemes with ample line bundles. Assume that $f$ is proper and $g$ has finite Tor dimension, and that $\mathcal{O}_X, \mathcal{O}_{Y'}$ are Tor-independent over $\mathcal{O}_Y$. Then $g^* \circ f_* = f'_* \circ g'^*$ in $\mathrm{Hom}(G_i(X), G_i(Y'))$.*

**Proof.** We leave the proof to the reader (see Quillen's paper, Prop. (2.11) for a proof using the analogous formula in the derived category). The point is to prove that if $\mathcal{F} \in \mathcal{M}(X, f) \cap \mathcal{M}(X, g')$, then $f_*\mathcal{F} \in \mathcal{M}(Y, g)$, $g'^* \mathcal{F} \in \mathcal{M}(X', f')$, and we have an isomorphism $g^* f_* \mathcal{F} \xrightarrow{\cong} f'_* g'^* \mathcal{F}$; the interested reader can give a direct proof of this using a suitable Čech complex, along the lines of the previous proof.

**Closed Subschemes.** Let $i : Z \longrightarrow X$ be a closed subscheme, $j : U \longrightarrow X$ the open complement. Let $I_Z$ denote the (coherent) sheaf of ideals of $Z$ in $\mathcal{O}_X$. We can identify $\mathcal{M}(Z)$ with the full subcategory of $\mathcal{M}(X)$ consisting of sheaves annihilated by $I_Z$, via the functor $i_*$ (see (D.29), (D.43)).

**Proposition (5.14).** *If $I_Z$ is nilpotent, then $G_i(Z) \cong G_i(X)$.*

**Proof.** Immediate from the dévissage theorem (4.8).

**Proposition (5.15).** *There is a long exact sequence*

$$\longrightarrow G_{i+1}(U) \longrightarrow G_i(Z) \xrightarrow{i_*} G_i(X) \xrightarrow{j^*} G_i(U) \longrightarrow G_{i-1}(Z)$$
$$\longrightarrow \cdots \longrightarrow G_0(X) \longrightarrow G_0(U) \longrightarrow 0.$$

## 5. The K-Theory of Rings and Schemes

**Proof.** Let $\mathcal{B} \subset \mathcal{M}(X)$ be the Serre subcategory of sheaves $\mathcal{F}$ with $\mathcal{F}|_U = 0$; then $\mathcal{M}(Z) \subset \mathcal{B}$ is a homotopy equivalence by dévissage, and the quotient $\mathcal{M}(X)/\mathcal{B}$ is equivalent to $\mathcal{M}(U)$. The result follows from the localization theorem (4.9).

**(5.16) Naturality.** The exact sequence of Prop. (5.15) has certain naturality properties. For example, if

$$Z \xrightarrow{i} Z' \xrightarrow{i} X$$

are closed subschemes, we have a diagram

$$\begin{array}{ccccccccc}
\cdots \to & G_{i+1}(X-Z) & \to & G_i(Z) & \xrightarrow{(i'_\circ i)_*} & G_i(X) & \xrightarrow{j^*} & G_i(X-Z) & \to \cdots \\
& \downarrow{j'^*} & & \downarrow{i_*} & & \| & & \downarrow{j'^*} & \\
\cdots \to & G_{i+1}(X-Z') & \to & G_i(Z') & \xrightarrow{i'_*} & G_i(X) & \xrightarrow{(j \circ j')^*} & G_i(X-Z') & \to \cdots
\end{array}$$

where

$$X - Z' \xrightarrow{j'} X - Z \xrightarrow{j} X$$

are the corresponding open immersions. Also, a flat map $f : X' \longrightarrow X$ induces a map from the sequence for $(X, Z)$ to that for $(X', f^{-1}Z)$.

Finally, if $U, V \subset X$ are open subschemes, there is a Mayer–Vietoris sequence

$$\cdots \to G_{i+1}(U \cap V) \longrightarrow G_i(U \cup V) \longrightarrow G_i(U) \oplus G_i(V) \longrightarrow G_i(U \cap V) \to \cdots .$$

Indeed, we may assume without loss of generality that $U \cup V = X$; then if $Y = X - U$, $Z = X - V$, then $Y \cap Z = \emptyset$ and $X - (Y \cup Z) = U \cap V$. Since $\mathcal{M}(Y \cup Z) \cong \mathcal{M}(Y) \times \mathcal{M}(Z)$, as $Y$ and $Z$ are disjoint, we have $G_i(Y \cup Z) \cong G_i(Y) \oplus G_i(Z)$. The Mayer–Vietoris sequence follows by a standard diagram chase from

$$\begin{array}{ccccccc}
\to G_i(Y) \oplus G_i(Z) & \to & G_i(X) \oplus G_i(X) & \to & G_i(U) \oplus G_i(V) & \to & G_{i-1}(Y) \oplus G_{i-1}(Z) \to \\
\| & & \downarrow & & \downarrow & & \| \\
\to G_i(Y \cup Z) & \to & G_i(X) & \to & G_i(U \cap V) & \to & G_{i-1}(Y \cup Z) \to
\end{array}$$

(where the top row is the direct sum of the localization sequences for $Y, Z$).

**Proposition (5.17)** (Homotopy property). *Let $f : P \longrightarrow X$ be a flat map whose fibers (see (D.31)) are affine spaces (e.g., a geometric vector bundle (D.59)). Then*

$$f^* : G_i(X) \longrightarrow G_i(P) \quad \text{is an isomorphism for all} \quad i.$$

**Proof.** For any morphism $T \longrightarrow X$ let $P_T = P \times_X T$, so that all fibers of $P_T \xrightarrow{f_T} T$ are also affine spaces and $f_T$ is flat. In particular, if $Z \subset X$ is a closed subscheme, $U = X - Z$, we have a diagram

$$
\begin{array}{ccccccccc}
\cdots \to & G_i(Z) & \to & G_i(X) & \to & G_i(U) & \to & G_{i-1}(Z) & \to \cdots \\
& \downarrow & & \downarrow & & \downarrow & & \downarrow & \\
\cdots \to & G_i(P_Z) & \to & G_i(P) & \to & G_i(P_U) & \to & G_{i-1}(P_Z) & \to \cdots
\end{array}
$$

By the 5-lemma, the proposition holds for any one of $X, Z, U$ if it is known to hold for the other two.

By Noetherian induction, we may assume that the proposition is valid for all proper closed subschemes $Z \subset X$, and we will prove it for $X$. If $X$ is non-reduced, we can take $Z = X_{\text{red}}$, so that $P_Z = P_{\text{red}}$, and the proposition holds for $X$ if it holds for $Z$, by Prop. (5.14). Next, if $X$ is reducible, say $X = Z_1 \cup Z_2$, then the proposition holds for $Z_1$ and $X - Z_1 = Z_2 - (Z_1 \cap Z_2)$ (since it holds for $Z_2$ and $Z_1 \cap Z_2$), hence it holds for $X$. Thus, we are reduced to the case when $X$ is reduced and irreducible.

Now we take the direct limit of the above diagrams as $Z$ runs over all proper closed subschemes of $X$ (ordered by inclusion; use the functoriality properties (5.16)). Since direct limits are exact, we obtain a diagram (where $k(X)$ is the function field of $X$)

$$
\begin{array}{ccccccccc}
\cdots \to & \varinjlim G_i(Z) & \to & G_i(X) & \to & G_i(k(X)) & \to & \varinjlim G_{i-1}(Z) & \to \cdots \\
& \downarrow \cong & & \downarrow & & \downarrow & & \downarrow \cong & \\
\cdots \to & \varinjlim G_i(P_Z) & \to & G_i(P) & \to & G_i(P_{k(X)}) & \to & \varinjlim G_{i-1}(P_Z) & \to \cdots
\end{array}
$$

where we have identified $\varinjlim G_i(U)$ with $G_i(k(X))$ (and similarly computed $\varinjlim G_i(P_U)$) by Lemma (5.9). Thus we are reduced to the case $X = \operatorname{Spec} k$, where $k$ is a field, and $P \cong \operatorname{Spec} k[t_1, \ldots, t_n]$ is affine space of some dimension over $k$; this follows from Theorem (5.2).

**Proposition (5.18)** (Projective Bundles). *Let $\mathcal{E}$ be a vector bundle of rank $r$ over $X$, $\mathbb{P}(\mathcal{E}) = \operatorname{Proj}(S(\mathcal{E}))$ the associated projective bundle, where $S(\mathcal{E})$ is the symmetric algebra of $\mathcal{E}$. Let $f : \mathbb{P}(\mathcal{E}) \longrightarrow X$ be the structure map. Then we have an isomorphism of $K_0(\mathbb{P}(\mathcal{E}))$-modules*

$$K_0(\mathbb{P}(\mathcal{E})) \otimes_{K_0(X)} G_i(X) \xrightarrow{\cong} G_i(\mathbb{P}(\mathcal{E}))$$

*for each $i \geq 0$, given by $y \otimes x \mapsto y \cdot f^*x$. Further, if $z \in K_0(\mathbb{P}(\mathcal{E}))$ is the*

## 5. The K-Theory of Rings and Schemes

class of $\mathcal{O}_{\mathbb{P}(\mathcal{E})}(-1)$, then the above isomorphism can be rewritten as

$$G_i(X)^r \xrightarrow{\cong} G_i(\mathbb{P}(\mathcal{E})),$$

$$(x_i)_{0 \leq i < r} \mapsto \sum_{i=0}^{r-1} z^i \cdot f^* x_i.$$

**Proof.** One knows that $K_0(\mathbb{P}(\mathcal{E}))$ is a free $K_0(X)$ module of rank $r$ with basis $1, z, \ldots, z^{r-1}$ (we prove a stronger result, Theorem (5.29), later; but this fact is "classical"— see e.g., Manin, Lectures on the $K$-functor in algebraic geometry, *Russian Math. Surveys* **24** (5) (1969) 1–89). Hence it suffices to prove the second formulation of the result.

Using the localization sequence and Noetherian induction as above, we easily reduce to the case $X = \operatorname{Spec} k$, where $k$ is a field, and $\mathcal{E}$ corresponds to a vector space $E$ of dimension $r$ over $k$. Then $S(\mathcal{E})$ corresponds to $S(E) \cong k[t_0, \ldots, t_{r-1}]$, the polynomial ring in $r$ variables, and $\mathbb{P}(\mathcal{E}) \cong \mathbb{P}_k^{r-1}$.

The standard correspondence (see (D.46)) between finite graded $S(E)$-modules and coherent sheaves on $\mathbb{P}(E)$ identifies $\mathcal{M}(\mathbb{P}(E))$ with the quotient of the Abelian category of finite positively graded $S(E)$-modules $\mathcal{M}grS(E)$, by the Serre subcategory $\mathcal{M}fgrS(E)$ of modules of finite length. By dévissage, the Serre subcategory has the same $K$-groups as the subcategory $\mathcal{M}gr(k)$ of positively graded finite dimensional $k$-vector spaces, considered as $S(E)$-modules annihilated by $E \cdot S(E)$. The $K$-groups of $\mathcal{M}grS(E)$ are computed by Prop. (5.4) as

$$K_i(\mathcal{M}gr(S(E)) \cong G_i(k) \otimes_{\mathbb{Z}} \mathbb{Z}[t]$$

where $t$ acts by shifting the grading by $-1$. Clearly $K_i(\mathcal{M}grk) \cong G_i(k) \otimes_{\mathbb{Z}} \mathbb{Z}[t]$ since $\mathcal{M}grk \cong \varinjlim \mathcal{M}(k)^n$, where $\mathcal{M}(k)^n \subset \mathcal{M}grk$ is identified with the subcategory of graded finite dimensional vector spaces which vanish in degrees $\geq n$; again $t \in \mathbb{Z}[t]$ acts by shifting the grading. Hence the localization sequence gives a diagram (whose commutativity defines $h$)

$$\begin{array}{ccccccc}
\cdots \longrightarrow & K_i(\mathcal{M}grk) & \longrightarrow & K_i(\mathcal{M}grS(E)) & \longrightarrow & G_i(\mathbb{P}(E)) \longrightarrow \cdots \\
 & \uparrow \cong & & \uparrow \cong & & \\
 & G_i(k) \otimes_{\mathbb{Z}} \mathbb{Z}[t] & \xrightarrow{h} & G_i(k) \otimes_{\mathbb{Z}} \mathbb{Z}[t] & &
\end{array}$$

where $h$ is $\mathbb{Z}[t]$-linear. Hence to compute $h$ it suffices to compute the composite $G_i(k) \longrightarrow K_i(\mathcal{M}grk) \xrightarrow{i_*} K_i(\mathcal{M}grS(E))$, which is induced by the functor associating to a $k$-vector space $W$ the $S(E)$-module obtained by letting $E \cdot S(E)$ act trivially on $W$; this is graded by assuming it to be concentrated in degree 0. The Koszul resolution for the $S(E)$-module $k$

gives an exact sequence

$$0 \longrightarrow S(E)(-r) \otimes_k \bigwedge^r E \otimes_k W \longrightarrow \cdots \longrightarrow S(E)(-1) \otimes_k E \otimes_k W$$
$$\longrightarrow S(E) \otimes_k W \longrightarrow W \longrightarrow 0.$$

Regarding this as an exact sequence of functors $\mathcal{M}(k) \longrightarrow \mathcal{M}gr S(E)$, we see that $h$ is just multiplication by $\lambda_{-t}(E) \in \mathbb{Z}[t]$, where $\lambda_{-t}(E) = \sum (-1)^i \dim(\bigwedge^i E) \cdot t^i$. Since $\lambda_{-t}(E)$ is a monic polynomial in $t$ of degree $r$, $h$ is injective and $\sum_{j=0}^{r-1} G_i(k) t^j$ maps isomorphically onto the cokernel. Thus the map

$$G_i(k)^r \longrightarrow G_i(\mathbb{P}(E)), (x_0, \ldots, x_{r-1}) \mapsto \sum z^i x_i$$

is an isomorphism, since the operation of shifting the grading by $-1$ corresponds to tensoring with $\mathcal{O}_{\mathbb{P}(E)}(-1)$, i.e., to the action of $z \in K_0(\mathbb{P}(E))$ under the module structure.

**(5.19) Filtration by Codimension of Support.** (See (D.30), (D.45).) Let $X$ be a Noetherian scheme, $Z \subset X$ a closed subscheme; we define the codimension of $Z$ in $X$ to be $\operatorname{codim}(Z, X) = \inf_{z \in Z} (\dim \mathcal{O}_{z,X})$. If $\mathcal{F} \in \mathcal{M}(X)$, the support of $\mathcal{F}$ is the subscheme of $X$ with ideal sheaf $\operatorname{Ann} \mathcal{F}$, the annihilator of $\mathcal{F}$ in $\mathcal{O}_X$. Let $\mathcal{M}^p(X) \subset \mathcal{M}(X)$ be the Serre subcategory consisting of those coherent sheaves $\mathcal{F}$ whose support is a subscheme of codimension $\geq p$ in $X$. From Lemma (3.8) and Prop. (5.14), $K_i(\mathcal{M}^p(X)) = \varinjlim G_i(Z)$, where $Z$ runs over closed subsets of codimension $\geq p$ in $X$. If $f : X' \longrightarrow X$ is flat, then $f^* \mathcal{M}^p(X) \subset \mathcal{M}^p(X')$; indeed it suffices to note that if $Z \subset X$ is a closed subscheme with $\operatorname{codim}(Z, X) \geq p$, then $\operatorname{codim}(f^{-1}(Z), X') \geq p$, which follows from the following statement about local rings — if $R \longrightarrow S$ is a flat homomorphism of local rings such that the maximal ideal of $R$ generates an ideal in $S$ primary to the maximal ideal of $S$, then $\dim R = \dim S$. (Compare with (D.40).)

If $i \mapsto X_i$ is a filtered inverse system of Noetherian schemes with affine, flat transition morphisms, such that the inverse limit $X$ is Noetherian, then $K_q(\mathcal{M}^p(X)) = \varinjlim K_q(\mathcal{M}^p(X_i))$. Indeed, since $K_q(\mathcal{M}^p(X)) = \varinjlim G_q(Z)$, where $Z$ runs over subschemes of codimension $\geq p$ in $X$, and a similar result holds for $K_q(\mathcal{M}^p(X_i))$ for each $i$, it suffices to prove that if $Z \subset X$ is a closed subscheme of codimension $\geq p$, and $f_i : X \longrightarrow X_i$ is the canonical map, then for some $i$ and some subscheme $Z' \subset X_i$ of codimension $\geq p$ in $X_i$, we have $f_i^{-1}(Z') = Z$. If $i \in I$ is fixed (where $I$ is the indexing category), and $U_i \subset X_i$ is an affine open subset, then for any $j \in I \setminus i$, $(j \to i)^{-1}(U_i) = U_j \subset X_j$ is affine open, and if $U_j = \operatorname{Spec} A_j$, then $f_i^{-1}(U_i) = U = \operatorname{Spec} A \subset X$ is affine open, with $A = \varinjlim A_j$. The ideal of

## 5. The $K$-Theory of Rings and Schemes

$Z \cap U$ is finitely generated, since $X$ is Noetherian, so that $Z \cap U = f_j^{-1}(Z_j')$ for some closed subscheme $Z_j' \subset U_j$. Since $X_i$ has a finite cover by affine open subsets, the claim follows.

Define a decreasing filtration on $G_i(X) = K_i(\mathcal{M}(X))$ by

$$F^p G_i(X) = \text{image}\bigl(K_i(\mathcal{M}^p(X)) \longrightarrow K_i(\mathcal{M}(X))\bigr).$$

This is called the *filtration by codimension of support*, and is a finite filtration provided $X$ has finite Krull dimension.

**Theorem (5.20).** *Let $X^p \subset X$ be the set of points of codimension $p$ in $X$. There is a spectral sequence* (of cohomological type)

$$E_1^{p,q} = E_1^{p,q}(X) = \coprod_{x \in X^p} K_{-p-q}(k(x)) \Longrightarrow G_{-p-q}(X),$$

*which is convergent when $X$ has finite Krull dimension, such that the induced filtration of $G_n(X)$ is the filtration by codimension of support. The spectral sequence is contravariant for flat morphisms; further, if $i \mapsto X_i$ is a filtered inverse system of Noetherian schemes with affine, flat transition morphisms whose inverse limit $X$ is Noetherian, then the spectral sequence for $X$ is the direct limit of the spectral sequences for the $X_i$.*

**Remark.** This spectral sequence is sometimes referred to in the literature as the BGQ (Brown–Gersten–Quillen) spectral sequence. In the statement of the above theorem, we interpret the notation to mean $K_n = 0$ for $n < 0$. Thus the spectral sequence is concentrated in degrees $p \geq 0$, $p + q \leq 0$; in particular it is a 4th quadrant spectral sequence of cohomological type.

**Proof of Theorem (5.20).** Consider the filtration by Serre subcategories

$$\mathcal{M}(X) \supset \mathcal{M}^1(X) \supset \mathcal{M}^2(X) \supset \cdots \supset \mathcal{M}^p(X) \cdots.$$

There is an equivalence of categories

$$\mathcal{M}^p(X)/\mathcal{M}^{p+1}(X) \cong \coprod_{x \in X^p} \mathcal{A}(\mathcal{O}_{x,X})$$

where $\mathcal{A}(\mathcal{O}_{x,X})$ is the category of $\mathcal{O}_{x,X}$-modules of finite length. By dévissage, if $k(x)$ is the residue field of $\mathcal{O}_{x,X}$, then

$$K_q(k(x)) \cong G_q(k(x)) \cong K_q(\mathcal{A}(\mathcal{O}_{x,X})),$$

for any $x \in X$. Hence we have localization sequences

$$\longrightarrow K_i(\mathcal{M}^{p+1}(X)) \longrightarrow K_i(\mathcal{M}^p(X)) \longrightarrow \coprod_{x \in X^p} K_i(k(x))$$

$$\longrightarrow K_{i-1}(\mathcal{M}^{p+1}(X)) \longrightarrow \cdots$$

giving rise to a spectral sequence by the method of exact couples (see Appendix C). The functoriality assertions follow immediately from the functorial properties of the filtration $\{\mathcal{M}^p\}$ noted earlier.

**Lemma (5.21).** *The following conditions on a Noetherian scheme $X$ are equivalent:*

(i) $\forall p \geq 0$, $\mathcal{M}^{p+1}(X) \longrightarrow \mathcal{M}^p(X)$ *induces 0 on $K$-groups.*

(ii) $\forall q \leq 0$, $E_2^{p,q}(X) = 0$ *for $p \neq 0$, and the edge homomorphism $G_{-q}(X) \to E_2^{0,q}(X)$ is an isomorphism.*

(iii) $\forall n \geq 0$, *the sequence of Abelian groups*

$$0 \longrightarrow G_n(X) \xrightarrow{e} \coprod_{x \in X^0} K_n(k(x)) \xrightarrow{d_1} \coprod_{x \in X^1} K_{n-1}(k(x)) \xrightarrow{d_1} \cdots$$

*is exact, where $d_1$ is the differential on the $E_1$ terms of the spectral sequence, and $e$ is obtained by functoriality from the (flat) morphisms $\operatorname{Spec} \mathcal{O}_{x,X} \longrightarrow X$, for $x \in X^0$, and the isomorphism $G_q(\mathcal{O}_{x,X}) \cong G_q(k(x)) \cong K_q(k(x))$.*

**Proof.** This follows easily from the construction of the spectral sequence. For any $p \geq 0$, $i \geq 1$, the differential $d_1$ on $\coprod_{x \in X^p} K_i(k(x))$ is obtained from the following diagram whose rows are exact localization sequences:

$$\to \coprod_{x \in X^p} K_i(k(x)) \longrightarrow K_{i-1}(\mathcal{M}^{p+1}(X)) \longrightarrow K_{i-1}(\mathcal{M}^p(X)) \to$$

$$\bigg\| d_1$$

$$\to K_{i-1}(\mathcal{M}^{p+2}(X)) \longrightarrow K_{i-1}(\mathcal{M}^{p+1}(X)) \longrightarrow \coprod_{x \in X^{p+1}} K_{i-1}(k(x)) \to$$

Now if (i) holds, the various localization sequences break up into short exact sequences

$$0 \longrightarrow K_i(\mathcal{M}^p(X)) \longrightarrow \coprod_{x \in X^p} K_i(k(x)) \longrightarrow K_{i-1}(\mathcal{M}^{p+1}(X)) \longrightarrow 0.$$

These sequences splice together to give the exact sequences in (iii). Since the sequences in (iii) are constructed from the complexes of $E_1$ terms, whose cohomology groups are precisely the $E_2$ terms, clearly (ii) is a reformulation of (iii). Hence (i) $\Rightarrow$ (iii) $\Leftrightarrow$ (ii).

Now assume (iii). We prove by induction on $p \geq 0$ that

$$K_i(\mathcal{M}^{p+1}(X)) \to K_i(\mathcal{M}^p(X))$$

is 0, for all $i$. To start the induction, we note that the injections

$$e : G_i(X) \longrightarrow \coprod_{x \in X^0} K_i(k(x))$$

fit into the localization sequence

$$\cdots \to K_i(\mathcal{M}^1(X)) \to K_i(\mathcal{M}(X)) \xrightarrow{e} \coprod_{x \in X^0} K_i(k(x)) \to K_{i-1}(\mathcal{M}^1(X)) \to \cdots$$

$$\| \quad G_i(X)$$

Hence this breaks up into short exact sequences

$$0 \longrightarrow K_i(\mathcal{M}(X)) \xrightarrow{e} \coprod_{x \in X^0} K_i(k(x)) \longrightarrow K_{i-1}(\mathcal{M}^1(X)) \longrightarrow 0,$$

and $K_i(\mathcal{M}^1(X)) \longrightarrow K_i(\mathcal{M}(X))$ is 0 for all $i \geq 0$.

Suppose we already know, by induction, that for $0 \leq p' < p$, the localization sequence of $\mathcal{M}^{p'+1}(X) \subset \mathcal{M}^{p'}(X)$ splits into short exact sequences

$$0 \longrightarrow K_i(\mathcal{M}^{p'}(X)) \longrightarrow \coprod_{x \in X^{p'}} K_i(k(x)) \longrightarrow K_{i-1}(\mathcal{M}^{p'+1}(X)) \longrightarrow 0,$$

so that $K_i(\mathcal{M}^{p'+1}(X)) \longrightarrow K_i(\mathcal{M}^{p'}(X))$ is 0 for all $i$. The differential

$$d_1 : \coprod_{x \in X^{p-2}} K_{i+1}(k(x)) \longrightarrow \coprod_{x \in X^{p-1}} K_i(k(x))$$

factors as the composite

$$\coprod_{x \in X^{p-2}} K_{i+1}(k(x)) \twoheadrightarrow K_i(\mathcal{M}^{p-1}(X)) \rightarrowtail \coprod_{x \in X^{p-1}} K_i(k(x)),$$

whose cokernel is

$$\coprod_{x \in X^{p-1}} K_i(k(x)) \twoheadrightarrow K_{i-1}(\mathcal{M}^p(X)).$$

Thus, in the factorization

$$d_1 : \coprod_{x \in X^{p-1}} K_i(k(x)) \longrightarrow K_{i-1}(\mathcal{M}^p(X)) \longrightarrow \coprod_{x \in X^p} K_{i-1}(k(x))$$

we see that by the exactness of the sequence in the hypothesis (iii) of the lemma,

$$K_{i-1}(\mathcal{M}^p(X)) \longrightarrow \coprod_{x \in X^p} K_{i-1}(k(x))$$

must be injective, for all $i$. Thus the localization sequence for $\mathcal{M}^{p+1}(X) \subset \mathcal{M}^p(X)$ breaks up into short exact sequences, and $K_i(\mathcal{M}^{p+1}(X)) \longrightarrow K_i(\mathcal{M}^p(X))$ is 0 for all $i$. We have assumed $p \geq 1$ in the above argument, but a minor variant works for $p = 0$.

**Proposition (5.22).** *Let $\mathcal{G}_{n,X}$ denote the (Zariski) sheaf on $X$ associated to the presheaf $U \mapsto G_n(U)$. Assume that $\operatorname{Spec}\mathcal{O}_{x,X}$ satisfies the equivalent conditions of Lemma (5.21) for each $x \in X$. Then there are canonical isomorphisms*
$$E_2^{p,q} \cong H^p(X, \mathcal{G}_{-q,X})$$
(where the $E_2$ groups are those obtained from the spectral sequence of Theorem (5.20)).

**Proof.** For each open set $U \subset X$, form the complex given in Lemma (5.21) (iii); as $U$ runs over all open subsets of $X$, we may view these complexes as defining a complex of presheaves for the Zariski topology on $X$. The associated complexes of sheaves have the form
$$0 \longrightarrow \mathcal{G}_{n,X} \longrightarrow \coprod_{x \in X^0} (i_x)_* K_n(k(x)) \longrightarrow \coprod_{x \in X^1} (i_x)_* K_{n-1}(k(x)) \longrightarrow \cdots$$
for $n \geq 0$, where $i_x : \operatorname{Spec} k(x) \longrightarrow X$ is the canonical map, and for $x \in X^p$, $K_{n-p}(k(x))$ is regarded as a constant sheaf on $\operatorname{Spec} k(x)$. The stalk of the above complex (for a given $n$) at $x \in X$ is just the corresponding complex for the scheme $\operatorname{Spec}\mathcal{O}_{x,X}$, since $\varinjlim_{x \in U} G_n(U) = G_n(\mathcal{O}_{x,X})$, and the spectral sequence commutes with filtered inverse limits of schemes (with affine flat transition maps). By hypothesis, this complex of stalks is exact for each $x \in X$ and $n \geq 0$. Hence the above complex of sheaves gives a resolution of $\mathcal{G}_{n,X}$ by flasque sheaves, which are known to be acyclic for the Zariski topology. Hence the associated complex of global sections, which is a complex of $E_1$ terms, computes the cohomology groups $H^i(X, \mathcal{G}_{n,X})$. But the cohomology groups of the complexes of $E_1$ terms are precisely the $E_2$ terms of the spectral sequence.

**Gersten's Conjecture.** *The equivalent conditions of Lemma (5.21) are valid if $X = \operatorname{Spec} R$, where $R$ is a regular local ring.*

**Theorem (5.23).** *Gersten's conjecture holds for $R = k[[k_1, \ldots, x_n]]$, the ring of formal power series in $n$ variables, and for $R$ equal to the ring of convergent power series in $n$ variables over a field $k$ complete with respect to a non-trivial valuation.*

**Proof.** First consider the case $R = k[[x_1, \ldots, x_n]]$. We prove that
$$\mathcal{M}^{p+1}(R) \longrightarrow \mathcal{M}^p(R)$$
induces 0 on $K$-groups (where $\mathcal{M}^p(R)$ stands for $\mathcal{M}^p(\operatorname{Spec} R)$). Clearly
$$K_i(\mathcal{M}^{p+1}(R)) = \varinjlim_t K_i(\mathcal{M}^p(R/tR))$$

where $t$ runs over non-zero non-units of $R$. Hence it suffices to prove that $\mathcal{M}^p(R/tR) \longrightarrow \mathcal{M}^p(R)$ induces 0 on $K$-groups for any such $t$.

By the Weierstrass preparation theorem, after a change of coordinates, we can assume that $A = k[[x_1,\ldots,x_{n-1}]]$ is such that the composite $A \longrightarrow R \longrightarrow R/tR$ is injective, and $R/tR$ is a finite $A$-module. Let $B = R \otimes_A R/tR$; since $R \cong A[[x_n]]$, $B \cong (R/tR)[[x_n]]$. There is a natural surjection of $(R/tR)$-algebras $\phi : B = R \otimes_A R/tR \longrightarrow R/tR$; if $\phi(x_n) = a \in R/tR$, then the kernel of $\phi$ is generated by $(x_n - a)$ (here $x_n \in R \subset R \otimes_A R/tR$). Thus, given any $(R/tR)$-module $M$ we have an exact sequence

$$0 \longrightarrow B \otimes_{R/tR} M \xrightarrow{(x_n-a)} B \otimes_{R/tR} M \longrightarrow M \longrightarrow 0$$

of $B$-modules, where $(x_n - a)$ denotes multiplication by $x_n - a$. Considering these as $R$-modules, if $M \in \mathcal{M}^p(R/tR)$, then the above sequence yields an exact sequence of exact functors

$$\mathcal{M}^p(R/tR) \longrightarrow \mathcal{M}^p(R).$$

Since the first two terms correspond to isomorphic functors, which yield the same map on $K$-groups, the last term, corresponding to the inclusion functor $\mathcal{M}^p(R/tR) \longrightarrow \mathcal{M}^p(R)$, induces 0 on $K$-groups.

The above argument also works when $R$ is a convergent power series ring, since the Weierstrass preparation theorem holds in that case too.

**Theorem (5.24)** (Quillen). *Let $R$ be a regular semi-local ring, which is a localization of a finitely generated algebra over a field $k$. Then the equivalent conditions of Lemma (5.21) hold for $R$.*

**Proof.** We will only prove the result in the special case when $\operatorname{Spec} R$ is smooth over an infinite field $k$ (see (D.67)). We refer the reader to Quillen's paper *Higher Algebraic K-Theory I* for the proof in the general case.

Let $A$ be a finitely generated $k$-algebra, and $S$ a finite set of primes of $A$, such that $R$ is the semi-local ring of $S$ on $X = \operatorname{Spec} A$. Since $R$ is smooth, we may assume that $A$ is smooth over $k$. Without loss of generality we may take $R$, $A$ to be domains.

We want to prove that $\mathcal{M}^{p+1}(R) \longrightarrow \mathcal{M}^p(R)$ induces 0 on $K$-groups. Clearly

$$K_i(\mathcal{M}^{p+1}(R)) = \varinjlim_{f} K_i(\mathcal{M}^{p+1}(A_f)),$$

where $f$ runs over all elements of $A$ which do not vanish at $S$. Replacing $A$ by any one such $A_f$, it suffices to prove that $\mathcal{M}^{p+1}(A) \longrightarrow \mathcal{M}^p(R)$ induces 0 on $K$-groups. Now

$$K_i(\mathcal{M}^{p+1}(A)) = \varinjlim_{t} K_i(\mathcal{M}^p(A/tA))$$

where $t$ runs over non-zero divisors in $A$. Hence it suffices to prove that for each non-zero divisor $t \in A$, there exists $f \in A$ such that $f$ does not vanish on $S$, and

$$\mathcal{M}^p(A/tA) \longrightarrow \mathcal{M}^p(A_f), \quad M \mapsto M_f,$$

induces 0 on $K$-groups. We now use:

**Normalization Lemma (5.25).** *Let $A$ be a smooth, finitely generated $k$-algebra of dimension $r$; let $S \subset \operatorname{Spec} A$ be a finite set, and $t \in A$ a non-zero divisor. Then there is a polynomial subring $B = k[x_1, \ldots, x_{r-1}] \subset A$ such that*

(i) *$A/tA$ is finite over $B$*

(ii) *$A$ is smooth over $B$ at the points of $S$.*

**Proof.** Let $X = \operatorname{Spec} A$; then there is an embedding $X \subset \mathbb{A}_k^N$ as a smooth, closed subvariety. Let $Y = \operatorname{Spec}(A/tA) \subset X$; then $\dim Y = r - 1$. Since $k$ is infinite, the "general" linear projection $\mathbb{A}^N \longrightarrow \mathbb{A}^{r-1}$ restricts to a finite morphism $Y \longrightarrow \mathbb{A}^{r-1}$, by the Noether normalization lemma (see (D.69)). Since $X$ is smooth, the "general" linear projection $X \longrightarrow \mathbb{A}^{r-1}$ is smooth at $S$, since $k$ is infinite, by Bertini's theorem (see (D.68)). This proves the lemma.

Now let $B' = A/tA$, $A' = A \otimes_B B'$, so that there is a map of $B'$-algebras $s : A' \twoheadrightarrow B'$, giving a diagram

$$s\left(\begin{array}{ccc} A' & \xleftarrow{v} & A \\ \big\downarrow u' & & \big\uparrow u \\ B' & \longleftarrow & B \end{array}\right.$$

Let $S' = v^{-1}(S)$ be the set of primes of $A'$ lying over $S$. Since $A$ is smooth over $B$ on $S$, $A'$ is smooth over $B'$ on $S'$. Since the relative dimension of $A'$ over $B'$ is 1, if $I = \ker s$, then $I$ is locally principal at the points of $S'$ (if $x \in S' \subset \operatorname{Spec} A'$, $y \in \operatorname{Spec} B'$ the image under $(u')^* : \operatorname{Spec} A' \longrightarrow \operatorname{Spec} B'$, then to see that $I$ is principal near $x$ we may replace $A', B'$ by their respective complete local rings at $x, y$, by Nakayama's lemma; now we are reduced to the situation $(A'_x)^\wedge \cong (B'_y)^\wedge[[z]]$, the ring of formal power series, where the claim is obvious—see the proof of (5.23)). Thus $I$ is principal in a neighborhood of $S'$. Since $A'$ is finite over $A$, and $S' = v^{-1}(S)$, we can find $f \in A$ such that $f$ does not vanish on $S$, and $I_f = I \cdot A'_f$ is principal. Since $A'$ is smooth over $B'$ on $S'$, it is smooth in a neighborhood of $S'$, and we may assume $f$ above has been chosen so that in addition $A'_f$ is smooth (hence flat) over $B'$.

## 5. The K-Theory of Rings and Schemes

Given any finite $(A/tA) = B'$-module $M$ we have an exact sequence of finite $A'_f$ modules

$$0 \longrightarrow I_f \otimes_{B'} M \longrightarrow A'_f \otimes_{B'} M \longrightarrow M_f \longrightarrow 0.$$

Since $A'_f$ is flat over $B'$, if $M \in \mathcal{M}^p(B')$, then $A'_f \otimes M \cong I_f \otimes M$ lies in $\mathcal{M}^p(A'_f)$. Since $A'_f$ is finite over $A_f$ we can then regard the above sequence as an exact sequence in $\mathcal{M}^p(A_f)$, giving an exact sequence of exact functors

$$\mathcal{M}^p(A/tA) \longrightarrow \mathcal{M}^p(A_f).$$

Since $I_f$ is principal, $I_f \cong A'_f$, and so $M \mapsto I_f \otimes M$, $M \mapsto A'_f \otimes M$ yield isomorphic functors, giving the same map on $K$-groups. Hence $M \mapsto M_f$ induces 0 on $K$-groups.

**Proposition (5.26).** *Let $X$ be a scheme of finite type over a field $k$. Then the image of the differential*

$$d_1 : \coprod_{x \in X^{p-1}} K_1(k(x)) \longrightarrow \coprod_{x \in X^p} K_0(k(x)) \cong \coprod_{x \in X^p} \mathbb{Z}$$

*in the spectral sequence of Theorem (5.20), consists precisely of the group of codimension $p$ cycles rationally equivalent to $0$ (in the sense of Fulton's book Intersection Theory). Hence $E_2^{p,-p} \cong CH^p(X)$, the Chow group of codimension $p$ cycles modulo rational equivalence.*

**Corollary (5.27)** (Bloch's formula). *Let $X$ be a regular scheme of finite type over a field $k$. Then there are natural isomorphisms*

$$H^p(X, \mathcal{K}_{p,X}) \cong CH^p(X), \quad p \geq 0$$

*where $\mathcal{K}_{p,X}$ is the sheaf associated to the presheaf (for the Zariski topology) $U \mapsto K_p(U)$. We have a flasque resolution*

$$0 \longrightarrow \mathcal{K}_{p,X} \longrightarrow \coprod_{x \in X^0} (i_x)_* K_p(k(x)) \longrightarrow \coprod_{x \in X^1} (i_x)_* K_{p-1}(k(x))$$

$$\longrightarrow \cdots \coprod_{x \in X^p} (i_x)_* K_0(k(x)) \longrightarrow 0$$

*and isomorphisms $E_2^{p,q} \cong H^p(X, \mathcal{K}_{-q,X})$ for the terms of the spectral sequence (5.20).*

**Proof of Corollary (5.27).** By Theorem (5.24) and Prop. (5.22), we have isomorphisms

$$E_2^{p,q} \cong H^p(X, \mathcal{G}_{-q,X}),$$

and from the proof of (5.22) we have a flasque resolution as above for $\mathcal{G}_{p,X}$. But $G_p(U) \cong K_p(U)$ for every open set $U \subset X$, since $X$ is regular; hence $\mathcal{G}_{p,X} \cong \mathcal{K}_{p,X}$. The formula for $CH^p(X)$ now follows from Prop. (5.26).

**Proof of (5.26).** For any $y \in X^{p-1}$ and $x \in X^p$ such that $x \in \overline{\{y\}}$, we have a natural map
$$\operatorname{ord}_{xy} : k(y)^* \longrightarrow \mathbb{Z},$$
defined as follows: let $Y = \overline{\{y\}}$ with the reduced structure, and let $R = \mathcal{O}_{x,Y}$. Then $R$ is a 1-dimensional Noetherian local domain with quotient field $k(y)$. Given $\alpha \in k(y)^*$, choose $a, b \in R - \{0\}$ with $\frac{a}{b} = \alpha$, and define
$$\operatorname{ord}_{xy}(\alpha) = \ell(R/aR) - \ell(R/bR), \quad \ell = \text{length}.$$
One sees (cf. Fulton, *Intersection Theory*, Appendix A) that this gives a well-defined homomorphism $\operatorname{ord}_{xy}$ with integer values. Combining the maps $\operatorname{ord}_{xy}$ for all $x, y$ we obtain a map
$$\operatorname{ord} : \coprod_{y \in X^{p-1}} k(y)^* \longrightarrow \coprod_{x \in X^p} \mathbb{Z}.$$
By definition, the cokernel of 'ord' is the Chow group $CH^p(X)$ of codimension $p$ cycles modulo rational equivalence. Hence, we need to show that 'ord' and $d_1$ have the same image.

Let $(d_1)_{xy} : k(y)^* \longrightarrow \mathbb{Z}$ be the $(xy)$-component of the differential $d_1$, for each $y \in X^{p-1}$, $x \in X^p$. Fix $y \in X^{p-1}$, and let $Y = \overline{\{y\}}$. The closed immersion $Y \longrightarrow X$ gives an exact functor $\mathcal{M}(Y) \hookrightarrow \mathcal{M}(X)$, such that $\mathcal{M}^i(Y) \subset \mathcal{M}^{p-1+i}(X)$ for all $i$. Hence we have a map of spectral sequences (5.20)
$$E_r^{i,j}(Y) \longrightarrow E_r^{i+p-1, j+1-p}(X),$$
which increases the filtration degree by $p - 1$; in particular we have a diagram

$$\begin{array}{ccc}
E_1^{p-1,-p}(X) & \longrightarrow & E_1^{p,-p}(X) \cong \coprod_{x \in X^p} \mathbb{Z} \\
\uparrow & & \uparrow \\
K_1(k(y)) \cong E_1^{0,-1}(Y) & \longrightarrow & E_1^{1,-1}(Y) \cong \coprod_{x \in Y^1} \mathbb{Z}
\end{array}$$

Thus $(d_1)_{xy} = 0$ unless $x \in Y$.

Next, if we fix $x_0 \in Y$, and let $R = \mathcal{O}_{x_0, Y}$, then the flat map $\operatorname{Spec} R \to Y$ induces a contravariant map of spectral sequences, yielding a diagram

$$\begin{array}{ccc}
K_1(k(y)) \cong E_1^{0,-1}(Y) & \xrightarrow{d_1} & E_1^{1,-1}(y) \cong \coprod_{x \in Y^1} \mathbb{Z} \\
\cong \downarrow & \downarrow & \downarrow p_0 \\
K_1(k(y)) \cong E_1^{0,-1}(R) & \xrightarrow{d_1} & E_1^{1,-1}(R) = \mathbb{Z}
\end{array}$$

where $p_0$ is projection onto the summand corresponding to $x_0$.

Hence we are reduced to proving:

**Lemma (5.28).** *Let $R$ be an equicharacteristic Noetherian local domain of dimension 1 with quotient field $F$ and residue field $k$, and let*

$$\longrightarrow G_1(R) \longrightarrow K_1(F) \xrightarrow{\partial} K_0(k) \longrightarrow G_0(R)$$

*be the localization sequence associated to the closed immersion $\operatorname{Spec} k \longrightarrow \operatorname{Spec} R$. Then $\partial : K_1(F) \longrightarrow K_0(k)$ is isomorphic to $\operatorname{ord} : F^* \longrightarrow \mathbb{Z}$ (i.e., there are functorial isomorphisms $K_1(F) \cong F^*$, $K_0(k) \cong \mathbb{Z}$ under which $\partial$ corresponds to $\operatorname{ord}$).*

**Proof.** By Theorem (5.1), there is a functorial isomorphism

$$F^* \cong \pi_1(BGL(F)^+) \cong K_1(F),$$

while by Theorem (4.0), there is a functorial isomorphism

$$K_0(k) \cong \mathbb{Z}.$$

With respect to these isomorphisms, we show $\partial = \pm \operatorname{ord}$ up to a universal choice of signs.

We have an isomorphism

$$R^* \cong \pi_1(BGL(R)^+) \cong K_1(R)$$

such that

$$\begin{array}{ccc} K_1(R) & \longrightarrow & K_1(F) \\ \cong \uparrow & & \uparrow \cong \\ R^* & \hookrightarrow & F^* \end{array}$$

commutes. Since $K_1(R) \longrightarrow K_1(F)$ factors through $G_1(R)$, we see that $\partial(x) = 0$ for $x \in R^* \subset F^*$; also $\operatorname{ord}(x) = \ell(R/xR) = 0$. So it suffices to show $\partial(x) = \pm \operatorname{ord}(x)$ for all $x \in R - (R^* \cup \{0\})$, for a universal choice of the sign; we fix such an $x$.

Let $k_0$ be the prime field; then there is a homomorphism $k_0[t] \to R$ mapping $t$ to $x$, where $k_0[t]$ is the polynomial ring. Since $x \neq 0$ and $x$ is a non-unit, this is flat. By the naturality of the localization sequence for flat maps, we have a diagram

$$\begin{array}{ccccccc} \longrightarrow & K_1(k_0[t]) & \longrightarrow & K_1(k_0[t, t^{-1}]) & \xrightarrow{\partial} & K_0(k_0) & \longrightarrow \\ & \downarrow & & \downarrow u & & \downarrow v & \\ \longrightarrow & G_1(R) & \longrightarrow & K_1(F) & \xrightarrow{\partial} & K_0(k) & \longrightarrow \end{array}$$

and a diagram

$$\begin{array}{ccc} K_1(k_0[t, t^{-1}]) & \cong \pi_1(BGL(k_0[t, t^{-1}])^+) \cong & k_0[t, t^{-1}]^* \\ \downarrow u & & \\ K_1(F) & \cong \pi_1(BGL(F)^*) \cong & F^* \end{array}$$

such that $u(t) = x$. The map $v$ is induced by the functor sending a $k_0$-vector space $V$ to the $R$-module of finite length $(R/xR) \otimes_{k_0} V$, and using dévissage to identify the $K$-groups of the categories of finite torsion $R$-modules and of finite dimensional $k$-vector spaces. Hence under the identifications $K_0(k_0) = \mathbb{Z}$, $K_0(k) = \mathbb{Z}$, $v$ is just multiplication by $\ell(R/xR) = \mathrm{ord}(x)$.

Hence it suffices to prove that $\partial(t) = \pm 1$ in the top row. But from Cor. (5.5)(ii),
$$K_1(k_0[t,t^{-1}]) \cong K_1(k_0[t]) \oplus K_0(k_0)$$
with the latter summand being identified with
$$\ker(K_1(k_0[t,t^{-1}]) \longrightarrow K_1(k_0)), \quad t \mapsto 1:$$
Under $K_1(k_0[t,t^{-1}]) \cong k_0[t,t^{-1}]^*$, the summand $K_0(k_0)$ is thus the cyclic subgroup generated by $t$, i.e., $\partial(t)$ is a generator of $K_0(k_0) = \mathbb{Z}$. Hence $\partial(t) = \pm 1$. To check that the sign is universal, we compare with the localization sequence for $\mathbb{Z}[t] \hookrightarrow \mathbb{Z}[t,t^{-1}]$ (which is possible because $k_0[t]$ has finite Tor-dimension over $\mathbb{Z}[t]$).

**Projective Bundles and Severi–Brauer Schemes.** Let $S$ be an arbitrary scheme, $\mathcal{E}$ a vector bundle on $S$ (i.e., locally free $\mathcal{O}_S$-module) of rank $r$, and $X = \mathbb{P}(\mathcal{E}) = \mathrm{Proj}(S(\mathcal{E}))$, where $S(\mathcal{E})$ is the symmetric algebra of $\mathcal{E}$ over $\mathcal{O}_S$ (see (D.58), (D.59)). Let $\mathcal{O}_X(1)$ be the canonically defined line bundle on $X$. We had earlier computed $G_*(\mathbb{P}(\mathcal{E}))$; our present goal is to prove:

**Theorem (5.29).** *If $S$ is quasi-compact, then one has isomorphisms*
$$K_q(S)^{\oplus r} \xrightarrow{\cong} K_q(X),$$
*for all $q \geq 0$, given by*
$$(a_i)_{0 \leq i < r} \mapsto \sum_{i=0}^{r-1} z^i \cdot f^* a_i,$$
*where $z \in K_0(X)$ is the class of $\mathcal{O}_X(-1)$, and $f : X \longrightarrow S$ is the structure morphism. Equivalently,*
$$K_q(X) \cong K_0(X) \otimes_{K_0(S)} K_q(S), \quad \text{for all } q \geq 0.$$

In the following discussion, let "$\mathcal{O}_X$-module" stand for "quasi-coherent $\mathcal{O}_X$-module", and similarly interpret "$\mathcal{O}_S$-module".

**Lemma (5.30).** a) *For any $\mathcal{O}_X$-module $\mathcal{F}$, $R^q f_* \mathcal{F}$ is an $\mathcal{O}_S$-module, and is 0 if $q \geq r$.*

b) *For any $\mathcal{O}_X$-module $\mathcal{F}$ and any vector bundle $\mathcal{E}'$ on $S$, we have isomorphisms*
$$(R^q f_* \mathcal{F}) \otimes_{\mathcal{O}_S} \mathcal{E}' \cong R^q f_* (\mathcal{F} \otimes_{\mathcal{O}_X} f^* \mathcal{E}'), \text{ for all } q \geq 0.$$

c) *For any $\mathcal{O}_S$-module $\mathcal{N}$ we have*
$$R^q f_* (\mathcal{O}_X(n) \otimes_{\mathcal{O}_X} f^* \mathcal{N})$$
$$= \begin{cases} 0, & \text{if } q \neq 0, r-1 \\ S^n(\mathcal{E}) \otimes_{\mathcal{O}_S} \mathcal{N}, & \text{if } q = 0 \\ S^{-n-r}(\mathcal{E})^* \otimes_{\mathcal{O}_S} \bigwedge^r \mathcal{E} \otimes_{\mathcal{O}_S} \mathcal{N}, & \text{if } q = r-1 \end{cases}$$
(where $*$ denotes the dual vector bundle, and we define $S^m(\mathcal{E}) = 0$ for $m < 0$).

d) *If $\mathcal{F}$ is an $\mathcal{O}_X$-module of finite type, and $S$ is affine, then $\mathcal{F}$ is a quotient of $\mathcal{O}_X(-n)^{\oplus k}$ for some $n, k > 0$ ($\mathcal{F}$ is of finite type if it is locally finitely generated).*

**Proof.** These are all standard facts about projective bundles. Clearly a), b), c) are local on the base, so we may assume $S = \operatorname{Spec} A$, and $\mathcal{E}$, $\mathcal{E}'$ are trivial bundles. Then $R^q f_* \mathcal{F}$ is the $\mathcal{O}_S$-module associated to the $A$-module $H^q(X, \mathcal{F})$, since the cohomology of $\mathcal{F}$ on $X$ can be computed using the Čech complex corresponding to the standard affine open cover of $\mathbb{P}_S^{r-1}$ (see the proof of (D.63)), and the formation of this Čech complex commutes with localization on $S = \operatorname{Spec} A$ (i.e., the Čech complex for $\mathcal{F}$ on $X_f = X \times_S \operatorname{Spec} A_f$ is obtained from the Čech complex for $\mathcal{F}$ on $X$ by tensoring with $A_f$). This proves a), and b) follows since $\mathcal{E}'$ is trivial. To prove c), we note that if $\mathcal{F} = \mathcal{O}_X(n)$, the terms of the Čech complex are flat $A$-modules; hence if $\mathcal{N}$ is the $\mathcal{O}_S$-module associated to the $A$-module $N$, then
$$H^q(X, \mathcal{O}_X(n) \otimes f^* \mathcal{N}) \cong H^q(X, \mathcal{O}_X(n)) \otimes_A N.$$

To prove d), since $S = \operatorname{Spec} A$ is affine, $\mathcal{E}$ is associated to a finitely generated projective $A$-module, so that there is a surjection $\mathcal{O}_S^{\oplus r} \twoheadrightarrow \mathcal{E}$, giving an embedding $X \subset \mathbb{P}_S^{r-1}$, which satisfies $\mathcal{O}_{\mathbb{P}^{m-1}}(1)|_X \cong \mathcal{O}_X(1)$. Hence we are reduced to the case $X = \mathbb{P}_S^{r-1} = \mathbb{P}_A^{r-1}$. If $\{U_i\}_{0 \leq i < r}$ is the standard affine open cover, with $U_i = \operatorname{Spec} A_i$, then $\mathcal{F}|_{U_i}$ is associated to a *finite* $A_i$-module. But any section of $\mathcal{F}$ on $U_i$ extends to a global section of $\mathcal{F}(n)$ for some $n$ (D.48), corresponding to a morphism $\mathcal{O}_X(-n) \longrightarrow \mathcal{F}$. We can choose a finite set of sections of $\mathcal{F}|_{U_i}$ which generate $\mathcal{F}|_{U_i}$, for each $i$, and choose $n$ so that all of these extend to sections of $\mathcal{F}(n)$; the induced map $\mathcal{O}_X(-n)^{\oplus m} \longrightarrow \mathcal{F}$ is clearly onto.

**Definition.** An $\mathcal{O}_X$-module $\mathcal{F}$ is *regular* if $R^q f_* \mathcal{F}(-q) = 0$ for all $q > 0$. For example, $\mathcal{O}_X(n) \otimes f^* \mathcal{N}$ is regular for $n \geq 0$, and any $\mathcal{O}_S$-module $\mathcal{N}$.

The idea of the proof of (5.29) is to show that any regular vector bundle on $X$ has a canonical resolution by twisted pullbacks of vector bundles on $S$.

**Lemma (5.31).** *Let* $0 \longrightarrow \mathcal{F}' \longrightarrow \mathcal{F} \longrightarrow \mathcal{F}'' \longrightarrow 0$ *be an exact sequence of* $\mathcal{O}_X$-*modules.*

a) *If* $\mathcal{F}'(n)$ *and* $\mathcal{F}''(n)$ *are regular, then so is* $\mathcal{F}(n)$.
b) *If* $\mathcal{F}(n)$ *and* $\mathcal{F}'(n+1)$ *are regular, then so is* $\mathcal{F}''(n)$.
c) *If* $\mathcal{F}(n+1)$ *and* $\mathcal{F}''(n)$ *are regular, and if* $f_*\mathcal{F}(n) \twoheadrightarrow f_*\mathcal{F}''(n)$, *then* $\mathcal{F}'(n+1)$ *is regular.*

**Proof.** Immediate from the definition of regularity and the long exact sequence of higher direct images.

**Lemma (5.32).** *If* $\mathcal{F}$ *is regular,* $\mathcal{F}(n)$ *is regular for* $n > 0$.

**Proof.** It suffices to prove that $\mathcal{F}(1)$ is regular, by induction. We have the Koszul exact sequence of vector bundles on $X$

$$0 \longrightarrow \mathcal{O}_X(-r) \otimes \bigwedge^r f^*\mathcal{E} \longrightarrow \cdots \longrightarrow \mathcal{O}_X(-1) \otimes f^*\mathcal{E} \longrightarrow \mathcal{O}_X \longrightarrow 0.$$

Tensoring with $\mathcal{F}$ we obtain the exact sequence

$$(*) \quad 0 \longrightarrow \mathcal{F}(-r) \otimes \bigwedge^r f^*\mathcal{E} \longrightarrow \cdots \longrightarrow \mathcal{F}(-1) \otimes f^*\mathcal{E} \longrightarrow \mathcal{F} \longrightarrow 0 \cdots.$$

Since $\mathcal{F}$ is regular, Lemma (5.30)(b) (projection formula) yields

$$R^q f_*(\mathcal{F}(-p) \otimes_{\mathcal{O}_X} (\bigwedge^p f^*\mathcal{E}(n))) \cong (R^q f_*\mathcal{F}(n-p)) \otimes_{\mathcal{O}_S} \bigwedge^p \mathcal{E}$$

so that $\mathcal{F} \otimes \bigwedge^p f^*\mathcal{E} \cong (\mathcal{F}(-p) \otimes \bigwedge^p f^*\mathcal{E})(p)$ is regular. Splitting $(*)$ into short exact sequences

$$0 \longrightarrow Z_p \longrightarrow \mathcal{F}(-p) \otimes \bigwedge^p f^*\mathcal{E} \longrightarrow Z_{p-1} \longrightarrow 0,$$

(with $Z_r = 0$, $Z_0 = \mathcal{F}$) we see, by descending induction on $p$ and Lemma (5.31)(b), that $Z_p(p+1)$ is regular for each $p \geq 0$. In particular $Z_0(1) = \mathcal{F}(1)$ is regular.

**Lemma (5.33).** *If* $\mathcal{F}$ *is regular, the natural map* $f^*f_*\mathcal{F} \longrightarrow \mathcal{F}$ *is onto.*

**Proof.** From the proof of (5.32) above, we have an exact sequence

$$0 \longrightarrow Z_1 \longrightarrow \mathcal{F}(-1) \otimes f^*\mathcal{E} \longrightarrow \mathcal{F} \longrightarrow 0$$

## 5. The K-Theory of Rings and Schemes

where $Z_1(2)$ is regular; hence $Z_1(n)$ is regular for all $n \geq 2$, and $R^1 f_* Z_1(n) = 0$ for all $n \geq 1$. Hence we have exact sequences

$$0 \longrightarrow f_* Z_1(n) \longrightarrow f_* \mathcal{F}(n-1) \otimes_{\mathcal{O}_S} \mathcal{E} \longrightarrow f_* \mathcal{F}(n) \longrightarrow 0, \quad n \geq 1.$$

Thus the natural map of graded $S(\mathcal{E})$-modules

$$f_* \mathcal{F} \otimes_{\mathcal{O}_S} S(\mathcal{E}) \longrightarrow \bigoplus_{n \geq 0} f_* \mathcal{F}(n)$$

is onto. The lemma follows by taking the associated sheaves on $\mathbb{P}(\mathcal{E})$ (in fact, the lemma would follow from the weaker statement that the above map of graded $S(\mathcal{E})$-modules is a surjection in sufficiently large degrees).

**Lemma (5.34).** *Any regular $\mathcal{O}_X$-module $\mathcal{F}$ has a resolution*

$$0 \longrightarrow \mathcal{O}_X(1-r) \otimes f^* T_{r-1}(\mathcal{F}) \longrightarrow \cdots \longrightarrow \mathcal{O}_X(-1) \otimes f^* T_1(\mathcal{F})$$
$$\longrightarrow f^* T_0(\mathcal{F}) \longrightarrow \mathcal{F} \longrightarrow 0$$

*where $T_i(\mathcal{F})$ are $\mathcal{O}_S$-modules determined up to isomorphism by $\mathcal{F}$. Further, $\mathcal{F} \longrightarrow T_i(\mathcal{F})$ is an exact functor from the category of regular $\mathcal{O}_X$-modules to the category of $\mathcal{O}_S$-modules.*

**Proof.** We first prove uniqueness. Given a resolution as in the lemma, since $\mathcal{F}$ is regular, the sequence obtained by twisting by $\mathcal{O}_X(n)$,

$$0 \longrightarrow \mathcal{O}_X(n+1-r) \otimes f^* T_{r-1}(\mathcal{F}) \longrightarrow \cdots \longrightarrow \mathcal{O}_X(n) \otimes f^* T_0(\mathcal{F}) \longrightarrow \mathcal{F}(n) \longrightarrow 0$$

consists of $\mathcal{O}_X$-modules which are acyclic for $R^i f_*$, $i \geq 1$, provided $n \geq 0$. Applying $f_*$ thus yields exact sequences

$$0 \longrightarrow S^{n+1-r}(\mathcal{E}) \otimes T_{r-1}(\mathcal{F}) \longrightarrow \cdots \longrightarrow S^n(\mathcal{E}) \otimes T_0(\mathcal{F}) \longrightarrow f_* \mathcal{F}(n) \longrightarrow 0$$

(as before we use the convention $S^n(\mathcal{E}) = 0$ for $n < 0$). In particular, for $0 \leq n < r$ we have

$$0 \longrightarrow T_n(\mathcal{F}) \longrightarrow \mathcal{E} \otimes_{\mathcal{O}_S} T_{n-1}(\mathcal{F}) \longrightarrow \cdots \longrightarrow f_* \mathcal{F}(n) \longrightarrow 0.$$

This shows that $T_n(\mathcal{F})$ is uniquely determined, by induction, up to isomorphism.

This also tells us how to define $T_n(\mathcal{F})$. We define inductively a sequence $Z_n = Z_n(\mathcal{F})$ of $\mathcal{O}_X$-modules, and $T_n = T_n(\mathcal{F})$ a sequence of $\mathcal{O}_S$-modules, by taking

$$Z_{-1} = \mathcal{F}, \quad T_n = f_*(Z_{n-1}(n)),$$

and letting $Z_n = \ker(\mathcal{O}_X(-n) \otimes f^* T_n \longrightarrow Z_{n-1})$. Clearly $Z_n$, $T_n$ are additive functors. We will prove, by induction on $n$, that $Z_n(n+1)$ is regular, this being given when $n = -1$ since $\mathcal{F}$ is regular.

If $Z_{n-1}(n)$ is regular, then by (5.33)
$$f^*T_n = f^*(f_*Z_{n-1}(n)) \twoheadrightarrow Z_{n-1}(n),$$
and by definition the kernel of this map is $Z_n(n)$. Thus we have an exact sequence

$(*)_n \qquad 0 \longrightarrow Z_n(n) \longrightarrow f^*T_n \longrightarrow Z_{n-1}(n) \longrightarrow 0 \cdots.$

By (5.31)c) and (5.32), $Z_n(n+1)$ is regular, giving the inductive step; the sequences $\mathcal{O}_X(-n) \otimes (*)_n$ splice together to give the resolution in the statement of the lemma (modulo the injectivity of the first map on the left). We also obtain
$$f_*Z_n(n) = 0 \text{ for all } n \geq 0, \text{ since } f_*f^*T_n = T_n = f_*Z_{n-1}(n).$$

From the sequence $\mathcal{O}_X(1) \otimes (*)_n$ we see that if $Z_{n-1}$ and $T_n$ are already known to be exact functors on the category of regular $\mathcal{O}_X$-modules, then so is $Z_n(n+1)$ (and hence also $Z_n$); since $f_*$ is exact on the category of regular $\mathcal{O}_X$-modules, $T_{n+1} = f_*Z_n(n+1)$ is also an exact functor.

Lastly, we show $Z_{r-1} = 0$. From $\mathcal{O}_X(n) \otimes (*)_{n+q}$ we have exact sequences
$$R^{q-1}f_*Z_{n+q-1}(n) \longrightarrow R^q f_*Z_{n+q}(n) \longrightarrow R^q f_*\mathcal{O}_X(-q) \otimes f^*T_{n+q}.$$
Starting from $f_*Z_n(n) = 0$, these sequences inductively show that $R^q f_*Z_{n+q}(n) = 0$ for all $q, n \geq 0$. Thus $R^q f_*Z_{r-1}(r-1-q) = 0$ for all $q \geq 0$ (this is a special case of the previous statement if $r-1-q = n \geq 0$, but is trivial if $r-1-q < 0$, i.e., if $q > r-1$). This means precisely that $Z_{r-1}(r-1)$ is regular. But $f_*Z_{r-1}(r-1) = 0$, so by (5.33), $Z_{r-1}(r-1) = 0$, i.e., $Z_{r-1} = 0$.

**Lemma (5.35).** *Assume that $S$ is quasi-compact. Then for any vector bundle $\mathcal{F}$ on $X$, there exists an integer $n_0$ such that for all $\mathcal{O}_S$-modules $\mathcal{N}$ and $n \geq n_0$, we have*

(a) $R^q f_*(\mathcal{F}(n) \otimes f^*\mathcal{N}) = 0$ *for all* $q > 0$.
(b) $f_*(\mathcal{F}(n) \otimes f^*\mathcal{N}) \cong (f_*\mathcal{F}(n)) \otimes_{\mathcal{O}_S} \mathcal{N}$
(c) $f_*\mathcal{F}(n)$ *is a vector bundle on $S$.*

**Proof.** By quasi-compactness, the existence of $n_0$ is local on the base $S$, so we may assume $S$ is affine. Then by (5.30)(d), $\mathcal{F}$ is a quotient of a vector bundle $\mathcal{L} = \mathcal{O}_X(-m)^{\oplus k}$ for some $m, k > 0$, and we have an exact sequence of vector bundles
$$0 \longrightarrow \mathcal{F}' \longrightarrow \mathcal{L} \longrightarrow \mathcal{F} \longrightarrow 0.$$

Further, (5.30) implies the lemma for $\mathcal{L}$. Since
$$0 \longrightarrow \mathcal{F}'(n) \otimes f^*\mathcal{N} \longrightarrow \mathcal{L}(n) \otimes f^*\mathcal{N} \longrightarrow \mathcal{F}(n) \otimes f^*\mathcal{N} \longrightarrow 0$$
is exact, we have a sequence
$$R^q f_*(\mathcal{L}(n) \otimes f^*\mathcal{N}) \longrightarrow R^q f_*(\mathcal{F}(n) \otimes f^*\mathcal{N}) \longrightarrow R^{q+1} f_*(\mathcal{F}'(n) \otimes f^*\mathcal{N}).$$
Hence (a) follows by descending induction on $q$, being trivially valid for $q \geq r$ (note that $\mathcal{F}'$ also satisfies the hypothesis of the lemma). Using (a), if $n \geq n_0$, we have a diagram with exact rows (for any $\mathcal{O}_S$-module $\mathcal{N}$)

$$\begin{array}{ccccccc}
(f_*\mathcal{F}'(n)) \otimes \mathcal{N} & \xrightarrow{v} & (f_*\mathcal{L}(n)) \otimes \mathcal{N} & \rightarrow & (f_*\mathcal{F}(n)) \otimes \mathcal{N} & \rightarrow & 0 \\
\downarrow{u'} & & \downarrow{\cong} & & \downarrow{u} & & \\
0 \rightarrow f_*(\mathcal{F}'(n) \otimes f^*\mathcal{N}) & \rightarrow & f_*(\mathcal{L}(n) \otimes f^*\mathcal{N}) & \rightarrow & f_*(\mathcal{F}(n) \otimes f^*\mathcal{N}) & \rightarrow & 0.
\end{array}$$

Hence $u$ is onto; a similar argument applied to $\mathcal{F}'$ shows that $u'$ is onto, so that $u$ is an isomorphism. Again, the same argument applied to $\mathcal{F}'$ shows $u'$ is an isomorphism. Thus $\ker v = \text{Tor}_1^{\mathcal{O}_S}(f_*\mathcal{F}(n), \mathcal{N}) = 0$, for any $\mathcal{N}$, i.e., $f_*\mathcal{F}(n)$ is a *flat* $\mathcal{O}_S$-module. Since it is a quotient of $f_*\mathcal{L}(n)$, it is of finite type; applying a similar argument to $\mathcal{F}'$, we see that $f_*\mathcal{F}(n)$ is finitely presented. Since a finitely presented flat module over a ring is projective, $f_*\mathcal{F}(n)$ is a vector bundle.

**Lemma (5.36).** *If $\mathcal{F}$ is a vector bundle on $X$ with $R^q f_*\mathcal{F}(n) = 0$ for all $q > 0$, $n \geq 0$, then $f_*\mathcal{F}(n)$ is a vector bundle on $S$ for all $n \geq 0$.*

**Proof.** Since the assertion is local on $S$, we may assume $S$ is affine; now by (5.35)c) the result holds if $n \geq n_0$. Applying the functor $\mathcal{H}om_{\mathcal{O}_X}(-, \mathcal{F}(n))$ to the Koszul exact sequence in the proof of (5.32) yields an exact sequence
$$0 \longrightarrow \mathcal{F}(n) \longrightarrow \mathcal{F}(n+1) \otimes f^*\mathcal{E}^* \longrightarrow \mathcal{F}(n+2)$$
$$\otimes \bigwedge^2 f^*\mathcal{E}^* \rightarrow \cdots \rightarrow \mathcal{F}(n+r) \otimes \bigwedge^r f^*\mathcal{E}^* \longrightarrow 0.$$
For $n \geq 0$ all of these sheaves are acyclic for $R^q f_*$, $q > 0$; hence on applying $f_*$ we have an exact sequence
$$0 \rightarrow f_*\mathcal{F}(n) \rightarrow (f_*\mathcal{F}(n+1)) \otimes_{\mathcal{O}_S} \mathcal{E}^* \rightarrow \cdots \rightarrow (f_*\mathcal{F}(n+r)) \otimes_{\mathcal{O}_S} \bigwedge^r \mathcal{E}^* \rightarrow 0.$$
Hence by descending induction on $n$, $f_*\mathcal{F}(n)$ is a vector bundle on $S$ for $n \geq 0$.

**Lemma (5.37).** *If $\mathcal{F}$ is a regular vector bundle on $X$, then $T_i(\mathcal{F})$, $0 \leq i < r$ are vector bundles on $S$.*

**Proof.** As in the proof of uniqueness in (5.34) we have sequences
$$0 \longrightarrow T_n \longrightarrow \mathcal{E} \otimes_{\mathcal{O}_S} T_{n-1} \longrightarrow \cdots \longrightarrow f_*\mathcal{F}(n) \longrightarrow 0, \quad 0 \leq n < r.$$
Hence the result follows from (5.36), by induction on $n$.

Recall that $\mathcal{P}(X)$ denotes the category of vector bundles on $X$, and $K_q(X) = K_q(\mathcal{P}(X))$ by definition. Let $\mathcal{P}_n \subset \mathcal{P}(X)$ be the full subcategory consisting of bundles $\mathcal{F}$ satisfying $R^i f_*\mathcal{F}(k) = 0 \ \forall i > 0$ and for all $k \geq n$. Let $\mathcal{R}_n \subset \mathcal{P}(X)$ denote the full subcategory of bundles $\mathcal{F}$ such that $\mathcal{F}(n)$ is regular. Both $\mathcal{P}_n, \mathcal{R}_n$ are closed under extensions in $\mathcal{P}(X)$, and so are exact categories; we have inclusions $\mathcal{P}_{n-1} \subset \mathcal{P}_n$, $\mathcal{R}_{n-1} \subset \mathcal{R}_n$ and $\mathcal{R}_n \subset \mathcal{P}_n$ and further $\cup \mathcal{P}_n = \varinjlim \mathcal{P}_n = \mathcal{P}(X) = \cup \mathcal{R}_n = \varinjlim \mathcal{R}_n$.

**Lemma (5.38).** *For all $n$, the inclusions $\mathcal{R}_n \subset \mathcal{P}_n \subset \mathcal{P}(X)$ induce isomorphisms $K_q(\mathcal{R}_n) \cong K_q(\mathcal{P}_n) \cong K_q(\mathcal{P}(X)) = K_q(X)$.*

**Proof.** For any vector bundle $\mathcal{F}$, we have an exact sequence, which is functorial in $\mathcal{F}$,
$$0 \longrightarrow \mathcal{F} \longrightarrow \mathcal{F}(1) \otimes f^*\mathcal{E}^* \longrightarrow \cdots \longrightarrow \mathcal{F}(r) \otimes f^* \bigwedge^r \mathcal{E}^* \longrightarrow 0.$$
For each $p > 0$, $\mathcal{F} \mapsto \mathcal{F}(p) \otimes f^* \bigwedge^p \mathcal{E}^*$ gives exact functors
$$\phi_p : \mathcal{P}_n \longrightarrow \mathcal{P}_{n-1}, \quad \psi_p : \mathcal{R}_n \longrightarrow \mathcal{R}_{n-1}.$$
If $i : \mathcal{P}_{n-1} \longrightarrow \mathcal{P}_n$ is the inclusion, then we have an exact sequence of functors (id = identity) $\mathcal{P}_n \longrightarrow \mathcal{P}_n$,
$$0 \longrightarrow \mathrm{id}_{\mathcal{P}_n} \longrightarrow i \circ \phi_1 \longrightarrow i \circ \phi_2 \longrightarrow \cdots \longrightarrow i \circ \phi_r \longrightarrow 0,$$
and an exact sequence of functors $\mathcal{P}_{n-1} \longrightarrow \mathcal{P}_{n-1}$,
$$0 \longrightarrow \mathrm{id}_{\mathcal{P}_{n-1}} \longrightarrow \phi_1 \circ i \longrightarrow \cdots \longrightarrow \phi_r \circ i \longrightarrow 0.$$
Thus $i_* : K_q(\mathcal{P}_{n-1}) \longrightarrow K_q(\mathcal{P}_n)$ and $\sum_{p>0}(-1)^{p-1}(\phi_p)_* : K_q(\mathcal{P}_n) \longrightarrow K_q(\mathcal{P}_{n-1})$ are inverse maps.

Similarly, if $j : \mathcal{R}_{n-1} \longrightarrow \mathcal{R}_n$ is the inclusion, then
$$j_* : K_q(\mathcal{R}_{n-1}) \longrightarrow K_q(\mathcal{R}_n)$$
and
$$\sum_{p>0}(-1)^{p-1}(\psi_p)_* : K_q(\mathcal{R}_n) \longrightarrow K_q(\mathcal{R}_{n-1})$$
are inverse maps.

Thus $K_q(\mathcal{P}_n) \cong \varinjlim K_q(\mathcal{P}_n) \cong K_q(\mathcal{P}(X)) = K_q(X)$, and similarly $K_q(\mathcal{R}_n) \cong \varinjlim K_q(\mathcal{R}_n) \cong K_q(\mathcal{P}(X)) = K_q(X)$.

## 5. The K-Theory of Rings and Schemes

**Proof of (5.29).** Let $u_n : K_q(S) \longrightarrow K_q(\mathcal{P}_0) \cong K_q(X)$ be the homomorphism induced by the exact functor $\mathcal{N} \mapsto \mathcal{O}_X(-n) \otimes f^*\mathcal{N}$, where $0 \leq n < r$ (these inequalities ensure that the functor has values in $\mathcal{P}_0$, by (5.30)). Let

$$u : K_q(S)^{\oplus r} \longrightarrow K_q(X) \text{ be given by}$$

$$u : (x_n)_{0 \leq n < r} \mapsto \sum_{n=0}^{r-1} u_n(x_n).$$

Then Theorem (5.29) states precisely that $u$ is an isomorphism.

If $\mathcal{F} \in \mathcal{P}_0$, then $f_*\mathcal{F}(n)$ is a vector bundle on $S$ for any $n \geq 0$, and $R^i f_*\mathcal{F}(n) = 0 \ \forall i > 0$, by (5.36). Hence we have maps $v_n : K_q(\mathcal{P}_0) \longrightarrow K_q(S)$, induced by $\mathcal{F} \mapsto f_*\mathcal{F}(n)$, for each $n \geq 0$.

The composite $v_n \circ u_m : K_q(S) \longrightarrow K_q(S)$ is induced by the functor

$$\mathcal{N} \mapsto (f_*\mathcal{O}_X(n-m)) \otimes \mathcal{N} \begin{cases} = 0 & \text{if } m > n \\ = \mathcal{N} & \text{if } m = n. \end{cases}$$

Thus if $v = (v_0, \ldots, v_{r-1}) : K_q(\mathcal{P}_0) \longrightarrow K_q(S)^{\oplus r}$, then $v \circ u$ is described by a triangular matrix with all diagonal entries equal to 1; hence $v \circ u$ is invertible, and so $u$ is injective.

On the other hand, by (5.34) and (5.37) we have exact functors $T_n : \mathcal{R}_0 \longrightarrow \mathcal{P}(S)$, for $0 \leq n < r$, which induce maps $t_n : K_q(\mathcal{R}_0) \longrightarrow K_q(S)$. Let $t = (\ldots, (-1)^n t_n, \ldots) : K_q(X) \longrightarrow K_q(S)^{\oplus r}$. The exact sequence

$$0 \longrightarrow \mathcal{O}_X(1-r) \otimes f^*T_{r-1} \to \cdots \to f^*T_0 \longrightarrow \mathcal{F} \longrightarrow 0$$

can be interpreted as an exact sequence of functors $\mathcal{R}_0 \to \mathcal{P}_0$, so that $u \circ t : K_q(\mathcal{R}_0) \longrightarrow K_q(\mathcal{P}_0)$ equals the isomorphism induced by the inclusion $\mathcal{R}_0 \subset \mathcal{P}_0$. Hence $u$ is surjective.

**$\mathbb{P}^1$ over a Ring.** Let $A$ be a not necessarily commutative ring, $t$ a (commuting) indeterminate and let

$$i_+ : A[t] \longrightarrow A[t, t^{-1}], \quad i_- : A[t^{-1}] \longrightarrow A[t, t^{-1}]$$

be the natural inclusions of the polynomial ring in the Laurent polynomials. We *define* $\mathcal{M}(\mathbb{P}_A^1)$ to be the Abelian category of triples $M = (M_+, M_-, \theta)$ where $M_+$ is a left $A[t]$-module, $M_-$ a left $A[t^{-1}]$-module, and $\theta : i_+^* M_+ \to i_-^* M_-$ is an isomorphism of $A[t, t^{-1}]$-modules. Similarly define $\mathcal{P}(\mathbb{P}_A^1)$ to be the full subcategory of $\mathcal{M}(\mathbb{P}_A^1)$ consisting of triples $M$ with $M_+ \in \mathcal{P}(A[t])$, $M_- \in \mathcal{P}(A[t^{-1}])$.

**Theorem (5.39).** Let $h_n : \mathcal{P}(A) \longrightarrow \mathcal{P}(\mathbb{P}_A^1)$ be given by

$$P \mapsto (A[t] \otimes_A P, A[t^{-1}] \otimes_A P, \theta_n)$$

where $\theta_n$ = multiplication by $t^n$ on $A[t,t^{-1}] \otimes_A P$. Then we have isomorphisms
$$K_q(A)^{\oplus 2} \xrightarrow{\cong} K_q(\mathcal{P}(\mathbb{P}_A^1)), \quad (x,y) \mapsto (h_0)_*x + (h_1)_*y,$$
and we have the relations $(h_{n-1})_* - 2(h_n)_* + (h_{n+1})_* = 0$ for all $n$.

**Proof.** For any $M = (M_+, M_-, \theta)$ define
$$M(n) = (M_+, M_-, t^{-n}\theta), \quad \forall n.$$
Thus $h_n$ is induced by $P \longrightarrow h_0(P)(-n)$. For any triple $M = (M_+, M_-, \theta)$ define $X_0, X_1 \in \text{Hom}(M, M(1))$ by
$$X_0\big|_{M_+} = 1, \quad X_0\big|_{M_-} = t^{-1},$$
$$X_1\big|_{M_+} = t, \quad X_1\big|_{M_-} = 1$$
(here '$t$' denotes multiplication by $t$). Then one checks easily that there is an exact sequence, functorial in $M$,
$$0 \longrightarrow M(m) \xrightarrow{\alpha} M(m+1)^{\oplus 2} \xrightarrow{\beta} M(m+2) \longrightarrow 0$$
where $\alpha = (X_0, X_1)$ and $\beta(x,y) = X_1 x - X_0 y$. In particular taking $m = -1-n$ and $M = h_0(P)$, $P \in \mathcal{P}(A)$, we have an exact sequence of functors $\mathcal{P}(A) \longrightarrow \mathcal{P}(\mathbb{P}_A^1)$
$$0 \longrightarrow h_{n+1} \longrightarrow h_n^{\oplus 2} \longrightarrow h_{n-1} \longrightarrow 0$$
giving the relation $(h_{n-1})_* - 2(h_n)_* + (h_{n+1})_* = 0$.

Next, for any triple $M = (M_+, M_-, \theta)$, we define $f_*M, R^1f_*M \in \mathcal{M}(A)$ to be the cohomology modules of the 2-term complex
$$0 \longrightarrow M_+ \oplus M_- \xrightarrow{\psi} i_-^*M_- \longrightarrow 0,$$
where $\psi(x,y) = \theta(1 \otimes x) - 1 \otimes y$. We now define a triple $M$ to be regular if $R^1f_*M(-1) = 0$. With this definition, one can check that the various steps in the proof of (5.29) go through in the given situation also. The details are left to the reader.

**Severi–Brauer Schemes.** Let $S$ be a scheme, and let $X$ be a Severi–Brauer scheme over $S$ of relative dimension $r - 1$. By definition $X$ is locally isomorphic to $\mathbb{P}_S^{r-1}$ as an $S$-scheme, in the étale topology on $S$, i.e., for some faithfully flat étale morphism $S' \longrightarrow S$, if $X' = X \times_S S'$, then $X' \cong \mathbb{P}_{S'}^{r-1}$ as $S'$-schemes. Let $f : X \longrightarrow S$ be the structure map.

If there exists a line bundle $\mathcal{L} \in \text{Pic}\, X$ which restricts to $\mathcal{O}(-1)$ on each geometric fiber, then $X = \mathbb{P}(\mathcal{E})$, where $\mathcal{E} = f_*\mathcal{L}^{-1}$. In general, $\mathcal{L}$ will exist only étale locally on $S$. However, there is a canonically defined vector bundle $\mathcal{J}$ of rank $r$ on $X$, which restricts to $\mathcal{O}(-1)^{\oplus r}$ on each geometric fiber of $f$. The idea is as follows (compare (D.70)).

Let $Y = \mathbb{P}_S^{r-1}$, and let $GL_{r,S}$ act on $\mathcal{O}_S^{\oplus r}$ in the usual way. Then the induced action on $Y = \mathbb{P}(\mathcal{O}_S^{\oplus r})$ factors through the quotient group scheme $G = PGL_{r,S} = GL_{r,S}/\mathbb{G}_{m,S}$. Since $\mathbb{G}_{m,S}$ acts trivially on $\mathcal{O}_Y(-1) \otimes f^*f_*\mathcal{O}_Y(1) \cong \mathcal{O}_Y(-1)^{\oplus r}$, $G$ acts on this vector bundle, compatibly with its action on $Y$. Since $X/S$ is étale locally on $S$ isomorphic to $Y/S$, and $G$ is the group scheme of automorphisms of $Y/S$, we know that $X = Y \times^G T$, where $T$ is a torsor (principal $G$-bundle) for $G$ over $S$ which is étale locally trivial, and $Y \times^G T$ is the associated fiber space over $S$ with fiber $Y$ (i.e., $X = (Y \times T)/G$ where $G$ acts by $g(y,t) = (gy, gt)$). At any rate, by descent theory, $\mathcal{O}_Y(-1) \otimes f^*f_*\mathcal{O}_Y(1)$ descends to a vector bundle $\mathcal{J}$ of rank $r$ on $X$.

The construction of $\mathcal{J}$ is compatible with base change; further, $\mathcal{J} = \mathcal{O}_X(-1) \otimes f^*\mathcal{E}$ if $X = \mathbb{P}(\mathcal{E})$ for a vector bundle $\mathcal{E}$ on $S$. In general, there is a faithfully flat étale morphism $g: S' \longrightarrow S$ giving a Cartesian square

$$\begin{array}{ccc} X' & \xrightarrow{g'} & X \\ f' \downarrow & & \downarrow f \\ S' & \xrightarrow{g} & S \end{array}$$

such that $X' \cong \mathbb{P}_{S'}^{r-1}$, and further $g'^*\mathcal{J} \cong \mathcal{O}_{X'}(-1) \otimes f'^*f'_*\mathcal{O}_{X'}(1)$.

Let $\mathcal{A}$ be the sheaf of non-commutative $\mathcal{O}_S$-algebras

$$\mathcal{A} = f_* \mathcal{E}nd_{\mathcal{O}_X}(\mathcal{J})^{\mathrm{op}}$$

(where the superscript "op" denotes the opposite algebra). As $g$ is flat, $g^*f_* \cong f'_*g'^*$. Hence

$$g^*\mathcal{A}^{\mathrm{op}} \cong f'_* \mathcal{E}nd_{\mathcal{O}_{X'}}(\mathcal{O}_{X'}(-1) \otimes f^*f_*\mathcal{O}_{X'}(1))$$
$$\cong f'_*\mathcal{M}_r(\mathcal{O}_{X'}) = \mathcal{M}_r(\mathcal{O}_{S'})$$

where $\mathcal{M}_r$ denotes the algebra of $r \times r$ matrices. Hence $\mathcal{A}$ is an Azumaya algebra of rank $r^2$. Further, the natural map

$$f^*\mathcal{A} = f^*f_*\mathcal{E}nd(\mathcal{J})^{\mathrm{op}} \longrightarrow \mathcal{E}nd(\mathcal{J})^{\mathrm{op}}$$

is an isomorphism, as is easily seen by applying $g'^*$ (note that $g'$ is also a faithfully flat étale map).

Conversely, if $\mathcal{A}$ is an Azumaya algebra over $S$ of degree $r$, so that the underlying sheaf of $\mathcal{O}_X$-modules of $\mathcal{A}$ is locally free of rank $r^2$, let $f: X \longrightarrow S$ be the scheme over $S$ parametrizing locally free quotient $\mathcal{O}_S$-modules of rank $r$ of $\mathcal{A}$, whose kernels are right ideals of $\mathcal{A}$, so that $X$ is a closed subscheme of the Grassmannian of locally free quotients of rank $r$ of $\mathcal{A}$. Then $X$ is a Severi–Brauer scheme, and the vector bundle $\mathcal{J}$ on

$X$ is identified with the dual of the restriction of the universal quotient bundle on the Grassmannian. These facts are checked by passing to $S'$, where $g : S' \longrightarrow S$ is faithfully flat and étale, and $g^*\mathcal{A} \cong \mathcal{M}_r(\mathcal{O}_{S'})$.

Let $\mathcal{J}_n = \mathcal{J}^{\otimes n}$, $\mathcal{A}_n = \mathcal{A}^{\otimes n}$, so that $\mathcal{J}_n$ is a vector bundle of rank $r^n$ on $X$, and $\mathcal{A}_n$ is an Azumaya algebra on $S$ of rank $(r^n)^2$. Then $\mathcal{A}_n \cong f_*\mathcal{E}nd_{\mathcal{O}_X}(\mathcal{J}_n)^{\mathrm{op}}$, and $f^*\mathcal{J}_n \cong \mathcal{E}nd_{\mathcal{O}_X}(\mathcal{J}_n)^{\mathrm{op}}$ (as can be verified by pulling back to $X' \cong \mathbb{P}^{r-1}_{S'}$).

Let $\mathcal{P}(\mathcal{A}_n)$ be the exact category of vector bundles on $S$ which are left $\mathcal{A}_n$-modules. Since $\mathcal{J}_n$ is a right $f^*(\mathcal{A}_n)$-module, which is locally (on $X$) a direct summand of $f^*(\mathcal{A}_n)$, we have an exact functor

$$\psi_n : \mathcal{P}(\mathcal{A}_n) \longrightarrow \mathcal{P}(X), \quad M \mapsto \mathcal{J}_n \otimes_{f^*(\mathcal{A}_n)} f^*(M).$$

Define $K_i(\mathcal{A}_n) = K_i(\mathcal{P}(\mathcal{A}_n))$.

**Theorem (5.40).** *If $S$ is quasi-compact, we have isomorphisms*

$$\coprod_{n=0}^{r-1} K_i(\mathcal{A}_n) \xrightarrow{\cong} K_i(X), \quad (x_n)_{0 \leq n < r} \mapsto \sum_{n=0}^{r-1} (\psi_n)_*(x_n).$$

**Proof** (Sketch). This is again just a modification of the proof of (5.29). Fix $g : S' \longrightarrow S$, a faithfully flat étale morphism such that $X' = X \times_S S' \cong \mathbb{P}^{r-1}_S$, and let $g' : X' \longrightarrow X$, $f' : X' \longrightarrow S'$ be the resulting maps.

We define an $\mathcal{O}_X$-module $\mathcal{F}$ to be *regular* if $g'^*\mathcal{F}$ is regular on $X'$. For regular $\mathcal{F}$, define inductively

$$Z_{-1}(\mathcal{F}) = \mathcal{F}, \quad T_n(\mathcal{F}) = f_*\mathcal{H}om_{\mathcal{O}_X}(\mathcal{J}_n, Z_{n-1}(\mathcal{F})),$$

and

$$Z_n(\mathcal{F}) = \ker(\mathcal{J}_n \otimes f^*T_n(\mathcal{F}) \longrightarrow Z_{n-1}(\mathcal{F})) \quad (\text{where } \mathcal{J}_0 = \mathcal{O}_X).$$

Then we obtain a resolution

$$(\alpha) \quad 0 \longrightarrow \mathcal{J}_{r-1} \otimes_{f^*(\mathcal{A}_{r-1})} f^*T_{r-1}(\mathcal{F}) \to \cdots \to f^*T_0(\mathcal{F}) \longrightarrow \mathcal{F} \longrightarrow 0 \cdots$$

where $T_n(\mathcal{F})$ has the natural left $\mathcal{A}_n$-module structure; this resolution pulls back to that of (5.34) on $X'$ for $g'^*\mathcal{F}$.

Similarly, the canonical epimorphism $\mathcal{J} \twoheadrightarrow \mathcal{O}_X$ gives rise to a Koszul complex

$$0 \longrightarrow \bigwedge^r \mathcal{J} \to \cdots \to \mathcal{J} \longrightarrow \mathcal{O}_X \longrightarrow 0$$

which pulls back to the Koszul resolution on $X'$ used in (5.32), (5.36), (5.38). In particular, the analogue of (5.38) is valid: let $\mathcal{R}_n(X) \subset \mathcal{P}_n(X) \subset \mathcal{P}(X)$ be the full subcategories of bundles $\mathcal{F}$ such that $g'^*\mathcal{F}$ lies in the corresponding subcategory of $\mathcal{P}(X')$; then

$$K_i(\mathcal{R}_n(X)) \cong K_i(\mathcal{P}_n)) \cong K_i(\mathcal{P}(X)) = K_i(X).$$

Now let $u_n : K_i(\mathcal{A}_n) \longrightarrow K_i(X)$ be given by $u_n = (\psi_n)_*$, where $\psi_n(\mathcal{M}) = \mathcal{J}_n \otimes_{f^*(\mathcal{A}_n)} \mathcal{M}$. Let $u : \coprod_{0 \leq n < r} K_i(\mathcal{A}_n) \longrightarrow K_i(X)$ be $u : (x_n)_{0 \leq n < r} \mapsto \sum u_n(x_n)$. Let $v_m : K_i(X) \longrightarrow K_i(\mathcal{A}_m)$ be induced by the functor $V_m : \mathcal{P}_0(X) \longrightarrow \mathcal{P}(\mathcal{A}_m)$, $V_m(\mathcal{F}) = f_* \mathcal{H}om(\mathcal{J}_m, \mathcal{F})$. Then $v_m \circ u_n$ is induced by the functor $V_m \circ \psi_n : \mathcal{P}(\mathcal{A}_n) \longrightarrow \mathcal{P}(\mathcal{A}_m)$, given by

$$N \mapsto f_* \mathcal{H}om(\mathcal{J}_m, \mathcal{J}_n \otimes_{f^*(\mathcal{A}_n)} N).$$

But we compute that

$$\mathcal{H}om_{\mathcal{O}_X}(\mathcal{J}_m, \mathcal{J}_n \otimes_{f^*(\mathcal{A}_n)} N) = \mathcal{H}om_{\mathcal{O}_X}(\mathcal{J}_m, \mathcal{J}_n) \otimes_{f^*(\mathcal{A}_n)} f^*N$$
$$\begin{cases} = 0 & \text{if } m < n, \text{ by applying } g'^* \\ = f^*N & \text{if } m = n. \end{cases}$$

Hence if $v = (v_0, \ldots, v_{r-1}) : K_i(X) \longrightarrow \coprod_{0 \leq n < r} K_i(\mathcal{A}_n)$, then $v \circ u$ has a triangular matrix with diagonal entries equal to 1. Hence $u$ is injective.

Similarly, if $t_n : K_i(X) \longrightarrow K_i(\mathcal{A}_n)$ is induced by $T_n : \mathcal{R}_0(X) \longrightarrow \mathcal{P}(\mathcal{A}_n)$, and

$$t = (\ldots, (-)^n t_n, \ldots) : K_i(X) \longrightarrow \coprod_{0 \leq n < r} K_i(\mathcal{A}_n),$$

then the resolution $(\alpha)$ obtained above shows that $u \circ t$ is the isomorphism $K_i(\mathcal{R}_0) \longrightarrow K_i(\mathcal{P}_0)$ induced by $\mathcal{R}_0 \subset \mathcal{P}_0$. Hence $u$ is surjective.

There is a different technique for proving lemma (5.34), which is perhaps longer than that given earlier, but yields an explicit description of the functors $T_i$. This point of view is closely related to the following paper of A. A. Beilinson: "Coherent sheaves on $\mathbb{P}^n$ and problems of linear algebra", J. Funct. Anal. Appl. 12 (1978), pp. 214–215.

The idea is that on $X = \mathbb{P}(\mathcal{E})$, we have an exact sequence of locally free $\mathcal{O}_X$-modules

$$0 \to \mathcal{S} \to f^*\mathcal{E} \to \mathcal{O}_X(1) \to 0$$

and a dual exact sequence

$$0 \to \mathcal{O}_X(-1) \to f^*\mathcal{E}^* \to \mathcal{S}^* \to 0.$$

Hence if $p_i : X \times_S X \to X$, $i = 1, 2$ are the projections, then one has a surjection

$$p_1^* f^* \mathcal{E} \otimes p_2^* f^* \mathcal{E}^* \to p_1^* \mathcal{O}_X(1) \otimes p_2^* \mathcal{S}^*.$$

Now if $q : X \times_S X \to S$ is the structure morphism, then there is a natural isomorphism

$$q^* \mathcal{E}nd(\mathcal{E}) \cong p_1^* f^* \mathcal{E} \otimes p_2^* f^* \mathcal{E}^*,$$

and the identity endomorphism of $\mathcal{E}$ yields a global section

$$\xi \in \Gamma(X \times_S X, p_1^* \mathcal{O}_X(1) \otimes p_2^* \mathcal{S}^*).$$

We claim that the scheme of zeroes of $\xi$ is the diagonal $\Delta_X \subset X \times_S X$. This is easily proved locally on $S$ using homogeneous coordinates, for example. Or else, it can be deduced from the universal property of a projective bundle (for any $S$-scheme $g : T \to S$, there is a bijection between $S$-morphisms $T \to \mathbb{P}(\mathcal{E})$ and isomorphism classes of invertible quotient sheaves of $g^*\mathcal{E}$). We leave the details to the reader. Granting this, we deduce that the structure sheaf of the diagonal has a resolution by a Koszul complex

$$\cdots \to \bigwedge^i (p_1^*\mathcal{O}_X(-1) \otimes p_2^*\mathcal{S}) \to \cdots p_1^*\mathcal{O}_X(-1) \otimes p_2^*\mathcal{S} \to \mathcal{O}_{X\times_S X} \to \mathcal{O}_{\Delta_X} \to 0.$$

Further, if $\mathcal{I}$ is the ideal sheaf of the diagonal, then

$$\mathcal{I}/\mathcal{I}^2 \cong (p_1^*\mathcal{O}_X(-1) \otimes p_2^*\mathcal{S}) \otimes \mathcal{O}_{\Delta_X}$$

(this is a general property of ideals resolved by a Koszul complex). But $\mathcal{I}/\mathcal{I}^2 \cong \Omega^1_{X/S}$, so that $\mathcal{S} \cong \Omega^1_{X/S}(1)$, and our Koszul complex assumes the form

$$\cdots \to p_1^*\mathcal{O}_X(-i) \otimes p_2^*\Omega^i_{X/S}(i) \to \cdots$$
$$\to p_1^*\mathcal{O}_X(-1) \otimes p_2^*\Omega^1_{X/S}(1) \to \mathcal{O}_{X\times_S X} \to \mathcal{O}_{\Delta_X} \to 0.$$

Now (5.34) follows from:

**Proposition (5.41)** *Let $f : X \to S$ be as above, $\mathcal{F}$ an $\mathcal{O}_X$-module.*

(a) *There is a convergent spectral sequence of $\mathcal{O}_X$-modules (the Beilinson spectral sequence), concentrated in the range $p \leq 0, q \geq 0$,*

$$E_1^{p,q} = \mathcal{O}_X(p) \otimes f^*\left(R^q f_*\left(\Omega^{-p}_{X/S}(-p) \otimes \mathcal{F}\right)\right),$$

*with*

$$E_\infty^{p,q} = \begin{cases} 0 & \text{if } (p,q) \neq (0,0) \\ \mathcal{F} & \text{if } (p,q) = (0,0). \end{cases}$$

(b) *Suppose $R^i f_*\left(\Omega^j_{X/S}(j) \otimes \mathcal{F}\right) = 0$ for all $i > 0, j \geq 0$. Then there is a resolution of $\mathcal{F}$ given by the $E_1$ terms along the line $q = 0$*

$$\cdots \to \mathcal{O}_X(-i) \otimes f^*T_i(\mathcal{F}) \to \cdots \to f^*T_0(\mathcal{F}) \to \mathcal{F} \to 0,$$

*where*

$$T_i(\mathcal{F}) = f_*\left(\Omega^i_{X/S}(i) \otimes \mathcal{F}\right).$$

(c) *Let $\mathcal{F}$ be a regular $\mathcal{O}_X$-module; then $R^i f_*\left(\Omega^j_{X/S}(j) \otimes \mathcal{F}\right) = 0$ for all $i > 0, j \geq 0$, and $T_i$ is an exact functor on the category of regular $\mathcal{O}_X$-modules for each $i \geq 0$.*

## 5. The K-Theory of Rings and Schemes

**Proof.** (Sketch) We have a canonical identification $p_{1*}(\mathcal{O}_{\Delta_X} \otimes p_2^*\mathcal{F}) \cong \mathcal{F}$, and $R^i p_{1*}(\mathcal{O}_{\Delta_X} \otimes \mathcal{F}) = 0$ for $i > 0$. On the other hand, the resolution of the diagonal yields a resolution on $X \times_S X$

$$\cdots \to p_1^*\mathcal{O}_X(-i) \otimes p_2^*\left(\Omega^i_{X/S}(i) \otimes \mathcal{F}\right) \to \cdots$$

$$\cdots \to p_1^*\mathcal{O}_X(-1) \otimes p_2^*\left(\Omega^1_{X/S}(1) \otimes \mathcal{F}\right) \to p_2^*\mathcal{F} \to (\mathcal{O}_{\Delta_X} \otimes p_2^*\mathcal{F}) \to 0.$$

Hence there is a spectral sequence for the hyperderived functors of $f_*$, concentrated in degrees $-r \le p \le 0, r \ge q \ge 0$ (i.e., set $p = -i$ in the resolution)

$$E_1^{p,q} = R^q p_{1*}\left(p_1^*\mathcal{O}_X(p) \otimes p_2^*(\Omega^{-p}_{X/S}(-p) \otimes \mathcal{F})\right) \Rightarrow R^{p+q}p_{1*}(\mathcal{O}_{\Delta_X} \otimes p_2^*\mathcal{F}).$$

Now $p_1 : X \times_S X \to X$ is itself a projective bundle, associated to $f^*\mathcal{E}$, so that by (5.30)(b), we may rewrite the $E_1$ terms as

$$E_1^{p,q} = \mathcal{O}_X(-p) \otimes R^q p_{1*} p_2^*\left(\Omega^{-p}_{X/S}(-p) \otimes \mathcal{F}\right).$$

Further, since $f$ is flat, we have a canonical isomorphism

$$R^q p_{1*} p_2^*\left(\Omega^{-p}_{X/S}(-p) \otimes \mathcal{F}\right) \cong f^*\left(R^q f_*\left(\Omega^{-p}_{X/S}(-p) \otimes \mathcal{F}\right)\right).$$

This gives the expression for the $E_1$ terms in the statement of the proposition.

The vanishing condition in (b) means that $E_1^{p,q} = 0$ for $q > 0$, and the spectral sequence is concentrated on the line $q = 0$. Further, $E_2^{-p,0} = E_\infty^{-p,0} = 0$ for all $p > 0$, so that the complex $E_1^{-p,0}$ is a resolution of $E_2^{0,0} = E_\infty^{0,0} = \mathcal{F}$.

Now we prove (c). One first proves that if $\mathcal{F}$ is regular, then

$$R^j f_*\left(\Omega^i_{X/S}(i+1-j) \otimes \mathcal{F}\right) = 0 \ \forall \ i \ge 0, \ j > 0.$$

This is done by descending induction on $i$; the case $i = r$ is the definition of regularity, since $\Omega^r_{X/S} \cong \mathcal{O}_X(-r-1)$. By taking exterior powers of the Euler exact sequence

$$0 \to \Omega^1_{X/S}(1) \to f^*\mathcal{E} \to \mathcal{O}_X(1) \to 0$$

(which we constructed earlier, when we identified $\mathcal{S}$ with $\Omega^1_{X/S}(1)$), we obtain exact sequences of locally free $\mathcal{O}_X$-modules

$$0 \to \Omega^i_{X/S}(i) \to f^*(\bigwedge^i \mathcal{E}) \to \Omega^{i-1}_{X/S}(i) \to 0.$$

Tensoring with $\mathcal{F}(-j)$ and applying $R^j f_*$, we obtain an exact sequence of $\mathcal{O}_S$-modules

$$R^j f_*\left(f^*(\bigwedge^i \mathcal{E}) \otimes \mathcal{F}(-j)\right) \to R^j f_*\left(\Omega^{i-1}(i) \otimes \mathcal{F}(-j)\right) \to$$

$$\to R^{j+1}f_*\left(\Omega^i_{X/S}(i) \otimes \mathcal{F}(-j)\right)$$

where $R^jf_*\left(f^*(\bigwedge^i \mathcal{E}) \otimes \mathcal{F}(-j)\right) \cong \bigwedge^i \mathcal{E} \otimes R^jf_*\mathcal{F}(-j) = 0$ by the regularity of $\mathcal{F}$, and $R^{j+1}f_*\left(\Omega^i_{X/S}(i) \otimes \mathcal{F}(-j)\right) = 0$ by induction.

In particular, setting $i = 0$, we see that $\mathcal{F}(1)$ is regular, and hence $\mathcal{F}(i)$ is regular for all $i \geq 0$ (compare (5.32)). Hence $R^jf_*\left(\Omega^i_{X/S} \otimes \mathcal{F}(k)\right) = 0$ for $i \geq 0, j > 0$, and $k \geq i+1-j$. This includes the case $k = i$, which is what was needed. The exactness of $T_i$ on the category of regular $\mathcal{O}_X$-modules is immediate from the vanishing of $R^1f_*\Omega^i_{X/S}(i) \otimes \mathcal{F}$ for regular $\mathcal{F}$.

The above argument for (c) was shown to me by Donu Arapura.

Resolutions for the diagonal are known, which yield computations of the $K$-groups, in certain other cases (Grassmannians, their twisted forms, and quadrics); this point of view is seen in papers of Kapranov, Panin and a joint paper of mine with Levine and Weyman. Panin noted that in fact for $K$-theoretic consequences, one only needs that the class of the diagonal in $K_0$ has a certain form, which would be implied by the existence of an appropriate resolution of the structure sheaf of the diagonal. A recent reference putting much of this work in perspective is the following paper of Panin: A.I. Panin, On Algebraic $K$-Theory of Generalized Flag Fibre Bundles and Some of their Twisted Forms, in *Advances in Soviet Mathematics*, Vol. 4, ed. A. A. Suslin, Amer. Math. Soc. (1991).

# 6. Proofs of the Theorems of Chapter 4

We first prove:

**Theorem (4.0).** *There is a natural isomorphism $K_0(\mathcal{C}) \cong \pi_1(B Q \mathcal{C}, \{0\})$ for any small exact category $\mathcal{C}$ and null object $0 \in \mathcal{C}$.*

Here, $K_0(\mathcal{C})$ is the quotient of the free Abelian group on the objects of $\mathcal{C}$ modulo the relations $[M] = [M'] + [M'']$ for all exact sequences

$$0 \longrightarrow M' \longrightarrow M \longrightarrow M'' \longrightarrow 0$$

in $\mathcal{C}$. We begin by proving a lemma. We use the notation and terminology of Chapter 3.

**Lemma (6.1.).** *The category of covering spaces of the classifying space $B\mathcal{C}$ of a small category $\mathcal{C}$ is naturally equivalent to the category of functors $F : \mathcal{C} \longrightarrow \underline{\text{Set}}$ (where $\underline{\text{Set}}$ is the category of sets) such that $F(u)$ is a bijection for any morphism $u$ of $\mathcal{C}$.*

**Proof.** Let $p : E \longrightarrow B\mathcal{C}$ be a covering space. For any object $X \in \mathcal{C}$, let $E(X)$ be the fiber over $X \in B\mathcal{C}$ (where $X$ is regarded as a 0-simplex in $N\mathcal{C}$, and hence determines a 0-cell in $B\mathcal{C}$). Given a morphism $u : X_1 \longrightarrow X_2$, we may regard $u$ as a 1-simplex in $N\mathcal{C}$, which determines a path $Bu$ in $B\mathcal{C}$ joining $X_1$ to $X_2$. Since $p$ is a covering, it has the unique path lifting property, which gives a bijection $(Bu)_* : E(X_1) \longrightarrow E(X_2)$, by associating to a point $y \in E(X_1)$ the second end-point of the unique path in $E$ which lifts $Bu$ and begins at $y$. Hence $X \mapsto E(X)$, $u \mapsto (Bu)_*$ determine a functor $\mathcal{C} \longrightarrow \underline{\text{Set}}$ carrying all arrows of $\mathcal{C}$ into bijections.

Conversely, if $F : \mathcal{C} \longrightarrow \underline{\text{Set}}$ is a "morphism inverting" functor (i.e., $F(u)$ is a bijection for each $u \in \text{Mor}(\mathcal{C})$), let $F \setminus \mathcal{C}$ be the category of pairs $(X, x)$ with $X \in \text{Ob}\,\mathcal{C}$, $x \in F(X)$, where a morphism $(X, x) \longrightarrow (X', x')$ is a morphism $u : X \longrightarrow X'$ such that $F(u)(x) = x'$. The forgetful functor $F \setminus \mathcal{C} \longrightarrow \mathcal{C}$, $(X, x) \mapsto X$ gives a map on classifying spaces $p_F$ :

$B(F \setminus C) \longrightarrow BC$, with fibers $p_F^{-1}(X) = F(X)$ for any object $X \in C$. We claim that $p_F$ is a covering space. If we prove this, then for any morphism $u : X \longrightarrow X'$ in $C$ and any $x \in F(X)$, if $x' = F(u)(x)$, then $u$ determines a morphism $(X, x) \longrightarrow (X', x')$ in $F \setminus C$, which gives the unique path in $B(F \setminus C)$ lifting $Bu$ and beginning at $x \in p_F^{-1}(X)$. Thus the above two constructions are inverse to each other, and give the desired equivalence of categories.

To check that $p_F : B(F \setminus C) \longrightarrow BC$ is a covering space, we use the criterion (A.48) of Appendix A. Thus, we must show that the map of simplicial sets $N(F \setminus C) \longrightarrow NC$ is a simplicial covering, i.e., if $\Delta(n)$ is the simplicial set $\Delta(n)(\underline{p}) = \mathrm{Hom}_{\Delta}(\underline{p}, \underline{n})$, (so that $|\Delta(n)| = \Delta_n$, the standard $n$-simplex), then given any diagram of maps of simplicial sets

$$\begin{array}{ccc} \Delta(0) & \longrightarrow & N(F \setminus C) \\ \downarrow & & \downarrow \\ \Delta(n) & \longrightarrow & NC \end{array}$$

we must show that there is a unique map $\Delta(n) \longrightarrow N(F \setminus C)$ (of simplicial sets) making the diagram commute. Of course, a map $\sigma : \Delta(n) \longrightarrow NC$ is just an $n$-simplex $\sigma \in N_n C$, so we must show that if $\sigma \in N_n C$ is an $n$-simplex of $NC$, $\sigma_0 \in N_0(F \setminus C)$ a 0-simplex lying over the $i$th vertex of $\sigma$, then there exists a unique $n$-simplex $\tau \in N_n(F \setminus C)$ which maps to $\sigma$, such that $\sigma_0$ is the $i$th vertex of $\tau$. Assume that $\sigma$ is given by the diagram in $C$

$$M_0 \xrightarrow{u_1} M_1 \xrightarrow{u_2} \cdots \xrightarrow{u_n} M_n;$$

the $i$th vertex of $\sigma$ is given by the object $M_i$, so that $\sigma_0$ is given by an object $(M_i, x_i) \in F \setminus C$, where $x_i \in F(M_i)$. We have bijections

$$F(M_0) \xrightarrow{F(u_1)} F(M_1) \xrightarrow{F(u_2)} F(M_2) \longrightarrow \cdots \xrightarrow{F(u_n)} F(M_n)$$

giving composite bijections $f_j : F(M_j) \longrightarrow F(M_i)$ for each $j$ ($f_i$ is the identity) such that $f_j = f_{j+1} \circ F(u_{j+1})$. Let $x_j \in F(M_j)$ be the unique element satisfying $f_j(x_j) = x_i$. Then clearly $x_i = f_j(x_j) = f_{j+1}(F(u_{j+1})(x_j))$, so that $x_{j+1} = F(u_{j+1})(x_j)$. Thus we have a diagram in $F \setminus C$, giving $\tau \in N_n(F \setminus C)$,

$$(M_0, x_0) \xrightarrow{\tilde{u}_1} (M_1, x_1) \xrightarrow{\tilde{u}_2} \cdots \xrightarrow{\tilde{u}_n} (M_n, x_n)$$

where $\tilde{u}_j$ is the morphism induced by $u_j$. One sees at once that $\tau$ is the unique $n$-simplex lifting $\sigma$ whose $i$th-vertex is $(M_i, x_i)$. This proves that $p_F : B(F \setminus C) \longrightarrow BC$ is a covering, and finishes the proof of (6.1).

Now let $C$ be a (small) exact category. Recall that if $M, N \in C$, an arrow $i : M \longrightarrow N$ is called an *admissible monomorphism* if we have an

exact sequence in $\mathcal{C}$ (for some $P \in \mathcal{C}$)
$$0 \longrightarrow M \xrightarrow{i} N \longrightarrow P \longrightarrow 0.$$
Similarly, $q: M \longrightarrow N$ is an *admissible epimorphism* if there is an exact sequence
$$0 \longrightarrow P \longrightarrow M \xrightarrow{q} N \longrightarrow 0$$
for some $P \in \mathcal{C}$. We write $i: M \rightarrowtail N$ to denote that $i$ is an admissible monomorphism, and $q: M \twoheadrightarrow N$ to denote that $q$ is an admissible epimorphism.

Next, we recall the construction of morphisms in the category $\mathcal{QC}$. A morphism $M \to N$ in $\mathcal{QC}$ is an equivalence class of diagrams $M \leftarrow M' \rightarrowtail N$, where $M \leftarrow M'' \rightarrowtail N$ is an equivalent diagram if and only if there is an isomorphism $u: M' \to M''$ making the following diagram commute:

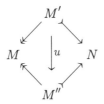

Composition of morphisms is defined as follows. Given diagrams $M \leftarrow M' \rightarrowtail N$, $N \leftarrow N' \rightarrowtail P$ the composite morphism $M \to P$ in $\mathcal{QC}$ is represented by the diagram $M \leftarrow M' \times_N N' \rightarrowtail P$. We have the diagram (where the square is a pullback)

$$\begin{array}{ccccc} M' \times_N N' & \rightarrowtail & N' & \rightarrowtail & P \\ \downarrow & & \downarrow & & \\ M' & \rightarrowtail & N & & \\ \downarrow & & & & \\ M & & & & \end{array}$$

In particular, if $i: M \rightarrowtail N$, we have an associated arrow $i_! : M \to N$ in $\mathcal{QC}$, given by $M \underset{1_M}{\leftarrow} M \underset{i}{\rightarrowtail} N$, where $1_M$ is the identity of $M$. Similarly, if $q: M \twoheadrightarrow N$, we have an associated arrow $q^! : N \to M$ in $\mathcal{QC}$, given by $N \underset{q}{\leftarrow} M \underset{1_M}{\rightarrowtail} M$.

If $f: M \to N$ is an arbitrary arrow in $\mathcal{QC}$, given by the diagram $M \underset{q}{\leftarrow} M' \underset{i}{\rightarrowtail} N$, then in fact $f = i_! \circ q^!$ as is immediate from the above description of composition of morphisms in $\mathcal{QC}$. We can form the pushout square in $\mathcal{C}$:

$$\begin{array}{ccc} M' & \stackrel{i}{\rightarrowtail} & N \\ q \downarrow & & \downarrow q' \\ M & \stackrel{i'}{\rightarrowtail} & N' \end{array}$$

which will also be Cartesian (i.e., pullback); such a square is called *bicartesian*. Then we claim that $f = q'^! \circ i'_!$ in $Q\mathcal{C}$, from the diagram giving the composition on the right:

$$\begin{array}{ccccc} M' & \stackrel{i}{\rightarrowtail} & N & \stackrel{1_N}{\rightarrow} & N \\ q \downarrow & & \downarrow q' & & \\ M & \stackrel{i'}{\rightarrowtail} & N' & & \\ 1_M \downarrow & & & & \\ M & & & & \end{array}$$

Thus, the assigments $i \mapsto i_!$, $q \mapsto q^!$ have the following properties:

(i) if $i, i'$ are composable admissible monomorphisms (i.e., range $i' =$ domain $i$) then $(i \circ i')_! = i_! \circ i'_!$; similarly, if $q, q'$ are composable, admissible epimorphisms, then $(q \circ q')^! = q'^! \circ q^!$.

(ii) if

$$\begin{array}{ccc} M' & \stackrel{i}{\rightarrowtail} & N \\ q \downarrow & & \downarrow q' \\ M & \rightarrowtail & N' \end{array}$$

is a bicartesian square, where $i, i'$ are admissible monomorphisms, $q, q'$ are admissible epimorphisms, then $i_! \circ q' = q'^! \circ i'_!$.

In fact, (i) and (ii) characterize $Q\mathcal{C}$, in the following sense.

**Lemma (6.2).** *Let $\mathcal{C}$ be an exact category, $\mathcal{D}$ a category. Assume that*

(i) *for each $M \in \mathcal{C}$, we are given an object $F(M) \in \mathcal{D}$,*

(ii) *for each admissible mono $i : M' \rightarrowtail M$, we are given an arrow $F_1(i) : F(M') \longrightarrow F(M)$, such that $F_1(i \circ i') = F_1(i) \circ F_1(i')$ if $i, i'$ are composable; similarly, for each admissible epi $q : M \twoheadrightarrow N$, we are given $F_2(q) : F(N) \longrightarrow F(M)$, such that $F_2(q \circ q') = F_2(q') \circ F_2(q)$ if $q, q'$ are composable*

(iii) *if*

$$\begin{array}{ccc} M' & \stackrel{i}{\rightarrowtail} & N \\ q \downarrow & & \downarrow q' \\ M & \stackrel{i'}{\rightarrowtail} & N' \end{array}$$

## 6. Proofs of the Theorems of Chapter 4

*is a bicartesian square, then*

$$F_1(i) \circ F_2(q) = F_2(q') \circ F_1(i').$$

*Then there is a well-defined functor* $F : Q\mathcal{C} \to \mathcal{D}$ *given by* $M \mapsto F(M)$,

$$(M \xleftarrow{q} M' \xrightarrowtail{i} N) \mapsto F_1(i) \circ F_2(q).$$

**Proof.** If $M \xleftarrow{q} M' \xrightarrowtail{i} N$, $M \xleftarrow{q_1} M'' \xrightarrowtail{i_1} N$ are equivalent diagrams giving a morphism $M \longrightarrow N$ in $Q\mathcal{C}$, we have an isomorphism $u : M' \longrightarrow M''$ such that $q = q_1 \circ u$, $i = i_1 \circ u$. Regarding $u$ as admissible epi, we get $F_2(q) = F_2(u) \circ F_2(q_1)$, and regarding $u$ as admissible mono, we get $F_1(i) = F_1(i_1) \circ F_1(u)$. From the bicartesian square

$$\begin{array}{ccc} M' & \xrightarrowtail{u} & M'' \\ u \downarrow & & \downarrow 1_{M''} \\ M'' & \xrightarrowtail{M''} & M'' \end{array}$$

we have $F_1(u) \circ F_2(u) = F_2(1_{M''}) \circ F_1(1_{M''}) = 1_{F(M'')}$. Hence $F_1(i) \circ F_2(q) = F_1(i_1) \circ F_2(q_1)$. Thus, $F_1(i) \circ F_2(q)$ depends only on the arrow in $Q\mathcal{C}$, and not on the particular diagram which represents it.

Next, if $M \xleftarrow{q_1} M' \xrightarrowtail{i_1} N$, $N \xleftarrow{q_2} N' \xrightarrowtail{i_2} P$ are given and $M \xleftarrow{q} M' \times_N N' \xrightarrowtail{i} P$ represents the composite arrow in $Q\mathcal{C}$, we have a diagram (where the square is bicartesian)

$$\begin{array}{ccccc} M' \times_N N' & \xrightarrowtail{i} & N' & \xrightarrowtail{i_2} & P \\ q' \downarrow & & \downarrow q_2 & & \\ M' & \xrightarrowtail{i_1} & N & & \\ q_1 \downarrow & & & & \\ M & & & & \end{array}$$

and $q = q_1 \circ q'$, $i = i_2 \circ i'$. Then

$$F_1(i) \circ F_2(q) = F_1(i_2) \circ F_1(i') \circ F_2(q') \circ F_2(q_1)$$
$$= F_1(i_2) \circ F_2(q_2) \circ F_1(i_1) \circ F_2(q_1).$$

This proves that $(M \xleftarrow{q} M' \xrightarrowtail{i} N) \mapsto F_1(i) \circ F_2(q)$ is compatible with composition in $Q\mathcal{C}$, and so yields a well-defined functor $Q\mathcal{C} \longrightarrow \mathcal{D}$. This proves the lemma.

**Corollary (6.3).** *For any exact category* $\mathcal{C}$, *there is an isomorphism* $Q\mathcal{C} \cong Q\mathcal{C}^{\mathrm{op}}$.

**Proof.** For any arrow $f$ in $\mathcal{C}$, let $\bar{f}$ be the corresponding arrow in $\mathcal{C}^{\mathrm{op}}$. Then if $i : M \rightarrowtail N$ is admissible mono, $\bar{i} : N \twoheadrightarrow M$ is admissible epi, and if $q : M \twoheadrightarrow N$ is admissible epi, $\bar{q} = N \rightarrowtail M$ is admissible mono. If

$$\begin{array}{ccc} M' & \stackrel{i}{\rightarrowtail} & N \\ {\scriptstyle q}\downarrow & & \downarrow {\scriptstyle q'} \\ M & \stackrel{}{\underset{i'}{\rightarrowtail}} & N' \end{array}$$

is bicartesian, then

$$\begin{array}{ccc} N' & \stackrel{\bar{q}'}{\rightarrowtail} & N \\ {\scriptstyle i'}\downarrow & & \downarrow {\scriptstyle \bar{i}} \\ M & \stackrel{}{\underset{\bar{q}}{\rightarrowtail}} & M' \end{array}$$

is bicartesian in $\mathcal{C}^{\mathrm{op}}$. Thus $i_! \circ q^! \mapsto \bar{q}'_! \circ \bar{i}'^!$ gives a functor (which is the identity on objects) $Q\mathcal{C} \longrightarrow Q\mathcal{C}^{\mathrm{op}}$, inducing a bijection

$$\mathrm{Hom}_{Q\mathcal{C}}(M, N) \longrightarrow \mathrm{Hom}_{Q\mathcal{C}^{\mathrm{op}}}(M, N).$$

This is the desired isomorphism.

**Proof of (4.0).** Let $\mathcal{C}$ be an exact category, $0 \in \mathcal{C}$ a null object. The category of covering spaces of $BQ\mathcal{C}$ is equivalent to the category $\mathcal{F}$ of functors $F : Q\mathcal{C} \longrightarrow \underline{\mathrm{Set}}$ such that $F(u)$ is a bijection for every arrow $u$ of $Q\mathcal{C}$. Let $\mathcal{F}' \subset \mathcal{F}$ be the full subcategory consisting of functors $F : Q\mathcal{C} \longrightarrow \underline{\mathrm{Set}}$ with $F(M) = F(0)$, $F(i_!) = 1_{F(0)}$ for any admissible mono $i : M' \rightarrowtail M$ in $\mathcal{C}$. If $F \in \mathcal{F}$ is an arbitrary functor, let $\bar{F} \in \mathcal{F}'$ be the functor given by $\bar{F}(M) = F(0)$, and if $M \stackrel{q}{\twoheadleftarrow} M' \stackrel{i}{\rightarrowtail} N$ represents an arrow $u : M \rightarrow N$ in $Q\mathcal{C}$, let $\bar{F}(u) = F(i_{M'!})^{-1} \circ F(q^!) \circ F(i_{M!})$ (where for any $M \in \mathcal{C}$, we have $i_M : 0 \rightarrowtail M$, $q_M : M \twoheadrightarrow 0$). Clearly $M \mapsto F(i_{M!})$ gives a natural transformation $\bar{F} \rightarrow F$ which is an isomorphism of functors, since $F(i_{M!})$ is an isomorphism in the category $\underline{\mathrm{Set}}$. Thus every object of $\mathcal{F}$ is isomorphic to an object of $\mathcal{F}'$, so that $\mathcal{F}$ is equivalent to $\mathcal{F}'$.

It suffices to show that $\mathcal{F}'$ is equivalent to the category of $K_0(\mathcal{C})$-sets (a $K_0(\mathcal{C})$-set is a set on which $K_0(\mathcal{C})$ acts through permutations). Indeed, by (6.1) and Chapter 3, Ex. (3.10), the category of coverings of $BK_0(\mathcal{C})$ (the classifying space of the group $K_0(\mathcal{C})$) is clearly equivalent to the category of $K_0(\mathcal{C})$-sets. But $\widetilde{BK_0(\mathcal{C})}$, the universal cover of $BK_0(\mathcal{C})$, is an initial object in the category of pointed coverings of $BK_0(\mathcal{C})$, and the automorphism group of $\widetilde{BK_0(\mathcal{C})}$ in the category of coverings is just $K_0(\mathcal{C})$, the fundamental group of $BK_0(\mathcal{C})$. Hence the category of pointed

## 6. Proofs of the Theorems of Chapter 4

coverings of $BQC$ also has an initial object (which must be the universal cover) whose automorphism group in the category of coverings is $K_0(C)$.

If $S$ is a $K_0(C)$-set, $\psi : K_0(C) \longrightarrow \text{Aut}(S)$ the permutation representation, let $F_S : QC \longrightarrow \underline{\text{Set}}$ be the functor defined by Lemma (6.2) and the assignments $F_S(M) = S \, \forall \, M \in C$, $(F_S)_1(i_!) = 1_S$, the identity map, and $(F_S)_2(q^!) = \psi([\ker q]) \in \text{Aut}(S)$. If

$$\begin{array}{ccc} M' & \stackrel{i}{\rightarrowtail} & N \\ q \downarrow & & \downarrow q' \\ M & \underset{i'}{\rightarrowtail} & N' \end{array}$$

is a bicartesian square, then $\ker q \cong \ker q'$, so that $\psi(\ker q) = \psi(\ker q')$, and the conditions of (6.2) hold. One sees at once that $F_S \in \mathcal{F}'$.

Conversely, if $F \in \mathcal{F}'$, let $\psi_F : K_0(C) \longrightarrow \text{Aut}(F(0))$ be given by $\psi_F([M]) = F(q_M^!)$. To see that this gives a well-defined homomorphism on $K_0(C)$, if we have an exact sequence in $C$

$$0 \longrightarrow M' \underset{i}{\rightarrowtail} M \underset{q}{\twoheadrightarrow} M'' \longrightarrow 0$$

we have a bicartesian square

$$\begin{array}{ccc} M' & \stackrel{i}{\rightarrowtail} & M \\ q_{M'} \downarrow & & \downarrow q \\ 0 & \underset{M''}{\rightarrowtail} & M'' \end{array}$$

so that $q^! \circ i_{M''!} = i_! \circ q_M^!$, so that $F(q_{M'}^!) = F(q^!)$; also $q_M^! = q^! \circ q_{M''}^!$, so that $F(q_M^!) = F(q_{M''}^!) = F(q_{M'}^!) \circ F(q_{M''}^!)$. In particular, by considering the split exact sequences

$$0 \longrightarrow M' \longrightarrow M' \oplus M'' \longrightarrow M'' \longrightarrow 0$$
$$0 \longrightarrow M'' \longrightarrow M' \oplus M'' \longrightarrow M' \longrightarrow 0$$

we see that $F(q_{M'}^!)$, $F(q_{M''}^!) \in \text{Aut}(F(0))$ commute. Hence $\psi_F$ is well defined.

Clearly $(S, \psi) \mapsto F_S$, $F \mapsto (F(0), \psi_F)$ give the desired equivalence of categories. This proves Theorem (4.0).

Our next goal is to prove two technical results on classifying spaces of (small) categories, which are the basic homotopy theoretic tools needed to prove the remaining results of Chapter 4. We begin with a result, called "Theorem A" by Quillen, which gives a criterion for a functor $f : C \longrightarrow \mathcal{D}$ to be a homotopy equivalence. For any functor $f : C \longrightarrow \mathcal{D}$, and any object $Y \in \mathcal{D}$, let $Y \setminus f$ be the category whose objects are pairs $(X, v)$, $X \in \text{Ob}\,C$, $v : Y \longrightarrow f(X)$ an arrow in $\mathcal{D}$, where a morphism $(X, v) \longrightarrow (X', v')$ is a morphism $w : X \longrightarrow X'$ such that the triangle below commutes:

Given an arrow $u : Y' \longrightarrow Y$ in $\mathcal{D}$, we have a functor $u^* : Y \setminus f \to Y' \setminus f$ given by $u^*(X, v) = (X, v \circ u)$.

**Theorem A.** *If $f : \mathcal{C} \longrightarrow \mathcal{D}$ is a functor, such that $Y \setminus f$ is contractible, for each $Y \in \mathcal{D}$, then $f$ is a homotopy equivalence.*

If $f : \mathcal{C} \longrightarrow \mathcal{D}$ is a functor, the fiber $f^{-1}(Y)$ over $Y \in \mathcal{D}$ is the subcategory of $\mathcal{C}$ whose objects $X$ satisfy $f(X) = Y$, with morphisms $v : X \longrightarrow X'$ being precisely those morphisms in $\mathcal{C}$ such that $f(v) = 1_Y$, the identity morphism. There is a naturally defined functor $f^{-1}(Y) \longrightarrow Y \setminus f$, for any $Y \in \mathcal{D}$, given by $X \mapsto (X, 1_Y)$. We say that $f$ makes $\mathcal{C}$ *prefibered* over $\mathcal{D}$ if for each $Y \in \mathcal{D}$, $f^{-1}(Y) \to Y \setminus f$ has a right adjoint. If this is to hold, then for any $(X, v) \in Y \setminus f$, we have an object $v^* X \in f^{-1}(Y)$ (so that $v^* : f^{-1}(f(X)) \longrightarrow f^{-1}(Y)$ is a functor) such that

$$\mathrm{Hom}_{Y \setminus f}(-, (X, v)) \cong \mathrm{Hom}_{f^{-1}(Y)}(-, v^* X).$$

Thus if $v : Y \longrightarrow Y'$ is any arrow in $\mathcal{D}$, we have a functorial *base-change* $v^* : f^{-1}(Y') \longrightarrow f^{-1}(Y)$. We say that $f : \mathcal{C} \longrightarrow \mathcal{D}$ makes $\mathcal{C}$ *fibered* over $\mathcal{D}$ if for $Y \xrightarrow{v} Y' \xrightarrow{v'} Y''$ in $\mathcal{D}$, the canonically defined natural transformation $v^* \circ v'^* \longrightarrow (v' \circ v)^*$ is an isomorphism. Thus, we have the following corollary to Theorem A (recall (3.7) that a functor which has a right adjoint is a homotopy equivalence).

**Corollary (6.4).** *Let $f : \mathcal{C} \longrightarrow \mathcal{D}$ make $\mathcal{C}$ prefibered over $\mathcal{D}$. Suppose that for each $Y \in \mathcal{D}$, $f^{-1}(Y)$ is contractible; then $f$ is a homotopy equivalence.*

Since $B\mathcal{C}$ is naturally homeomorphic to $B\mathcal{C}^{\mathrm{op}}$, we can deduce "dual" versions of Theorem A and Corollary (6.4). For any functor $f : \mathcal{C} \longrightarrow \mathcal{D}$, and any $Y \in \mathcal{D}$, let $f \setminus Y$ denote the category of pairs $(X, v)$, where $X \in \mathcal{C}$, $v : f(X) \longrightarrow Y$ a morphism in $\mathcal{D}$, where a morphism $(X, v) \longrightarrow (X', v')$ is a morphism $w : X \longrightarrow X'$ in $\mathcal{C}$ such that the triangle below commutes:

## 6. Proofs of the Theorems of Chapter 4

**Theorem A** (dual version). *Let $f : \mathcal{C} \longrightarrow \mathcal{D}$ be a functor such that $f/Y$ is contractible for all $Y \in \mathcal{D}$. Then $f$ is a homotopy equivalence.*

Next, $f : \mathcal{C} \longrightarrow \mathcal{D}$ is said to make $\mathcal{C}$ *pre-cofibered* over $\mathcal{D}$ if the functor $f^{-1}(Y) \longrightarrow f/Y$, $X \mapsto (X, 1_Y)$, has a left adjoint, for every $Y \in \mathcal{D}$. This gives functorial co-base change arrows $u_* : f^{-1}(Y) \longrightarrow f^{-1}(Y')$ associated to morphism $v : Y \longrightarrow Y'$. $f$ makes $\mathcal{C}$ *cofibered* over $\mathcal{D}$ if the natural transformations $(u \circ v)_* \longrightarrow u_* \circ v_*$ are isomorphisms, for any $Y \xrightarrow{u} Y' \xrightarrow{v} Y''$ in $\mathcal{D}$.

**Corollary (6.4)** (dual version). *Let $f : \mathcal{C} \longrightarrow \mathcal{D}$ make $\mathcal{C}$ pre-cofibered over $\mathcal{D}$. Suppose $f^{-1}(Y)$ is contractible for each $Y \in \mathcal{C}$. Then $f$ is a homotopy equivalence.*

(**Note:** Below, we may refer to Theorem A, Corollary (6.4), or to their dual versions, simply as Theorem A; it will be clear from the context as to which result we mean).

We need two lemmas in the proof of Theorem A.

**Lemma (6.5).** *Let $i \mapsto X_i$ be a functor from a small category $I$ to the category of topological spaces, and let $g : X_I \longrightarrow BI$ be the space over $BI$ obtained by realizing the simplicial space*

$$p \mapsto \coprod_{i_0 \to \cdots \to i_p} X_{i_0}$$

*(where $i_0 \to \cdots \to i_p$ ranges over $p$-simplices in the simplicial set $NI$). If for every $i \to i'$ in $I$, $X_i \longrightarrow X_{i'}$ is a homotopy equivalence, then $g : X_I \longrightarrow BI$ is a quasi-fibration.*

(See Chapter 3 for the definition of a simplicial space; if $p \mapsto X_p$ is a simplicial space, its *realization* is obtained by putting the quotient topology on $(\coprod_{p \geq 0} X_p \times \Delta_p)/\sim$, where $\sim$ is the equivalence relation used to define the realization of the underlying simplicial set. See Appendix A, (A.29) for the definition and some properties of quasi-fibrations).

**Proof.** Since $BI = \varinjlim F_p$, where $F_p \subset BI$ is the $p$-skeleton (the realization of the simplicial subset of $NI$ generated by the non-degenerate simplices of dimension $\leq p$), it suffices to prove that $g^{-1}(F_p) \longrightarrow F_p$ is a quasi-fibration for each $p \geq 0$, by (A.35). We have a map of pushout squares, induced by $g$,

$$\begin{array}{ccccccc}
\coprod_{N'_p I} X_{i_0} \times \partial \Delta_p & \longrightarrow & \coprod_{N'_p I} X_{i_0} \times \Delta_p & & \coprod_{N'_p I} \partial \Delta_p & \longrightarrow & \coprod_{N'_p I} \Delta_p \\
\downarrow & & \downarrow & & \downarrow & & \downarrow \\
g^{-1}(F_{p-1}) & \longrightarrow & g^{-1}(F_p) & & F_{p-1} & \longrightarrow & F_p
\end{array}$$

where $N'_p I \subset N_p I$ is the set of non-degenerate $p$-simplices in $NI$. Let $U \subset F_p$ be the open set obtained by removing the barycenters of the $p$-cells (indexed by $N'_p I$), and let $V = F_p - F_{p-1}$. By (A.30), it suffices to prove that $U$, $V$ and $U \cap V$ are distinguished for $g$ (as $U \cup V = F_p$); this is clear for $V$ and for $U \cap V$, since $g$ is a product over $V$.

We may assume by induction that $F_{p-1}$ is distinguished for $g$. There is a fiber-preserving deformation $D_t$ of $g^{-1}(U)$ into $g^{-1}(F_{p-1})$ induced by the radial deformation of $\Delta_p$, with its barycenter removed, onto $\partial \Delta_p$. Let $d_t$ be the induced deformation of $U$ onto $F_{p-1}$. To apply (A.34), we need only show that if $x \in U$, $x' = d_1(x)$, then $D_1 : g^{-1}(x) \longrightarrow g^{-1}(x')$ induces an isomorphism on homotopy groups. Since $d_1$ is the identity when restricted to $F_{p-1}$, we may assume $x \in U \cap V$. Suppose $x$ lies in the interior of the simplex $i_0 \to i_1 \to \cdots \to i_p$, and suppose $d_1(x)$ lies in the interior of the $q$-face with vertices $i_{j_0}, i_{j_1}, \ldots, i_{j_q}$ (the interior of $\Delta_q$ is $\Delta_q - \partial \Delta_q$, where $\partial \Delta_q = \emptyset$ if $q = 0$). Then $g^{-1}(x) = X_{i_0}$, and $g^{-1}(d_1(x)) = g^{-1}(x') = X_k$, where $k = i_{j_0}$. The map $D_1 : X_{i_0} \longrightarrow X_k$ is the one induced by the edge $i_0 \to i_{j_0}$ of $i_0 \to i_1 \to \cdots \to i_p$. Indeed, $D_t$ is induced by the deformation of $\Delta_p \setminus \{b\}$ onto $\partial \Delta_p$, where $b \in \Delta_p$ is the barycenter, by first taking the product deformation of $X_{i_0} \times (\Delta_p - \{b\})$ onto $X_{i_0} \times \partial \Delta_p$, for each simplex in $N'_p I$. If $x'$ lies in the interior of a $q$-face of $\partial \Delta_p$ whose first vertex is $k$, then we identify $X_{i_0} \times \{x'\} \subset X_{i_0} \times \partial \Delta_p$ with the image of $X_{i_0}$ in $X_k$, where $X_{i_0} \longrightarrow X_k$ is induced by the arrow in $I$ corresponding to the edge joining $i_0$ to $k$ in $\partial \Delta_p$. (The geometric realization is constructed using such identifications.)

Now $X_{i_0} \to X_k$ is a homotopy equivalence, by hypothesis. Hence $D_1 : g^{-1}(x) \longrightarrow g^{-1}(x')$ induces isomorphisms on homotopy groups, proving (6.7).

The next lemma involves the notion of a *bisimplicial space*. A bisimplicial space is defined to be a functor $\Delta^{\mathrm{op}} \times \Delta^{\mathrm{op}} \longrightarrow$ Top where Top denotes the category of topological spaces. Thus for each ordered pair $(\underline{p}, \underline{q})$ of objects of $\Delta$, we are given a topological space $T_{pq}$, such that given any pair of morphisms $\underline{p} \longrightarrow \underline{p}'$, $\underline{q} \longrightarrow \underline{q}'$ in $\Delta$, we have a corresponding map of topological spaces $T_{p'q'} \longrightarrow T_{pq}$. In particular, for each fixed $\underline{p} \in \Delta$, $\underline{q} \mapsto T_{pq}$ is a simplicial space; similarly for each fixed $\underline{q} \in \Delta$, $\underline{p} \mapsto T_{pq}$ is a simplicial space. Thus, we can form the geometric realizations $|\underline{q} \mapsto T_{pq}|$

and $|q \mapsto T_{pq}|$, and obtain simplicial spaces
$$p \mapsto |q \mapsto T_{pq}|, \quad q \mapsto |p \mapsto T_{pq}|,$$
and hence their geometric realizations (which are topological spaces)
$$|p \mapsto |q \mapsto T_{pq}||, \quad |q \mapsto |p \mapsto T_{pq}||.$$
Finally, we have the diagonal simplicial space $p \mapsto T_{pp}$, with geometric realization $|p \mapsto T_{pp}|$.

**Lemma (6.8).** *There are natural homeomorphisms*
$$|p \mapsto T_{pp}| \cong |p \mapsto |q \mapsto T_{pq}|| \cong |q \mapsto |p \mapsto T_{pq}||.$$

**Proof.** Suppose first that $T$ is of the form $\Delta(r) \times \Delta(s) \times S$ for a given topological space $S$, i.e., $T$ is given by the functor $\Delta^{op} \times \Delta^{op} \to \underline{\text{Top}}$, $h^{r,s} \times S : (p,q) \longrightarrow \text{Hom}_\Delta(p,r) \times \text{Hom}_\Delta(q,s) \times S$, where the Hom-sets are regarded as discrete spaces. Then we claim that the diagonal realization has a natural homeomorphism

(1)... $\qquad |p \mapsto \text{Hom}_\Delta(p,r) \times \text{Hom}_\Delta(p,s) \times S| \cong \Delta_r \times \Delta_s \times S.$

Indeed, given *any* simplicial set $F : \Delta^{op} \longrightarrow \underline{\text{Set}}$ and a topological space $S$, if $F \times S : \Delta^{op} \longrightarrow \underline{\text{Top}}$, is the simplicial space $p \mapsto F(p) \times S$, where $F(p)$ is regarded as a discrete space, then from the definition of the geometric realization of a simplicial space, there is a continuous bijection

(2)... $\qquad\qquad |F \times S| \longrightarrow |F| \times S$

which is a homeomorphism if $|F|$ is compact. Thus, it suffices to verify that there is a natural homeomorphism
$$|h^{r,s}| \longrightarrow \Delta_r \times \Delta_s.$$
But $h^{r,s} : \Delta^{op} \to \underline{\text{Set}}$ is just the product simplicial set $\Delta(r) \times \Delta(s)$, so that the projections $h^{r,s} \to \Delta(r)$, $h^{r,s} \to \Delta(s)$ (which are maps of simplicial sets) induce maps $|h^{r,s}| \to \Delta_r$, $|h^{r,s}| \to \Delta_s$ and hence a map $|h^{r,s}| \to \Delta_r \times \Delta_s$. By Appendix A, (A.55), this is a homeomorphism. This gives the homeomorphism in (1).

Next, we have homeomorphisms (applying (2))
$$|p \mapsto |q \mapsto \text{Hom}_\Delta(p,r) \times \text{Hom}_\Delta(q,s) \times S||$$
$$\cong |p \mapsto \text{Hom}_\Delta(p,r) \times \Delta_s \times S| \cong \Delta_r \times \Delta_s \times S,$$
and similarly
$$|q \mapsto |p \mapsto \text{Hom}_\Delta(p,r) \times \text{Hom}_\Delta(q,s) \times S|$$
$$\cong |q \mapsto \Delta_r \times \text{Hom}_\Delta(q,s) \times S|$$
$$\cong \Delta_r \times \Delta_s \times S.$$

Thus, Lemma (6.8) holds for bisimplicial sets of the form $h^{r,s} \times S$.

Now let $T = \{T_{rs}\}$ be a general bisimplicial space. Given any arrow $(\underline{r}, \underline{s}) \longrightarrow (\underline{r}', \underline{s}')$ in $\Delta \times \Delta$, we have (i) a map of topological spaces $T_{r's'} \longrightarrow T_{rs}$, and (ii) a natural transformation of functors (i.e., a map of bisimplicial sets) $h^{r,s} \to h^{r',s'}$. Thus we have two maps of bisimplicial spaces

$$\coprod_{(\underline{r},\underline{s}) \to (\underline{r}',\underline{s}')} h^{r,s} \times T_{r's'} \rightrightarrows \coprod_{(\underline{r},\underline{s})} h^{r,s} \times T_{rs},$$

such that the direct limit of the above diagram is $T$, i.e., the direct limit in $\underline{\text{Top}}$ of the diagram

$$\coprod_{(\underline{r},\underline{s}) \to (\underline{r}',\underline{s}')} h^{r,s}(\underline{p},\underline{q}) \times T_{r's'} \rightrightarrows \coprod_{(\underline{r},\underline{s})} h^{r,s}(\underline{p},\underline{q}) \times T_{rs}$$

is $T_{pq}$, for every $(\underline{p},\underline{q}) \in \Delta \times \Delta$. We leave the proof of this claim to the reader (who may find it instructive to first prove the analogous claim for simplicial spaces; it is useful to observe that there is at least a map

$$g: \coprod_{(\underline{r},\underline{s})} h^{r,s}(\underline{p},\underline{q}) \times T_{rs} \longrightarrow T_{pq},$$

since if $f: (\underline{p},\underline{q}) \longrightarrow (\underline{r},\underline{s}) \in h^{r,s}(\underline{p},\underline{q})$, we have a map $f^*: T_{rs} \longrightarrow T_{pq}$ as part of the data defining the bisimplicial space $T$. One verifies that $g$ induces a map from the direct limit of the diagram to $T_{pq}$, which one proves is a homeomorphism).

Now Lemma (6.8) follows from the special case dealt with earlier, and the observation that all three realization functors commute with direct limits.

**Proof of Theorem A.** Let $S(f)$ be the category of triples $(X, Y, v)$ with $X \in \mathcal{C}$, $Y \in \mathcal{D}$, $v: Y \longrightarrow f(X)$ an arrow in $\mathcal{D}$; a morphism of triples $(X, Y, v) \longrightarrow (X', Y', v')$ is defined to be a pair of arrows $u: X \longrightarrow X'$, $w: Y' \longrightarrow Y$ in $\mathcal{C}, \mathcal{D}$ respectively, such that

$$\begin{array}{ccc} Y & \xrightarrow{v} & f(X) \\ w \uparrow & & \downarrow f(u) \\ Y' & \xrightarrow{v'} & f(X') \end{array}$$

commutes, i.e., $v' = f(u) \circ v \circ w$. We have functors $p_1: S(f) \longrightarrow \mathcal{C}$, $p_2: S(f) \longrightarrow \mathcal{D}^{\text{op}}$, given by $p_1(X, Y, v) = X$, $p_2(X, Y, v) = Y$.

Let $T(f)$ be the bisimplicial set given by $T_{pq}$ = set of pairs of diagrams $(Y_p \to \cdots \to Y_0 \to f(X_0), X_0 \to \cdots \to X_q)$ where $Y_p \to \cdots \to Y_0$ is a $p$-simplex in $N\mathcal{D}^{\text{op}}$, and $X_0 \to \cdots \to X_q$ is a $q$-simplex in $N\mathcal{C}$. The bisimplicial structure of $T(f)$ is induced by the simplicial structures of $N\mathcal{C}$

## 6. Proofs of the Theorems of Chapter 4

and $N\mathcal{D}^{\mathrm{op}}$ in the obvious way. We may regard $N\mathcal{C}$ as a bisimplicial set with $(N\mathcal{C})_{pq} = N_q\mathcal{C}$. Then

$$(Y_p \to \cdots \to Y_0 \to f(X_0), X_0 \to \cdots \to X_q) \mapsto (X_0 \to \cdots \to X_q)$$

yields a map of bisimplicial sets

$$(*) \ldots \qquad T(f)_{pq} \longrightarrow (N\mathcal{C})_{pq} = N_q\mathcal{C}.$$

The diagonal simplicial set of $T(f)$ is just $NS(f)$, the nerve of $S(f)$, and the map $NS(f) \longrightarrow N\mathcal{C}$ of simplicial sets given by $(*)$ is just the natural map on nerves induced by the functor $p_1$. Hence the realization of $(*)$ (in any of the equivalent senses of Lemma (6.8)) is the map $Bp_1 : BS(f) \longrightarrow B\mathcal{C}$.

On the other hand, we may compute the realization of $(*)$ by first realizing in the $p$-direction, to obtain a map of simplicial spaces in the variable $q$, and then forming the associated map between the realizations of these simplicial spaces. Realizing $(*)$ in the $p$-direction yields the map of simplicial spaces which, on the spaces of $q$-simplices, is

$$\coprod_{(X_0\to\cdots\to X_q)\in N_q\mathcal{C}} B(\mathcal{D}/f(X_0))^{\mathrm{op}} \longrightarrow \coprod_{(X_0\to\cdots\to X_q)\in N_q\mathcal{C}} (\text{point}) = N_q\mathcal{C}.$$

Here $N\mathcal{C}$ is regarded as a simplicial space with the discrete space $N_q\mathcal{C}$ of $q$-simplices; for any $Y \in \mathcal{D}$, $\mathcal{D}/Y$ is the category of pairs $(Y', v)$ where $Y' \in \mathcal{D}$, $v : Y' \to Y$ is a morphism in $\mathcal{D}$, with morphisms $(Y', v) \longrightarrow (Y'', w)$ being arrows $u : Y' \longrightarrow Y''$ such that

commutes (thus, $\mathcal{D}/Y = 1_\mathcal{D}/Y$, $1_\mathcal{D} : \mathcal{D} \longrightarrow \mathcal{D}$ the identity functor). Now $X \mapsto B(\mathcal{D}/f(X))^{\mathrm{op}}$ is a functor $\mathcal{C} \longrightarrow \underline{\text{Top}}$ from the small category $\mathcal{C}$ to the category of topological spaces, where if $u : X \longrightarrow X'$ is an arrow in $\mathcal{C}$, composition with $f(u) : f(X) \longrightarrow f(X')$ yields a functor $\mathcal{D}/f(X) \longrightarrow \mathcal{D}/f(X')$, and hence a map of topological spaces

$$B(\mathcal{D}/f(X))^{\mathrm{op}} \longrightarrow B(\mathcal{D}/f(X'))^{\mathrm{op}}.$$

The simplicial space obtained by realizing $T(f) = \{T_{pq}\}$ in the $p$-direction is clearly just the simplicial space associated to $\mathcal{C} \longrightarrow \underline{\text{Top}}$ by Lemma (6.7), with $I = \mathcal{C}$. Clearly $B(\mathcal{D}/Y)$ is contractible for any $Y \in \mathcal{D}$, since $(Y, 1_Y)$ is a final object of $\mathcal{D}/Y$. Hence the map

$$B(\mathcal{D}/f(X))^{\mathrm{op}} \longrightarrow B(\mathcal{D}/f(X'))^{\mathrm{op}},$$

induced by any given arrow $X \longrightarrow X'$, is a homotopy equivalence. Hence the hypotheses of (6.7) are satisfied, and $Bp_1 : BS(f) \longrightarrow BC$ is a quasi-fibration. Clearly, $p_1^{-1}(X) = (\mathcal{D}/f(X))^{\mathrm{op}}$ for any $X \in \mathcal{C}$, so that

$$(Bp_1)^{-1}\{X\} = B(\mathcal{D}/f(X))^{\mathrm{op}}$$

(where $X$ is regarded on the left as a 0-cell of $BC$), which is a contractible space. Thus $Bp_1$ induces isomorphisms on homotopy groups, from the definition of a quasi-fibration; since $BS(f)$, $BC$ are $CW$-complexes, $Bp_1$ is a homotopy equivalence, from the Whitehead theorem (see Appendix A, (A.9)). We note here that so far, the arguments given are valid for *any* functor $f : \mathcal{C} \longrightarrow \mathcal{D}$ between small categories.

Next, let $N\mathcal{D}^{\mathrm{op}}$ be regarded as a bisimplicial set with $(N\mathcal{D}^{\mathrm{op}})_{pq} = N_p \mathcal{D}^{\mathrm{op}}$. Then

$$(Y_p \to \cdots \to Y_0 \to f(X_0), X_0 \to \cdots \to X_q) \mapsto (Y_p \to \cdots \to Y_0)$$

gives a map of bisimplicial sets

$$T(f)_{pq} \longrightarrow (N\mathcal{D}^{\mathrm{op}})_{pq} = N_p \mathcal{D}^{\mathrm{op}},$$

whose diagonal realization is just $Bp_2 : BS(f) \longrightarrow B\mathcal{D}^{\mathrm{op}}$. On the other hand, if we first realize in the $q$-direction, we obtain a map of simplicial spaces, which is given on the space of $p$-simplices by

$$\coprod_{Y_0 \leftarrow \cdots \leftarrow Y_p} B(Y_0 \setminus f) \longrightarrow \coprod_{Y_0 \leftarrow \cdots \leftarrow Y_p} (\text{point}) = N_p \mathcal{D}^{\mathrm{op}}.$$

Since by the hypothesis of Theorem A, $B(Y \setminus f)$ is contractible for every $Y \in \mathcal{D}$, Lemma (6.7) applies again, showing (as above) that $Bp_2 : BS(f) \longrightarrow B\mathcal{D}^{\mathrm{op}}$ is a quasi-fibration with the contractible fibers $(Bp_2)^{-1}(Y) = B(Y \setminus f)$, and is hence a homotopy equivalence.

Finally, let $f' : S(f) \longrightarrow S(1_\mathcal{D})$ be the functor $f'(X, Y, v) = (f(X), Y, v)$. Then we have a commutative diagram of categories and functors

$$\begin{array}{ccccc}
\mathcal{D}^{\mathrm{op}} & \xleftarrow{p_2} & S(f) & \xrightarrow{p_1} & \mathcal{C} \\
\parallel & & \downarrow f' & & \downarrow f \\
\mathcal{D}^{\mathrm{op}} & \xleftarrow{p_2} & S(1_\mathcal{D}) & \xrightarrow{p_1} & \mathcal{D}
\end{array}$$

where all the arrows except $f, f'$ are known to be homotopy equivalences. This finishes the proof of Theorem A.

Given a functor $f : \mathcal{C} \longrightarrow \mathcal{D}$ between categories, we have the homotopy fiber $F(Bf)$ of the map $Bf : BC \longrightarrow B\mathcal{D}$ (over any given base point of

$B\mathcal{D}$), giving rise to a long exact homotopy sequence (we omit the base points)

$$\cdots \to \pi_i(F(Bf)) \longrightarrow \pi_i(B\mathcal{C}) \xrightarrow{Bf_*} \pi_i(B\mathcal{D}) \longrightarrow \pi_{i-1}(F(Bf)) \to \cdots.$$

However, $F(Bf)$ is not in general the classifying space of a category naturally associated to $f$. Suppose that we are given a category $\mathcal{C}'$, together with a functor $g: \mathcal{C}' \longrightarrow \mathcal{C}$, and a natural transformation from the composite functor $f \circ g: \mathcal{C}' \longrightarrow \mathcal{D}$ to the constant functor $\mathcal{C}' \longrightarrow \{Y\}$, for some object $Y \in \mathcal{D}$. Then we have the map $Bg: B\mathcal{C}' \longrightarrow B\mathcal{C}$, together with a homotopy from $Bf \circ g: B\mathcal{C}' \longrightarrow B\mathcal{D}$ to the constant map $B\mathcal{C}' \longrightarrow \{Y\}$, where we regard $Y$ as a 0-cell in $B\mathcal{D}$. Thus we have an induced factorization of $Bg$ through the homotopy fiber over $Y$, i.e., a map $h: B\mathcal{C}' \longrightarrow F(Bf, \{Y\})$. If this is known to be a homotopy equivalence, then we would have a long exact homotopy sequence

$$\cdots \to \pi_i(B\mathcal{C}') \xrightarrow{Bg_*} \pi_i(B\mathcal{C}) \xrightarrow{Bf_*} \pi_i(B\mathcal{D}) \longrightarrow \pi_{i-1}(B\mathcal{C}') \to \cdots.$$

Given a diagram of topological spaces and maps

$$\begin{array}{ccc} E' & \xrightarrow{h'} & E \\ {\scriptstyle g'}\downarrow & & \downarrow{\scriptstyle g} \\ B' & \xrightarrow{h} & B \end{array}$$

we say that the diagram is *homotopy Cartesian* (or a homotopy fiber product) if the induced map $E' \longrightarrow B' \times_B \mathbb{F}(I, B) \times_B E$ is a homotopy equivalence. Here $\mathbb{F}(I, B)$ is the path space of $B$, and

$$B' \times_B \mathbb{F}(I, B) \times_B E = \{(b', \gamma, e) \in B' \times \mathbb{F}(I, B) \times E \mid h(b') = \gamma(0),$$
$$g(e) = \gamma(1)\}$$

is the homotopy fiber product; the map $E' \longrightarrow B' \times_B \mathbb{F}(I, B) \times_B E$ is $e' \longrightarrow (g'(e'), \gamma_0, h'(e'))$ where $\gamma_0: I \longrightarrow \{g \circ h'(e')\}$ is the constant path. The homotopy fiber product is characterized by the property that to give a map $X \longrightarrow B' \times_B \mathbb{F}(I, B) \times_B E$ is equivalent to giving a pair of maps $g_X: X \longrightarrow E$, $h_X: X \longrightarrow B'$ together with a homotopy $H: X \times I \longrightarrow B$ between $h \circ g_X$ and $g \circ h_X$. In particular, the identity map of $B' \times_B \mathbb{F}(I, B) \times_B E$ is given by the natural projections to $B'$ and $E$, and the homotopy $(B' \times_B \mathbb{F}(I, B) \times_B E) \times I \longrightarrow B$, $((b', \gamma, e), t) \longrightarrow \gamma(t)$.

Suppose $B'$ is contractible; let $H': B' \times I \longrightarrow B'$ be a homotopy from the identity map of $B'$ to the constant map $B' \longrightarrow \{b'_0\}$, for a fixed base point $b'_0 \in B'$. Let $b_0 = g(b'_0) \in B$. Let

$$\tilde{H}: (B' \times_B \mathbb{F}(I, B) \times_B E) \times I \longrightarrow B' \times_B \mathbb{F}(I, B) \times_B E$$

be given by
$$\tilde{H}((b',\gamma,e),t) = (H'(b',t),\gamma_t,e)$$
where $\gamma_t \in \mathbb{F}(I,B)$ is the path

$$\gamma_t(s) = \begin{cases} g \circ H'(b', t - (1+t)s) & \text{if } 0 \leq s \leq \frac{t}{1+t} \\ \gamma((1+t)s - t) & \text{if } \frac{t}{1+t} \leq s \leq 1. \end{cases}$$

Then one verifies immediately that $\tilde{H}$ is a deformation retraction of $B' \times_B \mathbb{F}(I,B) \times_B E$ onto $\{b'_0\} \times_B \mathbb{F}(I,B) \times_B E$, which is naturally homeomorphic to the homotopy fiber $F(h,b_0)$. Thus, if the square is homotopy Cartesian, and $B'$ is contractible, there is a homotopy equivalence $E \longrightarrow F(h,b_0)$, from $E'$ to the homotopy fiber of $h$ over any point $b_0 \in g(B')$.

**Theorem B.** *Let $f : \mathcal{C} \longrightarrow \mathcal{D}$ be a functor between small categories such that for every arrow $u : Y \longrightarrow Y'$ of $\mathcal{D}$, the functor $u^* : Y' \setminus f \longrightarrow Y \setminus f$ is a homotopy equivalence. Then for any object $Y$ of $\mathcal{D}$ the square*

$$\begin{array}{ccc} BY \setminus f & \xrightarrow{Bj} & B\mathcal{C} \\ {\scriptstyle Bf'}\downarrow & & \downarrow{\scriptstyle Bf} \\ BY \setminus \mathcal{D} & \xrightarrow[Bj']{} & B\mathcal{D} \end{array}$$

*is homotopy Cartesian, where* $j(X,v) = X$, $j'(Y',v) = Y'$, $f'(X,v) = (f(X),v)$. *Since $Y \setminus \mathcal{D}$ is contractible, for any $X \in f^{-1}(Y)$ we have an exact homotopy sequence* $(\bar{X} = (X,1_Y))$

$$\cdots \to \pi_{i+1}(B\mathcal{D},\{Y\}) \longrightarrow \pi_i(B(Y \setminus f),\{X\}) \xrightarrow{Bj_*} \pi_i(B\mathcal{C},\{X\}) \xrightarrow{Bf_*}$$
$$\pi_i(B\mathcal{D},\{Y\}) \longrightarrow \pi_{i-1}(B(Y \setminus f),\{\bar{X}\}) \to \cdots.$$

As with Theorem A, there is a dual version of the above theorem, involving the categories $f/Y$; further, if $f$ is either prefibered or pre-cofibered, we have an analogous result with suitable hypothesis on the fibers $f^{-1}(Y)$ (if $f$ is prefibered, $u : Y' \longrightarrow Y$ an arrow in $\mathcal{D}$, we assume $u^* : f^{-1}(Y) \longrightarrow f^{-1}(Y')$ is a homotopy equivalence; then we deduce the result that

$$\begin{array}{ccc} Bf^{-1}(Y) & \longrightarrow & B\mathcal{C} \\ \downarrow & & \downarrow \\ \{Y\} & \longrightarrow & B\mathcal{D} \end{array}$$

is homotopy Cartesian). We leave it to the reader to formulate these statements in detail, and deduce them from the above form of Theorem B.

**Proof of Theorem B.** As in the proof of Theorem A, we consider the category $S(f)$ of triples $(X,Y,v)$ with $X \in \mathcal{C}$, $Y \in \mathcal{D}$, $v : Y \longrightarrow f(X)$ an

## 6. Proofs of the Theorems of Chapter 4

arrow in $\mathcal{D}$, together with the diagram $\mathcal{D}^{op} \xleftarrow{p_2} S(f) \xrightarrow{p_1} \mathcal{C}$. As remarked during the proof of Theorem A, $p_1$ is a homotopy equivalence for an arbitrary functor $f$. Since $u^* : Y' \setminus f \longrightarrow Y \setminus f$ is a homotopy equivalence for any arrow $u : Y \longrightarrow Y'$ in $\mathcal{D}$, by hypothesis, Lemma (6.7) shows that $B_{p_2} : BS(f) \longrightarrow B\mathcal{D}^{op}$ is a quasi-fibration, as in the proof of Theorem A, i.e., any fiber of $B_{p_2}$ is weak homotopy equivalent (and hence homotopy equivalent, since both spaces have the homotopy types of $CW$-complexes) to the corresponding homotopy fiber. If $Y \in \mathcal{D}$, $p_2^{-1}(Y) = Y \setminus f$; hence

$$\begin{array}{ccc} BY \setminus f & \longrightarrow & BS(f) \\ \downarrow & & \downarrow B_{p_2} \\ \{Y\} & \longrightarrow & B\mathcal{D}^{op} \end{array}$$

is homotopy Cartesian. Consider now the diagram ($f'$ is as in the proof of Theorem A)

$$\begin{array}{ccccc} BY \setminus f & \longrightarrow & BS(f) & \xrightarrow{Bp_1} & B\mathcal{C} \\ \downarrow & \text{\textcircled{1}} & Bf' \downarrow & \text{\textcircled{2}} & \downarrow Bf \\ BY \setminus \mathcal{D} & \longrightarrow & BS(1_\mathcal{D}) & \xrightarrow{Bp_1} & B\mathcal{D} \\ \cong \downarrow & \text{\textcircled{3}} & Bp_2 \downarrow \cong & & \\ \{Y\} & \longrightarrow & B\mathcal{D}^{op} & & \end{array}$$

where the arrows marked $Bp_1$ are homotopy equivalences. The square \textcircled{1}+\textcircled{3}(with vertices $B(Y \setminus f)$, $\{Y\}$, $BS(f)$, $B\mathcal{D}^{op}$) has been shown, above, to be homotopy Cartesian. The two vertical arrows marked $\cong$ are homotopy equivalences. Hence \textcircled{1} is homotopy Cartesian, and also \textcircled{1} + \textcircled{2} (since the maps $Bp_1$ are homotopy equivalences). This is precisely the conclusion of the theorem.

Our next goal is to prove Theorem (4.1) and its corollaries. We recall the statement below. Let $\mathcal{C}$ be an exact category, $\mathcal{E}$ the category of (short) exact sequences in $\mathcal{C}$. If $E, E', E'' \in \mathcal{E}$, we say that the diagram $0 \longrightarrow E' \longrightarrow E \longrightarrow E'' \longrightarrow 0$ in $\mathcal{E}$ is an exact sequence if the corresponding diagram in $\mathcal{C}$ has exact rows and columns. We have three exact functors $s, t, q$ from $\mathcal{E} \longrightarrow \mathcal{C}$ and natural transformations $s \longrightarrow t$, $t \longrightarrow q$ such that for any $E \in \mathcal{E}$,

$$0 \longrightarrow s(E) \longrightarrow t(E) \longrightarrow q(E) \longrightarrow 0$$

is the corresponding exact sequence in $\mathcal{C}$.

**Theorem (4.1).** *The functor $(s, q) : Q\mathcal{E} \longrightarrow Q\mathcal{C} \times Q\mathcal{C}$ is a homotopy equivalence.*

**Proof.** By Theorem A, it suffices to prove that the category $(s,q)/(M,N)$ is contractible for every $(M,N) \in Q\mathcal{C} \times Q\mathcal{C}$. Let $\mathcal{C}_1 = (s,q)/(M,N)$; by definition it is the category of triples $(E,u,v)$ with $E \in Q\mathcal{E}$, $u: s(E) \longrightarrow M$, $v: q(E) \longrightarrow N$ morphisms in $Q\mathcal{C}$.

Let $\mathcal{C}_2 \subset \mathcal{C}_1$ be the full subcategory of triples $(E,u,v)$ such that $u = j^!$ for some admissible epimorphism $j$, and let $\mathcal{C}_3 \subset \mathcal{C}_2$ be the full subcategory of triples $(E,u,v)$ such that $v = i_!$ for some admissible monomorphism $i$.

**Lemma (6.9).** *The inclusion functors* $\mathcal{C}_2 \subset \mathcal{C}_1$ *and* $\mathcal{C}_3 \subset \mathcal{C}_2$ *have left adjoints, and are thus homotopy equivalences.*

**Proof.** First consider $\mathcal{C}_2 \subset \mathcal{C}_1$. If $X = (E,u,v) \in \mathcal{C}_1$, we must show that there is a universal arrow $X \longrightarrow \bar{X}$ in $\mathcal{C}_1$, with $\bar{X} \in \mathcal{C}_2$.

Factor $u$ as $u = j^! \circ i_!$, where $i: s(E) \rightarrowtail M'$, $j: M \twoheadrightarrow M'$ are respectively an admissible mono and an admissible epi in $\mathcal{C}$ (by the discussion preceding (6.2), every arrow in $Q\mathcal{C}$ has such a factorization which is unique up to isomorphism). Define an exact sequence $i_*E$ in $\mathcal{C}$ by pushout:

$$
\begin{array}{ccccccccc}
E:0 & \longrightarrow & s(E) & \longrightarrow & t(E) & \longrightarrow & q(E) & \longrightarrow & 0 \\
& & \Big\downarrow i & & \Big\downarrow & & \Big\| & & \\
i_*E:0 & \longrightarrow & M' & \longrightarrow & T & \longrightarrow & q(E) & \longrightarrow & 0
\end{array}
$$

(since $\mathcal{C}$ is closed under extensions in an ambient Abelian category, $T \in \mathcal{C}$, and so $i_*E \in \mathcal{E}$). Let $\bar{X} = (i_*E, j^!v)$; clearly $\bar{X} \in \mathcal{C}_2$, and there is an evident map $X \longrightarrow \bar{X}$ given by the admissible monomorphism $E \rightarrowtail i_*E$ in $\mathcal{E}$, described by the above diagram.

We claim $X \longrightarrow \bar{X}$ is universal for maps $X \longrightarrow X'$ with $X' \in \mathcal{C}_2$. Indeed, given $X \longrightarrow X'$ in $\mathcal{C}_1$ where $X' = (E', j'^!, v')$ lies in $\mathcal{C}_2$, we represent the corresponding morphism $E \longrightarrow E'$ in $Q\mathcal{E}$ by the diagram $E \rightarrowtail E_0 \twoheadleftarrow E'$ in $\mathcal{E}$. By definition of the morphisms in $\mathcal{C}_1 = (s,q)/(M,N)$, we have a diagram in $Q\mathcal{C}$

$$
\begin{array}{ccc}
s(E) & \longrightarrow & s(E') \\
{}_u\searrow & & \swarrow_{j'^!} \\
& M &
\end{array}
$$

and hence a diagram in $\mathcal{C}$

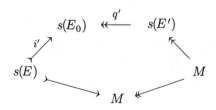

Since the factorization $u = q^! \circ i_!$ is unique up to isomorphism (by the definition of a morphism in $Q\mathcal{C}$), there is a (unique) isomorphism $s(E_0) \cong M$ making the diagram commute. Hence, without loss of generality we may assume $s(E_0) = M'$, $i' = i$, and $q' : s(E') \twoheadrightarrow M'$ is the unique arrow in $\mathcal{C}$ such that $q' \circ j' = q$. By the universal property of the pushout, $E \rightarrowtail E_0$ factors uniquely as $E \rightarrowtail i_*E \rightarrowtail E_0$. This gives a map $\bar{X} \longrightarrow X'$ in $\mathcal{C}_2$, associated to the diagram $i_*E \rightarrowtail E_0 \leftarrowtail E'$ in $\mathcal{E}$, such that $X \longrightarrow X'$ factors as $X \longrightarrow \bar{X} \longrightarrow X'$. If $E_0$ is replaced by an isomorphic exact sequence (so as to yield the same morphism $X \longrightarrow X'$), clearly we obtain the same morphism $i_*E \longrightarrow E'$ in $Q\mathcal{E}$, and hence the same morphism $\bar{X} \longrightarrow X'$ in $\mathcal{C}_2$. Lastly, the morphism $\bar{X} \longrightarrow X'$, factoring the given morphism $X \longrightarrow X'$, is unique; indeed, once we choose a diagram $E \rightarrowtail E_0 \leftarrowtail E'$ representing the underlying morphism $E \longrightarrow E'$ in $Q\mathcal{E}$, it suffices to show that there is a unique arrow $i_*E \rightarrowtail E_0$ in $\mathcal{E}$ factoring $E \rightarrowtail E_0$, which yields an arrow in $\mathcal{C}_2$. But by the universal property of the pushout, such morphisms $i_*E \rightarrowtail E_0$ are in bijection with morphisms $s(i_*E) \rightarrowtail s(E_0)$ factoring $s(E) \rightarrowtail s(E_0)$. However, as seen above, since $E \longrightarrow E'$ is an arrow in $Q\mathcal{E}$ arising from $X \longrightarrow X'$ in $\mathcal{C}_1$, for the arrow $s(i_*E) \longrightarrow s(E_0)$ to arise from an arrow in $\mathcal{C}_2$, we must have a diagram

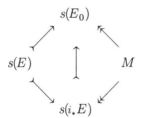

where all the arrows except the vertical one are already given. Thus the vertical arrow is uniquely determined, and is an isomorphism. This completes the proof that $\mathcal{C}_2 \subset \mathcal{C}_1$ has a left adjoint.

Next, consider the inclusion $\mathcal{C}_3 \subset \mathcal{C}_2$, and let $(E, u, v) \in \mathcal{C}_2$. We factor $v : q(E) \longrightarrow N$ as $v = i_! \circ j^!$, where $j : N' \twoheadrightarrow q(E)$, $i : N' \rightarrowtail N$, and define $j^*E$ by pullback:

$$\begin{array}{ccccccccc} j^*E : 0 & \longrightarrow & s(E) & \longrightarrow & T & \longrightarrow & N' & \longrightarrow & 0 \\ & & \| & & \downarrow & & \downarrow j & & \\ E : 0 & \longrightarrow & s(E) & \longrightarrow & t(E) & \longrightarrow & q(E) & \longrightarrow & 0. \end{array}$$

By an argument similar to the one above, we see that the map $(E, u, v) \longrightarrow (j^*E, u, i_!)$, induced by the admissible epi $j^*E \twoheadrightarrow E$ in $\mathcal{E}$, is universal, so that $(E, u, v) \longrightarrow (j^*E, u, i_!)$ is a left adjoint to the inclusion $\mathcal{C}_3 \subset \mathcal{C}_2$. This proves (6.9).

To finish the proof of (4.1), it suffices to prove that $\mathcal{C}_3$ is contractible, for any $M, N \in \mathcal{C}$. Let $(E, j^!, i_!) \in \mathcal{C}_3$, and consider $j_M : M \twoheadrightarrow 0$, $i_N : 0 \rightarrowtail N$, giving an object $(0, j_M^!, i_{N!}) \in \mathcal{C}_3$. We claim there is a unique morphism $(0, j_M^!, i_{N!}) \longrightarrow (E, j^!, i_!)$ in $\mathcal{C}_3$, so that $(0, j_M^!, i_{N!})$ is an initial object. A morphism $(0, j_M^!, i_{N!}) \longrightarrow (E, j^!, i_!)$ consists of a morphism $0 \longrightarrow E$ in $Q\mathcal{E}$ such that the induced diagrams below are commutative:

$$s(E) \xrightarrow{q^!} M \qquad q(E) \xrightarrow{i_!} N$$
$$\nwarrow \quad \nearrow_{q_M^!} \qquad\qquad \nwarrow \quad \nearrow_{i_{N!}}$$
$$0 \qquad\qquad\qquad 0$$

This forces $0 \longrightarrow s(E)$ to equal $q^!_{s(E)}$, and $0 \longrightarrow q(E)$ to equal $i_{q(E)!}$ and hence the morphism $0 \longrightarrow E$ in $Q\mathcal{E}$ must be the morphism given by the diagram

$$\begin{array}{ccccccccc}
0 & \longrightarrow & 0 & \longrightarrow & 0 & \longrightarrow & 0 & \longrightarrow & 0 \\
& & \uparrow & & \uparrow & & \| & & \\
0 & \longrightarrow & s(E) & \longrightarrow & s(E) & \longrightarrow & 0 & \longrightarrow & 0 \\
& & \| & & \downarrow u & & \downarrow & & \\
0 & \longrightarrow & s(E) & \xrightarrow{u} & t(E) & \longrightarrow & q(E) & \longrightarrow & 0
\end{array}$$

(this is immediate from a factorization $0 \twoheadleftarrow E_0 \rightarrowtail E$). Hence $(0, j_M^!, i_{N!})$ is an initial object of $\mathcal{C}_3$. Thus $\mathcal{C}_3$ is contractible.

Next, we recall the statement of Corollary (4.2).

**Corollary (4.2).** *Let $\mathcal{C}$ be an exact category, $\mathcal{E}$ the exact category of short exact sequences in $\mathcal{C}$, and let $s, t, q : \mathcal{E} \longrightarrow \mathcal{C}$ be the exact functors defined above. Then $t_* = s_* + q_* : K_i(\mathcal{E}) \longrightarrow K_i(\mathcal{C}) \; \forall i \geq 0$.*

**Proof.** Let $f : \mathcal{C} \times \mathcal{C} \longrightarrow \mathcal{E}$ be the exact functor given on objects by $f(M, N) = $ (the split exact sequence $0 \longrightarrow M \longrightarrow M \oplus N \longrightarrow N \longrightarrow 0$ in $\mathcal{E}$). Now the exact functor $(s, q) : \mathcal{E} \longrightarrow \mathcal{C} \times \mathcal{C}$ has the property that the induced functor $Q\mathcal{E} \longrightarrow Q(\mathcal{C} \times \mathcal{C}) = Q\mathcal{C} \times Q\mathcal{C}$ is a homotopy equivalence, by (4.1) proved above. Hence

$$(s, q)_* : K_i(\mathcal{E}) \longrightarrow K_i(\mathcal{C} \times \mathcal{C}) \cong K_i(\mathcal{C}) \oplus K_i(\mathcal{C})$$

is an isomorphism. The composite exact functor $(s, q) \circ f : \mathcal{C} \times \mathcal{C} \longrightarrow \mathcal{C} \times \mathcal{C}$ is the identity; hence $f_* : K_i(\mathcal{C} \times \mathcal{C}) \longrightarrow K_i(\mathcal{E})$ is an isomorphism $\forall i \geq 0$. Next, we have an exact functor $\oplus : \mathcal{C} \times \mathcal{C} \longrightarrow \mathcal{C}$ given by $\oplus(M, N) = M \oplus N$; if we fix a 0-object $0 \in \mathcal{C}$, then the functors $\mathcal{C} \longrightarrow \mathcal{C}$ given by $M \mapsto \oplus(M, 0)$ and $M \mapsto \oplus(0, M)$ are both isomorphic to the identity functor of $\mathcal{C}$. Since

6. Proofs of the Theorems of Chapter 4    109

the functors $\mathcal{C} \longrightarrow \mathcal{C} \times \mathcal{C}$ given by $M \mapsto (M, 0)$, $M \mapsto (0, M)$ represent $K_i(\mathcal{C} \times \mathcal{C})$ as a direct sum $K_i(\mathcal{C}) \oplus K_i(\mathcal{C})$, we deduce that

$$\oplus_* : K_i(\mathcal{C} \times \mathcal{C}) \longrightarrow K_i(\mathcal{C})$$

is identified with the addition map

$$K_i(\mathcal{C}) \oplus K_i(\mathcal{C}) \longrightarrow K_i(\mathcal{C}), \quad (x, y) \mapsto x + y.$$

Thus $\oplus_* \circ (s, q)_* : K_i(\mathcal{E}) \longrightarrow K_i(\mathcal{C})$ is just $s_* + q_*$. Now $t \circ f = \oplus \circ (s, q) \circ f : \mathcal{C} \times \mathcal{C} \longrightarrow \mathcal{C}$; hence $t_* \circ f_* = (s_* + q_*) \circ f_*$. Since $f_*$ is an isomorphism, $t_* = s_* + q_*$, which is what we wanted to prove.

**Corollary (4.3).** *Let $\mathcal{C}$, $\mathcal{D}$ be exact categories, $F, G, H : \mathcal{D} \longrightarrow \mathcal{C}$ exact functors, $F \longrightarrow G$, $G \longrightarrow H$ natural transformations, such that for any object $M \in \mathcal{D}$, $0 \longrightarrow F(M) \longrightarrow G(M) \longrightarrow H(M) \longrightarrow 0$ is an exact sequence in $\mathcal{C}$. Then $G_* = F_* + H_* : K_i(\mathcal{D}) \longrightarrow K_i(\mathcal{C})$ $\forall i \geq 0$.*

**Proof.** Let $\mathcal{E}$ be the exact category of short exact sequences in $\mathcal{C}$. Then we have a well-defined functor $L : \mathcal{D} \longrightarrow \mathcal{E}$ given by $L(M) = $ (the exact sequence $0 \longrightarrow F(M) \longrightarrow G(M) \longrightarrow H(M) \longrightarrow 0$). If $s, t, q : \mathcal{E} \longrightarrow \mathcal{C}$ are the exact functors defined above, clearly $F = s \circ L$, $G = t \circ L$, $H = g \circ L$. Since $t_* = s_* + q_*$, we deduce that $G_* = (t \circ L)_* = t_* \circ L_* = (s_* + q_*) \circ L_* = (s \circ L)_* + (q \circ L)_* = F_* + H_*$.

Corollaries (4.4), (4.5) follow easily from (4.3), and are left to the reader.

Next, we prove the resolution theorem. Let $\mathcal{M}$ be an exact category, $\mathcal{P} \subset \mathcal{M}$ a full additive subcategory which is closed under extensions in $\mathcal{M}$, so that $\mathcal{P}$ is an exact category, and $\mathcal{P} \subset \mathcal{M}$ an exact functor.

**Theorem (4.6) (Resolution Theorem).** *Let $\mathcal{P} \subset \mathcal{M}$ be as above. Assume that*

(i) *if $0 \longrightarrow M' \longrightarrow M \longrightarrow M'' \longrightarrow 0$ is exact in $\mathcal{M}$, and $M, M'' \in \mathcal{P}$, then $M' \in \mathcal{P}$*

(ii) *for each object $M \in \mathcal{M}$, there is a finite resolution in $\mathcal{M}$*

$$0 \longrightarrow P_n \longrightarrow P_{n-1} \longrightarrow \cdots \longrightarrow P_0 \longrightarrow M \longrightarrow 0$$

*with $P_i \in \mathcal{P}$ (and $n$ may depend on $M$).*

*Then $Q\mathcal{P} \longrightarrow Q\mathcal{M}$ is a homotopy equivalence; hence $K_i(\mathcal{P}) \cong K_i(\mathcal{M})$ $\forall i$.*

**Proof.** Let $\mathcal{M}_i \subset \mathcal{M}$ be the full additive subcategory whose objects are $M \in \mathcal{M}$ which have a resolution

$$0 \longrightarrow P_n \longrightarrow P_{n-1} \longrightarrow \cdots \longrightarrow P_0 \longrightarrow M \longrightarrow 0$$

with $P_j \in \mathcal{P}$ $\forall j$, and $n \leq i$. Clearly $\mathcal{P} = \mathcal{M}_0 \subset \mathcal{M}_1 \subset \cdots \subset \mathcal{M}$, and $\mathcal{M} = \varinjlim \mathcal{M}_i$.

**Lemma (6.10).** *Let* $0 \longrightarrow M' \longrightarrow M \longrightarrow M'' \longrightarrow 0$ *be an exact sequence in* $\mathcal{M}$. *Then*

(i)$_n$ $M \in \mathcal{M}_n, M'' \in \mathcal{M}_{n+1} \Longrightarrow M' \in \mathcal{M}_n$

(ii)$_n$ $M', M'' \in \mathcal{M}_{n+1} \Longrightarrow M \in \mathcal{M}_{n+1}$

(iii)$_n$ $M, M'' \in \mathcal{M}_{n+1} \Longrightarrow M' \in \mathcal{M}_{n+1}$.

**Proof.** We first prove the above statements for $n = 0$; then we verify that the three statements for $n - 1$ together imply the three statements for $n$, if $n \geq 1$. We use the following two constructions: given an admissible epi $P'' \twoheadrightarrow M''$ with $P'' \in \mathcal{P}$, we have the pullback diagram in $\mathcal{M}$ (with exact rows and columns)

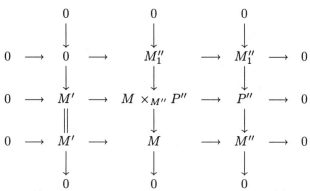

Next, if $P' \twoheadrightarrow M'$, $P'' \twoheadrightarrow M$ are admissible epimorphisms in $\mathcal{M}$ with $P', P'' \in \mathcal{P}$, then the composite $P'' \twoheadrightarrow M''$ is an admissible epimorphism. We have a diagram with exact rows and columns

$$\begin{array}{ccccccccc}
& & 0 & & 0 & & 0 & & \\
& & \downarrow & & \downarrow & & \downarrow & & \\
0 & \longrightarrow & K' & \longrightarrow & K & \longrightarrow & K'' & \longrightarrow & 0 \\
& & \downarrow & & \downarrow & & \downarrow & & \\
0 & \longrightarrow & P' & \longrightarrow & P' \oplus P'' & \longrightarrow & P'' & \longrightarrow & 0 \\
& & \downarrow & & \downarrow & & \downarrow & & \\
0 & \longrightarrow & M' & \longrightarrow & M & \longrightarrow & M'' & \longrightarrow & 0 \\
& & \downarrow & & \downarrow & & \downarrow & & \\
& & 0 & & 0 & & 0 & &
\end{array}$$

where the middle row is the split exact sequence.

6. Proofs of the Theorems of Chapter 4     111

If $M \in \mathcal{M}_0 = \mathcal{P}$, $M'' \in \mathcal{M}_1$, choose an exact sequence
$$0 \longrightarrow M_1'' \longrightarrow P'' \longrightarrow M'' \longrightarrow 0$$
with $M_1'', P'' \in \mathcal{P}$. Then from the middle column of the first (pullback) diagram above, $M \times_{M''} P'' \in \mathcal{P}$. Hence from the middle row of the same diagram, $M' \in \mathcal{P}$. This proves (i)$_0$.

If $M', M'' \in \mathcal{M}_1$, then from (i)$_0$, we see that $K', K'' \in \mathcal{P}$ in the second diagram above; hence $K \in \mathcal{P}$, and so from the middle vertical sequence, $M \in \mathcal{M}_1$. This proves (ii)$_0$.

If $M', M'' \in \mathcal{M}_1$, then from (i)$_0$, we have $K, K'' \in \mathcal{P}$ in the second diagram above; hence $K' \in \mathcal{P}$, and from the left hand vertical sequence, $M' \in \mathcal{M}_1$. This proves (iii)$_0$.

Now assume $n \geq 1$, and (i)$_{n-1}$, (ii)$_{n-1}$, (iii)$_{n-1}$ hold. Suppose $M \in \mathcal{M}_{n+1}$, $M'' \in \mathcal{M}_n$. Then from (i)$_{n-1}$, $M_1'' \in \mathcal{M}_{n-1}$ in the first diagram above (as $P'' \in \mathcal{P} \subset \mathcal{M}_{n-1}$). Hence from the middle row of the first diagram and (iii)$_{n-1}$, $M'' \in \mathcal{M}_n$. This proves (i)$_n$.

Next, suppose $M', M'' \in \mathcal{M}_{n+1}$. Then from (i)$_n$, $K', K'' \in \mathcal{M}_n$ in the second diagram above. Hence from (ii)$_{n-1}$, $K \in \mathcal{M}_n$, and so from the middle column of the same diagram, $M \in \mathcal{M}_{n+1}$. This proves (ii)$_n$.

Lastly, let $M, M'' \in \mathcal{M}_{n+1}$. Then from (i)$_n$, $K, K'' \in \mathcal{M}_n$ in the second diagram above. Hence from the top row of that diagram, and (iii)$_{n-1}$, $K \in \mathcal{M}_n$, so that $M' \in \mathcal{M}_{n+1}$. This proves (iii)$_n$, and completes the proof of (6.10).

Thus, it suffices to prove:

**Theorem (6.11).** *Let $\mathcal{P}$ be a full additive subcategory of an exact category $\mathcal{M}$ which is closed under extensions, such that*

(i) *for any exact sequence $0 \longrightarrow M' \longrightarrow M \longrightarrow M'' \longrightarrow 0$ in $\mathcal{M}$, $M \in \mathcal{P} \Longrightarrow M' \in \mathcal{P}$*

(ii) *for any $M'' \in \mathcal{M}$, there is an exact sequence as in (i) with $M \in \mathcal{P}$.*

*Then the natural functor $Q\mathcal{P} \longrightarrow Q\mathcal{M}$ is a homotopy equivalence.*

**Proof of (4.6).** Assuming (6.11), we see that $Q\mathcal{M}_n \subset Q\mathcal{M}_{n+1}$ is a homotopy equivalence for each $n \geq 0$. Thus we have homotopy equivalences
$$Q\mathcal{P} \cong \varinjlim Q\mathcal{M}_n \cong Q(\varinjlim \mathcal{M}_n) \cong Q\mathcal{M}.$$

**Proof of (6.11).** Note that $Q\mathcal{P} \subset Q\mathcal{M}$ is a subcategory which is in general not full. Let $\mathcal{C} \subset Q\mathcal{M}$ be the full subcategory with the same objects as $Q\mathcal{P}$. Then the inclusion of $Q\mathcal{P}$ in $Q\mathcal{M}$ factors as the composite $Q\mathcal{P} \underset{g}{\hookrightarrow} \mathcal{C} \underset{f}{\hookrightarrow} Q\mathcal{M}$; it suffices to prove that $f, g$ are homotopy equivalences.

To prove $g$ is a homotopy equivalence, it suffices to prove that $g/P$ is contractible $\forall P \in \mathcal{C}$, by Theorem A. Let $(P_1, u) \in g/P$, where $P_1 \in \mathcal{P}$, $u : P_1 \longrightarrow P$ the morphism in $\mathcal{C} \subset Q\mathcal{M}$ given by the diagram $P_1 \twoheadleftarrow P_1' \rightarrowtail P$; a morphism $(P_2, v) \longrightarrow (P_1, u)$ in $g/P$ is given by a diagram

where the top row defines $u$, the bottom row defines $v$, and the left vertical column defines a morphism in $Q\mathcal{P}$. The commutativity of

$$P_1 \xrightarrow{u} P$$
$$\uparrow \quad \nearrow v$$
$$P_2$$

in $Q\mathcal{M}$ is expressed by the existence of an isomorphism $w$ making the diagram commute. Since $P_2 \twoheadleftarrow P_3 \rightarrowtail P_1$ is an arrow in $Q\mathcal{P}$, $P_3 \rightarrowtail P_1$ and $P_3 \twoheadrightarrow P_2$ are admissible mono and epi in $\mathcal{P}$, respectively.

Given an object $(p_1, u) \in g/P$, with $u$ represented by $P_1 \twoheadleftarrow_q P_1' \rightarrowtail_i P$, we can associate to it the $\mathcal{M}$-admissible layer (image $i$, $i(\ker q)$) of subobjects of $P$, (where an $\mathcal{M}$-admissible layer $(P_1, P_2)$ of subobjects in $P$ is a pair of subobjects $P_2 \subset P_1 \subset P$, such that $P_2 \rightarrowtail P_1$, $P_1 \rightarrowtail P$ are admissible monomorphisms in $\mathcal{M}$). Given an arrow $(P_2, v) \longrightarrow (P_1, u)$, if $(P_1', P_1'')$ and $(P_2', P_2'')$ are the layers associated to $(P_1, u)$ and $(P_2, v)$, respectively, then there are inclusions $P_1'' \subset P_2'' \subset P_2' \subset P_1' \subset P$.

Thus, $g/P$ is equivalent to the partially ordered set (considered as a category in the usual way—see Appendix B) of $\mathcal{M}$-admissible layers $(P', P'')$ in $P$ such that $P'/P'' \in \mathcal{P}$, i.e., $P'' \hookrightarrow P'$ is $\mathcal{P}$-admissible. The partial order is given by $(P_1', P_1'') \leq (P_2', P_2'') \iff P_2'' \subset P_1'' \subset P_1' \subset P_2'$ (so that there is a (unique) morphism $(P_1', P_1'') \longrightarrow (P_2', P_2'')$ in $J$) and $P_1' \hookrightarrow P_2'$ is $\mathcal{P}$-admissible (so that all the inclusions $P_2'' \hookrightarrow P_1'' \hookrightarrow P_1' \hookrightarrow P_2'$ are $\mathcal{P}$-admissible).

We have inequalities (i.e., arrows) in $J$

$$(P', P'') \leq (P', 0) \geq (0, 0),$$

since $0 \hookrightarrow P'$, $P'' \hookrightarrow P'$ are $\mathcal{P}$-admissible. These can be viewed as natural transformations of functors $J \longrightarrow J$

$$1_J \longrightarrow h \longleftarrow 0$$

where $1_J$ is the identity, 0 is the constant functor with value $(0,0)$, and $h(P', P'') = (P', 0)$. Thus the identity functor of $J$ is homotopic to the constant functor 0, and so $J$ is contractible. This proves that $g$ is a homotopy equivalence.

To prove $f$ is a homotopy equivalence, it suffices to prove that $M \setminus f$ is contractible $\forall M \in \mathcal{QM}$. If $\mathcal{F} = M \setminus f$, then $\mathcal{F}$ consists of pairs $(P, u)$ with $P \in \mathcal{P}$, $u : M \longrightarrow P$ a morphism in $\mathcal{QM}$. Let $\mathcal{F}'$ be the full subcategory of pairs where $u = q^!$ for some admissible mono $q : P \twoheadrightarrow M$. If $X = (P, u) \in \mathcal{F}$, write $u = i_! \circ q^!$ where $q : \bar{P} \twoheadrightarrow M$, $i : \bar{P} \rightarrowtail P$. Since $\mathcal{P}$ is closed under taking subobjects, $\bar{P} \in \mathcal{P}$, so that $\bar{X} = (\bar{P}, q^!) \in \mathcal{F}'$, and $i_!$ gives a map $\bar{X} \longrightarrow X$ in $\mathcal{F}$. From the uniqueness (up to isomorphism) of the factorization $u = i_! \circ q^!$ we verify easily that $\bar{X} \longrightarrow X$ is universal among morphisms $X' \longrightarrow X$ in with $X' \in \mathcal{F}'$. Hence $\mathcal{F}' \hookrightarrow \mathcal{F}$ has a right adjoint, and so is a homotopy equivalence.

Now $\mathcal{F}'^{\mathrm{op}}$ is equivalent to the category of pairs $(P, j)$ where $P \in \mathcal{P}$, $j : P \twoheadrightarrow M$ an admissible epi in $\mathcal{M}$, where a morphism $(P, j) \longrightarrow (P', j')$ is an admissible epi $P \twoheadrightarrow P'$ in $\mathcal{M}$ making

$$\begin{array}{ccc} P & \xrightarrow{v} & f(X) \\ {}_{j}\searrow & & \swarrow_{j'} \\ & M & \end{array}$$

commute. There exists at least one admissible epi $j_0 : P_0 \twoheadrightarrow M$, by the hypothesis of (6.11); fix one such. Given any admissible epi $j : P \twoheadrightarrow M$ in $\mathcal{M}$, we have a diagram of admissible epis

$$\begin{array}{ccc} P_0 & \xrightarrow{j_0} & M \\ \uparrow & & \uparrow \\ P_0 \times_M P & \longrightarrow & P \end{array}$$

where $\ker(P_0 \times_M P \twoheadrightarrow P_0) \cong \ker(P \twoheadrightarrow M) \in \mathcal{P}$, so that $P_0 \times_M P \in \mathcal{P}$. Let $k : \mathcal{F}' \longrightarrow \mathcal{F}'$ be the functor $k(P, j) = (P_0 \times_M P, j_0 \circ j_1)$. Then the above can be viewed as giving natural transformations from $k$ to the identity functor of $\mathcal{F}'$, and to the constant functor with value $(P_0, j_0)$. Hence the identity functor of $\mathcal{F}'$ is homotopic to a constant functor, and so $\mathcal{F}'$ is contractible. This completes the proof of (6.11), and hence (4.6). The proof of Corollary (4.7) is left to the reader.

Next, we prove the dévissage theorem.

**Theorem (4.8).** *Let $\mathcal{A}$ be an Abelian category, $\mathcal{B} \subset \mathcal{A}$ a full Abelian subcategory which is closed under taking subobjects, quotients and finite* products in $\mathcal{A}$. Suppose each object $M \in \mathcal{A}$ has a finite filtration $0 = M_0 \subset M_1 \subset \cdots \subset M_n = M$ (where $n$ may depend on $M$) such that $M_i/M_{i-1} \in \mathcal{B} \ \forall i > 0$. Then $Q\mathcal{B} \longrightarrow Q\mathcal{A}$ is a homotopy equivalence; hence $K_i(\mathcal{B}) \cong K_i(\mathcal{A})$.

**Proof.** Let $f : Q\mathcal{B} \longrightarrow Q\mathcal{A}$ be the inclusion; by Theorem A, it suffices to prove that $f/M$ is contractible for each $M \in \mathcal{A}$. The category $f/M$ consists of pairs $(N, u)$ with $N \in \mathcal{B}$, $u : N \longrightarrow M$ a morphism in $Q\mathcal{A}$. Let $N \leftarrow M' \rightarrowtail M$ be a diagram giving $u$; there is a unique such diagram with $M' \subset M$, and we then have an associated layer $(M', M'')$ in $M$, where $M'' = \ker(M' \twoheadrightarrow N)$. Thus $f/M$ is equivalent to the partially ordered set $J(M)$ of $\mathcal{A}$-admissible layers $(M', M'')$ in $M$ with $M'/M'' \in \mathcal{B}$. Since $M$ has a finite filtration $0 = M_0 \subset M_1 \subset \cdots \subset M_n = M$ with $M_i/M_{i-1} \in \mathcal{B}$, it suffices to show that if $M' \subset M$ with $M/M' \in \mathcal{B}$, then $J(M') \longrightarrow J(M)$ is a homotopy equivalence.

If $(M_1, M_2) \in J(M)$, then $(M_1, M_2 \cap M') \in J(M)$, and $(M_1 \cap M', M_2 \cap M') \in J(M')$, since

$$(M_1 \cap M')/(M_2 \cap M') \subset M_1/(M_2 \cap M') \subset (M_1/M_2) \oplus (M/M') \in \mathcal{B}.$$

Let $r : J(M) \longrightarrow J(M')$, $s : J(M) \longrightarrow J(M)$ be the functors

$$r(M_1, M_2) = (M_1 \cap M', M_2 \cap M'), \quad s(M_1, M_2) = (M_1, M_2 \cap M'),$$

and let $i : J(M') \hookrightarrow J(M)$ be the inclusion. Then $r \circ i = 1_{J(M')}$. Also, we have inequalities in $J(M)$

$$(M_1 \cap M', M_2 \cap M') \leq (M_1, M_2 \cap M') \geq (M_1, M_2)$$

giving natural transformations $i \circ r \longleftarrow s \longrightarrow 1_{J(M)}$. Hence $i \circ r$ is homotopic to the identity, and so $i$ is a homotopy equivalence. This proves (4.8).

**Corollary (6.12).** *Let $\mathcal{A}$ be an Abelian category* (with a set of isomorphism classes) *such that every object of $\mathcal{A}$ has finite length. Let $\{X_j \mid j \in J\}$ be a set of representatives for the isomorphism classes of simple objects, and let $D_j = \mathrm{End}(X_j)^{\mathrm{op}}$. Then $K_i(\mathcal{A}) \cong \coprod_{j \in J} K_i(D_j)$.*

**Proof.** Let $\mathcal{B} \subset \mathcal{A}$ be the full Abelian subcategory of semi-simple objects. From (4.8), $K_i(\mathcal{B}) \cong K_i(\mathcal{A})$. Since $K_i$ commutes with finite products and filtered direct limits, we reduce to the case when $\mathcal{A} = \mathcal{B}$ has a single object $X$ up to isomorphism. But then $M \mapsto \mathrm{Hom}_{\mathcal{A}}(X, M)$ is an equivalence of $\mathcal{A}$ with $\mathcal{P}(D)$, $D = \mathrm{End}_{\mathcal{A}}(X)^{\mathrm{op}}$, where $\mathcal{P}(D)$ is the category of finite dimensional left $D$-vector spaces. This proves the corollary.

The last aim of the chapter is the proof of the localization theorem (see Appendix B for the construction of the quotient Abelian category of an Abelian category modulo a Serre subcategory).

**Theorem (4.9).** *Let $\mathcal{B}$ be a Serre subcategory of an Abelian category $\mathcal{A}$, with quotient Abelian category $\mathcal{A}/\mathcal{B}$; let $e : \mathcal{B} \longrightarrow \mathcal{A}$, $s : \mathcal{A} \longrightarrow \mathcal{A}/\mathcal{B}$ be the natural functors. Then there is a long exact sequence*

$$\cdots \longrightarrow K_{i+1}(\mathcal{A}/\mathcal{B}) \longrightarrow K_i(\mathcal{B}) \xrightarrow{e_*} K_i(\mathcal{A}) \xrightarrow{s_*} K_i(\mathcal{A}/\mathcal{B}) \longrightarrow K_{i-1}(\mathcal{B}) \cdots$$

$$\xrightarrow{e_*} K_0(\mathcal{A}) \xrightarrow{s_*} K_0(\mathcal{A}/\mathcal{B}) \longrightarrow 0.$$

*Further, the sequence is functorial for exact functors $(\mathcal{A}, \mathcal{B}) \longrightarrow (\mathcal{A}', \mathcal{B}')$.*

**Corollary (6.13).** *Let $A$ be a Dedekind domain with quotient field $F$. Then there is a long exact sequence*

$$\cdots \longrightarrow K_{i+1}(F) \longrightarrow \coprod_{\mathcal{M}} K_i(A/\mathcal{M}) \longrightarrow K_i(A) \longrightarrow K_i(F) \longrightarrow \cdots$$

$$\longrightarrow K_0(A) \longrightarrow K_0(F) \longrightarrow 0,$$

*where $\mathcal{M}$ runs over the maximal ideals of $A$.*

**Proof.** Let $\mathcal{A} = \mathcal{M}(A)$, $\mathcal{B} \subset \mathcal{A}$ the full subcategory of torsion $A$-modules (i.e., modules $M$ with $F \otimes_A M = 0$). Then $\mathcal{A}/\mathcal{B}$ is equivalent to the category $\mathcal{P}(F)$ of finite dimensional $F$-vector spaces, while $K_i(\mathcal{B}) \cong \coprod_{\mathcal{M}} K_i(A/\mathcal{M})$ from Corollary (6.12) above. Since $A$ is a Dedekind domain, any $M \in \mathcal{A}$ has a projective dimension $\leq 1$ over $A$, so that $K_i(\mathcal{A}) \cong K_i(\mathcal{P}(A)) = K_i(A)$ by the resolution of Theorem (4.6). Hence the result follows from (4.9).

**Proof of Theorem (4.9).** Fix a zero object $0 \in \mathcal{A}$, and let $0$ also denote its image in $\mathcal{A}/\mathcal{B}$, which is a zero object in $\mathcal{A}/\mathcal{B}$. Now $\mathcal{B}$ is the full subcategory of objects $M \in \mathcal{A}$ such that $s(M) \cong 0$ in $\mathcal{A}/\mathcal{B}$; since $0$ is a zero object, there is a unique such isomorphism. Thus the composite functor $Q\mathcal{B} \xrightarrow{Qe} Q\mathcal{A} \xrightarrow{Qs} Q(\mathcal{A}/\mathcal{B})$ is isomorphic to the constant functor with value $0$, and $Qe$ factors as

$$Q\mathcal{B} \longrightarrow 0 \setminus Qs \longrightarrow Q\mathcal{A}$$

$$M \mapsto (M, 0 \xrightarrow{\cong} s(M)), \; (N, u) \mapsto N.$$

By Theorem B, it suffices to prove the following claims, in order to prove (4.9):

(a) for every $u : V' \longrightarrow V$ in $Q(\mathcal{A}/\mathcal{B})$,

$$u^* : V \setminus Qs \longrightarrow V' \setminus Qs$$

is a homotopy equivalence, and

(b) the functor $QB \longrightarrow 0 \setminus Qs$ is a homotopy equivalence.

Indeed, (a) together with Theorem B implies that

$$\begin{array}{ccc} 0 \setminus Qs & \longrightarrow & Q\mathcal{A} \\ \downarrow & & \downarrow Qs \\ 0 \setminus Q(\mathcal{A}/\mathcal{B}) & \longrightarrow & Q(\mathcal{A}/\mathcal{B}) \end{array}$$

is homotopy Cartesian, where $0 \setminus Q(\mathcal{A}/\mathcal{B})$ has 0 as an initial object and is hence contractible. By (b), we deduce that the following square is homotopy Cartesian (which implies (4.9))

$$\begin{array}{ccc} Q\mathcal{B} & \xrightarrow{Qe} & Q\mathcal{A} \\ \downarrow & & \downarrow \\ 0 \setminus Q(\mathcal{A}/\mathcal{B}) & \longrightarrow & Q(\mathcal{A}/\mathcal{B}). \end{array}$$

To prove (a) for $u = i_! \circ q^!$ where $i$ is an admissible mono, and $q$ is admissible epi, it suffices to prove it when $u = i_!$ or $u = q^!$. Now by Corollary (6.3), we have isomorphisms

$$Q(\mathcal{A}/\mathcal{B}) \cong Q(\mathcal{A}/\mathcal{B})^{\mathrm{op}} \cong Q(\mathcal{A}^{\mathrm{op}}/\mathcal{B}^{\mathrm{op}}),$$

under which the roles of admissible monos and epis are interchanged. Thus $u = i_! \circ q^!$ gets transformed into $q'_! \circ i'^!$ where $q', i'$ are the dual arrows in $(\mathcal{A}/\mathcal{B})^{\mathrm{op}}$ to $q, i$ respectively (in $\mathcal{A}/\mathcal{B}$). Thus if we prove (a) for all $u = i_!$, for all quotients $\mathcal{A}/\mathcal{B}$ of an Abelian category by a Serre subcategory, we deduce (a) for all $u = q^!$ by passing to the opposite category $(\mathcal{A}/\mathcal{B})^{\mathrm{op}}$. Hence it suffices to prove (a) for all $u = i_!$. For any $V \in \mathcal{A}/\mathcal{B}$, let $i_V : 0 \rightarrowtail V$; then if $i : V' \rightarrowtail V$, $i_! \circ i_{V'!} = i_{V!}$. Hence it suffices to prove (a) for all $u = i_{V!}$, for all $V \in \mathcal{A}/\mathcal{B}$.

Let $\mathcal{F}_V$ be the full subcategory of $V \setminus Qs$ consisting of pairs $(M, u)$ such that $u : V \longrightarrow s(M)$ is an isomorphism in $Q(\mathcal{A}/\mathcal{B})$. Clearly $\mathcal{F}_0$ is isomorphic to $Q\mathcal{B}$; thus (b) is a particular case of:

**Lemma (6.14).** *The inclusion $\mathcal{F}_V \longrightarrow V \setminus Qs$ is a homotopy equivalence.*

**Proof.** Let $f : \mathcal{F}_V \longrightarrow V \setminus Qs$ be the inclusion functor. By Theorem A it suffices to prove that $f/(M, u)$ is contractible for any $(M, u)$ in $V \setminus Qs$. The arrow $u : V \longrightarrow s(M)$ in $Q(\mathcal{A}/\mathcal{B})$ corresponds to a unique diagram $V \xleftarrow{j} V_1 \xrightarrowtail{i} s(M)$ in $\mathcal{A}/\mathcal{B}$; let $V_0 = \ker j$. This gives an $\mathcal{A}/\mathcal{B}$-admissible layer $(i(V_1), i(V_0))$ in $s(M)$. Choosing $i$ to be an inclusion (there is a unique diagram representing $u$ with this property), $(V_1, V_0)$ is an $\mathcal{A}/\mathcal{B}$-admissible layer in $s(M)$.

6. Proofs of the Theorems of Chapter 4        117

An object of $f/(M,u)$ is a triple $(N,v,w)$ consisting of $(N,v) \in \mathcal{F}_V$ and a morphism $w : N \longrightarrow M$ in $\mathcal{QA}$ such that

$$\begin{array}{ccc} V & \xrightarrow{u} & s(M) \\ & \searrow v \quad \nearrow s(w) & \\ & s(N) & \end{array}$$

commutes in $Q(\mathcal{A}/\mathcal{B})$ (thus $w$ determines an arrow $(N,v) \longrightarrow (M,u)$ in $V \setminus Qs$). Since $(N,v) \in \mathcal{F}_V$, $v : V \xrightarrow{\cong} s(N)$. Let $N \leftarrow M_1 \rightarrowtail M$ be the unique diagram representing $w$ such that $M_1 \longrightarrow M$ is a subobject, and let $M_0 = \ker(M_1 \twoheadrightarrow N)$; thus we have an $\mathcal{A}$-admissible layer $(M_1, M_0)$ in $M$ associated to $w$. Since $v$ is an isomorphism, $(s(M_1), s(M_0)) = (V_1, V_0)$ as layers in $s(M)$. The layer $(M_1, M_0)$ determines $(N, v, w)$ up to isomorphism; hence $f \setminus (M, u)$ is equivalent to the partially ordered set $J$ of $\mathcal{A}$-admissible layers $(M_1, M_0)$ in $M$ such that $(s(M_1), s(M_0)) = (V_1, V_0)$, with $(M_1, M_0) \leq (M'_1, M'_0) \iff M'_0 \subset M_0 \subset M_1 \subset M'_1$.

We claim that $J$ is filtering (so that by (3.9), $J$ is contractible), i.e., $J$ is non-empty, and is directed (i.e., any two elements of $J$ have a common upper bound). Every subobject $V_1$ of $s(M)$ is (up to isomorphism) of the form $s(N)$ for some subobject $N$ of $M$; indeed $V_1 = s(N)$ for some object $N \in \mathrm{Ob}\,\mathcal{A} = \mathrm{Ob}(\mathcal{A}/\mathcal{B})$, and $V_1 \longrightarrow s(M)$ is induced by a morphism $N' \longrightarrow M'$ in $\mathcal{A}$, where $N' \subset N$, $M \twoheadrightarrow M'$, and $N/N'$, $\ker(M \twoheadrightarrow M') \in \mathcal{B}$. Then $N' \times_{M'} M \twoheadrightarrow N'$ has kernel in $\mathcal{B}$, so $s(N) \cong s(N') \cong s(N' \times_{M'} M)$. So $s(N) \cong V_1$ is the image of an arrow $u : N \longrightarrow M$ in $\mathcal{A}$, without loss of generality. Since $s(u)$ is mono, $\ker u \in \mathcal{B}$, and $N \twoheadrightarrow \mathrm{im}\,u$ maps to an isomorphism in $\mathcal{A}/\mathcal{B}$. Hence if we let $N = \mathrm{im}\,u$, then $N \subset M$ and $s(N) \cong V_1$. Thus any layer $(V_1, V_0)$ in $s(M)$ is of the form $(s(M_1), s(M_0))$ for some layer $(M_1, M_0)$ in $M$. Hence $J$ is non-empty.

If $(M_1 M_0)$, $(M'_1, M'_0)$ are layers in $J$, then clearly $(M_1 + M'_1, M_0 \cap M'_0) \in J$ is a common upper bound. Hence $J$ is directed, and hence contractible, proving (6.14).

To prove (a) above for $u = i_{V!}$, we must show that $(i_{V!})^* : V \setminus Qs \longrightarrow 0 \setminus Qs$ is a homotopy equivalence. This is done in a number of steps. First, we introduce some notation. For any $N \in \mathcal{A}$, let $\mathcal{E}_N$ be the category whose objects are pairs $(M, h)$ with $M \in \mathcal{A}$, $h : M \longrightarrow N$ a morphism in $\mathcal{A}$ such that $s(h)$ is an isomorphism, i.e., $\ker h$, $\mathrm{coker}\,h \in \mathcal{B}$. A morphism $(M, h) \longrightarrow (M', h')$ in $\mathcal{E}_N$ is defined to be a morphism $u : M \longrightarrow M'$ in $\mathcal{QA}$ such that if $u = i_! \circ j^!$, then the diagram in $\mathcal{A}$

$$\begin{array}{ccc} M_1 & \xrightarrow{i} & M' \\ j \downarrow & & \downarrow h' \\ M & \xrightarrow{h} & N \end{array}$$

commutes; this condition does not depend on the specific factorization $u = i_! \circ j^!$, since for an isomorphic factorization $u = i'_! \circ j'^!$, there is an isomorphism $w$ making the following diagram commute:

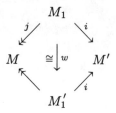

Associated to a morphism $(M, h) \longrightarrow (M, h')$, we have a morphism $\ker h \longrightarrow \ker h'$ in $Q\mathcal{B}$, given by the diagram in $\mathcal{B}$

$$\ker h \twoheadleftarrow j^{-1}(\ker h) \rightarrowtail \ker h'$$

corresponding to $u = i_! \circ j^!$; again, the morphism in $Q\mathcal{B}$ is independent of the factorization $u = i_! \circ j^!$. Thus, there is a well-defined functor $k_N : \mathcal{E}_N \longrightarrow Q\mathcal{B}$, $k_N(M, h) = \ker h$. We first prove that $k_N$ is a homotopy equivalence, in two steps.

Let $\mathcal{E}'_N \subset \mathcal{E}_N$ be the full subcategory of pairs $(M, h)$ where $h$ is an admissible epimorphism in $\mathcal{A}$, and let $k'_N : \mathcal{E}'_N \longrightarrow Q\mathcal{B}$ be the functor obtained by restricting $k_N$.

**Lemma (6.15).** $k'_N : \mathcal{E}'_N \longrightarrow Q\mathcal{B}$ *is a homotopy equivalence.*

**Proof.** By Theorem A, it suffices to prove $k'_N/T$ is contractible for any $T \in Q\mathcal{B}$. Let $\mathcal{C} = k'_N/T$, so that $\mathcal{C}$ is the category whose objects are triples $(M, h, u)$ with $(M, h) \in \mathcal{E}'_N$, and $u : \ker h \longrightarrow T$ a morphism in $Q\mathcal{B}$. Let $\mathcal{C}' \subset \mathcal{C}$ be the full subcategory of triples $(M, h, u)$ with $u = q^!$ for $q : T \twoheadrightarrow \ker h$ in $\mathcal{B}$. Given $X = (M, h, u) \in \mathcal{C}$, with $u = j^! \circ i_!$, $i : \ker h \rightarrowtail T_0$, $j : T \twoheadrightarrow T_0$, let $(i_*M, \bar{h}) \in \mathcal{E}'_N$ be defined by pushout along $i$,

$$\begin{array}{ccccc} \ker h & \rightarrowtail & M & \twoheadrightarrow & N \\ {\scriptstyle i}\downarrow & & \downarrow & & \parallel \\ T_0 & \rightarrowtail & i_*M & \underset{\bar{h}}{\twoheadrightarrow} & N \end{array}$$

Let $\bar{X} = (i_*M, \bar{h}, j^!) \in \mathcal{C}'$; there is an evident map $X \longrightarrow \bar{X}$ in $\mathcal{C}$. We claim this is universal among arrows $X \longrightarrow X'$ with $X' \in \mathcal{C}'$. This is easy to verify from the definitions of $\mathcal{C}$, $\mathcal{C}'$ and the uniqueness up to isomorphism of the factorization $u = j^! \circ i_!$, and is analogous to the argument given in the proof of (6.9); the details are left to the reader.

## 6. Proofs of the Theorems of Chapter 4

Thus, the inclusion of $\mathcal{C}'$ in $\mathcal{C}$ has a left adjoint, and is hence a homotopy equivalence. But $(N, 1_N, q_T^!)$ (where $1_N : N \longrightarrow N$ is the identity, $q_T : T \twoheadrightarrow 0$) is an initial object of $\mathcal{C}'$, from the diagram (in $\mathcal{A}$)

$$\begin{array}{ccccccc}
T & \twoheadrightarrow & \ker h & \longrightarrow & M & \xrightarrow{h} & N \\
& \searrow & \downarrow & & \downarrow h & & \| \\
& & 0 & \longrightarrow & N & \xrightarrow{1_N} & N
\end{array}$$

(note that we need $h$ to be an admissible epimorphism, for this argument to work, since $h$ must simultaneously give a morphism in $\mathcal{A}$ and $Q\mathcal{A}$). Thus $\mathcal{C}'$, and hence $\mathcal{C}$, are contractible. This proves the lemma.

**Lemma (6.16).** $k_N : \mathcal{E}_N \longrightarrow Q\mathcal{B}$ *is a homotopy equivalence.*

**Proof.** From the previous lemma, it suffices to check that $\mathcal{E}'_N \hookrightarrow \mathcal{E}_N$ is a homotopy equivalence. Let $\mathcal{I}$ be the ordered set of subobjects $I \subset N$ such that $N/I \in \mathcal{B}$. There is a functor $f : \mathcal{E}_N \longrightarrow \mathcal{I}$ given by $(M, h) \longrightarrow \operatorname{im} h$. We claim $f$ makes $\mathcal{E}_N$ fibered over $\mathcal{I}$; indeed the fiber $f^{-1}(I) = \mathcal{E}'_I$, and for any arrow $J \longrightarrow I$ in $\mathcal{I}$, corresponding to the inclusions $J \subset I \subset N$, the base change functor $\mathcal{E}'_I \longrightarrow \mathcal{E}'_J$ is given by

$$J \times_I ? : (M \twoheadrightarrow I) \mapsto (J \times_I M \twoheadrightarrow J).$$

We claim

$$\begin{array}{ccc}
\mathcal{E}'_I & \xrightarrow{J \times_I ?} & \mathcal{E}'_J \\
{}_{k'_I}\searrow & & \swarrow_{k'_J} \\
& Q\mathcal{B} &
\end{array}$$

commutes up to a natural transformation (so that the induced diagram of classifying spaces commutes up to homotopy). Indeed, there is a natural isomorphism

$$\ker(M \twoheadrightarrow I) \cong \ker(J \times_I M \twoheadrightarrow J).$$

Since $k'_I, k'_J$ are homotopy equivalences, by (6.15), $J \times_I ?$ is also a homotopy equivalence. By Theorem B,

$$\begin{array}{ccc}
\mathcal{E}'_N & \longrightarrow & \mathcal{E}_N \\
\downarrow & & \downarrow f \\
\{N\} & \longrightarrow & \mathcal{I}
\end{array}$$

is homotopy Cartesian. Since $N \in \mathcal{I}$ is a final object, $\mathcal{I}$ is contractible. Hence $\mathcal{E}'_N \longrightarrow \mathcal{E}_N$ is a homotopy equivalence, proving the lemma.

We now want to prove that if $s(N) \cong V$, then $\mathcal{F}_V$ is homotopy equivalent to $\mathcal{E}_N$; this will give homotopy equivalences $V \setminus Qs \cong \mathcal{F}_V \cong \mathcal{E}_N \cong Q\mathcal{B}$, so that the homotopy type of $V \setminus Qs$ is independent of $V$.

**Lemma (6.17).** *Let $g : N \longrightarrow N'$ be a map in $\mathcal{A}$ such that $s(g) : s(N) \longrightarrow s(N')$ is an isomorphism. Then $g_* : \mathcal{E}_N \longrightarrow \mathcal{E}_{N'}$, $(M, h) \longrightarrow (M, g \circ h)$ is a homotopy equivalence.*

**Proof.** Given $(M, h) \in \mathcal{E}_N$ there is a natural admissible monomorphism $\ker h \rightarrowtail \ker g \circ h$. This defines a natural transformation $k_N \longrightarrow k_{N'} \circ g_*$ of functors $\mathcal{E}_N \longrightarrow Q\mathcal{B}$. Since $k_N, k_{N'}$, are homotopy equivalences, so is $g_*$. This proves (6.17).

Given $V \in \mathcal{A}/\mathcal{B}$, let $\mathcal{I}_V$ be the category of pairs $(N, \phi)$ with $N \in \mathcal{A}$, $\phi : s(N) \xrightarrow{\cong} V$ an isomorphism in $\mathcal{A}/\mathcal{B}$, where a morphism $(N, \phi) \longrightarrow (N', \phi')$ is a morphism $g : N \longrightarrow N'$ in $\mathcal{A}$ such that the following diagram commutes:

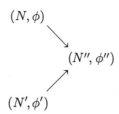

**Lemma (6.18).** *$\mathcal{I}_V$ is a filtering category for any $V \in \mathcal{A}/\mathcal{B}$.*

**Proof.** We must show (i) $\mathrm{Ob}\,\mathcal{I}_V$ is non-empty.

(ii) Given $(N, \phi), (N', \phi') \in \mathcal{I}_V$, there exists $(N'', \phi'') \in \mathcal{I}_V$ and a diagram in $\mathcal{I}_V$

$$(N, \phi) \searrow$$
$$(N'', \phi'')$$
$$(N', \phi') \nearrow$$

(iii) Given by two arrows $(N, \phi) \rightrightarrows (N', \phi')$ in $\mathcal{I}_V$, there is an arrow $(N', \phi') \longrightarrow (N'', \phi'')$ in $\mathcal{I}_V$ such that the two composite arrows $(N, \phi) \rightrightarrows (N'', \phi'')$ are equal (in earlier situations where we encountered filtering categories, these were based on partially ordered sets, so that there was at most one morphism between any two objects; hence (iii) was always satisfied and was not mentioned explicitly).

Since $\mathrm{Ob}\,\mathcal{A} = \mathrm{Ob}\,\mathcal{A}/\mathcal{B}$, clearly (i) holds; indeed there is a unique pair $(N, 1_{s(N)}) \in \mathcal{I}_V$. Given any two objects $(N_1, \phi_1), (N_2, \phi_2)$ in $\mathcal{I}_V$, where $\phi_i : s(N_i) \longrightarrow V$ is an isomorphism, then $\phi_i$ is induced by a map $N_i' \longrightarrow N/N_i''$ for subobjects $N_i' \subset N_i$, $N_i'' \subset N$ with $N_i/N_i'$, $N_i'' \in \mathcal{B}$.

## 6. Proofs of the Theorems of Chapter 4

Replacing $N_1'', N_2''$ by $N'' = N_1'' + N_2''$, we may assume $N_1'' = N_2'' = N''$. Consider the pushout diagram

$$\begin{array}{ccc} N_i & \longrightarrow & \tilde{N}_i \\ \uparrow & & \uparrow \\ N' & \longrightarrow & N/N'' \end{array}$$

(which defines $\tilde{N}_i$); then $N/N'' \hookrightarrow \tilde{N}_i$ with cokernel contained in $\mathcal{B}$, so that $s(N) \cong s(\tilde{N}_i)$. Finally, let $N$ be the pushout

$$\begin{array}{ccc} N/N'' & \longrightarrow & \tilde{N}_1 \\ \downarrow & & \downarrow \\ \tilde{N}_2 & \longrightarrow & \tilde{N} \end{array}$$

Then $N \longrightarrow \tilde{N}$ induces $s(N) \cong s(\tilde{N})$; let $\tilde{\phi}$ be the inverse isomorphism. By construction, we have a diagram

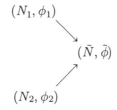

i.e., the composites $g_i : N_i \longrightarrow \tilde{N}_i \longrightarrow \tilde{N}$ yield commutative diagrams

$$\begin{array}{ccc} s(N_i) & \xrightarrow{s(g_i)} & s(\tilde{N}) \\ & \phi_i \searrow \quad \swarrow \tilde{\phi} & \\ & V & \end{array}$$

This follows from the diagrams

$$\begin{array}{ccccc} N_i & \longrightarrow & \tilde{N}_i & \longrightarrow & \tilde{N} \\ \uparrow & & \uparrow & & \\ N_i' & \longrightarrow & N/N'' & & \end{array}$$

where all arrows become isomorphisms on applying the functor $s$. This verifies (ii).

Finally, let $(N_1, \phi_1) \rightrightarrows (N_2, \phi_2)$ be two arrows in $\mathcal{I}_V$ given by $g_i : N_1 \longrightarrow N_2$, $i = 1, 2$. Let $N_3 = N_2/\text{im}(g_1 - g_2)$; then $s(g_1) = s(g_2) = \phi_2^{-1} \circ \phi_1$, so that $s(g_1 - g_2) = 0$, i.e., $\text{im}(g_1 - g_2) \in \mathcal{B}$, and $s(N_2) \cong s(N_3)$. Let $\phi_3 : s(N_3) \longrightarrow V$, induced by this isomorphism (and $\phi_2$). Clearly the two composite arrows $(N_1, \phi_1) \rightrightarrows (N_3, \phi_3)$ are equal. This proves (iii), and completes the proof of the lemma.

We claim the $\mathcal{E}_N$ give a directed system of categories indexed by $\mathcal{I}_V$. Indeed, associate $\mathcal{E}_N$ to $(N,\phi) \in \mathcal{I}_V$; if $g : (N,\phi) \longrightarrow (N',\phi')$ is an arrow in $\mathcal{I}_V$, associate to it the functor $g_* : \mathcal{E}_N \longrightarrow \mathcal{E}_{N'}$ of (6.17), $g_*(M,h) = (M, g \circ h)$. Then $(g_1 \circ g_2)_* = g_{1*} \circ g_{2*}$ follows from the associativity of composition of morphisms in a category.

Next, for each $(N,\phi) \in \mathcal{I}_V$, we have a functor $p_{(N,\phi)} = \mathcal{E}_N \longrightarrow \mathcal{F}_V$ given by $p_{(N,\phi)}(M,h) = (M, s(h)^{-1} \circ \phi^{-1})$, where we have isomorphisms (in $\mathcal{A}/\mathcal{B}$) $s(M) \xrightarrow{s(h)} s(N) \xrightarrow{\phi} V$. For any morphism $g : (N,\phi) \longrightarrow (N',\phi')$ in $\mathcal{I}_V$, we claim $p_{(N',\phi')} \circ g_* = p_{(N,\phi)}$. This follows from the identity.

$$s(g \circ h)^{-1} \circ \phi'^{-1} = s(h)^{-1} \circ s(g)^{-1} \circ \phi'^{-1} = s(h)^{-1} \circ \phi^{-1}.$$

Hence we obtain a functor on the direct limit category

$$p := \varinjlim_{\mathcal{I}_V}\{(N,\phi) \mapsto \mathcal{E}_N\} \longrightarrow \mathcal{I}_V.$$

**Lemma (6.19).** *$p$ is an isomorphism of categories* (i.e., $p$ is bijective on objects and on Hom-sets).

**Proof.** If $(M,\theta) \in \mathrm{Ob}\,\mathcal{F}_V$, where $\theta : V \longrightarrow s(M)$ is an isomorphism, then $(M,\theta) = p_{(M,\theta^{-1})}(M, 1_M)$. Hence $p$ is surjective on objects. If $p_{(N,\phi)}(M,h) = p_{(N,\phi)}(M',h')$, then clearly $M = M'$, and $s(h)^{-1} \circ \phi^{-1} = s(h')^{-1} \circ \phi^{-1}$, i.e., $s(h) = s(h')$. Hence if $N' = N/im(h - h')$, we obtain an arrow $g : (N,\phi) \longrightarrow (N',\phi')$ in $\mathcal{I}_V$ (where $\phi'$ is induced by $\phi$ and $s(N) \cong s(N')$) such that $g_*(M,h) = g_*(M,h')$. Hence $p$ is injective on objects (we have used condition (ii), verified in (6.18) for $\mathcal{I}_V$).

Next, let $(M,h) \rightrightarrows (M',h')$ be two morphisms in $\mathcal{E}_N$ which yield the same morphism on applying $p_{(N,\phi)}$. Then the two morphisms $M \rightrightarrows M'$ in $Q\mathcal{A}$ are the same, and so define the same arrow in $\mathcal{E}_N$. Thus $p_{(N,\phi)}$, and hence $p$, are injective on arrows.

Lastly, let $(M,h), (M',h') \in \mathcal{E}_N$, and let

$$t : p_{(N,\phi)}(M,h) \longrightarrow p_{(N,\phi)}(M',h')$$

be an arrow in $\mathcal{F}_V$. Then $t$ is given by an arrow $w : M \longrightarrow M'$ in $Q\mathcal{A}$ such that the diagram below commutes in $Q(\mathcal{A}/\mathcal{B})$:

$$\begin{array}{ccc} s(M) & \xrightarrow{s(w)} & s(M') \\ & \nwarrow_{s(h)^{-1} \circ \phi^{-1}} \quad \nearrow_{s(h')^{-1} \circ \phi^{-1}} & \\ & V & \end{array}$$

In particular, $s(w)$ is an isomorphism. Hence if we factor $w = i_! \circ j^!$ where $j : M'' \twoheadrightarrow M$, $i : M'' \rightarrowtail M'$, then $s(j), s(i)$ are isomorphisms. For $w$ to yield a morphism in $\mathcal{E}_N$, the diagram below must commute:

$$\begin{array}{ccc} M'' & \xrightarrow{i} & M' \\ j \downarrow & & \downarrow h' \\ M & \xrightarrow{h} & N \end{array}$$

But $s(w) \circ s(h)^{-1} \circ \phi^{-1} = s(h') \circ \phi^{-1}$, since $w$ yields the morphism $t$ in $\mathcal{E}_N$. Hence

$$\begin{array}{ccc} s(M'') & \xrightarrow{s(i)} & s(M') \\ s(j) \downarrow & & \downarrow s(h') \\ s(M) & \xrightarrow{s(h)} & s(N) \end{array}$$

commutes, all arrows being isomorphisms. Thus $\text{im}(h \circ j - h' \circ i) \in \mathcal{B}$. Let $N' = N/\text{im}(h \circ j - h' \circ i)$, so that $s(N) \cong s(N')$, and let $\phi' : V \cong s(N')$ be induced by this isomorphism (and $\phi$). Let $g : (N, \phi) \longrightarrow (N', \phi')$ be the natural map. Then $w : M \longrightarrow M'$ yields an arrow $\tilde{w} : (M, g \circ h) \longrightarrow (M', g \circ h')$ in $\mathcal{E}_{N'}$, such that

$$p_{(N',\phi')}(\tilde{w}) = t : p_{(N',\phi')}(M, g \circ h) \longrightarrow p_{(N',\phi')}(M', g \circ h').$$

Hence $p$ is surjective on arrows. This proves (6.19).

**Lemma (6.20).** *For any $(N, \phi) \in \mathcal{I}_V$, $p_{(N,\phi)} : \mathcal{E}_N \longrightarrow \mathcal{F}_V$ is a homotopy equivalence.*

**Proof.** From (6.17), for any arrows $g : (N, \phi) \longrightarrow (N', \phi')$ in $\mathcal{I}_V$, $g_* : \mathcal{E}_N \longrightarrow \mathcal{E}_{N'}$ is a homotopy equivalence. Hence for any $(N, \phi)$,

$$\mathcal{E}_N \longrightarrow \varinjlim_{\mathcal{I}_V} \{(N', \phi') \mapsto \mathcal{E}_{N'}\}$$

is a homotopy equivalence, by (3.8). Hence the lemma follows from (6.19).

We now prove $(i_{V!})^* : V \setminus Qs \longrightarrow 0 \setminus Qs$ is a homotopy equivalence. Choose $(N, \phi) \in \mathcal{I}_V$, and consider the diagram

$$\begin{array}{ccc} \mathcal{E}_N & \xrightarrow{p_{(N,\phi)}} & \mathcal{F}_V \hookrightarrow V \setminus Qs \\ k_N \downarrow & & \downarrow (i_{V!})^* \\ Q\mathcal{B} & \xrightarrow{\cong} & \mathcal{F}_0 \hookrightarrow 0 \setminus Qs. \end{array}$$

This does not commute in general; the two composite functors are

$$(M, h) \xrightarrow{\cong \circ k_N} (\ker h, 0 \xrightarrow{\cong} s(\ker h))$$

$$(M, h) \xrightarrow{(i_{V!})^* \circ p_{(N,)}} (M, i_{s(M)!} : 0 \rightarrowtail s(M)).$$

If $i : \ker h \rightarrowtail M$, then $i_!$ gives a natural transformation between these two functors, so the induced diagram of classifying spaces commutes up to homotopy. Since all sides except $(i_{V!})^*$ are known to be homotopy equivalences, $(i_{V!})^*$ is one too. This completes the proof of the localization theorem (4.9), modulo the naturality of the resulting long exact sequence of $K$-groups.

If $f : (\mathcal{A}, \mathcal{B}) \longrightarrow (\mathcal{A}', \mathcal{B}')$ is an exact functor, where $\mathcal{B} \subset \mathcal{A}$, $\mathcal{B}' \subset \mathcal{A}'$ are Serre subcategories, then we have a commutative diagram of categories and functors

$$\begin{array}{ccccc} Q\mathcal{B} & \xrightarrow{Qe} & Q\mathcal{A} & \xrightarrow{Qs} & Q(\mathcal{A}/\mathcal{B}) \\ \downarrow{Qg} & & \downarrow{Qf} & & \downarrow{Qs(f)} \\ Q\mathcal{B}' & \xrightarrow{Qe'} & Q\mathcal{A}' & \xrightarrow{Qs'} & Q(\mathcal{A}'/\mathcal{B}') \end{array}$$

where $e' : \mathcal{B}' \longrightarrow \mathcal{A}'$, $s' : \mathcal{A}' \longrightarrow \mathcal{A}'/\mathcal{B}'$, and $g = f|_{\mathcal{B}}$. From the right hand square, we have a diagram of topological spaces

$$\begin{array}{ccccc} F(BQs) & \longrightarrow & BQ\mathcal{A} & \xrightarrow{BQs} & BQ(\mathcal{A}/\mathcal{B}) \\ \downarrow & & \downarrow{BQf} & & \downarrow{BQs(f)} \\ F(BQs') & \longrightarrow & BQ\mathcal{A}' & \xrightarrow{BQs'} & BQ(\mathcal{A}'/\mathcal{B}') \end{array}$$

including a diagram of long exact sequences of homotopy groups; here we fix a zero object $0 \in \mathrm{Ob}\,\mathcal{A} = \mathrm{Ob}\,Q\mathcal{A}$, and its images in $Q(\mathcal{A}/\mathcal{B})$, $Q\mathcal{A}'$, $Q(\mathcal{A}'/\mathcal{B}')$ as the respective base points, and compute the homotopy fibers over these points. The localization sequences are obtained from homotopy equivalences

$$\alpha : BQ\mathcal{B} \longrightarrow F(BQs), \quad \alpha' : BQ\mathcal{B}' \longrightarrow F(BQs')$$

obtained using the functors $Qe$, $Qe'$ respectively, so that we need to check that

$$\begin{array}{ccc} BQ\mathcal{B} & \xrightarrow{\alpha} & F(BQs) \\ \downarrow{BQg} & & \downarrow \\ BQ\mathcal{B}' & \xrightarrow{\alpha'} & F(BQs') \end{array}$$

commutes.

The composite functor $Q(s \circ e) : Q\mathcal{B} \longrightarrow Q(\mathcal{A}/\mathcal{B})$ maps every $M \in Q\mathcal{B}$ to an object isomorphic to $0 \in Q(\mathcal{A}/\mathcal{B})$; hence there is a canonical isomorphism of functors between $Q(s \circ e)$ and the constant functor $Q\mathcal{B} \longrightarrow \{0\}$. This gives a null-homotopy of $BQ(s \circ e) = BQs \circ BQe$. The data, consisting of the map $BQe$, together with the null-homotopy of $BQs \circ BQe$ to the constant map from $BQ\mathcal{B}$ to the base point $0 \in BQ(\mathcal{A}/\mathcal{B})$, determine the map $\alpha$. The map $\alpha'$ is similarly determined. The composite map

$$BQ\mathcal{B} \xrightarrow{\alpha} F(BQs) \longrightarrow F(BQs')$$

is given by (i) the map $BQf \circ BQe : BQ\mathcal{B} \longrightarrow BQ\mathcal{A}'$, and (ii) the null-homotopy of $BQs' \circ (BQf \circ BQe) = BQs(f) \circ (BQe \circ BQs)$ induced by the null-homotopy of $BQe \circ BQs$ (used to determine $\alpha$). On the other hand, the composite
$$BQ\mathcal{B} \xrightarrow{BQg} BQ\mathcal{B}' \xrightarrow{\alpha'} F(BQs')$$
is induced by (i)' the map $BQe' \circ BQg : BQ\mathcal{B} \longrightarrow BQ\mathcal{A}'$, and (ii)'' the null-homotopy of $BQs' \circ (BQe' \circ BQg) = (BQs' \circ BQe') \circ BQg$ induced by the null-homotopy (used to define $\alpha'$) of $BQs' \circ BQe'$. However, $f \circ e = e' \circ g$, so the maps in (i), (i)' are equal. The homotopies in (ii), (ii)' are both induced by natural transformations from the functors $Qs' \circ Qf \circ Qe = Qs' \circ Qe' \circ Qg : Q\mathcal{B} \longrightarrow Q(\mathcal{A}'/\mathcal{B}')$ to the constant functor $Q\mathcal{B} \longrightarrow \{0\} \in Q(\mathcal{A}'/\mathcal{B}')$. These natural transformations are equal, since there is a *unique* isomorphism between any two zero objects in $\mathcal{A}'/\mathcal{B}'$, and hence in $Q(\mathcal{A}'/\mathcal{B}')$. This proves the claimed naturality.

# 7. Comparison of the Plus and $Q$-Constructions

Let $S$ be an Abelian monoid, i.e., $S$ has a commutative, associative binary operation with a two-sided identity. We say that $S$ acts on a set $X$ if there is a homomorphism of monoids $S \longrightarrow \text{Hom}_{\underline{\text{Set}}}(X, X)$; if $s \in S$, the corresponding map of sets $X \longrightarrow X$ is called translation by $s$. We say that $S$ acts *invertibly* on $X$ if each translation is bijective.

If $X$ is a set on which $S$ acts, let $S^{-1}X = (S \times X)/S$ where $S$ acts diagonally on the product ($S$ acts on itself by left translation; here $(S \times X)/S$ denotes the quotient of $S \times X$ by the equivalence relation $(s, x) \equiv (t, y) \Leftrightarrow$ for some $u, v \in S$, we have $(us, ux) = (vt, vy)$). Define a new action of $S$ on $S^{-1}X$ by $t(s, x) = (s, tx)$ for any $s, t \in S$, $x \in X$, and let $X \longrightarrow S^{-1}X$ be given by $x \mapsto (1, x)$. Then $X \longrightarrow S^{-1}X$ is a map of $S$-sets, and $S$ acts invertibly on $S^{-1}X$ (the inverse of translation by $t$ is given by $(s, x) \longrightarrow (ts, x)$), and $X \longrightarrow S^{-1}X$ is a universal arrow from $X$ to a set on which $S$ acts invertibly. If $S$ acts on itself by left translation, then $S^{-1}S$ is a group under the product $(s, t) \cdot (u, v) = (su, tv)$; $S \longrightarrow S^{-1}S$ is a homomorphism of monoids that is universal for homomorphisms from $S$ to groups. Thus, $S^{-1}S$ is the Grothendieck group of the commutative monoid $S$.

The above notions are generalized to categories, as follows (we retain the convention that the categories under consideration are equivalent to full small sub-categories, and one such equivalence has been fixed). A *monoidal category* $\mathcal{S}$ is a category $\mathcal{S}$ together with a functor $+ : \mathcal{S} \times \mathcal{S} \longrightarrow \mathcal{S}$ and an object $0 \in \mathcal{S}$, such that there are natural isomorphisms $(A + B) + C \cong A + (B + C)$, and $0 + A \cong A \cong A + 0$, for all $A, B, C \in \mathcal{S}$. These isomorphisms are required to be "coherent", i.e., the following diagrams must commute, for all $A, B, C, D \in \mathcal{S}$:

$$A + (B + (C + D)) \cong (A + B) + (C + D) \cong ((A + B) + C) + D$$
$$\cong \searrow \qquad \qquad \swarrow \cong$$
$$A + ((B + C) + D) \cong (A + (B + C)) + D$$

$$A + (0 + C) \cong (A + 0) + C$$
$$\cong \searrow \qquad \swarrow \cong$$
$$A + C$$

## 7. Comparison of the Plus and $Q$-Constructions

(Here, "natural isomorphism" means a natural transformation giving an isomorphism of functors. One knows that the commutativity of the above diagrams implies the commutativity of all similar diagrams; see MacLane's book *Categories for the Working Mathematician*).

A (left) *action* of a monoidal category $\mathcal{S}$ on a category $\mathcal{X}$ is a functor $+ : \mathcal{S} \times \mathcal{X} \longrightarrow \mathcal{X}$, together with natural isomorphisms $A + (B + F) \cong (A + B) + F$, $0 + F \cong F$ for all $A, B \in \mathcal{S}$, $F \in \mathcal{X}$. Diagrams analogous to the above two diagrams must commute (e.g., in the pentagon, the diagram with $D$ replaced by $F \in \mathcal{X}$ must commute, for all $A, B, C \in \mathcal{S}$). A *monoidal functor* between two monoidal categories $\mathcal{S}, \mathcal{T}$ is a functor $f : \mathcal{S} \longrightarrow \mathcal{T}$, together with natural isomorphisms $f(A + B) \cong f(A) + f(B)$, $f(0) \cong 0$, such that the following diagrams commute:

$$\begin{array}{ccccc} f((A+B)+C) & \cong & f(A+B) + f(C) & \cong & (f(A) + f(B)) + f(C) \\ \cong \downarrow & & & & \downarrow \cong \\ f(A+(B+C)) & \cong & f(A) + f(B+C) & \cong & f(A) + (f(B) + f(C)) \end{array}$$

$$\begin{array}{ccc} f(0+A) \cong f(0) + f(A) & & f(A+0) \cong f(A) + f(0) \\ \cong \downarrow \qquad \downarrow \cong & & \cong \downarrow \qquad \downarrow \cong \\ f(A) \cong 0 + f(A) & & f(A) \cong f(A) + 0 \end{array}$$

A functor $f : \mathcal{X} \longrightarrow \mathcal{Y}$ between categories with $\mathcal{S}$-actions *preserves the action* if there is a natural isomorphism $A + f(F) \cong f(A + F)$ for all $A \in \mathcal{S}, F \in \mathcal{X}$ such that appropriate diagrams commute.

If $\mathcal{S}$ is a monoidal category acting on a category $\mathcal{X}$, we say that $\mathcal{S}$ acts *invertibly* on $\mathcal{X}$ if each translation $\mathcal{X} \longrightarrow \mathcal{X}$, $F \mapsto A + F$, for $A \in \mathcal{S}$, is a homotopy equivalence. We try to imitate the construction of $X \longrightarrow S^{-1}X$ for a commutative monoid, to obtain a functor $f : \mathcal{X} \longrightarrow \mathcal{S}^{-1}\mathcal{X}$ of categories with $\mathcal{S}$-action such that $f$ preserves the action, $\mathcal{S}$ acts invertibly on $\mathcal{S}^{-1}\mathcal{X}$, and $f$ is "universal", at least in some homotopy theoretic sense.

If $\mathcal{S}$ is a monoidal category that acts on a category $\mathcal{X}$, let $\langle \mathcal{S}, \mathcal{X} \rangle$ be the category with the same objects as $\mathcal{X}$, such that an arrow $F \longrightarrow G$ in $\langle \mathcal{S}, \mathcal{X} \rangle$ is an equivalence class of pairs $(A, A + F \longrightarrow G)$, where $A \in \mathcal{S}$, and $A + F \longrightarrow G$ is an arrow in $\mathcal{X}$; $'(A, A + F \longrightarrow G), (A', A' + F \longrightarrow G)$ are equivalent if there is an isomorphism $u : A \longrightarrow A'$ in $\mathcal{S}$ such that

$$\begin{array}{c} A + F \xrightarrow[\cong]{u+1} A' + F \\ \searrow \qquad \swarrow \\ G \end{array}$$

commutes. $\langle \mathcal{S}, \mathcal{X} \rangle$ is somewhat analogous to the "quotient" of $\mathcal{X}$ modulo $\mathcal{S}$.

Let $\mathcal{X}$ be a category with $\mathcal{S}$-action, and let $\mathcal{S}$ act diagonally on $\mathcal{S} \times \mathcal{X}$, where $\mathcal{S}$ acts on itself by left translation. We define $\mathcal{S}^{-1}\mathcal{X} = \langle \mathcal{S}, \mathcal{S} \times \mathcal{X} \rangle$. Suppose $\mathcal{S}$ is commutative up to natural isomorphism, i.e., there are natural isomorphisms $A + B \cong B + A$ for all $A, B \in \mathcal{S}$ such that appropriate diagrams commute. Then the $\mathcal{S}$ action on $\mathcal{S} \times \mathcal{X}$ given by $A + (B, F) = (B, A + F)$ induces an $\mathcal{S}$-action on $\mathcal{S}^{-1}\mathcal{X}$. To see that commutativity is needed, we note that if $(D, (D + B, D + F) \longrightarrow (C, G))$ represents a morphism in $\mathcal{S}^{-1}\mathcal{X}$, we would want a naturally associated morphism in $\mathcal{S}^{-1}\mathcal{X}$ represented by $(D, (D + B, D + (A + F)) \longrightarrow (C, A + G))$. On the other hand, we have a morphism $(D + B, A + (D + F)) \longrightarrow (C, A + G)$ in $\mathcal{S} \times \mathcal{X}$. Thus if $\mathcal{S}$ is commutative up to isomorphism, $D + (A + F) \cong (D + A) + F \cong (A + D) + F \cong A + (D + F)$.

We observe that if $\mathcal{S}$ is commutative, then the above $\mathcal{S}$-action on $\mathcal{X}$ is invertible. Indeed, if $A \in \mathcal{S}$, then the functor $(B, F) \longrightarrow (B, A + F)$ has homotopy inverse $(B, F) \longrightarrow (A + B, F)$, since both composite functors equal $(B, F) \longrightarrow (A + B, A + F)$, and the arrow in $\mathcal{S}^{-1}\mathcal{X}$ given by $(A, A + (B, F) \cong (A + B, A + F))$ gives a natural transformation from the identity functor to the functor $(B, F) \longrightarrow (A + B, A + F)$.

If every arrow in $\mathcal{S}$ is an isomorphism, we claim that $0 \in \langle \mathcal{S}, \mathcal{S} \rangle$ is an initial object, so that $\langle \mathcal{S}, \mathcal{S} \rangle$ is contractible. To see this, if $A \in \mathcal{S}$, there is an arrow $(A, A + 0 \xrightarrow{\cong} A)$ from 0 to $A$ in $\langle \mathcal{S}, \mathcal{S} \rangle$; if $(B, B + 0 \xrightarrow{f} A)$ is any arrow in $\langle \mathcal{S}, \mathcal{S} \rangle$ then there is an arrow $u : B \longrightarrow A$ given by

$$\begin{array}{ccc} B + 0 & \cong & B \\ {\scriptstyle f} \downarrow & \swarrow {\scriptstyle u} & \\ A & & \end{array}$$

By the naturality of $A + 0 \cong A$, the square below commutes:

$$\begin{array}{ccc} B + 0 & \xrightarrow{1+u} & A + 0 \\ {\scriptstyle \cong} \downarrow & & \downarrow {\scriptstyle \cong} \\ B & \xrightarrow{u} & A \end{array}$$

Thus we have a commutative triangle

$$\begin{array}{ccc} B + 0 & \xrightarrow{u+1} & A + 0 \\ {\scriptstyle f} \searrow & & \swarrow {\scriptstyle \cong} \\ & A & \end{array}$$

and since $u$ is an isomorphism (since we assumed all arrows in $\mathcal{S}$ are isomorphisms), $(B, B + 0 \xrightarrow{f} A)$ and $(A, A + 0 \cong A)$ define the same arrow. Hence $0 \in \langle \mathcal{S}, \mathcal{S} \rangle$ is an initial object, as claimed.

Next, suppose that (in addition to all arrows being isomorphisms) all translations $S \longrightarrow S$, $B \longrightarrow A + B$, are faithful (one-one on morphisms in $S$). Then given any arrow $B \longrightarrow B'$ in $\langle S, S \rangle$ represented by $(A, A + B \xrightarrow{f} B')$, we claim that $A$ is determined up to a unique isomorphism. Indeed, in any case $A$ is determined up to isomorphism; so we must show that if $u : A \longrightarrow A$ is an automorphism such that the diagram in $S$

$$A + B \xrightarrow{u + 1_B} A + B$$
$$\phantom{A+B}\searrow_{f} \quad \swarrow_{f}$$
$$B'$$

commutes, then $u = 1_A$, the identity map on $A$. But since $f$ is an isomorphism, $u + 1_B = 1_{A+B} = 1_A + 1_B$; hence $u$ and $1_A$ become equal after translation by $B$, so that $u = 1_A$.

Thus, if $\rho : S^{-1}\mathcal{X} \longrightarrow \langle S, S \rangle$ is given by $\rho((B, F)) = B$ on objects,

$$\rho((A, (A + B, A + F)) \xrightarrow[(f,g)]{} (B', F'))) = (A, A + B \xrightarrow{f} B'),$$

then $\rho$ is a functor making $S^{-1}\mathcal{X}$ cofibered over $\langle S, S \rangle$. Indeed, given an arrow in $\langle S, S \rangle$ represented by $(A, A + B \longrightarrow B')$ we associate the functor $\rho^{-1}(B) \longrightarrow \rho^{-1}(B')$, $(B, F) \longrightarrow (B', A + F)$; there is a natural transformation from the identity functor on $\rho^{-1}(B)$ to the above functor, which (as $B$ varies) gives an equivalence $\rho/B' \longrightarrow \rho'(B')$ for any $B' \in S$. An arrow in $\rho^{-1}(B)$ is an arrow $(B, F) \longrightarrow (B, F')$ in $S^{-1}\mathcal{X}$ that covers the identity arrow of $B$ in $\langle S, S \rangle$. The identity arrow of $B$ is given by $(0, 0+B \cong B)$. Hence $F \longrightarrow (B, F)$ gives an isomorphism of categories $\mathcal{X} \longrightarrow \rho^{-1}(B)$, for any $B \in S$. Under this identification, if $(A, A + B \longrightarrow B')$ is an arrow in $\langle S, S \rangle$ then the associated cobase-change arrow $\rho^{-1}(B) \longrightarrow \rho^{-1}(B')$ becomes translation by $A$ on $\mathcal{X}$. We now have the tools to prove

**Theorem (7.1).** *Let $S$ be a monoidal category, commutative up to isomorphism, such that all arrows in $S$ are isomorphisms, and the functor $S \longrightarrow S$, $B \longrightarrow A + B$, given by translation by $A$, is faithful, for each $A \in S$. Let $S$ act on $\mathcal{X}$. Then $S$ acts invertibly on $\mathcal{X} \iff \mathcal{X} \longrightarrow S^{-1}\mathcal{X}$ is a homotopy equivalence.*

**Proof.** If $\mathcal{X} \longrightarrow S^{-1}\mathcal{X}$ is a homotopy equivalence, then since $S$ acts invertibly on $S^{-1}\mathcal{X}$ and the functor preserves the $S$-action, $S$ must act invertibly on $\mathcal{X}$. Conversely, if $S$ acts invertibly on $\mathcal{X}$, so that all translations on $\mathcal{X}$ are homotopy equivalences, then the cobase-change arrows of $\rho$ are homotopy-equivalences. Hence by Theorem B

$$\begin{array}{ccc} \mathcal{X} & \longrightarrow & \mathcal{S}^{-1}\mathcal{X} \\ \downarrow & & \downarrow \\ pt. & \longrightarrow & \langle \mathcal{S}, \mathcal{S} \rangle \end{array}$$

is homotopy Cartesian. Since $\langle \mathcal{S}, \mathcal{S} \rangle$ is contractible, $\mathcal{X} \longrightarrow \mathcal{S}^{-1}\mathcal{X}$ is a homotopy equivalence. This proves the theorem.

If $\mathcal{S}$ is a monoidal category as in (7.1) above, then $\pi_0(\mathcal{S}) = \pi_0(B\mathcal{S})$ is an Abelian monoid. If $\mathcal{S}$ acts on $\mathcal{X}$, $\pi_0(\mathcal{S})$ acts on $H_p(\mathcal{X}, \mathbb{Z}) = H_p(B\mathcal{X}, \mathbb{Z})$ for each $p \geq 0$. Since $\mathcal{S}$ acts invertibly on $\mathcal{S}^{-1}\mathcal{X}$, $\pi_0(\mathcal{S})$ acts invertibly on $H_*(\mathcal{S}^{-1}\mathcal{X}, \mathbb{Z})$. The natural map $H_*(\mathcal{S}, \mathbb{Z}) \longrightarrow H_*(\mathcal{S}^{-1}\mathcal{X}, \mathbb{Z})$ thus induces a map on localized modules $\pi_0(\mathcal{S})^{-1} H_*(\mathcal{X}, \mathbb{Z}) \longrightarrow H_*(\mathcal{S}^{-1}\mathcal{X}, \mathbb{Z})$.

**Theorem (7.2).** *Under the above conditions, the natural maps*

$$\pi_0(\mathcal{S})^{-1} H_p(\mathcal{X}, \mathbb{Z}) \longrightarrow H_p(\mathcal{S}^{-1}\mathcal{X}, \mathbb{Z})$$

*are isomorphisms $\forall p \geq 0$.*

**Proof.** Consider the double complex

$$E^0_{p,q} = \coprod_{(B_0 \to \cdots \to B_p) \in N_p\langle \mathcal{S}, \mathcal{S} \rangle} \coprod_{N_q(\rho \backslash B_0)} \mathbb{Z}$$

(which can be regarded as the double complex associated to the bisimplicial set $T_{pq} = T_{pq}(\rho)$ introduced in the proof of Theorem A, with the functor $f : \mathcal{C} \longrightarrow \mathcal{D}$ replaced by $\rho : \mathcal{S}^{-1}\mathcal{X} \longrightarrow \langle \mathcal{S}, \mathcal{S} \rangle$). The natural $\mathcal{S}$-action on $\mathcal{S}^{-1}\mathcal{X}$ (via the $\mathcal{S}$-action on $\mathcal{X}$), together with the trivial $\mathcal{S}$-action on $\langle \mathcal{S}, \mathcal{S} \rangle$, yield an "action" of $\mathcal{S}$ on the double complex, in the following sense. Each object of $\mathcal{S}$ yields an automorphism of $E^0_{p,q}$ for each $p, q$, compatible with the differentials; an arrow $A \longrightarrow B$ in $\mathcal{S}$ gives a chain homotopy between the automorphisms induced by $A$ and $B$, respectively, with respect to the differential in the $q$-direction (in fact, for any given $B_0 \in \langle \mathcal{S}, \mathcal{S} \rangle$, $\mathcal{S}$ "acts" on the (single) complex given in degree $q$ by $\coprod_{N_q(\rho \backslash B_0)} \mathbb{Z}$, in the above sense).

We may rewrite $E^0_{p,q}$ as

$$E^0_{p,q} = \coprod_{(F_0 \to \cdots \to F_q) \in N_q(\mathcal{S}^{-1}\mathcal{X})} \coprod_{N_p(\rho(F_q) \backslash \langle \mathcal{S}, \mathcal{S} \rangle)} \mathbb{Z}.$$

Hence, the homology groups in the $p$-direction (i.e., the $E^1$ terms for one of the spectral sequences for the double complex) are direct sums of groups $H_p(\rho(F_q) \backslash \langle \mathcal{S}, \mathcal{S} \rangle, \mathbb{Z})$. But $\rho(F_q) \backslash \langle \mathcal{S}, \mathcal{S} \rangle$ has $\rho(F_q)$ as an initial object, and

## 7. Comparison of the Plus and Q-Constructions

is hence contractible. Hence we have

$$E^1_{p,q} = \begin{cases} \coprod_{N_q(S^{-1}\mathcal{X})} \mathbb{Z} & \text{if } p = 0. \\ 0 & \text{if } p > 0. \end{cases}$$

Hence this spectral sequence degenerates at $E^2$, with

$$E^2_{0,q} \cong E^\infty_{0,q} \cong H_q(S^{-1}\mathcal{X}, \mathbb{Z}),$$

$E^\infty_{p,q} = 0$ for $p > 0$. Also, the automorphism induced by any object of $\mathcal{S}$ is just the automorphism of $H_q(S^{-1}\mathcal{X}, \mathbb{Z})$ induced by translation on $S^{-1}\mathcal{X}$, i.e., the action of $\mathcal{S}$ is induced by the natural (invertible) action of the monoid $\pi_0(\mathcal{S})$ on $H_q(S^{-1}\mathcal{X}, \mathbb{Z})$.

The other spectral sequence for the double complex has $E^1$ terms obtained by computing the homology in the $q$-direction of the $E^0_{p,q}$ terms. Thus we get

$$E^1_{p,q} = \coprod_{B_0 \to \cdots \to B_p} H_q(\rho \setminus B_0, \mathbb{Z}) \cong \coprod_{B_0 \to \cdots \to B_p} H_q(\rho^{-1}(B_0), \mathbb{Z})$$

$$\cong \coprod_{B_0 \to \cdots \to B_p} H_q(\mathcal{X}, \mathbb{Z})$$

from the cofibered structure of $\rho$, and the natural isomorphism $\mathcal{X} \cong \rho^{-1}(B_0)$. Let $\overline{H_q(\mathcal{X})} : \langle \mathcal{S}, \mathcal{S} \rangle \longrightarrow \mathcal{A}b$ denote the functor $B \mapsto H_q(\mathcal{X}, \mathbb{Z})$, $(A, A+B \longrightarrow B') \longrightarrow$ (endomorphism of $H_q(\mathcal{X}, \mathbb{Z})$ induced by translation by $A$) (here $\mathcal{A}b$ denotes the category of Abelian groups). For any functor $F : \mathcal{C} \longrightarrow \mathcal{A}b$, define $H_p(\mathcal{C}, F)$ to be the $p$th homology of the complex

$$C_p(\mathcal{C}, F) = \coprod_{(A_0 \to \cdots \to A_p) \in N_p\mathcal{C}} F(A_0)$$

with the natural differential induced from the simplicial structure of $N\mathcal{C}$. Then we may write the $E^2$ terms in the above spectral sequence as

$$E^2_{p,q} = H_p(\langle \mathcal{S}, \mathcal{S} \rangle, \overline{H_q(\mathcal{X})}).$$

The action of $\mathcal{S}$ on $E^1_{p,q}$ is given by the action on $H_q(\mathcal{X}, \mathbb{Z})$ induced by translation on $\mathcal{X}$. Thus, the action of $\mathcal{S}$ corresponds to the natural $\pi_0(\mathcal{S})$-module action as endomorphisms of the functor $\overline{H_q(\mathcal{X})}$. Hence, the spectral sequence

$$E^2_{p,q} = H_p(\langle \mathcal{S}, \mathcal{S} \rangle, \overline{H_q(\mathcal{X})}) \Longrightarrow H_{p+q}(S^{-1}\mathcal{X}, \mathbb{Z})$$

can be viewed as spectral sequence of $\pi_0(\mathcal{S})$-modules.

Since localization of modules over a commutative ring with respect to a multiplicative set is an exact functor, we may invert the $\pi_0(\mathcal{S})$-action on the above spectral sequence to obtain a new spectral sequence of $\pi_0(\mathcal{S})^{-1}$ $\pi_0(\mathcal{S})$-modules

$$E^1_{p,q} = \coprod_{N_p\langle\mathcal{S},\mathcal{S}\rangle} \pi_0(\mathcal{S})^{-1}H_q(\mathcal{X}) \Longrightarrow \pi_0(\mathcal{S})^{-1}H_{p+q}(\mathcal{S}^{-1}\mathcal{X},\mathbb{Z})$$
$$= H_{p+q}(\mathcal{S}^{-1}\mathcal{X},\mathbb{Z}).$$

Now $0 \in \langle\mathcal{S},\mathcal{S}\rangle$ is an initial object, so that we have a unique arrow $0 \longrightarrow B$ in $\langle\mathcal{S},\mathcal{S}\rangle$ for any $B \in \langle\mathcal{S},\mathcal{S}\rangle$, given by $(B, B+0 \cong B)$. The corresponding cobase-change functor $\rho^{-1}(0) \longrightarrow \rho^{-1}(B)$ is identified with translation by $B$ on $\mathcal{X} \cong \rho^{-1}(0) \cong \rho^{-1}(B)$. Now translation by $B$ induces an automorphism of $\pi_0(\mathcal{S})^{-1}H_q(\mathcal{X})$, so that we have a canonical isomorphism

$$\pi_0(\mathcal{S})^{-1}H_q(\rho^{-1}(0),\mathbb{Z}) \cong \pi_0(\mathcal{S})^{-1}H_q(\rho^{-1}(B),\mathbb{Z}) \text{ for any } B \in \langle\mathcal{S},\mathcal{S}\rangle.$$

Thus we may rewrite the complex of $E^1_{p,q}$ terms as

$$E^1_{p,q} = \left(\coprod_{N_p\langle\mathcal{S},\mathcal{S}\rangle} \mathbb{Z}\right) \otimes_{\mathbb{Z}} \pi_0(\mathcal{S})^{-1}H_q(\rho^{-1}(0),\mathbb{Z}).$$

Since $\langle\mathcal{S},\mathcal{S}\rangle$ is contractible, the complex

$$C_p(\langle\mathcal{S},\mathcal{S}\rangle,\mathbb{Z}) = \coprod_{N_p\langle\mathcal{S},\mathcal{S}\rangle} \mathbb{Z}$$

has homology groups $H_p(\langle\mathcal{S},\mathcal{S}\rangle,\mathbb{Z}) = 0$ unless $p = 0$, and $H_0(\langle\mathcal{S},\mathcal{S}\rangle,\mathbb{Z}) = \mathbb{Z}$. Thus $E^2_{p,q} = 0$ unless $p = 0$, and $E^2_{0,q} = \pi_0(\mathcal{S})^{-1}H_q(\mathcal{X})$. Hence the localized spectral sequence degenerates at $E^2$, giving isomorphisms

$$\pi_0(\mathcal{S})^{-1}H_q(\mathcal{X},\mathbb{Z}) \cong H_q(\mathcal{S}^{-1}\mathcal{X},\mathbb{Z}).$$

To see that this isomorphism is just the map induced by localizing

$$H_q(\mathcal{X},\mathbb{Z}) \longrightarrow H_q(\mathcal{S}^{-1}\mathcal{X},\mathbb{Z})$$

given by $\mathcal{X} \longrightarrow \mathcal{S}^{-1}\mathcal{X}$, compare the spectral sequence above with the trivial spectral sequence

$$E^1_{p,q} = \begin{cases} H_p(\mathcal{X},\mathbb{Z}) & \text{if } q = 0 \\ 0 & \text{if } q \neq 0 \end{cases}$$
$$E^1_{p,q} \Longrightarrow H_{p+q}(\mathcal{X},\mathbb{Z}),$$

which we regard as the analogous spectral sequence for the functor $\mathcal{X} \longrightarrow \{0\}$. The comparison is done using the diagrams of categories and functors

$$\begin{array}{ccc} \mathcal{X} & \longrightarrow & \mathcal{S}^{-1}\mathcal{X} \\ \downarrow & & \downarrow \\ \{0\} & \longrightarrow & \langle\mathcal{S},\mathcal{S}\rangle \end{array}$$

This completes the proof of (7.2).

## 7. Comparison of the Plus and $Q$-Constructions 133

**(7.3). The Functorial Version of the Plus Construction.**

Let $\mathcal{P}$ be an exact category in which all short exact sequences split, and let $\mathrm{Iso}(\mathcal{P})$ be the subcategory with the same objects as $\mathcal{P}$, whose arrows are all the isomorphisms of $\mathcal{P}$. Then the direct sum $\oplus : \mathrm{Iso}(\mathcal{P}) \times \mathrm{Iso}(\mathcal{P}) \longrightarrow \mathrm{Iso}(\mathcal{P})$ makes $\mathcal{S} = \mathrm{Iso}(\mathcal{P})$ into a monoidal category in a natural way. Then $B\mathcal{S}^{-1}\mathcal{S}$ is an $H$-space, with the multiplication $B\mathcal{S}^{-1}\mathcal{S} \times B\mathcal{S}^{-1}\mathcal{S} \longrightarrow B\mathcal{S}^{-1}\mathcal{S}$ induced by the functor $\mathcal{S}^{-1}\mathcal{S} \times \mathcal{S}^{-1}\mathcal{S} \longrightarrow \mathcal{S}^{-1}\mathcal{S}$, $((A, B), (C, D)) \mapsto (A \oplus C, B \oplus D)$.

In particular, let $R$ be a ring, $\mathcal{P} = \mathcal{P}(R)$, the category of finitely generated projective (left) $R$-modules. One checks easily that $\pi_0(\mathcal{S}^{-1}\mathcal{S}) \cong K_0(R)$ (this follows from the fact that $\pi_0(\mathcal{S}^{-1}\mathcal{S}) = \pi_0(F_1 B\mathcal{S}^{-1}\mathcal{S})$, where $F_1 B\mathcal{S}^{-1}\mathcal{S}$ is the 1-skeleton of $B\mathcal{S}^{-1}\mathcal{S}$).

If $A \in \mathcal{P}$, let $\underline{\mathrm{Aut}}(A)$ be the category with 1 object $A$, and arrows given by the group $\mathrm{Aut}(A)$ of automorphisms of $A$ as an object of $\mathcal{P}$. Thus $\underline{\mathrm{Aut}}(A)$ is the full subcategory of $\mathcal{S}$ with the single object $A$. There is a functor $\underline{\mathrm{Aut}}(A) \longrightarrow \mathcal{S}^{-1}\mathcal{S}$ given by $A \longrightarrow (A, A)$, $u \mapsto (0, 0+(A, A) \xrightarrow{\bar{u}} (A, A))$ for $u \in \mathrm{Aut}(A)$, where $\bar{u} : (0 \oplus A, 0 \oplus A) \longrightarrow (A, A)$ is given by $\bar{u} = (u_1, u_2)$, $u_1 : 0 \oplus A \longrightarrow A$ the natural isomorphism, and $u_2 : 0 \oplus A \longrightarrow A$ being the composite $u_2 = u \circ u_1$. The arrow $(R, R + (A, A) \longrightarrow (A \oplus R, A \oplus R))$ yields a natural transformation of functors $\underline{\mathrm{Aut}}(A) \longrightarrow \mathcal{S}^{-1}\mathcal{S}$, making the diagram

$$\begin{array}{ccc} B\,\underline{\mathrm{Aut}}(A) & \longrightarrow & B\,\underline{\mathrm{Aut}}(A \oplus R) \\ & \searrow \quad \swarrow & \\ & B\mathcal{S}^{-1}\mathcal{S} & \end{array}$$

commute up to homotopy. Thus, there is a map, well defined up to homotopy,

$$BGL(R) = \varinjlim_n B\,\underline{\mathrm{Aut}}(R^{\oplus n}) \longrightarrow B\mathcal{S}^{-1}\mathcal{S}.$$

In fact, the map factors through the identity component $B(\mathcal{S}^{-1}\mathcal{S})_0$.

The map $BGL(R) \longrightarrow B\mathcal{S}^{-1}\mathcal{S}$ can be concretely realized as follows. Let $\mathcal{S}_n$ be the connected component of $\mathcal{S}$ containing $R^{\oplus n}$ (so that $\mathcal{S}_n$ is the full subcategory of $\mathcal{S}$ with objects $A \cong R^{\oplus n}$). Let $\mathbb{N}$ be the totally ordered set of non-negative integers (regarded as a category, using the ordering). Let $\mathcal{L}$ denote the category of pairs $(n, B)$ with $n \in \mathbb{N}$, $B \in \mathcal{S}_n$, where an arrow $(n, B) \longrightarrow (n + m, C)$ is an isomorphism $T_m(B) \longrightarrow C$, where $T_0(B) = B$, $T_1(B) = R \oplus B$, $T_m(B) = T_1 \circ T_{m-1}(B)$ for $m > 1$ (there are no arrows $(n, B) \longrightarrow (n', C)$ with $n' < n$). Let $f : \mathcal{L} \longrightarrow \mathbb{N}$ be the functor $f(n, B) = n$. Then $f^{-1}(n) \cong \mathcal{S}_n$, and $f$ makes $\mathcal{L}$ cofibered over $\mathbb{N}$ with cobase change maps $(n, B) \longrightarrow (n+m, T_m(B))$ for any arrow $n \longrightarrow n+m$ in $\mathbb{N}$. Clearly $\mathcal{L} \cong \varinjlim(f \backslash n)$, and $f^{-1}(n) \hookrightarrow (f \backslash n)$ is a homotopy equivalence for each $n \in \mathbb{N}$. Further $\underline{\mathrm{Aut}}(R^{\oplus n}) \longrightarrow \mathcal{S}_n$ is a homotopy equivalence for each $n \in \mathbb{N}$, so that we have homotopy equivalences $BGL_n(R) \longrightarrow B(f \backslash n)$, such that the diagram

$$\begin{array}{ccc} BGL_n(R) & \longrightarrow & BGL_{n+1}(R) \\ \downarrow & & \downarrow \\ B(f \backslash n) & \longrightarrow & B(f \backslash n+1) \end{array}$$

commutes up to homotopy. Thus

$$\pi_i(B\mathcal{L}) = \varinjlim \pi_i(B(f \backslash n)) = \varinjlim \pi_i(BGL_n(R))$$

i.e., $\pi_1(B\mathcal{L}) \cong GL(R)$, $\pi_i(B\mathcal{L}) = 0$ for $i \neq i$. Thus $B\mathcal{L}$ is another model for the homotopy type of $BGL(R)$. Finally, we note that $(n, B) \longrightarrow (R^{\oplus n}, B)$ yields a functor $g : \mathcal{L} \longrightarrow (\mathcal{S}^{-1}\mathcal{S})$, if we regard $R^{\oplus n}$ as $T_n(0)$. We claim that $Bg : B\mathcal{L} \longrightarrow B(\mathcal{S}^{-1}\mathcal{S})_0$ induces an isomorphism on integral homology groups.

To prove the claim, we first note that if $e \in \pi_0(B\mathcal{S})$ is the class of the component containing $R$, then $\pi_0(B\mathcal{S})[e^{-1}] \cong \pi_0(B\mathcal{S}^{-1}\mathcal{S}) \cong K_0(R)$, since $\pi_0(B\mathcal{S})$ is the set of isomorphism classes of objects of $\mathcal{P}(R)$, made into a monoid by direct sum, and $P, Q \in \mathcal{P}(R)$ have the same image in $K_0(R)$ if and only if $R^{\oplus n} \oplus P \cong R^{\oplus n} \oplus Q$ for some $n \in \mathbb{N}$. Hence, from (7.2),

$$H_p(\mathcal{S}^{-1}\mathcal{S}) \cong \pi_0(\mathcal{S}^{-1})H_p(\mathcal{S}) \cong H_p(\mathcal{S})[e^{-1}].$$

Now

$$H_p(\mathcal{S}^{-1}\mathcal{S}) \cong H_p((\mathcal{S}^{-1}\mathcal{S})_0) \times K_0(R),$$

and

$$H_p(\mathcal{S}) \cong \coprod_{[P] \in \pi_0(B\mathcal{S})} H_p(B\underline{\mathrm{Aut}}(P)) \cong \coprod_{[P] \in \pi_0(B\mathcal{S})} H_p(\mathcal{S}_P)$$

where $\mathcal{S}_P$ is the connected component of $\mathcal{S}$ containing $P$. The map

$$H_p(\mathcal{S}_n) \longrightarrow H_p(\mathcal{S}^{-1}\mathcal{S})$$

is induced by the composite functor $\mathcal{S}_n \subset \mathcal{S} \longrightarrow \mathcal{S}^{-1}\mathcal{S}$, $A \mapsto (0, A)$. If $\mathcal{S}_n = f^{-1}(n) \subset \mathcal{L} \longrightarrow (\mathcal{S}^{-1}\mathcal{S})_0 \subset (\mathcal{S}^{-1}\mathcal{S})$ is the functor induced by $g$, given by $A \mapsto (R^{\oplus n}, A)$, then we have the functor obtained by translating this functor by $R^{\oplus n}$ (which corresponds on homology to multiplication by $e^n$),

$$A \mapsto (R^{\oplus n}, R^{\oplus n} \oplus A).$$

But we have an arrow in $\mathcal{S}^{-1}\mathcal{S}$

$$(R^{\oplus n}, R^{\oplus n} \oplus (0, A)) \longrightarrow (R^{\oplus n}, R^{\oplus n} \oplus A))$$

## 7. Comparison of the Plus and $Q$-Constructions

which yields a natural transformation of functors $\mathcal{S}_n \longrightarrow \mathcal{S}^{-1}\mathcal{S}$. Thus, we have a commutative diagram

$$\begin{array}{ccccc} H_p(\mathcal{S}_n) & \longrightarrow & H_p(\mathcal{S}) & \longrightarrow & H_p(\mathcal{S}^{-1}\mathcal{S}) \\ \downarrow & & & & \uparrow {\cdot e^n} \\ H_p(\mathcal{L}) & \xrightarrow{g_*} & H_p((\mathcal{S}^{-1}\mathcal{S})_0) & \longrightarrow & H_p(\mathcal{S}^{-1}\mathcal{S}) \end{array}$$

where $\cdot e^n$ denotes multiplication by $e^n$. Thus, the map

$$H_p(\mathcal{S}_n) \longrightarrow H_p((\mathcal{S}^{-1}\mathcal{S})_0) \subset H_p(\mathcal{S})[e^{-1}]$$

induced by $g$ is $x \mapsto x \cdot e^{-n}$. We claim that in fact

$$H_p((\mathcal{S}^{-1}\mathcal{S})_0) \cong \sum H_p(\mathcal{S}_n) \cdot e^{-n} \subset H^p(\mathcal{S})[e^{-1}].$$

Since $H_p((\mathcal{S}^{-1}\mathcal{S})_0) \subset H_p(\mathcal{S}^{-1}\mathcal{S}) \cong H_p(\mathcal{S})[e^{-1}]$, every $y \in H_p((\mathcal{S}^{-1}\mathcal{S})_0)$ can be written as $y = x \cdot e^{-n}$ with $x \in H_p(\mathcal{S})$; since

$$H_p(\mathcal{S}) = \coprod_{[P] \in \pi_0(\mathcal{S})} H_p(\mathcal{S}_P),$$

we must have $x \in \sum_{[P]} H_p(\mathcal{S}_P)$ where $[P]$ runs over the classes in $\pi_0(\mathcal{S})$ such that $[P] \cdot e^{-n} = 0$ in $K_0(R)$, i.e., $[P] \cong [R^{\oplus n}]$ in $K_0(R)$. Since $x \cdot e^{-n} = (x \cdot e^m) \cdot e^{-(m+n)}$ and $x \mapsto x \cdot e^m$ is induced by the functor $\mathcal{S} \longrightarrow \mathcal{S}$ given by translation by $R^{\oplus m}$, we can write $y = x \cdot e^{-n}$ with $x \in H_p(\mathcal{S}_n)$, for some $n$. Thus

$$H_p((\mathcal{S}^{-1}\mathcal{S})_0) \cong \sum H_p(\mathcal{S}_n) \cdot e^{-n}$$

as claimed. But it is easy to see (from the definition of localization) that

$$\sum H_p(\mathcal{S}_n) \cdot e^{-n} \subset \left( \coprod_{n \in \mathbb{N}} H_p(\mathcal{S}_n) \right)[e^{-1}]$$

is canonically isomorphic to $\varinjlim H_p(\mathcal{S}_n)$, where the maps

$$H_p(\mathcal{S}_n) \longrightarrow H_p(\mathcal{S}_{n+1})$$

in the direct system are all $\cdot e$. Hence, from the diagrams

$$\begin{array}{ccc} H_p(\mathcal{S}_n) & \xrightarrow{\cdot e} & H_p(\mathcal{S}_{n+1}) \\ & \searrow \quad \swarrow & \\ & H_p((\mathcal{S}^{-1}\mathcal{S})_0) & \end{array}$$

and

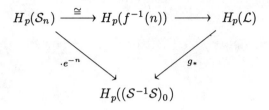

we deduce that

$$\varinjlim H_p(\mathcal{S}_n) \xrightarrow{\cong} H_p(\mathcal{L})$$
$$\cong \searrow \quad \swarrow g_*$$
$$H_p((\mathcal{S}^{-1}\mathcal{S})_0)$$

commutes, i.e., $g_*$ is an isomorphism.

Thus, we have an isomorphism of integral homology groups

$$H_p(BGL(R)) \cong H_p(\mathcal{L}) \xrightarrow[g_*]{} H_p((\mathcal{S}^{-1}\mathcal{S})_0).$$

Since $(\mathcal{S}^{-1}\mathcal{S})_0$ is an $H$-group, $\pi_1((\mathcal{S}^{-1}\mathcal{S})_0)$ is Abelian, and is thus isomorphic to $H_1((\mathcal{S}^{-1}\mathcal{S})_0) \cong H_1(BGL(R)) \cong GL(R)/E(R)$ (see (A.47), (A.52)). Hence from the universal property of the plus construction, we have a diagram

$$BGL(R) \cong B\mathcal{L} \xrightarrow{h} BGL(R)^+$$
$$Bg \searrow \quad \swarrow Bg^+$$
$$B(\mathcal{S}^{-1}\mathcal{S})_0$$

where $Bg_*^+ : H_p(BGL(R)^+) \longrightarrow H_p(B(\mathcal{S}^{-1}\mathcal{S})_0)$ is an isomorphism, since $h$, $Bg$ are homology isomorphisms. But $BGL(R)^+$, $B(\mathcal{S}^{-1}\mathcal{S})_0$ are $H$-spaces with the homotopy types of $CW$-complexes, and so $Bg^+$ is a homotopy equivalence (A.54). Thus we have proved:

**Theorem (7.4).** $B(\mathcal{S}^{-1}\mathcal{S})$ *is homotopy equivalent to* $K_0(R) \times BGL(R)^+$, *where* $\mathcal{S} = \mathrm{Iso}(\mathcal{P}(R))$.

**(7.5) The Loop Space of $B\mathcal{QP}(R)$.** Our final goal in this chapter is to prove Theorem (5.1), i.e., that there is a natural homotopy equivalence

$$BGL(R)^+ \longrightarrow (\Omega B\mathcal{QP}(R))^0.$$

In fact we prove below (Theorem (7.7)) that there is a natural homotopy equivalence

$$\Omega B\mathcal{QP}(R) \xrightarrow{\cong} B(\mathcal{S}^{-1}\mathcal{S}),$$

which, combined with (7.4), yields the result.

## 7. Comparison of the Plus and Q-Constructions

Let $\mathcal{S}$ be a monoidal category as in (7.1), and let $f : \mathcal{X} \longrightarrow \mathcal{Y}$ be a functor between categories with $\mathcal{S}$-action, such that $f$ preserves the action. If $\mathcal{S}$ acts trivially on $\mathcal{Y}$, we say $\mathcal{S}$ acts fiber-wise with respect to $f$; $\mathcal{S}$ does indeed act on the fibers $f^{-1}(G)$, $G \in \mathcal{Y}$. If in addition, $f$ makes $\mathcal{X}$ fibered over $\mathcal{Y}$, we say that the $\mathcal{S}$-action is *Cartesian*. In this case, one can show (details left to the reader) that $\mathcal{S}^{-1} f : \mathcal{S}^{-1} \mathcal{X} \longrightarrow \mathcal{Y}$ is also fibered, with fibers $\mathcal{S}^{-1} f^{-1}(G)$, $G \in \mathcal{Y}$, and base change maps induced by localizing those of $f$. Similarly, if $f$ makes $\mathcal{X}$ cofibered over $\mathcal{Y}$, we say that the $\mathcal{S}$-action is *co-Cartesian*; then $\mathcal{S}^{-1} f : \mathcal{S}^{-1} \mathcal{X} \longrightarrow \mathcal{Y}$ is still cofibered, with fibers $\mathcal{S}^{-1} f^{-1}(G)$, $G \in \mathcal{Y}$.

Now suppose $\mathcal{S}$ is as in (7.1), and $\mathcal{S}$ acts on $\mathcal{X}$. Assume

(i) every arrow in $\mathcal{X}$ is monic;

(ii) $\forall F \in \mathcal{X}$, the functor $\mathcal{S} \longrightarrow \mathcal{X}$, $B \mapsto B + F$, is faithful (i.e., injective on Hom-sets).

Then as in the proof of (7.1), one checks that the projection $\mathcal{S} \times \mathcal{X} \longrightarrow \mathcal{X}$ induces a cofibered functor $q : \mathcal{S}^{-1} \mathcal{X} \longrightarrow \langle \mathcal{S}, \mathcal{X} \rangle$, with fibers $q^{-1}(F) \cong \mathcal{S}$ ($\mathcal{S} \longrightarrow q^{-1}(F)$ is given by $A \mapsto (A, F)$), such that cobase-change maps become translations on $\mathcal{S}$.

Define a new $\mathcal{S}$-action on $\mathcal{S}^{-1} \mathcal{X}$, through the action of $\mathcal{S}$ on $\mathcal{S} \times \mathcal{X}$ on the first factor, $A + (B, F) = (A + B, F)$. Then this action is clearly co-Cartesian with respect to $q$. Localizing, we obtain a cofibered functor $\mathcal{S}^{-1} q : \mathcal{S}^{-1} \mathcal{S}^{-1} \mathcal{X} \longrightarrow \langle \mathcal{S}, \mathcal{X} \rangle$, each fiber of which is isomorphic to $\mathcal{S}^{-1} \mathcal{S}$, with base change maps given by translations, which are invertible on $\mathcal{S}^{-1} \mathcal{S}$. Thus Theorem B yields a homotopy Cartesian square (for each $F \in \langle \mathcal{S}, \mathcal{X} \rangle$)

$$\begin{array}{ccc} \mathcal{S}^{-1}\mathcal{S} & \longrightarrow & \mathcal{S}^{-1}\mathcal{S}^{-1}\mathcal{X} \\ \downarrow & & \downarrow \\ \{F\} & \longrightarrow & \langle \mathcal{S}, \mathcal{X} \rangle \end{array}$$

The functor $\mathcal{S}^{-1} \mathcal{S} \longrightarrow \mathcal{S}^{-1} \mathcal{S}^{-1} \mathcal{X}$ is $(A, B) \mapsto (A, (B, F))$. If $\langle \mathcal{S}, \mathcal{X} \rangle$ is contractible, then $\mathcal{S}^{-1} \mathcal{S} \longrightarrow \mathcal{S}^{-1} \mathcal{S}^{-1} \mathcal{X}$ is a homotopy equivalence.

**Lemma (7.6).** *Let $\mathcal{S}$ be as in (7.1), and let $\mathcal{X}$ satisfy (i), (ii) above. Suppose $\langle \mathcal{S}, \mathcal{X} \rangle$ is contractible. Then for each $F \in \mathcal{X}$, $\mathcal{S}^{-1} \mathcal{S} \longrightarrow \mathcal{S}^{-1} \mathcal{X}$ given by $(A, B) \mapsto (A, B + F)$ is a homotopy equivalence.*

**Proof.** As seen above, $\mathcal{S}^{-1} \mathcal{S} \longrightarrow \mathcal{S}^{-1} \mathcal{S}^{-1} \mathcal{X}$, $(A, B) \mapsto (A, (B, F))$ is a homotopy equivalence. Clearly $\mathcal{S}^{-1} \mathcal{S} \longrightarrow \mathcal{S}^{-1} \mathcal{S}$, $(A, B) \mapsto (B, A)$ is a homotopy equivalence (which is its own inverse). Hence $\mathcal{S}^{-1} \mathcal{S} \longrightarrow \mathcal{S}^{-1} \mathcal{S}^{-1} \mathcal{X}$, $(A, B) \mapsto (B, (A, F))$, is a homotopy equivalence.

On the other hand, we have a functor $S^{-1}\mathcal{X} \longrightarrow S^{-1}S^{-1}\mathcal{X}$, $(A, F) \mapsto (0, (A, F))$, which preserves the $S$-action given by $B + (A, F) = (B + A, F)$ on $S^{-1}\mathcal{X}$, and $B + (C, (A, F)) = (C, (B + A, F))$ on $S^{-1}S^{-1}\mathcal{X}$. But $B + (A, F) = (A, B + F)$ gives a functor $S^{-1}\mathcal{X} \longrightarrow S^{-1}\mathcal{X}$ which is a homotopy inverse for the above translation by $B$ on $S^{-1}\mathcal{X}$, since there is a natural transformation from the identity functor to either composite $(B, B + (A, F) \longrightarrow (B + A, B + F))$ (where we use the translation $B + (A, F) = (B + A, B + F)$, used in defining $S^{-1}\mathcal{X}$). Thus, the new $S$-action $B + (A, F) = (B + A, F)$ on $S^{-1}\mathcal{X}$ is invertible. Hence $S^{-1}\mathcal{X} \longrightarrow S^{-1}S^{-1}\mathcal{X}$, $(A, F) \mapsto (0, (A, F))$ is a homotopy equivalence, by (7.1).

Finally, consider the triangle

$$\begin{array}{ccc} S^{-1}S & \longrightarrow & S^{-1}S^{-1}\mathcal{X} \\ & \searrow & \nearrow \\ & S^{-1}\mathcal{X} & \end{array}$$

corresponding to $S^{-1}S \longrightarrow S^{-1}S^{-1}\mathcal{X}$, $(A, B) \mapsto (B, (A, F))$, $S^{-1}S \longrightarrow S^{-1}\mathcal{X}$, $(A, B) \mapsto (A, B + F)$, $S^{-1}\mathcal{X} \longrightarrow S^{-1}S^{-1}\mathcal{X}$, $(A, F) \mapsto (0, (A, F))$. The composite $S^{-1}S \longrightarrow S^{-1}\mathcal{X} \longrightarrow S^{-1}S^{-1}\mathcal{X}$ is $(A, B) \mapsto (A, (A, B+F))$. The arrows in $S^{-1}S^{-1}\mathcal{X}$

$$(B, B + (0, (A, B + F)) \xrightarrow{\cong} (B, (B + A, B + F)))$$

$$(0, 0 + (B, (A, F)) \xrightarrow{u} (B, (B + A, B + F)))$$

(where $u$ is induced by the arrow $(B, B + (A, F) \xrightarrow{\cong} (B + A, B + F))$ in $S^{-1}\mathcal{X}$) give a chain of natural transformations connecting the two functors $S^{-1}S \longrightarrow S^{-1}S^{-1}\mathcal{X}$. Hence the triangle commutes up to homotopy. Since two sides are homotopy equivalences, so is the third. This proves (7.6).

Now we can compute $\Omega BQP(R)$. More generally, let $\mathcal{P}$ be any exact category where all exact sequences split; let $S = \text{Iso}\,\mathcal{P}$.

**Theorem (7.7).** *There is a natural homotopy equivalence*

$$\Omega BQP \longrightarrow BS^{-1}S.$$

**Proof.** Let $\mathcal{E}$ be the following category. An object of $\mathcal{E}$ is a short exact sequence in $\mathcal{P}$. An arrow in $\mathcal{E}$

$$(0 \to A \to B \to C \to 0) \longrightarrow (0 \to A' \to B' \to C' \to 0)$$

## 7. Comparison of the Plus and $Q$-Constructions

is defined to be an equivalence class of diagrams

$$\begin{array}{ccccccccc} 0 & \to & A & \to & B & \to & C & \to & 0 \\ & & \uparrow & & \| & & \uparrow & & \\ 0 & \to & A' & \to & B & \to & C_1 & \to & 0 \\ & & \| & & \downarrow & \square & \downarrow & & \\ 0 & \to & A' & \to & B' & \to & C' & \to & 0 \end{array}$$

where the square marked $\square$ is a pullback; a diagram isomorphic to the above one with an isomorphism inducing the identity maps on $A$, $B$, $C$, $A'$, $B'$, $C'$ defines the same arrow in $\mathcal{E}$. There is a functor $f : \mathcal{E} \longrightarrow Q\mathcal{P}$, $(0 \to A \to B \to C \to 0) \mapsto C$. The fiber $f^{-1}(C) = \mathcal{E}_C$ is the category of short exact sequences $(0 \to A \to B \to C \to 0)$, with morphisms given by isomorphisms

$$\begin{array}{ccccccccc} 0 & \to & A & \to & B & \to & C & \to & 0 \\ & & \downarrow \simeq & & \downarrow \simeq & & \| & & \\ 0 & \to & A' & \to & B' & \to & C & \to & 0. \end{array}$$

We claim $f$ makes $\mathcal{E}$ fibered over $Q\mathcal{P}$. The base change maps are defined as follows: if $q : C' \twoheadrightarrow C$, then

$$(q^!)^* : \mathcal{E}_{C'} \longrightarrow \mathcal{E}_C$$

is given by

$$(0 \longrightarrow A \longrightarrow B \xrightarrow{u} C' \longrightarrow 0)$$
$$\mapsto (0 \longrightarrow u^{-1}(\ker q) \longrightarrow B \xrightarrow{q \circ u} C \longrightarrow 0);$$

if $i : C \rightarrowtail C'$, then $(i_!)^* : \mathcal{E}_{C'} \longrightarrow \mathcal{E}_C$ is

$$(0 \longrightarrow A \longrightarrow B \longrightarrow C' \longrightarrow 0)$$
$$\mapsto (0 \longrightarrow A \longrightarrow B \times_{C'} C \longrightarrow C \longrightarrow 0).$$

One checks that these definitions do make $f$ a fibered functor.

Let $\mathcal{S}$ act on $\mathcal{E}$ by

$$(A') + (0 \longrightarrow A \longrightarrow B \longrightarrow C \longrightarrow 0)$$
$$= (0 \longrightarrow A' \oplus A \longrightarrow A' \oplus B \longrightarrow C \longrightarrow 0).$$

One verifies that this $\mathcal{S}$-action is Cartesian with respect to $f$. Also, $\mathcal{S} \longrightarrow \mathcal{E}_0$ given by

$$A \mapsto (0 \longrightarrow A \xrightarrow{1_A} A \longrightarrow 0 \longrightarrow 0)$$

is an equivalence of categories.

**Lemma (7.8).** *For any $C \in \mathcal{P}$, $\langle \mathcal{S}, \mathcal{E}_C \rangle$ is contractible.*

**Proof.** We define a product $\mathcal{E}_C \times \mathcal{E}_C \longrightarrow \mathcal{E}_C$ by
$$(0 \to A \to B \to C \to 0) + (0 \to A' \to B' \to C \to 0)$$
$$= (0 \to A \oplus A' \to B \times_C B' \to C \to 0).$$
One checks easily that this makes $\langle \mathcal{S}, \mathcal{E}_C \rangle$ into an $H$-space, which is homotopy commutative and associate, with identity element
$$0 \longrightarrow 0 \longrightarrow C \xrightarrow{1_C} C \longrightarrow 0.$$

Next, since every short exact sequence in $\mathcal{P}$ is split, there exists an isomorphism
$$A + (0 \to 0 \to C \xrightarrow{1_C} C \to 0) \cong (0 \to A \to B \to C \to 0),$$
giving an arrow
$$(0 \to 0 \to C \xrightarrow{1_C} C \to 0) \mapsto (0 \to A \to B \to C \to 0)$$
in $\langle \mathcal{S}, \mathcal{E}_C \rangle$. Hence $\langle \mathcal{S}, \mathcal{E}_C \rangle$ is a connected, homotopy associative $H$-space, and is hence an $H$-group (see (A.47)).

Finally, for any $(0 \to A \to B \to C \to 0) \in \mathcal{E}_C$, the diagonal map $B \longrightarrow B \times_C B$ provides a natural isomorphism
$$A + (0 \to A \to B \to C \to 0) \cong (0 \to A \oplus A \to B \times_C B \to C \to 0).$$
This can be viewed as giving a natural transformation from the identity functor on $\langle \mathcal{S}, \mathcal{E}_C \rangle$ to the functor corresponding to multiplication by 2 in the $H$-group structure. Thus, in the group (under pointwise multiplication) of homotopy classes of self maps of $B\langle \mathcal{S}, \mathcal{E}_C \rangle$, we have the identity $x = x^2$. Thus, this group is trivial, i.e., $B\langle \mathcal{S}, \mathcal{E}_C \rangle$ is contractible. This proves (7.8).

**Lemma (7.9).** *The square*
$$\begin{array}{ccc} S^{-1}\mathcal{S} & \longrightarrow & S^{-1}\mathcal{E} \\ \downarrow & & \downarrow {\scriptstyle S^{-1}f} \\ \{0\} & \longrightarrow & Q\mathcal{P} \end{array}$$
*is homotopy Cartesian.*

**Proof.** By Theorem B, it suffices to prove that the base change functors for the fibered functor $S^{-1}f$ are homotopy equivalences. Since every arrow in $Q\mathcal{P}$ can be factored as $q^! \circ i_!$ where $i$ is admissible mono, $q$ is admissible epi, and for $i : C \rightarrowtail C'$, $q : C' \twoheadrightarrow C$ we have $i_{C''!} = i_! \circ i_{C!}$, $q^!_{C'} = q^! \circ q^!_C$ (where $i_C : 0 \rightarrowtail C$, $q_C : C \twoheadrightarrow 0$, etc.), it suffices to prove this for arrows $i_{C!}$ and $q^!_C$.

## 7. Comparison of the Plus and Q-Constructions

The base change $(q_C^!):^* (S^{-1}\mathcal{E}_C \longrightarrow S^{-1}\mathcal{E}_0$ is given by

$$(A', (0 \to A \to B \to C \to 0)) \mapsto (A', (0 \to B \xrightarrow{1_B} B \to 0 \to 0)).$$

Identifying $\mathcal{E}_0$ with $\mathcal{S}$, this is just $S^{-1}\mathcal{E}_C \longrightarrow S^{-1}\mathcal{S}$,

$$(A', (0 \to A \to B \to C \to 0) \mapsto (A', B).$$

Since $\langle \mathcal{S}, \mathcal{E}_C \rangle$ is contractible, (7.6) implies that $S^{-1}\mathcal{S} \longrightarrow S^{-1}\mathcal{E}_C$ given by

$$(A', A) \mapsto (A', 0 \longrightarrow A \longrightarrow A \oplus C \longrightarrow C \longrightarrow 0)$$

is a homotopy equivalence. The composite with $(q_C^!)^*$ is the functor $S^{-1}\mathcal{S} \longrightarrow S^{-1}\mathcal{S}$ given by $(A', A) \longrightarrow (A', A \oplus C)$. This is just translation by $C$ on $S^{-1}\mathcal{S}$, which is a homotopy equivalence, since $\mathcal{S}$ acts invertibly on $S^{-1}\mathcal{S}$. Hence $(q_C^!)$ is a homotopy equivalence.

Similarly $(i_{C!})^* : S^{-1}\mathcal{E}_C \longrightarrow S^{-1}\mathcal{E}_0$ is given by

$$(A', 0 \longrightarrow A \longrightarrow B \longrightarrow C \longrightarrow 0)$$
$$\mapsto (A', 0 \longrightarrow A \xrightarrow{1_A} A \longrightarrow 0 \longrightarrow 0),$$

which is identified with the functor $S^{-1}\mathcal{E}_C \longrightarrow S^{-1}\mathcal{S}$,

$$(A', 0 \longrightarrow A \longrightarrow B \longrightarrow C \longrightarrow 0) \mapsto (A', A).$$

Now the composite functor $S^{-1}\mathcal{S} \longrightarrow S^{-1}\mathcal{E}_C \longrightarrow S^{-1}\mathcal{S}$ is in fact the identity. This proves (7.9).

**Lemma (7.10).** $S^{-1}\mathcal{E}$ *is contractible.*

**Proof.** For any category $\mathcal{X}$, let $\text{Sub}\,\mathcal{X}$ denote the category whose objects are the arrows of $\mathcal{X}$, such that an arrow $f \to g$ in $\text{Sub}\,\mathcal{X}$ is a pair of arrows $h, k$ of $\mathcal{X}$ satisfying $k \circ f \circ h = g$.

$$\begin{array}{ccc} \bullet & \xrightarrow{f} & \bullet \\ h \uparrow & & \downarrow k \\ \bullet & \xrightarrow{g} & \bullet \end{array}$$

Then the functor $(f: M \longrightarrow N) \mapsto N$, $\text{Sub}\,\mathcal{X} \longrightarrow \mathcal{X}$ is an equivalence of categories, with inverse given by $N \mapsto (1_N : N \longrightarrow N)$.

Let $\mathcal{X}$ be the subcategory of $Q\mathcal{P}$ with the same objects, whose arrows are all the maps $i_!$, for admissible monos $i$ in $\mathcal{P}$. Then $0 \in \mathcal{X}$ is an initial object, so that $\mathcal{X}$, and hence $\text{Sub}\,\mathcal{X}$, are contractible.

There is a functor $\text{Sub}\,\mathcal{X} \longrightarrow \mathcal{E}$, given by $(i_! : M \longrightarrow N) \mapsto (0 \longrightarrow M \xrightarrow{i} N \longrightarrow \text{coker}\,i \longrightarrow 0)$. An arrow $i_{1!} \longrightarrow i_{2!}$ in $\text{Sub}\,\mathcal{X}$, for $i_j : M_j \rightarrowtail N_j$, $j = 1, 2$ consists of a square of admissible monomorphisms

$$\begin{array}{ccc} M_1 & \xrightarrow{i_1} & N_1 \\ {\scriptstyle i'_1}\uparrow & & \downarrow{\scriptstyle i'_2} \\ M_2 & \xrightarrow{i_2} & N_2 \end{array}$$

yielding the diagram

$$\begin{array}{ccccccccc} 0 & \longrightarrow & M_1 & \xrightarrow{i_1} & N_1 & \longrightarrow & \text{coker}\,i_1 & \longrightarrow & 0 \\ & & \uparrow & & \| & & \uparrow & & \\ 0 & \longrightarrow & M_2 & \xrightarrow{i_1 \circ i'_1} & N_1 & \longrightarrow & \text{coker}\,(i_1 \circ i'_1) & \longrightarrow & 0 \\ & & \| & & \downarrow & \square & \downarrow & & \\ 0 & \longrightarrow & M_2 & \xrightarrow{i_2} & N_2 & \longrightarrow & \text{coker}\,i_2 & \longrightarrow & 0. \end{array}$$

Thus, $\text{Sub}\,\mathcal{X} \longrightarrow \mathcal{E}$ is an equivalence of categories, with inverse

$$(0 \longrightarrow M \xrightarrow{i} N \longrightarrow P \longrightarrow 0) \mapsto (i_! : M \longrightarrow N).$$

Hence $\mathcal{E}$ is contractible. Thus $\mathcal{S}$ acts invertibly on $\mathcal{E}$; hence by (7.1), $\mathcal{E} \longrightarrow \mathcal{S}^{-1}\mathcal{E}$ is a homotopy equivalence. Hence $\mathcal{S}^{-1}\mathcal{E}$ is contractible. This proves (7.10).

From (7.9) and (7.10), we have a homotopy equivalence

$$B\mathcal{S}^{-1}\mathcal{S} \longrightarrow F(B\mathcal{S}^{-1}f, \{0\})$$

to the homotopy fiber of $B\mathcal{S}^{-1}f : B\mathcal{S}^{-1}\mathcal{E} \longrightarrow BQ\mathcal{P}$. But by (7.10), the natural inclusion $\Omega\,BQ\mathcal{P} \subset F(B\mathcal{S}^{-1}f, \{0\})$ (where $\Omega\,BQ\mathcal{P}$ is the space of loops in $BQ\mathcal{P}$ based at $\{0\}$) is a homotopy equivalence, where this inclusion is given by $\omega \mapsto (x_0, \omega)$ where $x_0$ is a base point of $B\mathcal{S}^{-1}\mathcal{E}$, $\omega : I \longrightarrow BQ\mathcal{P}$ a loop based at 0. Indeed, one sees easily that a deformation retraction of $B\mathcal{S}^{-1}\mathcal{E}$ into $\{x_0\}$ induces one of $F(B\mathcal{S}^{-1}f, \{0\})$ into $\Omega\,BQ\mathcal{P}$. Thus we have the chain of homotopy equivalences

$$B\mathcal{S}^{-1}\mathcal{S} \longrightarrow F(B\mathcal{S}^{-1}f, \{0\}) \longleftarrow \Omega\,BQ\mathcal{P}$$

which proves Theorem (7.7). As noted earlier, this gives the homotopy equivalence

$$BGL(R)^+ \cong \Omega\,BQ\mathcal{P}(R)_0$$

of Theorem (5.1).

## 7. Comparison of the Plus and $Q$-Constructions

We end this chapter with a brief discussion of the relation between the transfer (or norm) maps on $K$-groups defined using the plus and $Q$-constructions. The situation we consider is as follows: let $R$, $S$ be rings, with a homomorphism $R \to S$ making $S$ into a finitely generated projective $R$-module.

On the one hand, there is a 'direct image' functor $f_* : \mathcal{P}(S) \to \mathcal{P}(R)$ between the categories of finitely generated projective modules over $S$ and $R$ respectively, given by $f_*(P) = P$, regarded as an $R$-module via the homomorphism $R \to S$. This is exact, and so yields a functor $Q\mathcal{P}(S) \to Q\mathcal{P}(R)$, and a map on $K$-groups $f_* : K_i(S) \to K_i(R)$.

On the other hand, we may choose a finitely generated projective $R$-module $P_0$ such that there is an isomorphism $f_*(S) \oplus P_0 \cong R^{\oplus m}$ for some $m > 0$, which we fix. This yields a compatible family of isomorphisms $f_*(S^{\oplus n}) \oplus P_0^{\oplus n} \cong R^{\oplus mn}$, giving a compatible family of (injective) homomorphisms $GL_n(S) \to GL_{mn}(R)$, and hence a homomorphism $GL(S) \to GL(R)$. This clearly maps $E(S)$ into $E(R)$, hence yields a map of spaces $f'_* : BGL(S)^+ \to BLG(R)^+$. One checks by the methods of Chapter 2 that the homotopy class of this map is independent of the choices made, so that there is a well defined map on $K$-groups $f'_* : K_i(S) \to K_i(R)$ for each $i > 0$. For $i = 1, 2$ these are the transfer maps defined in Chapter 14 of Milnor's book *Introduction to Algebraic K-Theory*.

We will sketch a proof that $f_* = f'_*$ on $K_i(S)$ for all $i > 0$. Let $\mathcal{S}_S = \mathrm{Iso}(\mathcal{P}(S))$, $\mathcal{S}_R = \mathrm{Iso}(\mathcal{P}(R))$ be the monoidal categories associated to $\mathcal{P}(S)$, $\mathcal{P}(R)$ respectively; then the exact functor $f_* : \mathcal{P}(S) \to \mathcal{P}(R)$ induces a monoidal functor $f_* : \mathcal{S}_S \to \mathcal{S}_R$. Next, let $\mathcal{E}(S)$, $\mathcal{E}(R)$ be the categories obtained by the extension constructions over $\mathcal{P}(S)$, $\mathcal{P}(R)$ respectively. Again, there is a functor $f_* : \mathcal{E}(S) \to \mathcal{E}(R)$. These functors $f_*$ yield a map between the homotopy Cartesian squares of lemma (7.9) associated to the rings $S$ and $R$, respectively. Hence there is a homotopy commutative diagram

$$\begin{array}{ccc} \Omega BQ\mathcal{P}(S) & \xrightarrow{\simeq} & B\mathcal{S}_S^{-1}\mathcal{S}_S \\ f_* \downarrow & & \downarrow f_* \\ \Omega BQ\mathcal{P}(R) & \xrightarrow{\simeq} & B\mathcal{S}_R^{-1}\mathcal{S}_R \end{array}$$

showing that $f_* : B\mathcal{S}_S^{-1}\mathcal{S}_S \to B\mathcal{S}_R^{-1}\mathcal{S}_R$ induces the transfer map $f_*$ on $K_i$ for all $i$.

One now checks that the diagram

$$\begin{array}{ccc} BGL(S)^+ & \longrightarrow & B\mathcal{S}_S^{-1}\mathcal{S}_S \\ f'_* \downarrow & & \downarrow f_* \\ BGL(R)^+ & \longrightarrow & B\mathcal{S}_R^{-1}\mathcal{S}_R \end{array}$$

is homotopy commutative. This implies the claimed equality of the two transfer maps. We will use $f_*$ to denote either of them.

It is tempting to try to directly define products on $K$-groups using the $\mathcal{S}_R^{-1}\mathcal{S}_R$ construction, for which various functorial properties (like the projection formula) are obvious, and thus to avoid some of the work done in Chapter 2. There is an 'obvious' approach to doing this: one tries to use the tensor product of projective modules to define a functor $\mathcal{S}_R^{-1}\mathcal{S}_R \times \mathcal{S}_R^{-1}\mathcal{S}_R \to \mathcal{S}_R^{-1}\mathcal{S}_R$, which is given on objects by

$$(A, B) \times (C, D) \mapsto (A \otimes_R B \oplus B \otimes_R D, A \otimes_R D \oplus B \otimes_R C),$$

with brackets inserted at suitable places, and using the natural isomorphisms inherent in the structure of a monoidal category with tensor products to define what must happen on morphisms. However, it turns out that this fails for a somewhat subtle reason: it is impossible to make the assignment on morphisms compatible with composition! For details, see: R. Thomason, Beware the phony multiplication on Quillen's $\mathcal{A}^{-1}\mathcal{A}$, Proc. Amer. Math. Soc. 80 (1980) 569–573.

The method of monoidal categories can be used to prove that $\mathcal{S}^{-1}\mathcal{S}$ is an infinite loop space; see, for example, J.P. May (with contributions by F. Quinn, N. Ray, J. Tornehave): *$E_\infty$-Ring Spaces and $E_\infty$-Ring Spectra*, Lect. Notes in Math., No. 533, Springer-Verlag (1976).

# 8. The Merkurjev–Suslin Theorem

The main references in this chapter are:

[MS] A. S. Merkurjev, A. A. Suslin, $\mathcal{K}$-cohomology of Severi–Brauer varieties and the norm residue homomorphism, *Math. USSR Izv.* 21 (1983) 307–340 (English translation)

[CS] C. Soulé, $K_2$ et le groupe de Brauer (d'après A. S. Merkurjev et A. A. Suslin), *Séminaire Bourbaki* 601, Astérique 105–106, Soc. Math. France (1983).

[M] A. S. Merkurjev, $K_2$ of fields and the Brauer group, *Contemporary Math.* Vol. 55, Part II, Amer. Math. Soc. (1986).

[Mi] J. S. Milne, *Étale Cohomology*, Princeton Math. Ser. 33, Princeton (1980).

[AS] A. A. Suslin, Algebraic $K$-theory and the norm-residue homomorphism, *J. Soviet Math.* 30 (1985) 2556–2611.

[AS2] A. A. Suslin: Torsion in $K_2$ of fields, *K-Theory* 1 (1987) 5–29.

[S] J.-P. Serre, *Local Fields*, Grad. Texts in Math. No. 67, Springer-Verlag (1979).

[T] J. Tate, Relations between $K_2$ and Galois cohomology, *Invent. Math.* 36 (1976) 257–274.

Notation: If $F$ is a field, we will use multiplicative notation for the group operation on $F^*$, and on Steinberg symbols. Otherwise, we will use additive notation for the group operation in an Abelian group. If $A$ is an Abelian group, $n$ an integer, and $n_A : A \to A$ is multiplication by $n$, then we let $_nA = \ker n_A$ and $n \cdot A = \operatorname{im} n_A$.

## (8.1) The Galois Symbol

Let $F$ be a field, $\overline{F}$ a separable closure of $F$, $G = \operatorname{Gal}(\overline{F}/F)$. Let $n > 0$ be an integer relatively prime to char. $F$, and let $\mu_n$ denote the group of $n^{\text{th}}$ roots of unity in $\overline{F}^*$. We have the Kummer sequence of $G$-modules (with $(\cdot n)(a) = a^n$ for all $a \in \overline{F}^*$)

$$0 \to \mu_n \to \overline{F}^* \xrightarrow{\cdot n} \overline{F}^* \to 0$$

giving an exact sequence of Galois cohomology groups

$$F^* \xrightarrow{\cdot n} F^* \to H^1(F, \mu_n) \to \underset{\underset{0}{\|}}{H^1(F, \overline{F}^*)}$$

where the last group is 0 by Hilbert's Theorem 90. Thus we have an isomorphism
$$\chi_n = \chi_{n,F} : F^* \otimes_{\mathbb{Z}} \mathbb{Z}/n\mathbb{Z} \to H^1(F, \mu_n).$$
(We will abuse notation and also use $\chi_n$ to denote the surjection $F^* \to H^1(F, \mu_n)$.) The cup product gives a map
$$F^* \otimes_{\mathbb{Z}} F^* \to (F^* \otimes_{\mathbb{Z}} F^*) \otimes_{\mathbb{Z}} \mathbb{Z}/n\mathbb{Z} \to H^1(F, \mu_n) \otimes_{\mathbb{Z}} H^1(F, \mu_n) \to H^2(F, \mu_n^{\otimes 2}).$$

**Lemma 8.1** *If $a \in F^*$, $a \neq 1$, then $\chi_n(a) \cup \chi_n(1-a) = 0$.*

**Proof.** Let $T^n - a = \prod_{i=1}^k p_i(T)^{n_i}$ be the decomposition into prime factors over $F[T]$. Let $\alpha_i \in \overline{F}$ be a root of $p_i(T) = 0$, and $F_i = F(\alpha_i)$. Then
$$1 - a = \prod_{i=1}^k p_i(1)^{n_i} = \prod_{i=1}^k N_{F_i/F}(1 - \alpha_i)^{n_i}.$$
Hence
$$\chi_{n,F}(1-a) = \sum n_i N_{F_i/F} \circ \chi_{n,F_i}(1 - \alpha_i),$$
where $N_{F_i/F} : H^1(F_i, \mu_n) \to H^1(F, \mu_n)$ is the norm (i.e., the corestriction). By the projection formula (in Galois cohomology),
$$\chi_{n,F}(a) \cup \chi_{n,F}(1-a) = \sum n_i \chi_{n,F}(a) \cup (N_{F_i/F} \circ \chi_{n,F_i}(1 - \alpha_i)) =$$
$$\sum_i n_i N_{F_i/F} \left( \chi_{n,F_i}(a) \cup \chi_{n,F_i}(1 - \alpha_i) \right) =$$
$$\sum_i n_i N_{F_i/F} \left( \chi_{n,F_i}(\alpha_i^n) \cup \chi_{n,F_i}(1 - \alpha_i) \right) = 0.$$

**Corollary 8.2** *The map $F^* \otimes_{\mathbb{Z}} F^* \to H^2(F, \mu_n^{\otimes 2})$ given by*
$$a \otimes b \mapsto \chi_{n,F}(a) \cup \chi_{n,F}(b)$$
*is a Steinberg symbol (see (1.14)).*

**Definition:** Let
$$R_{n,F} : K_2(F) \otimes_{\mathbb{Z}} \mathbb{Z}/n\mathbb{Z} \to H^2(F, \mu_n^{\otimes 2})$$
be the homomorphism given by (8.2); it is called the *norm residue homomorphism* or the *Galois symbol* (see (1.16)).

Suppose the group of $n^{\text{th}}$ roots of unity $\mu_n \subset F$; let $\zeta \in \mu_n$ be a primitive $n^{\text{th}}$ root. Given $a, b \in F^*$, let $A_\zeta(a, b)$ be the central simple $F$-algebra with generators $X, Y$ subject to the relations $X^n = a$, $Y^n = b$,

$\zeta XY = YX$. One verifies easily that $A_\zeta(a,b)$ is indeed a central simple algebra over $F$ of degree $n$ (i.e., $\dim_F A_\zeta(a,b) = n^2$). $A_\zeta(a,b)$ is called a *cyclic algebra* since it is split (i.e., isomorphic to the algebra of $n \times n$ matrices) over the cyclic extension field $F(\sqrt[n]{b})$.

More generally, let $E/F$ be a cyclic Galois extension of degree $n$, and $\sigma \in \mathrm{Gal}\,(E/F)$ a generator. Given $b \in F^*$, consider the $F$-algebra $A(\sigma, b)$ which is an $E$-vector space of dimension $n$ with basis $1, Y, \ldots, Y^{n-1}$ and algebra structure given by the rules $Y^n = b$, $Yc = \sigma(c)Y$ for $c \in E$, and

$$Y^i Y^j = \begin{cases} Y^{i+j} & \text{if } 1 \le i, j \text{ and } i+j < n, \\ bY^{i+j-n} & \text{if } 1 \le i, j \le n-1 \text{ and } i+j \ge n. \end{cases}$$

If $E/F$ is a Kummer extension, so that $E = F(\sqrt[n]{a})$, this reduces to the earlier construction. We call $A(\sigma, b)$ the *cyclic algebra* over $F$ associated to the cyclic extension $E/F$, the generator $\sigma \in \mathrm{Gal}\,(E/F)$ and $b \in F^*$. For example, if char. $F = p > 0$, and $E = F[X]/(X^p - X - a)$ is an Artin-Schreier extension, $\sigma \in \mathrm{Gal}\,(E/F)$ given by $\sigma(X) = X+1$, then $A(\sigma, b)$ has generators $X, Y$ subject to the relations $X^p - X = a$, $Y^p = b$, $YX = XY + Y$.

If $E/F$ is a cyclic extension of degree $m$, $\sigma \in \mathrm{Gal}\,(E/F)$ a generator, $b \in F^*$, then there is another way of describing $[A(\sigma, b)] \in \mathrm{Br}\,(F)$, the Brauer class of $A(\sigma, b)$. There is a unique character $\chi: \mathrm{Gal}\,(E/F) \xrightarrow{\cong} \mathbb{Z}/m\mathbb{Z}$ given by $\chi(\sigma) = 1 \,(\mathrm{mod}\, m)$. Let $\overline{\chi}: \mathrm{Gal}\,(\overline{F}/F) \to \mathbb{Z}/m\mathbb{Z}$ be the induced character. From the Bockstein exact sequence (associated to the sequence of trivial $\mathrm{Gal}\,(\overline{F}/F)$-modules $0 \to \mathbb{Z} \xrightarrow{\cdot m} \mathbb{Z} \to \mathbb{Z}/m\mathbb{Z} \to 0$)

$$\cdots \to H^1(F, \mathbb{Z}) \to H^1(F, \mathbb{Z}/m\mathbb{Z}) \xrightarrow{\delta} H^2(F, \mathbb{Z}) \to \cdots$$

we have a class $\delta(\overline{\chi}) \in H^2(F, \mathbb{Z})$. The cup product

$$H^2(F, \mathbb{Z}) \otimes_\mathbb{Z} H^0(F, \overline{F}^*) \to H^2(F, \overline{F}^*) = \mathrm{Br}\,(F)$$

yields a class $\delta(\overline{\chi}) \cup (b) \in \mathrm{Br}\,(F)$. It can be shown (see A. Weil, *Basic Number Theory*, XII, §4.5) that $\delta(\overline{\chi}) \cup (b) = [A(\sigma, b)] \in \mathrm{Br}\,(F)$.

Now suppose $\mu_n \subset F$; fix a primitive $n^{\mathrm{th}}$ root $\zeta \in \mu_n$. Given a Kummer extension $E = F(\sqrt[n]{a})$ and a generator $\sigma \in \mathrm{Gal}\,(E/F)$, there is a unique $i(\sigma) \,(\mathrm{mod}\, n)$ such that $\sigma(\sqrt[n]{a}) = \zeta^{i(\sigma)} \sqrt[n]{a}$. This defines a character $\mathrm{Gal}\,(E/F) \to \mathbb{Z}/n\mathbb{Z}$, $\sigma \mapsto i(\sigma) \,(\mathrm{mod}\, n)$. Let $\overline{\chi}$ be the induced character of $\mathrm{Gal}\,(\overline{F}/F)$. Now $\delta(\overline{\chi}) \cup (b) \in {}_n\mathrm{Br}\,(F)$. But from Hilbert's Theorem 90 and the Kummer sequence, we have an exact sequence

$$0 \to H^2(F, \mu_n) \to \mathrm{Br}\,(F) \xrightarrow{\cdot n} \mathrm{Br}\,(F) \to 0,$$

so that ${}_n\mathrm{Br}\,(F) \cong H^2(F, \mu_n)$. Thus $\delta(\overline{\chi}) \cup (b) \cup (\zeta) \in H^2(F, \mu_n^{\otimes 2})$.

**Lemma 8.3** $R_{n,F}(\{a,b\}) = \delta(\overline{\chi}) \cup (b) \cup (\zeta)$.

**Proof.** See Serre's *Local Fields*, XIV, Prop. 5.

We denote the induced map

$$K_2(F) \otimes_{\mathbb{Z}} \mathbb{Z}/n\mathbb{Z} \to {}_n\text{Br}(F), \quad \{a,b\} \mapsto \delta(\overline{\chi}) \cup (b),$$

by $R'_{n,F}$.

**Lemma 8.4** *(a) Let $E/F$ be a cyclic extension of fields of degree $n$, $\sigma \in \text{Gal}(E/F)$ a generator, $\chi : \text{Gal}(E/F) \to \mathbb{Z}/n\mathbb{Z}$ the character given by $\sigma \mapsto 1$. Then the map $F^* \to \text{Br}(F)$, $b \mapsto \delta(\overline{\chi}) \cup (b)$ gives an isomorphism*

$$F^*/N_{E/F}E^* \cong \text{Br}(E/F) = \ker(\text{Br}(F) \to \text{Br}(E)).$$

*(b) Suppose $\mu_n \subset F$. Then*

$$R_{n,F}(\{a,b\}) = 0 \Leftrightarrow \{a,b\} \in n \cdot K_2(F)$$

$$\Leftrightarrow b = N_{E/F}(c) \text{ for some } c \in E^*, \text{ where } E = F(\sqrt[n]{a})$$

$$\Leftrightarrow b = N_{E'/F}(c) \text{ for some } c \in E'^*, \text{ where } E' = F[T]/(T^n - a).$$

**Proof.** (a) One knows (see Serre, *Local Fields*, X, §4, Cor. to Prop. 6) that $\text{Br}(E/F) = \ker(\text{Br}(F) \to \text{Br}(E)) \cong H^2(E/F, E^*)$. Thus we must show that if $\delta(\overline{\chi}) \in H^2(E/F, \mathbb{Z})$ is obtained using the boundary map $\delta$ in the exact sequence

$$\cdots \to H^1(E/F, \mathbb{Z}) \to H^1(E/F, \mathbb{Z}/n\mathbb{Z}) \xrightarrow{\delta} H^2(E/F, \mathbb{Z}) \to \cdots,$$

then $b \mapsto \delta(\overline{\chi}) \cup (b)$ gives an isomorphism $F^*/N_{E/F}E^* \cong \text{Br}(E/F)$. But $b \mapsto \delta(\overline{\chi}) \cup (b)$ is precisely the periodicity isomorphism for the Tate cohomology groups $\widehat{H}^0(E/F, E^*) \to \widehat{H}^2(E/F, E^*)$, for the cohomology groups of the cyclic group $\text{Gal}(E/F)$ (see Serre, *Local Fields*, VIII, §4).

(b) Clearly $b = N_{E/F}(c)$ for some $c \in E^* \Leftrightarrow b = N_{E'/F}(c')$ for some $c \in E'^*$. Hence it suffices to prove:

$$\{a,b\} \in n \cdot K_2(F) \Leftrightarrow R_{n,F}(\{a,b\}) = 0 \Leftrightarrow b = N_{E/F}(c) \text{ for some } c \in E^*.$$

From (8.3) and (8.4)(a) applied to $E/F$, we see that

$$\{a,b\} \in n \cdot K_2(F) \Rightarrow R_{n,F}(\{a,b\}) = 0 \Leftrightarrow b = N_{E/F}(c) \text{ for some } c \in E^*.$$

Hence it suffices to observe that if $N_{E/F} : K_2(E) \to K_2(F)$ is the transfer map (the map $f_*$ obtained from (5.11) for the morphism $f : \text{Spec } E \to \text{Spec } F$), then from the projection formula, we have

$$b = N_{E/F}(c) \Rightarrow \{a,b\} = N_{E/F}\{a,c\} = N_{E/F}\{\sqrt[n]{a}, c\}^n$$

which lies in $n \cdot K_2(F)$. Here, the projection formula we need is verified in Milnor's book *Introduction to Algebraic K-Theory*, Ch. 14, with a different definition of the transfer map on $K_2$: one regards $K_2(F)$ as $H_2(\mathrm{E}(F), \mathbb{Z})$, where $\mathrm{E}(F) \subset \mathrm{GL}(F)$ is the group of elementary matrices; now choose a basis for $E$ as an $F$-vector space, to get an embedding $\mathrm{GL}(E) \hookrightarrow \mathrm{GL}(F)$, and an induced embedding of commutator subgroups $\mathrm{E}(E) \hookrightarrow \mathrm{E}(F)$. The transfer map is the induced mapping on $H_2$. The inclusion $\mathrm{GL}(E) \hookrightarrow \mathrm{GL}(F)$ induces $BGL(E)^+ \to BGL(F)^+$, giving maps $f_* : K_i(E) \to K_i(F)$; this yields the above map for $i = 2$. As discussed at the end of Chapter 7, this agrees with the transfer map defined using (5.11).

With this background, we now state the theorem of Merkurjev and Suslin, which is the main goal of this chapter.

**Theorem 8.5** (Merkurjev–Suslin) *Let $F$ be a field, $n > 0$ an integer not divisible by char. $F$. Then the Galois symbol*

$$R_{n,F} : K_2(F) \otimes_{\mathbb{Z}} \mathbb{Z}/n\mathbb{Z} \to H^2(F, \mu_n^{\otimes 2})$$

*is an isomorphism.*

### (8.2) Proof of the Merkurjev–Suslin Theorem

We first make some preliminary reductions. We begin with a useful lemma due to Bass and Tate.

**Lemma 8.6** *Let $p$ be a prime number, and let $F$ be a field which has no non-trivial finite extensions of degree $< p$. Let $E/F$ be an extension of degree $p$. Then $K_2(E)$ is generated by symbols $\{a, b\}$ with $a \in F^*$, $b \in E^*$.*

**Proof.** Let $F = E(t)$, so that $t$ satisfies a monic polynomial over $F$ of degree $p$; thus every element of $E$ is a polynomial (over $F$) in $t$ of degree $< p$. Since $F$ has no non-trivial finite extensions of degree $< p$, any polynomial over $F$ of degree $< p$ factors into a product of linear factors. Hence every element of $E^*$ is a product of linear polynomials in $t$. Thus $K_2(E)$ is generated by symbols $\{at+b, ct+d\}$, $a, b, c, d \in F$, $(at+b), (ct+d) \in E^*$. If $ac = 0$ or $ad = bc$, then $\{at+b, ct+d\}$ lies in the image of $F^* \otimes_{\mathbb{Z}} E^* \to K_2(E)$ (use the relation $\{u, -u\} = 1$ in $K_2$, if $ad = bc$). If $ac(ad - bc) \neq 0$, write

$$\{at + b, ct + d\} =$$

$$\left\{\frac{c(at+b)}{bc-ad}, \frac{a(ct+d)}{ad-bc}\right\}\left\{\frac{c}{bc-ad}, \frac{a(ct+d)}{ad-bc}\right\}^{-1}\left\{at+b, \frac{a}{ad-bc}\right\}^{-1}.$$

But

$$\frac{c(at+b)}{bc-ad} + \frac{a(ct+d)}{ad-bc} = 1;$$

using $\{u, 1 - u\} = 1$, we deduce that $\{at + b, ct + d\} \in \text{image}(F^* \otimes_{\mathbb{Z}} E^*)$.

**Lemma 8.7** *Let $E/F$ be a finite separable extension of fields, $n$ a positive integer not divisible by char. $F$. Then we have a commutative diagram*

$$\begin{array}{ccc} K_2(E) & \stackrel{R_{n,E}}{\to} & H^2(E, \mu_n^{\otimes 2}) \\ N_{E/F} \downarrow & & \downarrow N_{E/F} \\ K_2(F) & \stackrel{R_{n,F}}{\to} & H^2(F, \mu_n^{\otimes 2}) \end{array}$$

(In fact the lemma is true without the separability assumption, but we do not need this.)

**Proof.** It suffices to consider the case when $n = p^k$ for some prime number $p$. If $F'/F$ is a finite extension of degree relatively prime to $p$, then since the composite

$$H^2(F, \mu_n^{\otimes 2}) \stackrel{\mathrm{Res}_{F'/F}}{\to} H^2(F', \mu_n^{\otimes 2}) \stackrel{N_{F'/F}}{\to} H^2(F, \mu_n^{\otimes 2})$$

is multiplication by $[F' : F]$, it is an isomorphism. Let $\widetilde{F}$ be a maximal algebraic extension of $F$ such that any finite sub-extension of $F$ contained in $\widetilde{F}$ has degree relatively prime to $p$. Then $E \otimes_F \widetilde{F} \cong \prod_{i=1}^{r} E_i$ for certain extension fields $E_i$ of $\widetilde{F}$ (this is true because $E/F$ is separable). Then we have a diagram

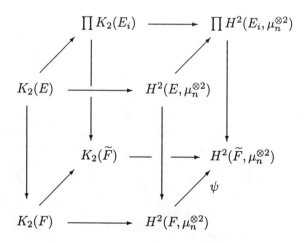

where the vertical arrows are norm maps, the horizontal maps are Galois symbols, and the maps joining the front and back faces are induced by the inclusions of fields. Since the map $\psi$ is injective, the lemma for $E/F$ follows from the lemma for each of the extensions $E_i/\widetilde{F}$.

## 8. The Merkurjev–Suslin Theorem

Thus, to prove the lemma, we may assume that $n = p^k$, and $F$ has no finite extensions of degree prime to $p$. Hence any finite separable extension of $F$ is a tower of cyclic Galois extensions of degree $p$. If $F \subset E \subset E'$ are finite extensions, then $N_{E'/F} = N_{E/F} \circ N_{E'/E}$ (where $N$ denotes the norm on either $K_2$ or $H^2$). So it suffices to consider the case when $E/F$ is cyclic of degree $p$.

Now by lemma (8.6), $K_2(E)$ is generated by symbols $\{a,b\}$ with $a \in F^*$, $b \in E^*$. But

$$R_{p,F} N_{E/F}\{a,b\} = R_{p,F}\{a, N_{E/F}b\} = \chi_{p,F}(a) \cup (\chi_{p,F} \circ N_{E/F}(b)) =$$

$$\chi_{p,F}(a) \cup (N_{E/F} \circ \chi_{p,E}(b)) = N_{E/F}(\chi_{p,E}(a) \cup \chi_{p,E}(b)) = N_{E/F} R_{p,E}\{a,b\}.$$

To prove (8.5), that $R_{n,F}$ is an isomorphism, it clearly suffices to do this when $n = p^k$ is a power of a prime number $p$. From the above lemma, we may also assume that $F$ contains a primitive $p^{\text{th}}$ root of unity. From now on, we will fix such a prime number $p$.

**Lemma 8.8** *Let $F$ be a field containing a primitive $p^{\text{th}}$ root of unity, such that $R_{p,F}$ is an isomorphism. Then $R_{p^k,F}$ is an isomorphism for all $k > 0$.*

**Proof.** By induction on $k$, we may assume $k > 1$, and $R_{p^l,F}$ is an isomorphism for $l < k$. We have a commutative diagram

$$\begin{array}{ccccccc} F^* \otimes_{\mathbb{Z}} \mu_p & \to & K_2(F) \otimes_{\mathbb{Z}} \mathbb{Z}/p^{k-1}\mathbb{Z} & \to & K_2(F) \otimes_{\mathbb{Z}} \mathbb{Z}/p^k\mathbb{Z} & \to & K_2(F) \otimes_{\mathbb{Z}} \mathbb{Z}/p\mathbb{Z} \to 0 \\ \chi_{p,F} \otimes_{\mathbb{Z}} 1 \downarrow & & R_{p^{k-1},F} \downarrow & & R_{p^k,F} \downarrow & & R_{p,F} \downarrow \\ H^1(F, \mu_p^{\otimes 2}) & \to & H^2(F, \mu_{p^{k-1}}^{\otimes 2}) & \to & H^2(F, \mu_{p^k}^{\otimes 2}) & \to & H^2(F, \mu_p^{\otimes 2}) \end{array}$$

in which the bottom row is exact, the top row is a complex which is exact except possibly at $K_2(F) \otimes_{\mathbb{Z}} \mathbb{Z}/p^{k-1}\mathbb{Z}$, and the vertical arrows other than $R_{p^k,F}$ are given to be isomorphisms. A diagram chase now implies that $R_{p^k,F}$ is also an isomorphism.

Hence we are reduced to proving that $R_{p,F}$ is an isomorphism, where $p$ is a prime number, and $F$ contains a primitive $p^{\text{th}}$ root of unity $\zeta$. The main idea of the proof in this case is to compare the Galois symbol for $F$ with that for an extension field $F(X)$, where $X$ is a Severi-Brauer variety over $F$ associated to a division algebra of degree $p$ over $F$, and $F(X)$ is its function field. Along the way, we also prove an analogue for $K_2$ of Hilbert's Theorem 90 (see Theorem (8.15)), which is of independent interest.

In Chapter 5, we gave Quillen's computation of the $K$-groups of a Severi-Brauer variety (see (5.29), (5.40)). Let $p$ be a prime and let $D$ be a central simple algebra over $F$ of degree $p$ (i.e., $D$ is a simple $F$-algebra with $\dim_F D = p^2$, and $D$ has centre $F$). Then either $D$ is a division algebra, or $D \cong M_p(F)$, the algebra of $p \times p$ matrices over $F$. Let $X$ be the Severi-Brauer variety over $F$ associated to $D$ (see Serre, *Local Fields*, Ch. X). We have an isomorphism, by (5.40),

$$K_i(X) \cong K_i(F) \oplus K_i(D) \oplus \cdots \oplus K_i(D^{\otimes j}) \oplus \cdots \oplus K_i(D^{\otimes p-1}), \quad \forall\, i \geq 0.$$

If $D = M_p(F)$, so that $D^{\otimes j} = M_{p^j}(F)$, we have canonical isomorphisms $K_i(D^{\otimes j}) \cong K_i(F)$. Indeed, for any $m > 0$, if we fix an isomorphism $M_m(F) \cong \mathrm{End}_F(W)$, where $W$ is an $F$-vector space of dimension $m$, then $W$ can be regarded as a left $M_m(F)$-module. Then $W$ is an indecomposable projective module, and we have a (Morita) equivalence of categories $\mathcal{P}(F) \to \mathcal{P}(M_m(F))$, $V \mapsto W \otimes_F V$, where $V \in \mathcal{P}(F)$ is a finite dimensional $F$-vector space, and $\mathcal{P}(M_m(F))$ is the category of finitely generated projective $M_m(F)$ modules. Hence $Q\mathcal{P}(F) \to Q\mathcal{P}(M_m(F))$ is an equivalence of categories, giving the desired isomorphisms $K_i(F) \cong K_i(M_m(F))$. Thus in the situation when $D = M_p(F)$, (5.40) gives an isomorphism

$$K_i(X) \cong K_i(F) \oplus K_i(D) \oplus \cdots \oplus K_i(D^{\otimes j}) \oplus \cdots \oplus K_i(D^{\otimes p-1}) = K_i(F)^{\oplus p}.$$

On the other hand, $X = \mathbb{P}_F^{p-1}$, so that (5.29) gives an isomorphism

$$K_i(F)^{\oplus p} \xrightarrow{\cong} K_i(\mathbb{P}_F^{p-1}),$$

$$(a_j)_{0 \leq j \leq p-1} \mapsto \sum_j a_j \cdot \zeta^j, \quad \zeta = [\mathcal{O}_{\mathbb{P}^{p-1}}(-1)] \in K_0(\mathbb{P}_F^{p-1}).$$

We claim that the above two isomorphisms $K_i(F)^{\oplus p} \to K_i(\mathbb{P}_F^{p-1})$ coincide. To see this, note that if $\pi : \mathbb{P}_F^{p-1} \to \mathrm{Spec}\, F$ is the structure map, then in the notation of (5.40), $\mathcal{J} = V \otimes_F \mathcal{O}_{\mathbb{P}^{p-1}}(-1)$, where $V = \Gamma(\mathbb{P}_F^{p-1}, \mathcal{O}_{\mathbb{P}^{p-1}}(1))$. Hence $\mathcal{J}^{\otimes n} = V^{\otimes n} \otimes_F \mathcal{O}_{\mathbb{P}^{p-1}}(-n)$, and $f_* \mathcal{E}nd\, (\mathcal{J}^{\otimes n})^{\mathrm{op}} \cong \mathrm{End}_F(V^{\otimes n})^{\mathrm{op}} \cong \mathrm{End}_F(V^{*\otimes n})$ (we identify coherent sheaves on $\mathrm{Spec}\, F$ with finite dimensional $F$-vector spaces). Since a simple left module for $\mathrm{End}_F(V^{*\otimes n})$ is $V^{*\otimes n}$, the map (given by (5.40))

$$K_i(F) \cong K_i(\mathrm{End}_F(V^{*\otimes n})) \to K_i(\mathbb{P}_F^{p-1})$$

is that given by the functor $\mathcal{P}(F) \to \mathcal{P}(\mathbb{P}_F^{p-1})$,

$$V_1 \mapsto \mathcal{J}^{\otimes n} \otimes_{f_* \mathrm{End}\,(V^{*\otimes n})} V^{*\otimes n} \otimes_F V_1.$$

For any $F$-vector space $W$, regarded as a right $\mathrm{End}_F(W)^{\mathrm{op}}$-module, we have a canonical isomorphism

$$W \otimes_{\mathrm{End}_F(W)^{\mathrm{op}}} W^* \cong F,$$

given by the trace mapping $W \otimes_F W^* \to F$. Hence the above functor $\mathcal{P}(F) \to \mathcal{P}(\mathbb{P}_F^{p-1})$ is just

$$V_1 \mapsto \mathcal{O}_{\mathbb{P}^{p-1}}(-n) \otimes_F V_1,$$

which corresponds to the map $K_i(F) \to K_i(\mathbb{P}_F^{p-1})$, $a \mapsto a \cdot \zeta^n$, where $\zeta = [\mathcal{O}_{\mathbb{P}^{p-1}}(-1)]$. This proves the claim.

If $D$ is a division algebra, $E \subset D$ a maximal subfield, then $D_E = E \otimes_F D \cong M_p(E)$. For any such $E$, if $X_E = X \times_{\operatorname{Spec} F} \operatorname{Spec} E$, then $X_E$ is the Severi-Brauer variety associated to $D_E$, i.e., $X_E \cong \mathbb{P}_E^{p-1}$. If $f: X_E \to X$, then we have a diagram

$$\begin{array}{ccc} K_i(X) & \xrightarrow{f^*} & K_i(X_E) \cong K_i(\mathbb{P}_E^{p-1}) \\ \cong \uparrow & & \uparrow \cong \\ \oplus_{j=0}^{p-1} K_i(D^{\otimes j}) & \to & \oplus_{j=0}^{p-1} K_i(D_E^{\otimes j}) \cong K_i(E)^{\oplus p} \end{array}$$

where the vertical arrows are the isomorphisms of (5.40). In particular, for $i = 0$, we see that $K_0(D^{\otimes j}) \cong \mathbb{Z}$, $K_0(E) \cong \mathbb{Z}$, and the natural map $K_0(D^{\otimes j}) \to K_i(E)$ is either (i) an isomorphism, if $j = 0$, or (ii) is identified with multiplication by $p$, if $1 \leq j < p$. To prove (ii), it suffices to show that $D^{\otimes j} = M_{p^{j-1}}(D_j)$ for some division algebra $D_j$ of degree $p$ over $F$. Now $[D^{\otimes j}] \in \operatorname{Br}(F)$ is an element of order $p$ if $1 \leq j \leq p - 1$, so that $D^{\otimes j} \cong M_{r_j}(D_j)$ for some division algebra $D_j$ over $F$ of degree $> 1$, such that $D_j$ is split by $E$. So it suffices to show that any divison algebra $D_j$ over $F$, which is split by $E$, has degree dividing $p$. This follows from the next lemma, shown to me by R. Sridharan.

**Lemma 8.9** *If $E/F$ is a finite extension of fields, $\widetilde{D}$ a division algebra over $F$ which is split by $E$, then $M_m(\widetilde{D})$ contains a maximal commutative subring isomorphic to $E$, for some $m > 0$ (thus $m \deg \widetilde{D} = [E : F]$).*

**Proof.** We note that $E$ also splits the opposite algebra $\widetilde{D}^{\operatorname{op}}$, so that there is an isomorphism of $E$-algebras $E \otimes_F \widetilde{D}^{\operatorname{op}} \to M_n(E)$, $n = \deg \widetilde{D}$; we can use this isomorphism to regard $E^{\oplus n}$ as a left-$E$ right-$\widetilde{D}$ bimodule. If $m$ is the dimension of $E^{\oplus n}$ as a right $\widetilde{D}$-vector space, then $E \hookrightarrow \operatorname{End}_{\widetilde{D}}(E^{\oplus n}) \cong M_m(\widetilde{D})$; by counting dimensions, $E$ is a maximal commutative subring — its dimension over the centre $F$ is the square root of the dimension of the algebra.

The following corollary, needed later, is immediate from the lemma.

**Corollary 8.10** *If $\widetilde{D}$ is a central division algebra over $F$ of degree $n$, the maximal subfields of $\widetilde{D}$ are, up to $F$-isomorphism, precisely the extensions of $F$ of degree $n$ which split $\widetilde{D}$.*

Next, consider the above map $f^* : K_1(X) \to K_1(X_E)$. The corresponding map $K_1(D^{\otimes j}) = K_1(D_j) \to K_1(E)$ is given as follows. By the corollary to Example (1.6), $K_1(D_j) \cong D_j/[D_j, D_j]$. Further, $K_1(E) \cong E^*$, and the map $K_1(D_j) \to K_1(E)$ is given by the composite

$$D_j^* \to (E \otimes_F D_j)^* \cong \mathrm{GL}_p(E) \xrightarrow{det} E^*$$

where $det$ denotes the determinant. By the definition of the *reduced norm* for a division algebra, which we denote $Nrd$, this factors as

$$D_j \xrightarrow{Nrd} F^* \hookrightarrow E^*.$$

We have the following result of Sh. Wang, On the commutator group of a simple algebra, *Amer. J. Math.* 72 (1950) 323–334.

**Lemma 8.11** *Let $D$ be a central division algebra over $F$ of prime degree $p$. Then $[D^*, D^*] = \ker(Nrd : D^* \to F^*)$.*

**Proof.** We prove the result in 3 steps. We recall that if $\alpha \in D^*$ and $K$ is a maximal subfield of $D$ containing $F(\alpha)$, then $Nrd(\alpha) = N_{K/F}(\alpha)$. Since $\mathrm{Br}(F) \neq 0$, $F$ is infinite.

(i) Suppose $\alpha \in D^*$ with $Nrd(\alpha) = 1$, and $\alpha \in K^* \subset D^*$ where $K$ is a maximal subfield of $D$ containing $F(\alpha)$, and $K$ is a cyclic Galois extension of $F$. If $\sigma \in \mathrm{Gal}(K/F)$ is a generator, then by Hilbert's Theorem 90, $N_{K/F}(\alpha) = 1 \Rightarrow \alpha = \sigma(\beta)/\beta$ for some $\beta \in K^*$. By the Skolem-Noether theorem, $\sigma : K \to K$ is given by conjugation by some element $\tau \in D^*$. Then

$$\alpha = \sigma(\beta)\beta^{-1} = \tau\beta\tau^{-1}\beta^{-1} \in [D^*, D^*].$$

(ii) Next, suppose $\alpha \in D^*$ with $Nrd(\alpha) = 1$, and for some extension field $E$ of degree $n$ of $F$, we have $\alpha \in [D_E^*, D_E^*]$. Regarding $E$ as an $F$-subalgebra of the matrix algebra $M_n(F)$, so that $D_E \subset M_n(D)$, we see that the $n \times n$ diagonal matrix

$$\alpha_n = \begin{bmatrix} \alpha & & 0 \\ & \ddots & \\ 0 & & \alpha \end{bmatrix} \in [D_E^*, D_E^*] \subset [\mathrm{GL}_n(D), \mathrm{GL}_n(D)].$$

Hence the Dieudonné determinant of $\alpha_n$ is trivial, i.e., $\alpha^n \in [D^*, D^*]$. In particular, since $D$ has prime degree $p$, $D$ has a splitting field $E$ of degree $p$ over $F$, so that $\alpha^p \in [D^*, D^*]$ (if $E$ splits $D$, then $[D_E^*, D_E^*] = [\mathrm{GL}_p(E), \mathrm{GL}_p(E)] = \mathrm{SL}_p(E)$, since $E$ is an infinite field).

(iii) By (ii), we may replace $F$ by a maximal algebraic extension $F'$ such that all finite extensions of $F$ in $F'$ have degree coprime to $p$. Indeed, if we write $F' = \varinjlim_i F_i$ where $F_i/F$ is finite of degree prime to $p$, then

$$\frac{\ker Nrd(D_{F'}^*) \to F'^*}{[D_{F'}^*, D_{F'}^*]} = \varinjlim_i \frac{\ker Nrd(D_{F_i}^*) \to F_i^*}{[D_{F_i}^*, D_{F_i}^*]},$$

and for each $i$,

$$\frac{\ker Nrd(D_F^*) \to F^*}{[D_F^*, D_F^*]} \hookrightarrow \frac{\ker Nrd(D_{F_i}^*) \to F_i^*}{[D_{F_i}^*, D_{F_i}^*]}.$$

So we may assume that any finite extension of $F$ has degree equal to a power of $p$; in particular, any separable extension of $F$ of degree $p$ is a cyclic Galois extension. Now let $\alpha \in D^*$ with $Nrd(\alpha) = 1$, and let $K$ be a maximal subfield of $D$ containing $F(\alpha)$. If $K/F$ is inseparable, then $1 = Nrd(\alpha) = N_{K/F}(\alpha) = \alpha^p$, so that $\alpha = 1$. If $K/F$ is separable, then it must be a cyclic Galois extension. But then $\alpha \in [D^*, D^*]$ by (i).

Thus, the map $f^* : K_1(X) \to K_1(X_E)$ is identified with

$$F^* \oplus Nrd(D_1^*) \oplus \cdots \oplus Nrd(D_j^*) \oplus \cdots \oplus Nrd(D_{p-1}^*) \hookrightarrow E^{\oplus p},$$

where $D^{\otimes j} \cong M_{p^{j-1}}(D_j)$. Now for any division algebra $D$ over $F$ of prime degree $p$, $Nrd(D^*) = \cup_{L/F}(L^*)$, where $L$ runs over the maximal subfields of $D$. Equivalently (by (8.10)), we may let $L$ run over all extensions of $F$ of degree $p$ which split $D$. Since $[D_j] = j[D]$ in $\mathrm{Br}\,(F)$, $D_j$ and $D$ have the same splitting fields. Hence $Nrd(D^*) = Nrd(D_j^*)$.

We will use the above computations of $K$-groups of Severi-Brauer varieties to study certain $K$-cohomology groups. Let $p$ be a prime number, $D$ a division algebra over $F$ of degree $p$. Let $E$ be a maximal subfield of $D$, so that $E/F$ is a splitting field of $D$. Let $X$ be the Severi-Brauer variety over $F$ associated to $D$, $f : X_E \to X$ the natural map.

**Proposition 8.12** *With the above notation, we have*

(a) *the Chow groups $CH^i(X) \cong CH^i(X_E) \cong \mathbb{Z}$ for $0 \le i \le p-1$, and $f^* : CH^i(X) \to CH^i(X_E)$ is identified with multiplication by $p$ for $i > 0$;*

(b) *$H^1(X, \mathcal{K}_{2,X}) \cong Nrd(D^*)$, $H^1(X_E, \mathcal{K}_{2,E}) \cong E^*$, and the map $f^* : H^1(X, \mathcal{K}_{2,X}) \to H^1(X_E, \mathcal{K}_{2,E})$ is identified with the inclusion $Nrd(D^*) \hookrightarrow E^*$.*

**Proof.** (a) We have seen that $K_0(X) \cong \mathbb{Z}^{\oplus p}$, $K_0(X_E) \cong \mathbb{Z}^{\oplus p}$ and $f^* : K_0(X) \to K_0(X_E)$ is identified with the map $\mathbb{Z}^{\oplus p} \to \mathbb{Z}^{\oplus p}$,

$$(n_0, n_1, \ldots, n_{p-1}) \mapsto (n_0, pn_1, pn_2, \ldots, pn_{p-1}).$$

By the projection formula, $f_* \circ f^*$ equals multiplication by the class $[f_*\mathcal{O}_{X_E}] = p[\mathcal{O}_X] \in K_0(X)$; hence the map $f_* : K_0(X_E) \to K_0(X)$ must be given by $\mathbb{Z}^{\oplus p} \to \mathbb{Z}^{\oplus p}$,

$$(n_0, n_1, \ldots, n_{p-1}) \mapsto (pn_0, n_1, \ldots, n_{p-1}).$$

If $\gamma = [\mathcal{O}_{X_E}(-1)] \in K_0(X_E)$, then the isomorphism $\mathbb{Z}^{\oplus p} \to K_0(X_E)$ is given by $(n_0, \ldots, n_{p-1}) \mapsto \sum n_i \gamma^i$. We claim that the filtration by codimension of support on $K_j(X_E)$ is given by $(\gamma - 1)^i K_j(X_E)$. This is the content of the next lemma.

**Lemma 8.13** *For any field $F$, and integers $i, j, m \geq 0$, we have*

$$F^i K_j(\mathbb{P}_F^m) = (\gamma_m - 1)^i K_j(\mathbb{P}_F^m),$$

*where $\gamma_m = [\mathcal{O}_{\mathbb{P}^m}(-1)] \in K_0(\mathbb{P}_F^m)$.*

**Proof.** If $f : \mathbb{P}_F^{m-i} \hookrightarrow \mathbb{P}_F^m$ is the inclusion of a linear subspace, then $f^* \gamma_m = \gamma_{m-i}$, and $f_*([\mathcal{O}_{\mathbb{P}^{m-i}}]) = (\gamma_m - 1)^i \in K_0(\mathbb{P}_F^m)$ by direct computation, using the fact that $f(\mathbb{P}_F^{m-i})$ is a complete intersection of $i$ hyperplanes in $\mathbb{P}_F^m$, so that its ideal sheaf is resolved by a Koszul complex. Hence by the projection formula,

$$f_* K_j(\mathbb{P}_F^{m-i}) = f_*(\sum_{s=0}^{m-i}(\gamma_{m-i} - 1)^s K_j(F)) = f_*(\sum_{s=0}^{m-i} f^*(\gamma_m - 1)^s K_j(F))$$

$$= \sum_{s=i}^{m}(\gamma_m - 1)^s K_j(F) = (\gamma_m - 1)^i K_j(\mathbb{P}_F^m).$$

Clearly $f_* K_j(\mathbb{P}_F^{m-i}) \subset F^i K_j(\mathbb{P}_F^m)$. Hence $(\gamma_m - 1)^i K_j(\mathbb{P}_F^m) \subset F^i K_j(\mathbb{P}_F^m)$.

We claim that, on the other hand, the natural composite map

$$\sum_{s=0}^{i-1}(\gamma_m - 1)^s K_j(F) \hookrightarrow K_j(\mathbb{P}_F^m) \to K_j(\mathbb{P}_F^m)/F^i K_j(\mathbb{P}_F^m)$$

is injective. To prove this, observe that

$$F^i K_j(\mathbb{P}_F^m) = \ker\left(K_j(\mathbb{P}_F^m) \to \varinjlim_{\mathrm{codim}\, Z \geq i} K_j(\mathbb{P}_F^m - Z)\right).$$

## 8. The Merkurjev–Suslin Theorem

Hence it suffices to show that if $Z \subset \mathbb{P}_F^m$ is any subscheme of codimension $\geq i$, then the composite map

$$\sum_{s=0}^{i-1} (\gamma_m - 1)^s K_j(F) \hookrightarrow K_j(\mathbb{P}_F^m) \to K_j(\mathbb{P}_F^m - Z)$$

is injective. But for any such $Z$, we can find a linear subspace $g : \mathbb{P}_F^{i-1} \hookrightarrow \mathbb{P}_F^m$ with $Z \cap \operatorname{im} g = \emptyset$. Hence it suffices to note that the composite

$$\sum_{s=0}^{i-1} (\gamma_m - 1)^s K_j(F) \hookrightarrow K_j(\mathbb{P}_F^m) \to K_j(\mathbb{P}_F^m - Z) \xrightarrow{g^*} K_j(\mathbb{P}_F^{i-1})$$

is an isomorphism. This proves the lemma.

Now $K_0(X_E) = K_0(E)[\gamma]/((\gamma-1)^p)$ is a free Abelian group with basis $(\gamma-1)^i$, $0 \leq i \leq p-1$. Thus

$$\operatorname{image}(F^i K_0(X_E) \to K_0(X_E) \otimes_{\mathbb{Z}} \mathbb{Z}/p\mathbb{Z}) = (\gamma-1)^i (\mathbb{Z}/p\mathbb{Z})[\gamma]/((\gamma-1)^p).$$

On the other hand,

$$\operatorname{image}(K_0(X) \otimes_{\mathbb{Z}} \mathbb{Z}/p\mathbb{Z} \to K_0(X_E) \otimes_{\mathbb{Z}} \mathbb{Z}/p\mathbb{Z}) = \mathbb{Z}/p\mathbb{Z} \subset (\mathbb{Z}/p\mathbb{Z})[\gamma]/((\gamma-1)^p).$$

Hence for $i > 0$,

$$f^* K_0(X) \cap F^i K_0(X_E) \subset p F^i K_0(X_E).$$

Now $f^* F^i K_0(X) \subset F^i K_0(X_E)$, and $f_* F^i K_0(X_E) \subset F^i K_0(X)$; hence

$$p F^i K_0(X_E) = f^* \circ f_* F^i K_0(X_E) \subset f^* F^i K_0(X) \subset f^* K_0(X) \cap F^i K_0(X_E).$$

This implies that

$$p F^i K_0(X_E) = f^* F^i K_0(X) = f^* K_0(X) \cap F^i K_0(X_E)$$

for $i > 0$. Thus for $0 \leq i \leq p-1$, we have $\operatorname{gr}^i_F K_0(X) \cong \operatorname{gr}^i_F K_0(X_E) \cong \mathbb{Z}$, and the induced map $f^* : \operatorname{gr}^i_F K_0(X) \to \operatorname{gr}^i_F K_0(X_E)$ is either an isomorphism ($i = 0$) or is identified with multiplication by $p$ on $\mathbb{Z}$; in any case the map is injective.

We may regard $F^i$ as the filtration associated to the BGQ-spectral sequence, so that

$$\operatorname{gr}^i_F K_0(X) \cong E_\infty^{i,-i}(X), \quad \operatorname{gr}^i_F K_0(X_E) \cong E_\infty^{i,-i}(X_E).$$

Thus $f^* : E_\infty^{i,-i}(X) \to E_\infty^{i,-i}(X_E)$ is injective. We have a commutative diagram

$$\begin{array}{ccccc} CH^i(X_E) & \cong E_2^{i,-i}(X_E) & \xrightarrow{\psi_i'} & E_\infty^{i,-i}(X_E) \\ f^* \uparrow & \uparrow f^* & & \uparrow f^* \\ CH^i(X) & \cong E_2^{i,-i}(X) & \xrightarrow{\psi_i} & E_\infty^{i,-i}(X) \end{array}$$

where the horizontal maps are the (surjective) cycle maps (associating to an irreducible subvariety of codimension $i$ the class in $K_0/F^{i+1}K_0$ of its structure sheaf). From the Riemann-Roch theorem 'without denominators' (see W. Fulton, *Intersection theory*, Ergeb. Math. Band 3 Folge 2, Springer-Verlag (1984), Ex. 15.1.5 and Ex. 15.3.6), we know that $\ker \psi_i$ is annihilated by $(i-1)!$. The map $\psi_i'$ is an isomorphism (being a surjection $\mathbb{Z} \to \mathbb{Z}$). Since $f_* : CH^i(X_E) \to CH^i(X)$ has the property that $f_* \circ f^* : CH^i(X) \to CH^i(X)$ is multiplication by $p = \deg f$, we have that $\ker f^* = \ker(\psi_i' \circ f^*) = \ker(f^* \circ \psi_i) = \ker \psi_i$ is annihilated by $p$. Since $(i-1)!$ and $p$ are relatively prime, $\psi_i$ is an isomorphism. Thus $CH^i(X) \cong \mathbb{Z}$, and $f^* : CH^i(X) \to CH^i(X_E)$ is identified with multiplication by $p$ for $0 < i < p$.

(b) Again, the strategy is to relate $H^1(X, \mathcal{K}_{2,X}) = E_2^{1,-2}(X)$ to $E_\infty^{1,-2}(X)$. We have

$$K_1(X) \cong F^* \oplus Nrd(D^*)^{\oplus p-1}, \quad K_1(X_E) \cong (E^*)^{\oplus p},$$

and $f^*$ is the map induced by the inclusions $F^* \subset E^*$, $Nrd(D^*) \subset E^*$. The isomorphism $E^{*\oplus p} \to K_1(X_E)$ is given by $(a_0, \ldots, a_{p-1}) \mapsto \sum \gamma^i \cdot a_i$, and for $i < p$,

$$F^i K_1(X_E) = (\gamma - 1)^i K_1(X_E) \cong (\gamma - 1)^i E^* + \cdots + (\gamma - 1)^{p-1} E^*.$$

We claim that for $0 < i < p$,

$$F^i K_1(X_E) \cap f^* K_1(X) = (\gamma - 1)^i Nrd(D^*) + \cdots + (\gamma - 1)^{p-1} Nrd(D^*).$$

Indeed, if

$$x = a_0 + \gamma a_1 + \cdots + \gamma^{p-1} a_{p-1} = (\gamma - 1)^i b_i + \cdots + (\gamma - 1)^{p-1} b_{p-1}$$

where $a_0 \in F^*$, $a_j \in Nrd(D^*)$ for $j > 0$, and $b_j \in E^*$, then comparing coefficients of $\gamma^j$, we see by descending induction that $b_j \in Nrd(D^*)$, so that $x \in (\gamma - 1)^i Nrd(D^*) + \cdots + (\gamma - 1)^{p-1} Nrd(D^*)$.

Next, if $L \subset D$ is any maximal subfield, then $X_L \cong \mathbb{P}_L^{p-1}$, and we have a fiber product diagram with finite flat maps

$$\begin{array}{ccc} X_{L \times E} & \xrightarrow{h} & X_E \\ k \downarrow & & \downarrow f \\ X_L & \xrightarrow{g} & X \end{array}$$

so that by (5.13), we have

$$f^* \circ g_* = h_* \circ k^* : K_1(X_L) \to K_1(X_E),$$

and both compositions map $F^i K_1(X_L)$ into $F^i K_1(X_E)$. If $\gamma_L = [\mathcal{O}_{X_L}(-1)] \in K_0(X_L)$, then $k^* \gamma_L = \gamma_{L \times E} = h^* \gamma$ in $K_0(X_{L \times E})$, where $\gamma_{L \times E} = [\mathcal{O}_{X_{L \times E}}(-1)]$. Hence

$$\begin{aligned} h_* \circ k^*(F^i K_1(X_L)) &= h_* \circ k^*(\gamma_L - 1)^i K_1(X_L) \\ &= h_*(h^*(\gamma - 1)^i \cdot k^* K_1(X_L)) \\ &= (\gamma - 1)^i \cdot h_* \circ k^* K_1(X_L) \\ &= (\gamma - 1)^i f^* \circ g_* K_1(X_L). \end{aligned}$$

Now $g_* : K_1(X_L) \to K_1(X)$ is a map $L^{*\oplus p} \to F^* \oplus Nrd(D^*)^{\oplus p-1}$.

**Lemma 8.14** *In the above situation, $g_* : K_1(X_L) \to K_1(X)$ is induced by the norm $N_{L/F} : L^* \to F^*$, where we note that $N_{L/F}(L^*) \subset Nrd(D^*)$.*

**Proof.** First note that from (5.13) applied to the diagram

$$\begin{array}{ccc} X_L & \xrightarrow{g} & X \\ \downarrow & & \downarrow \\ \operatorname{Spec} L & \to & \operatorname{Spec} F \end{array}$$

the composite $K_i(L) \to K_i(X_L) \xrightarrow{g_*} K_i(X)$ factors through the norm map $K_i(L) \to K_i(F)$.

Next, for each $j > 0$, the composite

$$K_i(L) \to K_i(X_L) \to K_i(X),$$

$$x \mapsto \gamma_L^j x \mapsto g_*(\gamma_L^j x),$$

is induced by the exact functor $A_j : \mathcal{P}(L) \to \mathcal{P}(X)$, where for a finite dimensional $L$-vector space $V_1$, we take

$$A_j(V_1) = g_*(\mathcal{O}_{X_L}(-j) \otimes_L V_1).$$

If $\mathcal{J}$ is the locally free $\mathcal{O}_X$-module of rank $p$ described in (5.40), then $g^* \mathcal{J} = V \otimes_L \mathcal{O}_{X_L}(-1)$, where $V = H^0(X_L, \mathcal{O}_{X_L}(1))$. Hence

$$\mathcal{O}_{X_L}(-j) = g^*(\mathcal{J}^{\otimes j} \otimes_{\operatorname{End}(V^{*\otimes j})} V^{*\otimes j})$$

$$\cong (g^* \mathcal{J}^{\otimes j}) \otimes_{D_L^{\otimes j}} V^{*\otimes j},$$

where we have identified $D_L = D \otimes_F L$ with $\operatorname{End}_L(V^*)$ (so that $M_{p^{j-1}}(D_j) \cong D_L^{\otimes j} \cong \operatorname{End}_L(V^{*\otimes j})$ for $1 \leq j < p$). This implies that

$$A_j(V_1) = \mathcal{J}^{\otimes j} \otimes_{D^{\otimes j}} B_j(V^{*\otimes j} \otimes_L V_1),$$

where $B_j : \mathcal{P}(L) \to \mathcal{P}(D^{\otimes j})$ is given by

$$B_j(V_1) = V^{*\otimes j} \otimes_L V_1,$$

regarded as a $D^{\otimes j}$-module using $D^{\otimes j} \hookrightarrow D_L^{\otimes j} \cong \mathrm{End}\,_L(V^{*\otimes j})$. However, we see easily that $V^{*\otimes j}$ is a simple $D_L^{\otimes j}$-module which is also a simple $D^{\otimes j} \cong M_{p^j-1}(D_j)$-module. Hence, $B_j$ is isomorphic to the composite of the change of rings functor $\mathcal{P}(L) \to \mathcal{P}(D_j)$ with the Morita equivalence functor $\mathcal{P}(D_j) \to \mathcal{P}(D^{\otimes j})$. In particular, the map

$$K_1(L) \to K_1(X),$$
$$x \mapsto g_*(\gamma_L^j x)$$

equals the composite

$$K_1(L) \to K_1(D_j) \cong K_1(D^{\otimes j}) \to K_1(X),$$

where the third map is that given in (5.40). Identifying $K_1(D_j)$ with $Nrd(D_j^*) = Nrd(D^*)$, we see that $K_1(L) \to K_1(D_j)$ is the map

$$L^* \xrightarrow{N_{L/F}} Nrd(D^*),$$

since the reduced norm on $D_j^*$ restricts to the field norm on $L^* \subset D_j^*$. This proves the claim.

Thus for $i > 0$, we have a sequence of inclusions

$$f^* \circ g_* F^i K_1(X_L) = (\gamma - 1)^i N_{L/F}(L^*) + \cdots + (\gamma - 1)^{p-1} N_{L/F}(L^*)$$
$$\subset f^* F^i K_1(X) \subset f^* K_1(X) \cap F^i K_1(X_E)$$
$$\subset (\gamma - 1)^i Nrd(D^*) + \cdots + (\gamma - 1)^{p-1} Nrd(D^*).$$

But $Nrd(D^*) = \cup_L N_{L/F}(L^*)$, where $L$ ranges over the maximal subfields (up to $F$-isomorphism) of $D$. Thus all but the first of the inclusions are equalities. We deduce that

$$\mathrm{gr}\,_F^i K_1(X) \cong Nrd(D^*)\ (i > 0), \quad \mathrm{gr}\,_F^0 K_1(X) \cong F^*,$$
$$\mathrm{gr}\,_F^i K_1(X_E) \cong E^* \ \forall\, i,$$

and $f^* : \mathrm{gr}\,_F^i K_1(X) \to \mathrm{gr}\,_F^i K_1(X_E)$ is identified with either the natural inclusion $F^* \hookrightarrow E^*$ (for $i = 0$), or the inclusion $Nrd(D^*) \hookrightarrow E^*$. In particular, $f^* : \mathrm{gr}\,_F^i K_1(X) \to \mathrm{gr}\,_F^i K_1(X_E)$ is *injective* for all $i$.

We use this fact for $i = 1$ to prove (8.12)(b). We have $H^1(X, \mathcal{K}_{2,X}) \cong E_2^{1,-2}(X)$. The relevant differentials for $E_r^{1,-2}(X)$ are

$$E_r^{1-r,r-3} \to E_r^{1,-2}, \quad E_r^{1,-2} \to E_r^{r+1,-1-r}.$$

Here $E_2^{1-r,r-3} = 0$ as $1 - r < 0$, while $E_2^{r+1,-1-r}(X) \cong E_\infty^{r+1,-1-r}(X) \cong CH^{r+1}(X)$ from (8.12)(a) proved above. Hence $E_2^{1,-2}(X) \cong E_\infty^{1,-2}(X) \cong Nrd(D^*)$; by a similar argument, $E_2^{1,-2}(X_E) \cong E_\infty^{1,-2}(X_E) \cong E^*$, and $f^* : H^1(X, \mathcal{K}_{2,X}) \to H^1(X_E, \mathcal{K}_{2,X_E})$ is identified with the inclusion $Nrd(D^*) \hookrightarrow E^*$.

We use the above computations of $K$-cohomology to prove the $K_2$-analogue of Hilbert's Theorem 90.

**Theorem 8.15** (Hilbert's Theorem 90 for $K_2$)  *Let $F$ be a field, $E/F$ a cyclic Galois extension, $\sigma \in \mathrm{Gal}\,(E/F)$ a generator. Then we have an exact sequence*
$$K_2(E) \xrightarrow{1-\sigma_*} K_2(E) \xrightarrow{N_{E/F}} K_2(F)$$
*(i.e., $x \in K_2(E)$ satisfies $N_{E/F}(x) = 0 \Leftrightarrow x = y - \sigma_*(y)$ for some $y \in K_2(E)$).*

We will need only the special case when $E/F$ is of prime degree, and give the proof in this case.

**Proof.** Let $p = [E : F]$. For any extension field $L$ of $F$, $L \otimes_F E$ is a cyclic Galois algebra over $L$. If $\sigma \in \mathrm{Gal}\,(E/F)$ is a generator, $\sigma$ determines a generator (which we also denote by $\sigma$) for $\mathrm{Gal}\,(L \otimes_F E/L)$. The action of $\sigma_*$ on $K_i(L \otimes_F E)$ is obtained from the functor $\sigma_* : \mathcal{P}(L \otimes_F E) \to \mathcal{P}(L \otimes_F E)$ on the category of finitely generated projective modules, $M \mapsto \sigma_* M$, where $\sigma_* M$ is the Abelian group $M$ with new module structure given by $\alpha \cdot m = \sigma^{-1}(\alpha)m$ (the expression on the right is computed using the original module structure). Next, the norm maps $N_{L \otimes_F E/L} : K_i(L \otimes_F E) \to K_i(L)$ are induced by the forgetful functor $\mathcal{P}(L \otimes_F E) \to \mathcal{P}(L)$ which associates to an $L \otimes_F E$-module the underlying $L$-vector space. Since $M$ and $\sigma_* M$ have the same $L$-vector space structure, $N_{L \otimes_F E/L} \circ \sigma_* = N_{L \otimes_F E/L}$ on $K_i(L \otimes_F E)$. Thus $N_{L \otimes_F E} \circ (1 - \sigma_*) = 0$ on $K_i(L \otimes_F E)$.

Let $N_L = N_{L \otimes_F E/L} : K_i(L \otimes_F E) \to K_i(L)$, and let
$$V(L) = \frac{\ker N_L}{(1 - \sigma_*)K_2(L \otimes_F E)}.$$

Thus, Hilbert's Theorem 90 for $K_2$ for $E/F$ is the assertion that $V(F) = 0$. If $L/F$ is a finite extension field of degree $d$, we have a norm map $N_{L/F} : V(L) \to V(F)$ such that the composite $V(F) \to V(L) \xrightarrow{N_{L/F}} V(F)$ is multiplication by $d$. Since $E \otimes_F E \cong E^{\oplus p}$, such that for any $i \geq 0$, $N_E : K_i(E \otimes_F E) \to K_i(E)$ is given by the addition map $K_i(E)^{\oplus p} \to K_i(E)$, and $\sigma_* : K_i(E \otimes_F E) \to K_i(E \otimes_F E)$ cyclically permutes the factors in $K_i(E \otimes_F E) \cong K_i(E)^{\oplus p}$, one sees easily that $V(E) = 0$. Hence $pV(F) = 0$.

Thus if $L$ is a finite extension of $F$ of degree relatively prime to $p$, then $V(F) \hookrightarrow V(L)$, so it suffices to prove the result for $L \otimes_F E/L$.

Note also that if $L = \varinjlim_i L_i$ for suitable extensions $L_i$ of $F$, then $V(L) = \varinjlim_i V(L_i)$ (this follows from (5.9), for example). Let $L$ be a maximal algebraic extension of $F$ such that any finite sub-extension has degree relatively prime to $p$. Then $V(F) \hookrightarrow V(L)$, and it suffices to prove the theorem for $L \otimes_F E/L$.

Thus to prove $V(F) = 0$, we may assume without loss of generality that every finite extension of $F$ has degree a power of $p$.

**Lemma 8.16** *For $b \in F^*$, let $X$ be the Severi-Brauer variety associated to the cyclic algebra $A(\sigma, b)$, and let $F(X)$ be its function field. Then $V(F) \to V(F(X))$ is injective.*

**Proof.** Since $E$ trivializes $A(\sigma, b)$, $X_E \cong \mathbb{P}_E^{p-1}$, and we claim that

$$H^0(X_E, \mathcal{K}_{2,E}) \cong K_2(E) \hookrightarrow K_2(E(X)),$$

where $E(X)$ is the function field of $X_E$.

Indeed, $H^0(\mathbb{P}_E^n, \mathcal{K}_2) \subset H^0(\mathbb{A}_E^n, \mathcal{K}_2) \subset K_2(E(\mathbb{P}^n))$ for any $n$, from (5.27); hence it suffices to prove that $H^0(\mathbb{A}_E^n, \mathcal{K}_2) \cong K_2(E) \hookrightarrow K_2(E(\mathbb{A}^n)) = K_2(E(\mathbb{P}^n))$. This is easy for $n = 1$, from $K_j(\mathbb{A}_E^1) \cong K_j(E)$ and the BGQ-spectral sequence, which degenerates at $E_2$ since $\dim \mathbb{A}_E^1 = 1$. In general, writing $\mathbb{A}_E^n = \mathbb{A}_E^{n-1} \times_E \mathbb{A}_E^1$,

$$H^0(\mathbb{A}_E^n, \mathcal{K}_2) \hookrightarrow H^0(\mathbb{A}_{E(\mathbb{A}^1)}^{n-1}, \mathcal{K}_2) = K_2(E(\mathbb{A}^1)) \subset K_2(E(\mathbb{A}^n)),$$

by the induction hypothesis for $\mathbb{A}^{n-1}$; now from the commutative diagram with exact rows

$$\begin{array}{ccccccc} 0 \to & H^0(\mathbb{A}_E^n, \mathcal{K}_2) & \to & K_2(E(\mathbb{A}^n)) & \to & \oplus_{x \in (\mathbb{A}_E^n)^1} E(x)^* \\ & \uparrow & & \uparrow & & \uparrow \\ 0 \to & K_2(E) & \to & K_2(E(\mathbb{A}^1)) & \to & \oplus_{x \in (\mathbb{A}_E^1)^1} E(x)^* \end{array}$$

we see that $H^0(\mathbb{A}_E^n, \mathcal{K}_2) = K_2(E) \hookrightarrow K_2(E(\mathbb{A}^n))$, as desired.

Now suppose $u \in K_2(E)$ such that $N_{E/F} u = 0$, and $u_{E(X)} = v - \sigma_* v$ for some $v \in K_2(E(X))$ (here $u_L \in K_2(L \otimes_F E)$ is the image of $u$ under $K_2(E) \to K_2(L \otimes_F E)$). We have the complex of $E_1$ terms of the BGQ-spectral sequence

$$0 \to K_2(E(X)) \xrightarrow{d_1} \oplus_{x \in X_E^1} E(x)^* \xrightarrow{d_1} \oplus_{x \in X_E^2} \mathbb{Z} \to 0.$$

Since $d_1(u_{E(X)}) = 0$, $d_1(v) = d_1(\sigma_* v) = \sigma_* d_1(v)$. Thus $d_1(v) = f^* w$ for some $w \in \oplus_{y \in X^1} F(y)^*$, under the natural map $f^* : E_1^{1,-2}(X) \to E_1^{1,-2}(X_E)$. Now $d_1 \circ d_1(v) = 0$, and

$$f^* : E_1^{2,-2}(X) = \oplus_{y \in X^2} \mathbb{Z} \to \oplus_{x \in X_E^2} \mathbb{Z} = E_1^{2,-2}(X_E)$$

is injective. Hence $d_1(w) = 0$, and $w$ determines a class in $E_2^{1,-2}(X) \cong H^1(X, \mathcal{K}_{2,X})$, which by construction, lies in

$$\ker(f^* : H^1(X, \mathcal{K}_{2,X}) \to H^1(X_E, \mathcal{K}_{2,X_E})).$$

## 8. The Merkurjev–Suslin Theorem

But by (8.12)(b), this map $f^*$ is injective. Hence $w = d_1(v')$ for some $v' \in K_2(F(X))$. Hence $u_{E(X)} = v - \sigma_* v = (v - f^* v') - \sigma_*(v - f^* v')$, and $d_1(v - f^* v') = d_1(v) - f^* w = 0$, so that $v - f^* v' \in H^0(X_E, \mathcal{K}_2) \cong K_2(E)$. Hence $u \in (1 - \sigma_*) K_2(E)$, as desired, proving (8.16).

We now inductively define a sequence of fields $F_m$, $m \geq 0$, as follows. Let $F_0 = F$; let $F_{2n+1}$ be the compositum of all the function fields $F_{2n}(X)$, where $X$ ranges over the Severi-Brauer varieties associated to the cyclic algebras $A(\sigma, b)$, $b \in F_{2n}^*$; let $F_{2n+2}$ be a maximal algebraic extension of $F_{2n+1}$ of degree relatively prime to $p$ (i.e., such that any finite subextension of $F_{2n+1}$ has degree relatively prime to $p$). Let $F_\infty = \varinjlim_n F_n$. Then $V(F) \to V(F_\infty)$ is injective, every finite extension of $F_\infty$ has degree a power of $p$, and $[A(\sigma, b)] = 0$ in $\mathrm{Br}(F_\infty)$ for all $b \in F_\infty^*$, i.e., by (8.4)(a), if $E_\infty = E F_\infty$, then $N_{E_\infty/F_\infty} : E_\infty^* \to F_\infty^*$ is onto.

Hence we may assume without loss of generality that every finite extension of $F$ has degree a power of $p$, and $N_{E/F} : E^* \to F^*$ is onto. We claim that under these conditions, $N_{E/F} : K_2(E) \to K_2(F)$ induces an isomorphism

$$\frac{K_2(E)}{\mathrm{im}\,(1 - \sigma_*)} \to K_2(F).$$

To see this, let $\varphi : F^* \otimes_\mathbb{Z} F^* \to K_2(E)/(\mathrm{im}\,(1-\sigma_*))$ be defined by $\varphi(a \otimes b) = \{\alpha, b\}$ where $\alpha \in E^*$ is a solution of $N_{E/F}\alpha = a$. Since any other solution $\alpha'$ satisfies $N_{E/F}\alpha'\alpha^{-1} = 1$, the usual version of Hilbert's Theorem 90 implies that $\{\alpha'\alpha^{-1}, b\} \in (1 - \sigma_*) K_2(E)$. Thus $\varphi$ is well defined and bilinear; further the composite $N_{E/F} \circ \varphi : F^* \otimes_\mathbb{Z} F^* \to K_2(F)$ is just the Steinberg symbol. By lemma (8.6) of Bass and Tate, $K_2(E)$ is generated by the symbols $\{\alpha, b\}$ with $\alpha \in E^*$, $b \in F^*$, so that $\varphi$ is surjective. If we prove that for any $a \in F^*$, $a \neq 1$, we have $\varphi(a \otimes (1 - a)) = 0$, then we would have (by (1.14)) an induced map $\widetilde{\varphi} : K_2(F) \to K_2(E)/\mathrm{im}\,(1 - \sigma_*)$ which is surjective, such that $N_{E/F} \circ \widetilde{\varphi}$ is the identity on $K_2(F)$, i.e., $\widetilde{\varphi}$, $N_{E/F}$ are inverse isomorphisms.

So it suffices to prove that if $a \in F^*$, $a \neq 1$ and $\alpha \in E^*$ such that $N_{E/F}\alpha = a$, then $\{\alpha, 1 - a\} \in (1 - \sigma_*) K_2(E)$. Let $T^p - a = \prod_i p_i(T)^{n_i}$ be the factorisation into irreducible factors over $F[T]$, and let $\alpha_i$ be a root of $p_i(T) = 0$, $F_i = F(\alpha_i)$, $E_i = E \otimes_F F_i$. Then

$$1 - a = \prod_i p_i(1)^{n_i} = \prod_i N_{E_i/E}(1 - \alpha_i)^{n_i},$$

$$\{\alpha, 1 - a\} = \prod_i N_{E_i/E}\{\alpha, 1 - \alpha_i\}^{n_i} = \prod_i N_{E_i/E}\{\alpha\alpha_i^{-1}, 1 - \alpha_i\}^{n_i}$$

(using $\{\alpha_i, 1-\alpha_i\} = 1$). Now $N_{E_i/F_i}(\alpha\alpha_i^{-1}) = 1$, so by Hilbert's Theorem 90, $\alpha\alpha_i^{-1} = \beta_i\sigma(\beta_i)^{-1}$ for some $\beta_i \in E_i^*$. Then

$$\{\alpha, 1-a\} = \prod_i N_{E_i/E}(1-\sigma_*)\{\beta_i, 1-\alpha_i\}^{n_i} \in (1-\sigma_*)K_2(E),$$

from the commutativity of

$$\begin{array}{ccc} K_2(E_i) & \overset{\sigma_*}{\to} & K_2(E_i) \\ N_{E_i/E} \downarrow & & \downarrow N_{E_i/E} \\ K_2(E) & \overset{\sigma_*}{\to} & K_2(E) \end{array}$$

(which follows by (5.13) from a corresponding commutative diagram of rings

$$\begin{array}{ccc} E_i & \overset{\sigma}{\to} & E_i \\ \uparrow & & \uparrow \\ E & \overset{\sigma}{\to} & E \end{array} \quad )$$

This completes the proof of (8.15).

We now proceed with the proof of (8.5), following [M].

**Lemma 8.17** *Let $E = F(\sqrt[p]{a})$ be a cyclic Galois extension of $F$ of degree $p$, and let $\sigma \in \mathrm{Gal}\,(E/F)$ be the generator such that $\sigma(\sqrt[p]{a}) = \zeta\sqrt[p]{a}$. Suppose $R_{p,F}$ is injective. Then:*

*(i) $\ker(K_2(F) \otimes_{\mathbb{Z}} \mathbb{Z}/p\mathbb{Z} \to K_2(E) \otimes_{\mathbb{Z}} \mathbb{Z}/p\mathbb{Z})$ consists of precisely the elements $\{a, b\}$ with $b \in F^*$*

*(ii) the square*

$$\begin{array}{ccc} K_2(F) \otimes_{\mathbb{Z}} \mathbb{Z}/p\mathbb{Z} & \overset{R'_{p,F}}{\to} & {}_p\mathrm{Br}\,(F) \\ \downarrow & & \downarrow \\ K_2(E) \otimes_{\mathbb{Z}} \mathbb{Z}/p\mathbb{Z} & \overset{R'_{p,E}}{\to} & {}_p\mathrm{Br}\,(E) \end{array}$$

*is a pullback*

*(iii) $R_{p,E}$ is injective.*

**Proof.** (i) Let $u \in \ker(K_2(F) \otimes_{\mathbb{Z}} \mathbb{Z}/p\mathbb{Z} \to K_2(E) \otimes_{\mathbb{Z}} \mathbb{Z}/p\mathbb{Z})$. Then $R'_{p,F}(u) \in \ker({}_p\mathrm{Br}\,(F) \to {}_p\mathrm{Br}\,(E))$. Hence by (8.4)(a), there exists $b \in F^*$ such that $R'_{p,F}(u) = R'_{p,F}(\{a, b\})$. Since $R_{p,F}$ (and hence $R'_{p,F}$) is injective, this means $u = \{a, b\} \pmod{p \cdot K_2(F)}$.

(ii) Let $u \in K_2(E) \otimes_{\mathbb{Z}} \mathbb{Z}/p\mathbb{Z}$, $v \in {}_p\mathrm{Br}\,(F)$ have the same image in ${}_p\mathrm{Br}\,(E)$. We must show that there is a unique $y \in K_2(F) \otimes_{\mathbb{Z}} \mathbb{Z}/p\mathbb{Z}$ such that $y_E = u$, and $R'_{p,F}(y) = v$ (here $y_E$ denotes the image of $y$ in $K_2(E) \otimes_{\mathbb{Z}} \mathbb{Z}/p\mathbb{Z}$). Since $R'_{p,F}$ is injective, such an element $y$ is unique, if it exists. Also, by (8.4)(b) and (i) of this lemma, we see that the two vertical arrows in the square have the same kernel. So it suffices to prove that $u \in \text{image}\,(K_2(F) \otimes_{\mathbb{Z}} \mathbb{Z}/p\mathbb{Z})$.

Now $R'_{p,F}(N_{E/F}(u)) = N_{E/F}(R'_{p,E}(u)) = N_{E/F}(v_E) = pv = 0$. Since $R'_{p,F}$ is injective, $N_{E/F}(u) = 0$ in $K_2(F) \otimes_{\mathbb{Z}} \mathbb{Z}/p\mathbb{Z}$.

We claim that

$$(*) \quad \cdots \quad \ker\bigl(N_{E/F} : K_2(E) \otimes_{\mathbb{Z}} \mathbb{Z}/p\mathbb{Z} \to K_2(F) \otimes_{\mathbb{Z}} \mathbb{Z}/p\mathbb{Z}\bigr) =$$
$$(1 - \sigma_*)K_2(E) \otimes_{\mathbb{Z}} \mathbb{Z}/p\mathbb{Z} + \text{image}\, K_2(F) \otimes_{\mathbb{Z}} \mathbb{Z}/p\mathbb{Z}.$$

If $z \in K_2(E) \otimes_{\mathbb{Z}} \mathbb{Z}/p\mathbb{Z}$ with $N_{E/F}(z) = 0$, and $\tilde{z} \in K_2(E)$ is a preimage of $z$, then $N_{E/F}(\tilde{z}) = p\,\tilde{w}$ for some $\tilde{w} \in K_2(F)$, so that $N_{E/F}(\tilde{z} - \tilde{w}_E) = 0$. Hence by (8.15), $\tilde{z} = \tilde{w}_E + (1 - \sigma_*)(\tilde{t})$ for some $\tilde{t} \in K_2(E)$. If $\tilde{w} \mapsto w \in K_2(F) \otimes_{\mathbb{Z}} \mathbb{Z}/p\mathbb{Z}$, $\tilde{t} \mapsto t \in K_2(E) \otimes_{\mathbb{Z}} \mathbb{Z}/p\mathbb{Z}$, then $z = w_E + (1 - \sigma_*)(t)$ in $K_2(E) \otimes_{\mathbb{Z}} \mathbb{Z}/p\mathbb{Z}$.

In particular, for our given $u$, we can write $u = w_E + (1 - \sigma_*)(t)$ with $w \in K_2(F) \otimes_{\mathbb{Z}} \mathbb{Z}/p\mathbb{Z}$ and $t \in K_2(E) \otimes_{\mathbb{Z}} \mathbb{Z}/p\mathbb{Z}$.

First suppose $p = 2$. If $N_{E/F}(t) = s$, then $s_E = (1+\sigma_*)(t) = (1-\sigma_*)(t)$. Hence $u = (w + s)_E$ is in the image of $K_2(F) \otimes_{\mathbb{Z}} \mathbb{Z}/p\mathbb{Z}$ as desired.

Now suppose $p$ is an odd prime. We claim by induction on $j$ that

$$(1 - \sigma_*)(t) \in \text{image}\, K_2(F) \otimes_{\mathbb{Z}} \mathbb{Z}/p\mathbb{Z} + (1 - \sigma_*)^j K_2(E) \otimes_{\mathbb{Z}} \mathbb{Z}/p\mathbb{Z}$$

for each $j = 1, 2, \ldots, p-1$. This is clear for $j = 1$. Suppose $1 \leq j < p-1$, and $(1 - \sigma_*)(t) - (1 - \sigma_*)^j(s)$ is in the image of $K_2(F) \otimes_{\mathbb{Z}} \mathbb{Z}/p\mathbb{Z}$. Then $(1 - \sigma_*)^{j+1}(s) = (1 - \sigma_*)^2(t)$. Hence

$$(1 - \sigma_*)^{j+1} R'_{p,E}(s) = (1 - \sigma_*)^2 R'_{p,E}(t) = (1 - \sigma_*) R'_{p,E}((1 - \sigma_*)(t))$$
$$= (1 - \sigma_*) R'_{p,E}(u - w_E) = (1 - \sigma_*)(v_E - (R'_{p,F}(w))_E) = 0.$$

Since $j + 1 \leq p - 1$, we get $(1 - \sigma_*)^{p-1} R'_{p,E}(s) = 0$. In the polynomial ring $\mathbb{Z}/p\mathbb{Z}[X]$, we have $1 + X + \cdots + X^{p-1} = (1 - X)^{p-1}$. Hence

$$\bigl(N_{E/F} \circ R'_{p,E}(s)\bigr)_E = (1 + \sigma_* + \cdots + \sigma_*^{p-1}) R'_{p,E}(s) = (1 - \sigma_*)^{p-1} R'_{p,E}(s) = 0.$$

Hence by (8.4)(a), $N_{E/F} \circ R'_{p,E}(s) = R'_{p,F}(\{a, b\})$, for some $b \in F^*$. We then have

$$R'_{p,F} \circ N_{E/F}(s) = N_{E/F} \circ R'_{p,E}(s) = R'_{p,F}(\{a, b\}) = R'_{p,F} \circ N_{E/F}(\{\sqrt[p]{a}, b\}),$$

where the last equality is because $p$ is odd. Since $R'_{p,F}$ is injective, by assumption, we have $N_{E/F}(s - \{\sqrt[p]{a}, b\}) = 0$. Hence by $(*)$, we must have

$$s - \{\sqrt[p]{a}, b\} \in (1 - \sigma_*) K_2(E) \otimes_{\mathbb{Z}} \mathbb{Z}/p\mathbb{Z} + \text{image}\, K_2(F) \otimes_{\mathbb{Z}} \mathbb{Z}/p\mathbb{Z},$$

so that
$$(1-\sigma_*)^j(s) - \{\zeta, b\} \in (1-\sigma_*)^{j+1} K_2(E) \otimes_{\mathbb{Z}} \mathbb{Z}/p\mathbb{Z}.$$
This proves the claim for $j+1$.

Now applying the claim for $j = p-1$, we get that
$$(1-\sigma_*)(t) - (1-\sigma_*)^{p-1}(s) \in \text{image } K_2(F) \otimes_{\mathbb{Z}} \mathbb{Z}/p\mathbb{Z}$$
for some $s \in K_2(E) \otimes_{\mathbb{Z}} \mathbb{Z}/p\mathbb{Z}$. But
$$(1-\sigma_*)^{p-1}(s) = (1 + \sigma_* + \cdots + \sigma_*^{p-1})(s)$$
$$= \left(N_{E/F}(s)\right)_E \in \text{image } K_2(F) \otimes_{\mathbb{Z}} \mathbb{Z}/p\mathbb{Z}.$$
Hence $(1-\sigma_*)(t) \in \text{image } K_2(F) \otimes_{\mathbb{Z}} \mathbb{Z}/p\mathbb{Z}$, and so $u \in \text{image } K_2(F) \otimes_{\mathbb{Z}} \mathbb{Z}/p\mathbb{Z}$.

(iii) This follows from (ii) and the injectivity of $R'_{p,F}$.

Our next goal is to determine the kernel of
$$K_2(F) \otimes_{\mathbb{Z}} \mathbb{Z}/p\mathbb{Z} \to K_2(F(\sqrt[p]{a})) \otimes_{\mathbb{Z}} \mathbb{Z}/p\mathbb{Z}$$
without first assuming (as we did in lemma (8.17)) that $R_{p,F}$ is injective. We will accomplish this in Proposition (8.23), by deducing it for an arbitrary field $F$ from a suitable 'generic' situation, where it is easier to prove. To do this, we need to first prove the injectivity of $R_{p,F}$ when $F$ is a pure transcendental extension of a finite field or of a number field. This is done in the next two lemmas. The reduction of the arbitrary field to the 'generic' situation is done using lemma (8.20). Note that if $F$ contains an algebraically closed subfield $F_0$, then (as explained later in more detail) we do not need the next lemma in proving Proposition (8.23), and so can avoid the reference to class field theory made below.

**Lemma 8.18** (Tate) *Let $F$ be a field which is algebraic over its prime subfield, and $p$ a prime such that $F$ contains a primitive $p^{\text{th}}$ root of unity. Then $R'_{p,F}$ is injective.*

**Proof.** We may assume without loss of generality that $F$ is a finite algebraic extension of its prime subfield. If $F$ is a finite field, then $K_2(F) = 0$. Hence the lemma is trivial in this case.

If $F$ is a number field, then the lemma follows from lemma 8.4(b), and the following two assertions:

(i) given $a, b, c, d \in F^*$ such that $R'_{p,F}(\{a,b\}) = R'_{p,F}(\{c,d\})$, there exist $x, y \in F^*$ such that
$$R'_{p,F}(\{a,b\}) = R'_{p,F}(\{x,b\}) = R'_{p,F}(\{x,y\}) = R'_{p,F}(\{c,y\}) = R'_{p,F}(\{c,d\})$$

## 8. The Merkurjev–Suslin Theorem

(ii) given $a, b, c, d \in F^*$, there exist $x, y, z \in F^*$ such that

$$R'_{p,F}(\{a,b\}) = R'_{p,F}(\{x,z\}) \text{ and } R'_{p,F}(\{c,d\}) = R'_{p,F}(\{y,z\}).$$

Indeed, granting these assertions, we first see, from (i) and (8.4)(b) above, that if $R'_{p,F}(\{a,b\}) = R'_{p,F}(\{c,d\})$, then $\{a,b\} = \{c,d\}$ in $K_2(F) \otimes_{\mathbb{Z}} \mathbb{Z}/p\mathbb{Z}$. Then (ii) implies that any element of $K_2(F) \otimes_{\mathbb{Z}} \mathbb{Z}/p\mathbb{Z}$ is the image of a symbol $\{x,y\}$ for some $x, y \in F^*$. Now (i) implies that $R'_{p,F}$ is injective.

We first prove (ii). We claim that since $F$ is a number field, any finite number of elements $\{\alpha_1, \ldots, \alpha_r\}$ of $_p\text{Br}(F)$ have a common cyclic splitting field $E$ with $[E : F] = p$. We will deduce this from the exact sequence describing the Brauer group of a number field

$$(\#) \qquad 0 \to \text{Br}(F) \to \bigoplus_{v \in M_F} \text{Br}(F_v) \xrightarrow{\sum_v \text{inv}_v} \mathbb{Q}/\mathbb{Z} \to 0$$

which is obtained from class field theory (see for example *Algebraic Number Theory*, J. W. S. Cassels and A. Fröhlich, eds., Academic Press Reprint (1990), Ch. VII, 9.6). Here $M_F$ denotes the set of places of $F$, and $\text{inv}_v : \text{Br}(F_v) \to \mathbb{Q}/\mathbb{Z}$ is the natural inclusion given by local class field theory ($\text{inv}_v$ is an isomorphism for non-Archimedean $v$, and is the inclusion $\mathbb{Z}/2\mathbb{Z} \to \mathbb{Q}/\mathbb{Z}$ if $F_v \cong \mathbb{R}$, the real field). Let $S$ be the finite set of places $v$ of $F$ such that some $\alpha_i$ has a non-zero image in $_p\text{Br}(F_v)$. For each $v \in S$, there is a cyclic Galois extension $E_v$ of $F_v$ of degree $p$ which splits the generator of $_p\text{Br}(F_v) \cong \mathbb{Z}/p\mathbb{Z}$. By Kummer theory, we have $E_v = F_v(\sqrt[p]{d_v})$ for some $d_v \in F_v^*$. We can find $d \in F^*$ which gives a sufficiently good $F_v$-approximation to $d_v$, for all $v \in S$. Then $d_v/d \in (F_v^*)^p$ for all $v \in S$, so $E_v = F_v(\sqrt[p]{d})$ for all $v \in S$; hence $E = F(\sqrt[p]{d})$ splits all the $\alpha_i$. This proves the claim. Since by lemma 8.4, $\ker(_p\text{Br}(F) \to {_p\text{Br}}(E))$ consists precisely of the elements $R'_{p,F}(\{a,d\})$ for $a \in F^*$, (ii) follows at once.

To prove (i), we again use the exact sequence $(\#)$. First observe that since $_p\text{Br}(F_v)$ is cyclic, (i) holds if we replace $F$ by $F_v$. Indeed, identifying $_p\text{Br}(F_v)$ with $\mathbb{Z}/p\mathbb{Z}$ (assuming $_p\text{Br}(F_v) \neq 0$), we may view $R'_{p,F_v}$ as a bilinear form on $F_v^* \otimes_{\mathbb{Z}} \mathbb{Z}/p\mathbb{Z}$. We may assume $R'_{p,F_v}(\{a,b\}) = R'_{p,F_v}(\{c,d\}) = \alpha \neq 0$. Then the linear forms $x \mapsto R'_{p,F_v}(\{x,b\})$ and $x \mapsto R'_{p,F_v}(\{x,d\})$ are both non-zero. If they are equal, or are linearly independent, then take $y = d$, and we can solve (for $x$) the system of linear equations $R'_{p,F_v}(\{x,b\}) = R'_{p,F_v}(\{x,d\}) = \alpha$. If the linear functionals are not equal but are dependent, then $p > 2$, and the bilinear form is *alternating* (any $t \in F_v^*$ is either a $p^{\text{th}}$ power or is a norm from $F_v(\sqrt[p]{t})$; hence $R'_{p,F_v}(\{t,t\}) = 0$). Now consider the linear forms $x \mapsto R'_{p,F_v}(\{x,c\})$ and $x \mapsto R'_{p,F_v}(\{x,d\})$. These are linearly independent since $R'_{p,F_v}(\{c,d\}) = \alpha \neq 0$. Now take

$y = cd$. Then $R'_{p,F_v}(\{c,y\}) = R'_{p,F_v}(\{c,d\}) = \alpha$, and the linear forms $x \mapsto R'_{p,F_v}(\{x,b\})$ and $x \mapsto R'_{p,F_v}(\{x,y\}) = R'_{p,F_v}(\{x,cd\})$ are linearly independent, so that we can solve $R'_{p,F_v}(\{x,b\}) = R'_{p,F_v}(\{x,y\}) = \alpha$ for $x \in F_v^* \otimes_{\mathbb{Z}} \mathbb{Z}/p\mathbb{Z}$.

Now let $\alpha = R'_{p,F}(\{a,b\}) = R'_{p,F}(\{c,d\})$. Let $S$ be the finite set of places of $F$ such that $\alpha \mapsto \alpha_v \neq 0 \in {}_p\mathrm{Br}\,(F_v)$. For each $v \in S$, we can then find $x_v, y_v \in F_v^*$ such that

$$R'_{p,F_v}(\{a,b\}) = R'_{p,F_v}(\{x_v,b\}) = R'_{p,F_v}(\{x_v,y_v\}) =$$
$$R'_{p,F_v}(\{c,y_v\}) = R'_{p,F_v}(\{c,d\}).$$

From the last equation, $y_v/d$ is a norm from $F_v(\sqrt[p]{c}) = E_v$, say $y_v/d = N_{E_v/F_v}(t_v)$. Choose $t \in F(\sqrt[p]{c}) = E$ such that $t/t_v \in (E_v^*)^p$ for each $v \in S$. Take $y = dN_{E/F}(t)$. Then

$$R'_{p,F}(\{c,d\}) = R'_{p,F}(\{c,y\}) = \alpha,$$

and

$$R'_{p,F_v}(\{c,y\}) = R'_{p,F_v}(\{c,y_v\})$$

for each $v \in S$. We now need to find $x \in F^*$ such that

$$R'_{p,F}(\{x,b\}) = R'_{p,F}(\{x,y\}) = \alpha.$$

By construction, we know that we can find a local solution $x_v \in F_v^*$ for each $v \in S$, and there is the trivial solution $x_v = 1$ for all $v \notin S$. Now we are reduced to (the case $r = 2$ of) the following assertion:
if $\alpha_1, \ldots, \alpha_r \in {}_p\mathrm{Br}\,(F)$, and $a_1, \ldots, a_r \in F^*$ such that for each place $v$ of $F$, there exists $x_v \in F_v^*$ satisfying $\{R'_{p,F_v}(\{a_i, x_v\}) = (\alpha_i)_v$ in ${}_p\mathrm{Br}\,(F_v)$, then there exists $x \in F^*$ satisfying $R'_{p,F}(\{a_i, x\}) = \alpha_i$ for each $i$.

This last assertion follows from class field theory; it is a reformulation of Cassels and Fröhlich, eds., *loc. cit.*, Ex.2.16.

**Lemma 8.19** (Bloch) *Let $L$ be a finitely generated pure transcendental extension of $F$. If $R_{p,F}$ is injective, then so is $R_{p,L}$.*

**Proof.** By induction, we may assume $L = F(t)$ where $t$ is an indeterminate. From the BGQ-spectral sequence for $\mathbb{A}_F^1$ (which degenerates at $E_2$) and the isomorphisms $K_i(F) \cong K_i(\mathbb{A}_F^1)$ for all $i$, we have $H^0(\mathbb{A}_F^1, \mathcal{K}_i) = K_i(F)$, and $H^j(\mathbb{A}_F^1, \mathcal{K}_i) = 0$ for $j > 0$, for all $i$. In particular there is an exact sequence

$$0 \to K_2(F) \to K_2(F(t)) \xrightarrow{\sum_x T_x} \oplus_{x \in (\mathbb{A}_F^1)^1} F(x)^* \to 0,$$

## 8. The Merkurjev–Suslin Theorem

where $T_x : K_2(F(t)) \to F(x)^*$ is the tame symbol associated to the discrete valuation $v_x$ determined by $x$ (see (9.12)). The corresponding sequence of $p$-torsion subgroups is

$$0 \to {}_pK_2(F) \to {}_pK_2(F(t)) \to \oplus_{x \in (\mathbb{A}_F^1)^1} \mu_p \to 0,$$

where the surjectivity of the map on the right is clear, because $T_x(\{a, \zeta\}) = \zeta^{v_x(a)}$ for any $p^{\text{th}}$ root of unity $\zeta$ and any $a \in F(t)^*$. Hence

$$0 \to K_2(F) \otimes_{\mathbb{Z}} \mathbb{Z}/p\mathbb{Z} \to K_2(F(t)) \otimes_{\mathbb{Z}} \mathbb{Z}/p\mathbb{Z} \to \oplus_{x \in (\mathbb{A}_F^1)^1} F(x)^*/F(x)^{*p} \to 0$$

is exact.

There is a corresponding exact sequence of Galois cohomology groups

$$0 \to H^2(F, \mu_p^{\otimes 2}) \to H^2(F(t), \mu_p^{\otimes 2}) \xrightarrow{\sum_x \partial_x} \oplus_{x \in (\mathbb{A}_F^1)^1} F(x)^*/F(x)^{*p} \to 0,$$

where $H^2(F, \mu_p^{\otimes 2})$ is identified with the étale cohomology group $H_{\text{ét}}^2(\mathbb{A}_F^1, \mu_p^{\otimes 2})$ (see Milne, *Étale Cohomology*, Ch. VI, Cor.4.20, for example). This exact sequence of Galois cohomology groups may be deduced from the Gysin exact sequence in étale cohomology, for the smooth $F$-variety $\mathbb{A}_F^1$ (see Milne's book, Ch. VI, 5.4(b)), as follows. For any open subset $U \subset \mathbb{A}_F^1$, and $Z = \mathbb{A}_F^1 - U$, we have a Gysin exact sequence

$$\cdots \to H_{et}^{i-2}(Z, \mu_p) \to H_{et}^i(\mathbb{A}_F^1, \mu_p^{\otimes 2}) \to H_{et}^i(U, \mu_p^{\otimes 2}) \to H_{et}^{i-1}(Z, \mu_p) \to$$
$$H_{et}^{i+1}(\mathbb{A}_F^1, \mu_p^{\otimes 2}) \to \cdots$$

Since $H^i(F, \mu_p^{\otimes 2}) \cong H_{et}^i(\mathbb{A}_F^1, \mu_p^{\otimes 2})$, the map $H_{et}^i(\mathbb{A}_F^1, \mu_p^{\otimes 2}) \to H_{et}^i(U, \mu_p^{\otimes 2})$ is injective for each $i$ (if $F$ is infinite, then $U$ has $F$-rational points; if $F$ is finite, $U$ has two closed points of relatively prime degree over $F$). Hence the long exact sequence above splits into short exact sequences for each $i$

$$0 \to H^i(F, \mu_p^{\otimes 2}) \to H_{et}^i(U, \mu_p^{\otimes 2}) \to H_{et}^{i-1}(Z, \mu_p) \to 0.$$

Since $Z$ consists of a finite set of closed points,

$$H_{et}^1(Z, \mu_p) = \oplus_{x \in Z} H^1(F(x), \mu_p) = \oplus_{x \in Z} F(x)^* \otimes_{\mathbb{Z}} \mathbb{Z}/p\mathbb{Z};$$

hence the above sequence for $i = 2$ reads

$$0 \to H^2(F, \mu_p^{\otimes 2}) \to H_{et}^2(U, \mu_p^{\otimes 2}) \to \oplus_{x \in Z} F(x)^* \otimes_{\mathbb{Z}} \mathbb{Z}/p\mathbb{Z} \to 0.$$

Taking the direct limit over all finite sets $Z$ of closed points, we obtain the desired sequence of Galois cohomology groups.

We claim that for each closed point $x \in \mathbb{A}_F^1$, the square

$$\begin{array}{ccc} K_2(F(t)) & \to & F(x)^* \otimes_\mathbb{Z} \mathbb{Z}/p\mathbb{Z} \\ R_{p,F(t)} \downarrow & & \| \\ H^2(F(t), \mu_p^{\otimes 2}) & \to & F(x)^* \otimes_\mathbb{Z} \mathbb{Z}/p\mathbb{Z} \end{array}$$

commutes up to a (universal) sign. Both composites clearly vanish on symbols $\{a, b\}$ with $a, b \in \mathcal{O}^*_{\mathbb{A}_F^1, x}$. So it suffices to compare the values of both composites on elements $\{a, b\}$ with $a \in \mathcal{O}^*_{\mathbb{A}_F^1, x}$, for a local parameter $b \in \mathcal{O}_{\mathbb{A}_F^1, x}$. Since $R_{p, F(t)}$ is defined via cup products using $\chi_{p, F(t)} : F(t)^* \otimes_\mathbb{Z} \mathbb{Z}/p\mathbb{Z} \xrightarrow{\cong} H^1(F(t), \mu_p)$, which is a boundary map in an exact cohomology sequence, the claim in this special case follows from the formula for the cup product of a boundary (see Milne's book, Ch. V, §1, and in particular, Prop. 1.16).

The claim yields a commutative diagram with exact rows

$$\begin{array}{ccccccc} 0 \to K_2(F) \otimes_\mathbb{Z} \mathbb{Z}/p\mathbb{Z} \to & K_2(F(t)) \otimes_\mathbb{Z} \mathbb{Z}/p\mathbb{Z} & \to & \oplus_{x \in (\mathbb{A}_F^1)^1} F(x)^*/F(x)^{*p} \to 0 \\ R_{p,F} \downarrow & R_{p,F(t)} \downarrow & & \cong \downarrow \\ 0 \to H^2(F, \mu_p^{\otimes 2}) \to & H^2(F(t), \mu_p^{\otimes 2}) & \xrightarrow{\sum_x \partial_x} & \oplus_{x \in (\mathbb{A}_F^1)^1} F(x)^*/F(x)^{*p} \to 0 \end{array}$$

A diagram chase immediately implies the lemma.

**Lemma 8.20** *Let $R = k[X_1, \ldots, X_n]$ be a polynomial ring over a field $k$, $T$ an integral domain which is a finite $R$-algebra, and $L$ the quotient field of $T$. Let $\mu : T \to F$ be a homomorphism of $k$-algebras to an extension field $F$ of $k$, and $\psi : R \to F$ the induced homomorphism. Let $P = \ker \psi$, $\widetilde{P} = \ker \mu$. Suppose $T_{\widetilde{P}}$ is a flat and unramified (i.e., étale) $R_P$-algebra. Then there are (non-canonical) homomorphisms*

$$\theta_1 : L^* \otimes_\mathbb{Z} \mathbb{Z}/p\mathbb{Z} \to F^* \otimes_\mathbb{Z} \mathbb{Z}/p\mathbb{Z},$$

$$\theta_2 : K_2(L) \otimes_\mathbb{Z} \mathbb{Z}/p\mathbb{Z} \to K_2(F) \otimes_\mathbb{Z} \mathbb{Z}/p\mathbb{Z},$$

*such that*

(i) $\theta_2\{a, b\} = \{\theta_1(a), \theta_1(b)\}$ *for all* $a, b \in L^*$

(ii) $\theta_1(y) = \mu(y)$ *for all* $y \in T^*_{\widetilde{P}}$.

**Proof.** Let $\mu \otimes 1 : T \otimes_k F \to F$ and $\psi \otimes 1 : R \otimes_k F \to F$ be the obvious extensions of $\mu$ and $\psi$. If $P' = \ker \psi \otimes 1$ and $Q = \ker \mu \otimes 1$, then $(T \otimes_k F)_Q$ is an étale algebra over $(R \otimes_k F)_{P'}$. In particular, $(T \otimes_k F)_Q$ is a regular local ring, and hence an integral domain. Hence there is a unique minimal prime $Q'$ of $T \otimes_k F$ contained in $Q$. We may now replace $R$ by $R \otimes_k F$, and $T$ by $T \otimes_k F/Q'$. Hence we are reduced to the case when $k = F$.

Next, if $\psi(X_i) = a_i$, we may change variables in $R$ to $Y_i = X_i - a_i$. Now under the given hypotheses, we have an isomorphism of completions $k[[Y_1, \ldots, Y_n]] \cong \widehat{R_P} \cong \widehat{T_{\widetilde{P}}}$. If $M$ is the quotient field of $k[[Y_1, \ldots, Y_n]]$, then we have a natural inclusion $L \hookrightarrow M$ giving a commutative diagram

$$\begin{array}{ccc} T_{\widetilde{P}} & \to & k[[Y_1, \ldots, Y_n]] \\ \downarrow & & \downarrow \\ L & \to & M \end{array}$$

Hence it suffices to construct homomorphisms

$$\theta_1 : M^* \otimes_{\mathbb{Z}} \mathbb{Z}/p\mathbb{Z} \to k^* \otimes_{\mathbb{Z}} \mathbb{Z}/p\mathbb{Z},$$

$$\theta_2 : K_2(M) \otimes_{\mathbb{Z}} \mathbb{Z}/p\mathbb{Z} \to K_2(k) \otimes_{\mathbb{Z}} \mathbb{Z}/p\mathbb{Z},$$

such that

(i) $\theta_2\{a, b\} = \{\theta_1(a), \theta_1(b)\}$

(ii) $\theta_1(y) \equiv y(0, 0, \ldots, 0) \pmod{k^{*p}}$ for all $y \in k[[Y_1, \ldots, Y_n]]^*$.

We may well-order the monomials in $Y_1, \ldots, Y_n$ using the lexicographic order: $Y_1^{p_1} Y_2^{p_2} \cdots Y_n^{p_n} < Y_1^{q_1} Y_2^{q_2} \cdots Y_n^{q_n}$ if for the first $i$ such that $p_i \neq q_i$, we have $p_i < q_i$. Then any non-zero element $z \in k[[Y_1, \ldots, Y_n]]$ is uniquely expressible as $z = a \cdot \text{in}(z) + z_1$, where $a \in k^*$, and $\text{in}(z)$ is the *initial monomial* of $z$, i.e., the smallest monomial (with respect to our chosen ordering) appearing in the power series $z$ with a non-zero coefficient. Here $a$ is the coefficient of $\text{in}(z)$ in $z$, so that every monomial appearing in $z_1$ is strictly larger than $\text{in}(z)$ with respect to the chosen ordering.

Then we have:

(a) if $z$ is a unit, we have $\text{in}(z) = 1$, and the unique expression is $z = z(0, \ldots, 0) \cdot 1 + z_1$

(b) for any non-zero $z', z'' \in k[[Y_1, \ldots, Y_n]]$, we have $\text{in}(z'z'') = \text{in}(z')\text{in}(z'')$.

Now an arbitrary element $y \in M^*$ can be (non-uniquely) expressed as $y = y'/y''$, where $y', y''$ are relatively prime (non-zero) elements in $k[[Y_1, \ldots, Y_n]]$. We can uniquely write $y' = a \cdot \text{in}(y') + y'_1$, $y'' = b \cdot \text{in}(y'') + y''_1$ where $a, b \in k^*$. Define

$$\theta_1(y) = \text{image of } a/b \text{ in } k^* \otimes_{\mathbb{Z}} \mathbb{Z}/p\mathbb{Z}.$$

We easily verify using (b) above that $\theta_1$ is well defined, and is a homomorphism. Now (a) implies that for $y \in k[[Y_1, \ldots, Y_n]]^*$, we have $\theta_1(y) = y(0, \ldots, 0) \pmod{k^{*p}}$.

We now define $\theta_2(\{y, z\}) = \{\theta_1(y), \theta_1(z)\}$. It only remains to show that this is well defined. This follows provided $\theta_2(\{y, 1-y\}) = 0$ for all $y \in M - \{0, 1\}$, which we now prove. If $y = y'/y''$ as above, then $1 - y = (y'' - y')/y''$. If $\text{in}(y'') < \text{in}(y')$, then $\text{in}(y'' - y') = \text{in}(y'')$, so $\theta_1(1-y) = 1$. On the other hand, if $\text{in}(y'') > \text{in}(y')$, then we similarly conclude that $\theta_1(1-y) = -\theta_1(y)$. Finally, suppose $\text{in}(y') = \text{in}(y'')$. In this case $\theta_1(1-y) = 1 - \theta_1(y)$. In all three cases, clearly $\{\theta_1(y), \theta_1(1-y)\}$ is trivial in $K_2(k) \otimes_{\mathbb{Z}} \mathbb{Z}/p\mathbb{Z}$.

**Remark.** The subring of $M$, whose non-zero elements are the fractions $y_1/y_2$ with $\text{in}(y_1) \leq \text{in}(y_2)$, is a valuation ring with quotient field $M$, and the maps $\theta_i$ are particular cases of specialization maps for $K$ groups associated to valuation rings.

**Lemma 8.21** *Let $F$ be a field, $p$ a prime $\neq \text{char}.F$. Let $f : F^* \to F^* \otimes_{\mathbb{Z}} \mathbb{Z}/p\mathbb{Z}$. Then*

$$K_2(F) \otimes_{\mathbb{Z}} \mathbb{Z}/p\mathbb{Z} = (F^* \otimes_{\mathbb{Z}} \mathbb{Z}/p\mathbb{Z})^{\otimes 2}/A,$$

*where $A$ is the subgroup generated by $f(x) \otimes f(y)$, where $x, y \in F^*$ range over all pairs such that $y$ is a norm from the field $F(\sqrt[p]{x})$.*

**Proof.** As in the proof of (8.1), we see that $f(a) \otimes f(1-a) \in A$ for all $a \in F^* - \{1\}$. Hence $(F^* \otimes_{\mathbb{Z}} \mathbb{Z}/p\mathbb{Z})^{\otimes 2}/A$ is a quotient of $K_2(F) \otimes_{\mathbb{Z}} \mathbb{Z}/p\mathbb{Z}$.

On the other hand, if $y = N_{F(\sqrt[p]{x})/F}(z)$, then in $K_2(F)$, we have $\{x, y\} = N_{F(\sqrt[p]{x})/F}\{x, z\} = N_{F(\sqrt[p]{x})/F}\{\sqrt[p]{x}, z\}^p \in p \cdot K_2(F)$. Hence $A$ is in the kernel of the (surjective) symbol map $(F^* \otimes_{\mathbb{Z}} \mathbb{Z}/p\mathbb{Z})^{\otimes 2} \to K_2(F) \otimes \mathbb{Z}/p\mathbb{Z}$.

In the next lemma, we require the following notation. For any positive integer $n$, let

$$\mathcal{A}_n = \{\alpha : \{1, 2, \ldots, n\} \to \{0, 1, \ldots, p-1\} \mid \alpha(i) \neq 0 \text{ for some } i\}.$$

For any $a_1, \ldots, a_n \in F^*$, and $\alpha \in \mathcal{A}_n$, let

$$a_\alpha = \prod_{i=1}^{n} a_i^{\alpha(i)},$$

$$F_\alpha = F(\sqrt[p]{a_\alpha}), \quad N_\alpha = N_{F_\alpha/F} : F_\alpha^* \to F^*.$$

**Lemma 8.22** *Let $a_1, \ldots, a_m \in F^*$ have $\mathbb{Z}/p\mathbb{Z}$-linearly independent images in $F^* \otimes_{\mathbb{Z}} \mathbb{Z}/p\mathbb{Z}$, and let $b_1, \ldots, b_m \in F^*$. Then the following are equivalent.*

8. The Merkurjev-Suslin Theorem   173

(i) $\prod_{i=1}^{m}\{a_i, b_i\}$ is trivial in $K_2(F) \otimes_{\mathbb{Z}} \mathbb{Z}/p\mathbb{Z}$.

(ii) There exist $a_{m+1}, \ldots, a_n \in F^*$, $x_\alpha \in F_\alpha^*$ for each $\alpha \in \mathcal{A}_n$, and $c_1, \ldots, c_n \in F^*$ such that
$$c_i^p = b_i \prod_{i=1}^{n} N_\alpha(x_\alpha)^{\alpha(i)}, \quad 1 \leq i \leq n,$$
where $b_{m+1} = \cdots = b_n = 1$.

**Proof.** (ii) $\Rightarrow$(i) is easy, and does not use the $\mathbb{Z}/p\mathbb{Z}$-linear independence of the $a_i$. Assuming (ii), we compute (in $K_2(F) \otimes_{\mathbb{Z}} \mathbb{Z}/p\mathbb{Z}$) that

$$\prod_{i=1}^{m}\{a_i, b_i\} = \prod_{i=1}^{n}\{a_i, b_i\} = \prod_{i=1}^{n}\prod_{\alpha \in \mathcal{A}_n}\{a_i, N_\alpha(x_\alpha)^{\alpha(i)}\}^{-1}$$
$$= \prod_{\alpha \in \mathcal{A}_n}\prod_{i=1}^{n}\{a_i^{\alpha(i)}, N_\alpha(x_\alpha)\}^{-1} = \prod_{\alpha \in \mathcal{A}_n}\{a_\alpha, N_\alpha(x_\alpha)\}^{-1},$$

which is trivial.

To prove (i) $\Rightarrow$(ii), we use the equivalent definition (8.21) of $K_2(F) \otimes_{\mathbb{Z}} \mathbb{Z}/p\mathbb{Z}$. Since $\prod_{i=1}^{n}\{a_i, b_i^{-1}\}$ is trivial in $K_2(F) \otimes_{\mathbb{Z}} \mathbb{Z}/p\mathbb{Z}$, we have $\sum_{i=1}^{n} f(a_i) \otimes f(b_i^{-1}) \in A$, i.e.,

$$\sum_{i=1}^{m} f(a_i) \otimes f(b_i^{-1}) = \sum_{j=1}^{r} f(e_j) \otimes f(g_j),$$

where $e_j, g_j \in F^*$, the $e_j$ are all distinct, and $g_j$ is a norm from $F(\sqrt[p]{e_j})$. Let
$$f(a_1), \ldots, f(a_m), f(a_{m+1}), \ldots, f(a_n)$$
be a $\mathbb{Z}/p\mathbb{Z}$-basis for the subgroup of $F^* \otimes_{\mathbb{Z}} \mathbb{Z}/p\mathbb{Z}$ generated by
$$f(a_1), \ldots, f(a_m), f(e_1), \ldots, f(e_r).$$
Then we can find $\alpha_j \in \mathcal{A}_n$, for $1 \leq j \leq r$, such that $f(e_j) = f(a_{\alpha_j})$, and $\alpha_j \neq \alpha_k$ for $j \neq k$. Clearly also $F(\sqrt[p]{e_j}) = F(\sqrt[p]{a_{\alpha_j}}) = F_{\alpha_j}$, so $g_j = N_{\alpha_j}(x_j)$ for some $x_j \in F_{\alpha_j}^*$. Define
$$x_\alpha = \begin{cases} x_{\alpha_j} & \text{if } \alpha = \alpha_j \text{ for some (necessarily unique) } j, \\ 1 & \text{otherwise.} \end{cases}$$

Then we have
$$\sum_{i=1}^{m} f(a_i) \otimes f(b_i^{-1}) = \sum_{\alpha \in \mathcal{A}_n} f(a_\alpha) \otimes f(N_\alpha(x_\alpha))$$
$$= \sum_{\alpha \in \mathcal{A}_n}\sum_{i=1}^{n} f(a_i) \otimes f(N_\alpha(x_\alpha)^{\alpha(i)}) = \sum_{i=1}^{n} f(a_i) \otimes f(\prod_{\alpha \in \mathcal{A}_n} N_\alpha(x_\alpha^{\alpha(i)})).$$

Since the $f(a_i)$ are $\mathbb{Z}/p\mathbb{Z}$-linearly independent in $F^* \otimes_{\mathbb{Z}} \mathbb{Z}/p\mathbb{Z}$, we must have
$$f(b_i \prod_{\alpha \in \mathcal{A}_n} N_\alpha(x_\alpha^{\alpha(i)})) = 1$$
for $1 \leq i \leq n$, where $b_i = 1$ for $i > m$, i.e.,
$$b_i \prod_{\alpha \in \mathcal{A}_n} N_\alpha(x_\alpha^{\alpha(i)}) = c_i^p$$
for some $c_i \in F^*$, for $1 \leq i \leq n$. This is what was to be proved.

**Proposition 8.23** *Let $F$ be a field, $p$ a prime $\neq$ char. $F$, such that $F$ contains a primitive $p^{\text{th}}$ root of unity $\zeta$. Let $a \in F^* - F^{*p}$, and $E = F(\sqrt[p]{a})$. Then*
$$\ker(K_2(F) \otimes_{\mathbb{Z}} \mathbb{Z}/p\mathbb{Z} \to K_2(E) \otimes_{\mathbb{Z}} \mathbb{Z}/p\mathbb{Z}) = \{\{a,b\} \mid b \in F^*\}.$$

**Proof.** This was proved in lemma (8.17) under the additional assumption that $R_{p,F}$ is injective. We now prove it in general.

Let $x = \prod_{i=1}^{m} \{a_i, b_i\}$ be an element of the kernel. To prove the result, we may change $x$ by multiplying it by $\{a,b\}$ for any $b \in F^*$, since such elements are certainly in the kernel. We may thus assume without loss of generality that $a_1, \ldots, a_m$ yield $\mathbb{Z}/p\mathbb{Z}$-linearly independent elements of $E^* \otimes_{\mathbb{Z}} \mathbb{Z}/p\mathbb{Z}$. This is because $\ker(F^* \otimes_{\mathbb{Z}} \mathbb{Z}/p\mathbb{Z} \to E^* \otimes_{\mathbb{Z}} \mathbb{Z}/p\mathbb{Z})$ is just the subgroup generated by the image of $a$, from Kummer theory.

Now by lemma (8.22), we can find

(i) $a_{m+1}, \ldots, a_n \in E^*$,

(ii) $c_1, \ldots, c_n \in E^*$,

(iii) $x_\alpha \in E_\alpha = E(\sqrt[p]{a_\alpha})$ for each $\alpha \in \mathcal{A}_n$,

such that for each $1 \leq i \leq n$, we have
$$c_i^p = b_i \prod_{\alpha \in \mathcal{A}_n} N_\alpha(x_\alpha)^{\alpha(i)}.$$

Here, recall that $\mathcal{A}_n$ is the set of non-zero functions $\{1 \ldots, n\} \to \{0, 1, \ldots, p-1\}$, and for $\alpha \in \mathcal{A}_n$, we have $a_\alpha = \prod_{i=1}^{n} a_i^{\alpha(i)}$. Also $N_\alpha : E_\alpha^* \to E^*$ is the norm, and we take $b_{m+1} = \cdots = b_n = 1$.

Let $\omega = \sqrt[p]{a}$. Then $1, \omega, \ldots, \omega^{p-1}$ is an $F$-basis for $E$, so that we can uniquely write
$$a_i = \sum_{j=0}^{p-1} a_{ij} \omega^j, \quad m+1 \leq i \leq n, \quad a_{ij} \in F,$$

8. The Merkurjev-Suslin Theorem 175

$$c_i = \sum_{j=0}^{p-1} c_{ij}\omega^j, \quad 1 \le i \le n, \quad c_{ij} \in F.$$

Further, for each $\alpha \in \mathcal{A}_n$, we have $[E_\alpha : F] = p^2$, with a basis given by $(\sqrt[p]{a_\alpha})^i \omega^j$, $0 \le i,j \le p-1$. Hence we can uniquely write

$$x_\alpha = \sum_{i,j=0}^{p-1} x_{ij\alpha}(\sqrt[p]{a_\alpha})^i \omega^j, \quad \forall\, \alpha \in \mathcal{A}_n, \quad x_{ij\alpha} \in F.$$

Let $F_0$ be the subfield of $F$ generated by the algebraic closure in $F$ of the prime subfield, and the elements $a$ and $\zeta$ (the given primitive $p^{\text{th}}$ root of unity). Consider the following independent variables over $F_0$:

$$\begin{array}{ll} A_i & (1 \le i \le m) \\ A_{ij} & (m+1 \le i \le n, 0 \le j \le p-1) \\ B_i & (1 \le i \le m) \\ C_{ij} & (1 \le i \le n, 0 \le j \le p-1) \\ X_{ij\alpha} & (0 \le i,j \le p-1,\ \alpha \in \mathcal{A}_n). \end{array}$$

We may then define the following rings and fields:

$$\begin{aligned} E_0 &= F_0(\sqrt[p]{a}) = F_0(\omega) \\ R &= F_0[A_i, B_i, A_{ij}, X_{ij\alpha}] \\ F_1 &= F_0(A_i, A_{ij}, B_i, X_{ij\alpha}) \text{ (the quotient field of } R) \\ S &= R[\sqrt[p]{a}] = R \otimes_{F_0} E_0 \\ E_1 &= F_1(\sqrt[p]{a}) = F_1 \otimes_{F_0} E_0 \text{ (the quotient field of } S). \end{aligned}$$

Also define

$$\begin{aligned} A_i &= \sum_{j=0}^{p-1} A_{ij}\omega^j \text{ for } m+1 \le i \le n \\ A_\alpha &= \prod_{i=1}^{n} A_i^{\alpha(i)}. \end{aligned}$$

Then $A_i, A_\alpha \in S$ for all $i, \alpha$. Let

$$E_{1,\alpha} = E_1(\sqrt[p]{A_\alpha}), \text{ for each } \alpha \in \mathcal{A}_n.$$

Finally, we may define elements

$$\begin{aligned} X_\alpha &= \sum_{i,j=0}^{p-1} X_{ij\alpha}(\sqrt[p]{A_\alpha})^i \omega^j \text{ for all } \alpha \in \mathcal{A}_n \text{ (thus } X_\alpha \in E_{1,\alpha}^*) \\ D_i &= B_i \prod_{\alpha \in \mathcal{A}_n} N_{1,\alpha}(X_\alpha)^{\alpha(i)}, \text{ for all } 1 \le i \le n \\ D &= \prod_{i=1}^{n} N_{E_1/F_1}(D_i). \end{aligned}$$

Here $N_{1,\alpha} : E_{1,\alpha}^* \to E_1^*$ is the norm map. Note that $D_i \in S$, and $D \in R$.

There is a homomorphism of $F_0$-algebras $\psi : R \to F$ given by $A_i \mapsto a_i$, $1 \le i \le m$, $A_{ij} \mapsto a_{ij}$, $m+1 \le i \le n$, $0 \le j \le p-1$, $B_i \mapsto b_i$, $1 \le i \le m$, $X_{ij\alpha} \mapsto x_{ij\alpha}$, $\alpha \in \mathcal{A}_n$, $0 \le i,j \le p-1$. There is an induced

$E_0$-homomorphism $\eta : S = R \otimes_{F_0} E_0 \to F \otimes_{F_0} E_0 = E$, which maps $A_i$ to $a_i$ for all $1 \leq i \leq n$, and maps $N_{1,\alpha}(X_\alpha)$ to $N_\alpha(x_\alpha)$. Let $P \subset R$ be the kernel of $\psi$. Then $R_P$ is a regular local ring, in which $A_i$, $1 \leq i \leq m$, $B_i$, $1 \leq i \leq m$, are units. Also, $Q = PS$ is a prime ideal, which is the kernel of $\eta$; the ring $S_Q$ is a regular local ring, in which $A_i$, $1 \leq i \leq n$, $B_i$, $1 \leq i \leq m$, $N_{1,\alpha}(X_\alpha)$, $\alpha \in \mathcal{A}_n$, $0 \leq i,j \leq p-1$ are units.

Let $T = R[C_{ij}]$ be the polynomial ring over $R$ in the variables $C_{ij}$, $1 \leq i \leq n$, $0 \leq j \leq p-1$. Thus $T$ is the polynomial ring over $F_0$ in all of the variables $A_i$, $1 \leq i \leq m$, $A_{ij}$, $m+1 \leq i \leq n$, $0 \leq j \leq p-1$, $B_i$, $1 \leq i \leq m$, $C_{ij}$, $1 \leq i \leq n$, $0 \leq j \leq p-1$, $X_{ij\alpha}$, $\alpha \in \mathcal{A}_n$, $0 \leq i,j \leq p-1$. The homomorphism $\psi : R \to F$ extends to a homomorphism $\mu : T \to F$, by letting $\mu \vert_R = \psi$, and $\mu(C_{ij}) = c_{ij}$. Let $T \otimes_{F_0} E_0 = U$, and let

$$C_i = \sum_{j=0}^{p-1} C_{ij} \omega^j \text{ for } 1 \leq i \leq n.$$

Then $\mu$ and $\eta$ have a common extension to a homomorphism $\nu : U \to E$, with $\nu(C_i) = c_i$. Clearly $U = S[C_{ij}]$ is a polynomial algebra over $S$.

If $\sigma \in \text{Gal}(E_1/F_1)$ is the generator such that $\sigma(\omega) = \zeta \omega$ (where $\zeta$ is the given primitive $p^{\text{th}}$ root of unity), then $S$ is a Galois $R$-algebra, $U$ is a Galois $T$-algebra, and $\sigma$ also generates $\text{Gal}(S/R)$ and $\text{Gal}(U/T)$. Clearly $\eta$ and $\nu$ are $\sigma$-equivariant, where we also identify $\sigma$ with the corresponding generator of $\text{Gal}(E/F) \cong \text{Gal}(E_0/F_0)$.

If we set

$$Z_{ik} = \sum_{j=0}^{p-1} C_{ij} \sigma^k(\omega^j) = \sum_{j=0}^{p-1} C_{ij} \zeta^{kj} \omega^j, \ 1 \leq i \leq n, 0 \leq k \leq p-1,$$

then $Z_{ik}$ are another set of independent variables generating the polynomial algebra $U$ over $S$, since $\det[\sigma^k(\omega^j)]_{0 \leq j,k \leq p-1} \neq 0$ (this is clear by direct computation using $\sigma^k(\omega^j) = \zeta^{kj} \omega^j$, or since the square of this determinant is the discriminant of the basis $1, \omega, \ldots, \omega^{p-1}$ for $E_0$ over $F_0$).

Let $I \subset U$ be the ideal generated by

$$f_i = \left( \sum_{j=0}^{p-1} C_{ij} \omega^j \right)^p - D_i, \ 1 \leq i \leq n.$$

In terms of the variables $Z_{ik}$, we may rewrite this as

$$f_i = Z_{i0}^p - D_i, \ 1 \leq i \leq n.$$

If we uniquely write

$$f_i = \sum_{j=0}^{p-1} g_{ij} \omega^j, \ 1 \leq i \leq n,$$

8. The Merkurjev-Suslin Theorem    177

let $J \subset T$ be the ideal generated by the $g_{ij}$. Then $JU$ is the ideal generated by
$$Z_{ik}^p - \sigma^k(D_i), \quad 1 \le i \le n, 0 \le k \le p-1.$$
Again, this is because
$$Z_{ik}^p - \sigma^k(D_i) = \sigma^k(f_i) = \sum_{j=0}^{p-1} g_{ij}\sigma^k(\omega^j),$$
so that the $\sigma^k(f_i)$ are $E_0$-linear combinations of the $g_{ij}$ obtained by applying the invertible $p \times p$ matrix $[\sigma^k(\omega^j)]$. Also note that $JU \cap T = J$ (this is true of any ideal $J$ of $T$).

We claim the elements $\sigma^k(D_i)$, $1 \le i \le n$, $0 \le k \le p-1$, yield $\mathbb{Z}/p\mathbb{Z}$-linearly independent elements of $E_1^* \otimes_\mathbb{Z} \mathbb{Z}/p\mathbb{Z}$ (recall $E_1$ is the quotient field of $S$). To see this, since $S$ is a polynomial algebra over the field $E_0$, it suffices to find an $E_0$-homomorphism $S \to M$ to some field $M$ such that the images of the $\sigma^k(D_i)$ are multiplicatively independent elements of $M^*$ modulo $p^{\text{th}}$ powers.

Let $M = E_0(Y_i)$, where $Y_i$, $1 \le i \le n$ are independent variables over $E_0$. Let $\alpha_l \in \mathcal{A}_n$ be the mapping given by $\alpha_l(l) = 1$, $\alpha_l(j) = 0$ for $j \ne l$. Consider the homomorphism $S \to M$ given by

(i) $A_i \mapsto Y_i$ for $1 \le i \le m$,

(ii) $A_{i0} \mapsto Y_i$ for $m+1 \le i \le n$, $A_{ij} \mapsto 0$ for $m+1 \le i \le n$, $1 \le j \le p-1$,

(iii) $B_i \mapsto 1$ for $1 \le i \le m$,

(iv) for $\alpha \in \mathcal{A}_n - \{\alpha_1, \ldots, \alpha_n\}$, we have $X_{00\alpha} \mapsto 1$, $X_{ij\alpha} \mapsto 0$ for $0 \le i, j \le p-1$, $(i,j) \ne (0,0)$,

(v) for $1 \le l \le n$, $X_{00\alpha_l} \mapsto 1$, $X_{10\alpha_l} \mapsto -1$, $X_{01\alpha_0} \mapsto 1$, $X_{ij\alpha_l} \mapsto 0$ if $1 \le i, j \le p-1$.

Then $A_i \mapsto Y_i$ for all $1 \le i \le n$, and $X_\alpha \mapsto 1$ for $\alpha \ne \alpha_l$ for all $l$, while
$$N_{1,\alpha_l}(X_{\alpha_l}) \mapsto \prod_{k=0}^{p-1}(1 + \omega - \sqrt[p]{Y_l}\zeta^k) = (1+\omega)^p - Y_l.$$

Hence $\sigma^k(D_i) \mapsto (1 + \zeta^k\omega)^p - Y_i$, $1 \le i \le n$, $0 \le k \le p-1$. Since $Y_1, \ldots, Y_n$ are independent variables over $E_0$, and for $0 \le k \le p-1$, the elements $(1+\zeta^k\omega)^p$ in $E_0^*$ are distinct, the images of the $\sigma^k(D_i)$ are multiplicatively independent, as desired.

By Kummer theory, this implies that $JU \subset U$ is a prime ideal, and the quotient field $E_2$ of $U/JU$ is a (solvable) Galois extension of $F_1$ (the quotient field of $R$) of degree $p^{np+1}$, generated by $\omega$ and the $p^{\text{th}}$ roots of $\sigma^k(D_i)$, $1 \leq i \leq n$, $0 \leq k \leq p-1$. Hence $J = JU \cap T$ is a prime ideal in $T$, and the quotient field $F_2$ of $T/J$ is an extension of $F_1$ (in $E_2$) of degree $p^{pn}$. If $G = \text{Gal}(E_2/F_1)$, $H = \text{Gal}(E_2/F_2)$, then $G$ has order $p^{pn+1}$, and $H \subset G$ has order $p$. Hence there is a chain of subgroups $H = H_0 \subset H_1 \subset \cdots \subset H_n = G$ where $H_{i-1}$ is a normal subgroup of $H_i$ of index $p$. There is a corresponding chain of subfields

$$F_1 = L_0 \subset L_1 \subset \cdots \subset L_{pn} = F_2$$

where each $F_i$ is a Galois extension of $L_{i-1}$ of degree $p$. Further, $(T/J)[D^{-1}]$ is a flat, unramified $R[D^{-1}]$ algebra (i.e., an étale algebra). Indeed, $S[D^{-1}]$ is faithfully flat over $R[D^{-1}]$ (since $S$ is faithfully flat over $R$), so it suffices to prove that $(U/JU)[D^{-1}]$ is an étale algebra over $S[D^{-1}]$ (see Milne, Étale Cohomology, I, 2.24). But this is true, since $(U/JU)[D^{-1}]$ is obtained by adjoining to $S[D^{-1}]$ the $p^{\text{th}}$ roots of a set of units of $S[D^{-1}]$.

Since $\nu(C_i) = c_i$, we have $\nu(D_i) = c_i^p = \nu(Z_{i0}^p)$. Hence $\nu(f_i) = 0$ for $1 \leq i \leq n$. Hence also $\nu(\sigma^k(f_i)) = 0$ for $1 \leq i \leq n$, $0 \leq k \leq p-1$. Thus $\mu$ factors through $T/J$, and $\nu$ factors through $U/JU$. The kernel of $\mu : T/J \to F$ is a prime ideal $\widetilde{P}$ contracting to $P \subset R$, such that $\widetilde{Q} = \widetilde{P}(U/JU) = \ker \nu : U/JU \to E$, and $\widetilde{Q}$ contracts to $Q \subset S$. Since $\nu(D_i) = c_i^p \neq 0$, so that $\nu(\sigma^j(D_i)) = \sigma^j(c_i)$ for $0 \leq j \leq p-1$, we have $\sigma^j(D_i) \notin \widetilde{Q}$ for all $i,j$, and hence $D \notin Q \cap R = P$. Thus $(T/J)_{\widetilde{P}}$ is localisation of $(T/J)[D^{-1}]$, and is hence an étale extension of $R_P$. Hence by lemma (8.20), there are homomorphisms

$$\theta_1 : F_2^* \to F^*,$$

$$\theta_2 : K_2(F_2) \otimes_{\mathbb{Z}} \mathbb{Z}/p\mathbb{Z} \to K_2(F) \otimes_{\mathbb{Z}} \mathbb{Z}/p\mathbb{Z}$$

such that

(i) $\theta_1(y) = \mu(y)$ for $y \in (T/J)_{\widetilde{P}}^*$

(ii) $\theta_2\{y_1, y_2\} = \{\theta_1(y_1), \theta_1(y_2)\}$ for all $y_1, y_2 \in F_2^*$.

In particular,

$$\theta_2(\prod_{i=1}^m \{A_i, B_i\}) = \prod_{i=1}^m \{a_i, b_i\} = x \in K_2(F) \otimes_{\mathbb{Z}} \mathbb{Z}/p\mathbb{Z};$$

also, for any $Y \in F_2^*$,

$$\theta_2\{a, Y\} = \{a, \theta_1(Y)\}.$$

In $E_2 = F_2(\sqrt[p]{a})$, we have that

$$(\sum_{j=0}^{p-1} C_{ij}\omega^j)^p = B_i \prod_{\alpha \in \mathcal{A}_n} N_{1,\alpha}(X_\alpha)^{\alpha(i)}, \quad 1 \leq i \leq n,$$

$$B_i = 1, \quad m+1 \leq i \leq n.$$

Here we abuse notation, and also use the symbol $C_{ij}$ to denote the image of $C_{ij}$ in $F_2$. Hence by lemma (8.22),

$$X = \prod_{i=1}^{m}\{A_i, B_i\} \in \ker\left(K_2(F_2) \otimes_\mathbb{Z} \mathbb{Z}/p\mathbb{Z} \to K_2(E_2) \otimes_\mathbb{Z} \mathbb{Z}/p\mathbb{Z}\right).$$

Now $F_1$ is a pure transcendental extension of $F_0$, which (we recall) is generated over an algebraic extension of the prime field by $\zeta$ and $a$; hence $F_1$ is a pure transcendental extension of a field which is algebraic over its prime subfield. By lemmas (8.18) and (8.19), the Galois symbol $R_{p,F_1}$ is injective.

Now by lemma (8.17) and induction, we see that the Galois symbol is injective for each of the fields $L_i$, and hence in particular for $F_2$. Hence again by lemma (8.17), we see that

$$\ker(K_2(F_2) \otimes_\mathbb{Z} \mathbb{Z}/p\mathbb{Z} \to K_2(E_2) \otimes_\mathbb{Z} \mathbb{Z}/p\mathbb{Z})$$

consist of symbols $\{a, Y\}$ with $Y \in F_2^*$. In particular, there exists $Y \in F_2^*$ such that $X = \{a, Y\}$. Then

$$x = \theta_2(X) = \theta_2\{a, Y\} = \{a, \theta_1(Y)\},$$

i.e., $x$ is a symbol $\{a, y\}$ for some $y \in F^*$, which proves the proposition.

Note that the reference to lemma (8.18), i.e., to class field theory, is not needed if $F$ contains an algebraically closed field, since lemma (8.18) is trivial for the algebraic closure of a prime field. This helps to make our discussion more self-contained, for that case.

We can now finish the proof of the main theorem (8.5). We first prove that for any field $F$ containing a primitive $p^{\text{th}}$ root of unity $\zeta$, the map $R_{p,F}$ is injective. Suppose $x = \prod_{i=1}^{m}\{a_i, b_i\}$, and $R_{p,F}(x) = 0$. We prove by induction on $m$ that $x = 0$ in $K_2(F) \otimes_\mathbb{Z} \mathbb{Z}/p\mathbb{Z}$. For $m = 1$, this follows at once from lemma (8.4)(b). We may assume $a_m$ is not a $p^{\text{th}}$ power in $F$, else $x$ is a product of $m - 1$ symbols in $K_2(F) \otimes_\mathbb{Z} \mathbb{Z}/p\mathbb{Z}$, and we are done by induction. Now let $E = F(\sqrt[p]{a_m})$. Then $x_E = \prod_{i=1}^{m-1}\{a_i, b_i\}$ in $K_2(E) \otimes_\mathbb{Z} \mathbb{Z}/p\mathbb{Z}$, and $R_{p,E}(x_E) = 0$, so by induction, we have $x_E = 0$. Hence by Proposition (8.23), $x = \{a_m, y\}$ in $K_2(F) \otimes_\mathbb{Z} \mathbb{Z}/p\mathbb{Z}$, for some $y \in F^*$; hence we are done again by the case $m = 1$.

Now we show $R_{p,F}$ is also surjective. By the usual norm (i.e., transfer) argument, we may assume that any finite extension of $F$ has degree a power of $p$. Let $u \in {}_p\text{Br}(F)$, and let $E$ be an extension of minimal degree splitting $u$. We work by induction on $[E:F]$. Let $L$ be the normal closure of $E$ over $F$. The Galois group $\text{Gal}(L/F)$ is a $p$-group, since every finite extension of $F$ has degree a power of $p$. Hence we can find an intermediate field $F \subset F_1 \subset E$ such that $[F_1:F] = p$. Then $u_{F_1} \in$ image $R'_{p,F_1}$ by induction, since $u_{F_1}$ is split by $E$, and $[E:F_1] < [E:F]$. By lemma (8.17)(ii), the square

$$\begin{array}{ccc} K_2(F) \otimes_{\mathbb{Z}} \mathbb{Z}/p\mathbb{Z} & \stackrel{R'_{p,F}}{\to} & {}_p\text{Br}(F) \\ \downarrow & & \downarrow \\ K_2(F_1) \otimes_{\mathbb{Z}} \mathbb{Z}/p\mathbb{Z} & \stackrel{R'_{p,F_1}}{\to} & {}_p\text{Br}(F_1) \end{array}$$

is a pullback. Hence $u_{F_1} \in$ image $R'_{p,F_1} \Rightarrow u \in$ image $R'_{p,F}$. This concludes the proof of Theorem (8.5).

### (8.3) Torsion in $K_2$

We now prove a theorem of Suslin, which describes the torsion subgroup of $K_2(F)$ in many situations.

**Theorem 8.24** *(a) Let $F$ be a field, $p$ a prime such that $F$ contains a primitive $p^{\text{th}}$ root of unity $\zeta$. Then the $p$-torsion subgroup of $K_2(F)$ is precisely the subgroup*

$$\{\{\zeta, a\} \mid a \in F^*\} \subset K_2(F).$$

*(b) In (a), if $F_0 \subset F$ is the algebraic closure in $F$ of the prime subfield, then for any $a \in F^*$ with $\{\zeta, a\} = 0$ in $K_2(F)$, there exists $b \in F^*$, $c \in F_0^*$ such that $a = b^p c$, and $\{\zeta, c\} = 0$ in $K_2(F_0)$.*

*(c) If $F$ has characteristic $p > 0$, then $K_2(F)$ has no $p$-torsion.*

**Corollary 8.25** *Let $F_1 \subset F_2$ be an inclusion of fields such that $F_1$ is algebraically closed in $F_2$. Then $K_2(F_1) \to K_2(F_2)$ is injective.*

**Proof.** We may assume without loss of generality that $F_1$ and $F_2$ are finitely generated extensions of the prime subfield; let $F_0 \subset F_1$ be the algebraic closure of the prime subfield in $F_2$. Then we can find a purely transcendental extension $F_3$ of $F_1$ in $F_2$ such that $F_2$ is finite algebraic over $F_3$. Now $K_2(F_1) \to K_2(F_3)$ is injective, since $K_2(F_3) = \varinjlim_f K_2(F_1[t_1, \ldots, t_d, f^{-1}])$, where $d$ is the relative transcendence degree of $F_3$ over $F_1$, and $f$ ranges over the monic polynomials, and $K_2(F_1) \to$

$K_2(F_1[t_1,\ldots,t_d,f^{-1}])$ is a split inclusion for any field $F_1$ (if $F_1$ is finite, $K_2(F_1) = 0$; if $F_1$ is infinite, $F_1 \to F_1[t_1,\ldots,t_d,f^{-1}]$ is a split inclusion of rings). On the other hand, by the usual transfer argument, $\ker(K_2(F_3) \to K_2(F_2))$ is an $n$-torsion group, where $n = [F_2 : F_3]$. Hence it suffices to prove that $K_2(F_1) \to K_2(F_2)$ is injective on $p$-torsion subgroups, for each prime $p$. By (c) of the Theorem (8.24), we may assume $F_1$ is not of characteristic $p$; hence by the usual tranfer argument, we may also assume $F_1$ contains a primitive $p^{\text{th}}$ root of unity. Now by (a) and (b) of Theorem (8.24), we are done.

**Proof of Theorem (8.24) (a):** Let $K = F(U)$, $L = F(U)[T]/(T^p - U) = F(T)$, so that $K$, $L$ are each isomorphic to the function field of $\mathbb{A}^1_F$, and the inclusion of fields $K \hookrightarrow L$ corresponds to a morphism $f : X \to Y$ over $F$, where $X = \operatorname{Spec} F[T] \cong \mathbb{A}^1_F$, and $Y = \operatorname{Spec} F[U] \cong \mathbb{A}^1_F$. Then $L/K$ is a cyclic Galois extension of degree $p$, which we may think of as a "universal Kummer extension over $F$". Let $\sigma \in \operatorname{Gal}(L/K)$ be the generator determined by $\sigma(T) = \zeta^{-1}T$. There is a commutative diagram

$$\begin{array}{ccc} K_2(L) & \xrightarrow{N_{L/K}} & K_2(K) \\ \uparrow & & \uparrow \\ K_2(F) & \xrightarrow{\cdot p} & K_2(F) \end{array}$$

where the (injective) vertical maps are induced by the inclusions of $F$ in $L$, $K$ respectively, and $\cdot p$ denotes multiplication by $p$. This follows immediately from the diagram of categories and functors

$$\begin{array}{ccc} \mathcal{P}(L) & \xrightarrow{N} & \mathcal{P}(K) \\ \alpha \uparrow & & \uparrow \beta \\ \mathcal{P}(F) & \xrightarrow{\cdot p} & \mathcal{P}(F) \end{array}$$

where $\alpha(V) = L \otimes_F V$, and $(\cdot p)(V) = V^{\oplus p}$, $\beta(V) = K \otimes_F V$ for any $V \in \mathcal{P}(F)$, and $N(W)$ is the underlying $K$-vector space of an $L$-vector space $W$. The choice of a basis for $N(L)$ gives a natural isomorphism of functors $N \circ \alpha \to \beta \circ (\cdot p)$.

Thus from Theorem (8.15),

$$_pK_2(F) = K_2(F) \cap \ker(N_{L/K} : K_2(L) \to K_2(K))$$
$$= K_2(F) \cap (1 - \sigma_*)(K_2(L)).$$

So it suffices to prove that

$$K_2(F) \cap (1 - \sigma_*)(K_2(L)) = \{\{\zeta, a\} \mid a \in F^*\} \subset K_2(F).$$

Let $u \in K_2(F)$, $v \in K_2(L)$ such that $u = v - \sigma_*(v)$ in $K_2(L)$. Consider the exact sequence

$$0 \to K_2(F) \to K_2(L) \xrightarrow{\partial} \oplus_{x \in X^1} F(x)^* \to 0,$$

obtained from the degeneration of the BGQ-spectral sequence for $\mathbb{A}_F^1$, and $H^0(\mathbb{A}_F^1, \mathcal{K}_2) = K_2(F)$, $H^1(\mathbb{A}_F^1, \mathcal{K}_2) = 0$ (see the proof of (8.19)). We have $\partial(u) = 0$, so that $\partial(v)$ is invariant under $\sigma$. If we write

$$\oplus_{x \in X^1} F(x)^* = \oplus_{y \in Y^1} \left( \oplus_{f(x)=y} F(x)^* \right),$$

then the subgroup of $\sigma$-invariant elements is

$$\oplus_{y \in Y^1} F(y)^*.$$

The map $f^* : E_1^{1,-2}(Y) \to E_1^{1,-2}(X)$ of $E_1$ terms of the BGQ-spectral sequences is given as follows: let $y_0 \in Y$ and $x_0 \in X$ be the respective origins, so that $X - \{x_0\} \to Y - \{y_0\}$ is unramified, and $f^{-1}(y_0) = x_0$. Then for $y \neq y_0$, $f^*$ induces the natural inclusion

$$F(y)^* \to \oplus_{f(x)=y} F(x)^*$$

as the subgroup of $\sigma$-invariants. If $y = y_0$, then $F(y_0) = F(x_0) = F$, and $\sigma$ acts trivially; however, $F(y_0)^* \to F(x_0)^*$ is multiplication by $p$, since $f^{-1}(y_0)$ is a fiber of multiplicity $p$ in $X$. Indeed, the terms $F(y_0)^*$, $F(x_0)^*$ are obtained by dévissage from $K_1$ groups of suitable categories of Artinian modules, and for any $\mathcal{O}_{Y,y_0}$-module $M$ of finite length, $f^*M$ has length equal to $p(\text{length}(M))$.

Thus we can find $w \in K_2(K)$ such that $\partial(w_L) - \partial(v)$ lies in the summand $F(x_0)^*$. Now we can find $a \in F^*$ such that $\partial(\{T, a\}) = \partial(w_L) - \partial(v)$, since (up to sign) $\partial$ is the sum of tame symbols associated to the discrete valuations given by points of $X^1$ (see (9.12)). Hence, after altering $w$ by adding an element of $K_2(F)$, we can write $v = w_L + \{T, a\}$. Hence $u = (1 - \sigma_*)(v) = \{\zeta, a\}$, as desired.

**Proof of (8.24)(c):** This is very similar to (a), using the "universal Artin-Schreier extension" $K = F(U) \hookrightarrow F(U)[T]/(T^p - T - U) = F(T) = L$. Since the corresponding morphism $\mathbb{A}_F^1 \to \mathbb{A}_F^1$ is unramified everywhere, the argument above shows that if $u \in K_2(F)$, $u = v - \sigma_*(v)$, then in fact $\partial(v) = \partial(w_L)$ for some $w \in K_2(K)$. Again, after altering $w$ by adding an element of $K_2(F)$, we may assume $\partial(v) = \partial(w_L)$; hence $u = v - \sigma(v) = 0$.

**Proof of (8.24)(b):** The proof uses continuous ($\ell$-adic) Galois cohomology, which we now discuss, as well as results of Tate [T].

# 8. The Merkurjev-Suslin Theorem

Given a topological group $G$, and a topological $G$-module $M$, let $\mathcal{D}^i(G, M)$ denote the group of continuous maps $G^{i+1} \to M$; then the $\mathcal{D}^i(G, M)$ can be made into a complex with differentials given by

$$df(g_0, \ldots, g_{i+1}) = \sum_{j=0}^{i+1} (-1)^j f(g_0, \ldots, \widehat{g_j}, \ldots, g_{i+1})$$

(here $\widehat{g_j}$ means that $g_j$ is omitted). $G$ acts on the complex $\mathcal{D}^*(G, M)$ by

$$(gf)(g_0, \ldots, g_i) = g \cdot f(g^{-1} g_0, \ldots, g^{-1} g_i),$$

and the subcomplex of $G$-invariants $\mathcal{D}^*(G, M)^G = C^*(G, M)$ is defined to be the standard complex of (continuous) cochains on $G$ with values in $M$. We may define $H^i(G, M) = H^i(C^*(G, M))$. If $F$ is a field, $\overline{F}$ a separable closure, $G = \mathrm{Gal}(\overline{F}/F)$, $M$ a continuous $G$-module with the discrete topology, then $H^i(G, M)$ is just $H^i(F, M)$, the Galois cohomology group considered earlier. If $M$ is a finitely generated $\mathbb{Z}_\ell$-module with its $\ell$-adic topology, which is a continuous $G$-module, we can consider the continuous ($\ell$-adic) cohomology groups $H^i(G, M)$. A notable example is given by $M = \mathbb{Z}_\ell(1) = \varprojlim_n \mu_{\ell^n}$, where $\ell$ is a prime $\neq \mathrm{char}. F$. The assertions below are proved in Tate's article [T].

(i) If $G$ is any topological group, and

$$0 \to M' \xrightarrow{\alpha} M \xrightarrow{\beta} M'' \to 0$$

is an exact sequence of continuous $G$-modules such that $M'$ has the subspace topology, and $\beta$ has a continuous section (which may not be either $G$-equivariant or a homomorphism), then there is a long exact sequence

$$\cdots \to H^i(G, M') \xrightarrow{\alpha_*} H^i(G, M) \xrightarrow{\beta_*} H^i(G, M'') \to \cdots$$

(ii) Next, if $M$ is a finite $\mathbb{Z}_\ell$-module with its $\ell$-adic topology, and $G$ a topological group acting continuously on $M$ through $\mathbb{Z}_\ell$-linear automorphisms, then we have an exact sequence

$$0 \to \varprojlim_n{}^1 H^{i-1}(G, M/\ell^n M) \to H^i(G, M) \to \varprojlim_n H^i(G, M/\ell^n M) \to 0,$$

where $\varprojlim_n{}^1$ is the first derived functor of $\varprojlim_n$. (Recall that if

$$\cdots \to G_n \xrightarrow{f_n} G_{n-1} \xrightarrow{f_{n-1}} \cdots \xrightarrow{f_2} G_1$$

is an inverse system of Abelian groups, and we let $A = \prod_{n=1}^{\infty} G_n$, and define $T : A \to A$ by $T(g_1, g_2, \ldots) = (g_1 - f_2(g_2), g_2 - f_3(g_3), \ldots)$, then we have $\ker T = \varprojlim_n G_n$ and $\operatorname{coker} T = \varprojlim_n {}^1 G_n$. Further, if each $G_n$ is finite, then $\varprojlim_n {}^1 G_n = 0$; more generally, this holds if the inverse system $\{G_n\}$ satisfies the *Mittag-Leffler condition*, that for each $n$, $\operatorname{im}(G_{n+m} \to G_n)$ is independent of $m$ for all large enough $m$.)

**Lemma 8.26** *Let $G$ be a compact topological group, $H \subset G$ a closed normal subgroup, such that $G \to G/H$ has a continuous section. For any topological $G$-module $M$, we have an exact sequence*

$$0 \to H^1(G/H, M^H) \to H^1(G, M) \to \overline{M} \to H^2(G/H, M^H) \to H^2(G, M)$$

*where $\overline{M} \subset H^1(H, M)^{G/H}$.*

**Proof.** This follows from an analogue of the Hochschild-Serre spectral sequence. A continuous section $s : G/H \to G$ gives a map of complexes $\bar{s} : \mathcal{D}^*(H, M) \to \mathcal{D}^*(G, M)$ which is $H$-equivariant, such that $\bar{s}$ and the restriction map $\mathcal{D}^*(G, M) \to \mathcal{D}^*(H, M)$ are chain homotopy inverses. If $\mathcal{E}^* = \mathcal{D}^*(G, M)^H$, then $\mathcal{E}^*$ computes the continuous cohomology $H^*(H, M)$; further $(\mathcal{E}^*)^{G/H} = \mathcal{D}^*(G, M)^G = C^*(G, M)$ computes the continuous cohomology $H^*(G, M)$. Now $\mathcal{E}^n \subset \mathcal{D}^n(G, M)$ is the subset of $H$-equivariant continuous maps $G^{n+1} \to M$. This can be topologised with the compact-open topology (see Appendix (A.1)), so that $\mathcal{E}^n$ becomes a continuous $G/H$-module (since $G$ is compact, the evaluation map is continuous). One checks that $\mathcal{E}^n$ is an induced $G/H$-module (i.e., $\mathcal{E}^n = \operatorname{Hom}(G/H, N)$ for some topological group $N$, where the group of continuous homomorphisms is given the compact-open topology); using this one shows that $H^i(G/H, \mathcal{E}^n) = 0$ for all $i > 0$, and $H^0(G/H, \mathcal{E}^n) = (\mathcal{E}^n)^{G/H} = C^n(G, M)$, for all $n$. Consider the double complex

$$A^{ij} = \mathcal{C}^i(G/H, \mathcal{E}^j).$$

Taking cohomology first in the $i$-direction, the above computation of $H^*(G/H, \mathcal{E}^n)$ shows that the first spectral sequence degenerates at $E_2$, and the cohomology groups of the total complex are $H^*(G, M)$. If $Z^j = \ker(\mathcal{E}^j \to \mathcal{E}^{j+1})$, then the second spectral sequence has $E_1$ terms

$$E_1^{i,j} = \mathcal{C}^i(G/H, Z^j)/(\operatorname{image} \mathcal{C}^i(G/H, \mathcal{E}^{j-1})).$$

In particular

$$E_1^{0,j} = H^j(\mathcal{E}^*) = H^j(H, M),$$
$$E_1^{i,0} = \mathcal{C}^i(G/H, Z^0) = \mathcal{C}^i(G/H, M^H).$$

The differential $d_1 : E_1^{0,j} \to E_1^{1,j}$ is a map

$$Z^j/B^j = H^j(H,M) \to \mathcal{C}^1(G/H, Z^j)/(\text{image}\,\mathcal{C}^1(G/H, \mathcal{E}^{j-1})),$$

where $B^j = \text{image}(\mathcal{E}^{j-1} \to \mathcal{E}^j)$. It is induced by the map

$$Z^j \to \mathcal{C}^1(G/H, Z^j), \quad z \mapsto f_z,$$

where $f_z : (G/H)^2 \to Z^j$ is given by

$$f_z(g_0 H, g_1 H) = g_0(z) - g_1(z).$$

If $f_z \mapsto 0$ in $E_1^{i,j}$, i.e., $f_z$ lifts to an element of $\mathcal{C}^1(G/H, \mathcal{E}^{j-1})$, then $g_0(z) - g_1(z) \in B^j$, for any $g_0, g_1 \in G$, so the class $[z] \in H^j(H,M)$ is $G/H$-invariant. Thus

$$E_2^{0,j} \subset H^j(H,M)^{G/H}.$$

Let $\overline{M} = E_2^{0,1} \subset H^1(H,M)^{G/H}$. Then the five-term exact sequence of low degree terms (for the above spectral sequence of the double complex) is

$$0 \to H^1(G/H, M^H) \to H^1(G,M) \to \overline{M} \to H^2(G/H, M^H) \to H^2(G,M),$$

which is the desired exact sequence. This proves (8.26).

Let $F$ be a field, finitely generated over the prime field, and $\ell$ a prime $\neq \text{char.}\,F$. One sees easily that $\varprojlim_n{}^1 H^0(F, \mu_{\ell^n}) = 0$, since $H^0(F, \mu_{\ell^n})$ is finite for each $n$. Thus we have an isomorphism (Kummer theory)

$$H^1(F, \mathbb{Z}_\ell(1)) \cong \varprojlim_n F^* \otimes_{\mathbb{Z}} \mathbb{Z}/\ell^n \mathbb{Z},$$

giving a character

$$\chi_{F, \mathbb{Z}_\ell} : F^* \to H^1(F, \mathbb{Z}_\ell(1)).$$

Tate shows (in [T]) that if $a \in F^*$, $a \neq 1$, then the cup product

$$\chi_{F, \mathbb{Z}_\ell}(a) \cup \chi_{F, \mathbb{Z}_\ell}(1-a) = 0 \in H^2(F, \mathbb{Z}_\ell(2)),$$

where $\mathbb{Z}_\ell(2) = \mathbb{Z}_\ell(1) \otimes_{\mathbb{Z}_\ell} \mathbb{Z}_\ell(1)$ (as a topological Galois module). Thus we obtain a homomorphism (Tate's $\ell$-adic Galois symbol)

$$K_2(F) \to H^2(F, \mathbb{Z}_\ell(2)).$$

In [T], Tate proves the following important result.

**Theorem 8.27** (Tate) *If $F$ is a number field, then for any prime $\ell$, the $\ell$-adic Galois symbol $K_2(F) \to H^2(F, \mathbb{Z}_\ell(2))$ is an isomorphism on $\ell$-primary components.*

We will now use this to prove (8.24)(b). We first prove:

**Lemma 8.28** *Let $F$ be a field which is finitely generated over the prime subfield, and $F_0 \subset F$ the algebraic closure of the prime subfield in $F$. Then the natural map*
$$H^i(F_0, \mathbb{Z}_\ell(2)) \to H^i(F, \mathbb{Z}_\ell(2))$$
*is an isomorphism for $i = 1$ and an injection for $i = 2$.*

**Proof.** (Sketch) Let $\overline{F}$ be a separable closure of $F$, and $\overline{F_0}$ the separable (= algebraic) closure of $F_0$ in $\overline{F}$. Since $F$ is finitely generated over $F_0$, and $F_0$ is algebraically closed in $F$, we can find a normal projective variety $X$ defined over $F_0$ whose function field (over $F_0$) is $F$. Let $\overline{X} = X_{\overline{F_0}}$, so that the compositum $\overline{F_0}F$ is the function field of $\overline{X}$ over $\overline{F_0}$. Then $\overline{F_0}F$ is a Galois extension of $F$ with Galois group isomorphic to $\mathrm{Gal}\,(\overline{F_0}/F_0)$. We can apply (8.26) with $G = \mathrm{Gal}\,(\overline{F}/F)$, $H = \mathrm{Gal}\,(\overline{F}/\overline{F_0}F)$, and $G/H \cong \mathrm{Gal}\,(\overline{F_0}/F_0)$, for the module $M = \mathbb{Z}_\ell(2)$. We then obtain an exact sequence of continuous Galois cohomology groups

$$0 \to H^1(F_0, \mathbb{Z}_\ell(2)) \to H^1(F, \mathbb{Z}_\ell(2)) \to \overline{M} \to H^2(F_0, \mathbb{Z}_\ell(2)) \to H^2(F, \mathbb{Z}_\ell(2)),$$

where $\overline{M} \subset H^0(F_0, H^1(\overline{F_0}F, \mathbb{Z}_\ell(2)))$. So it suffices to prove that

$$H^0(F_0, H^1(\overline{F_0}F, \mathbb{Z}_\ell(2))) = 0.$$

Let $\mathrm{Div}\,(\overline{X})$ denote the group of Cartier divisors on $\overline{X}$; being a subgroup of the (free Abelian) group of Weil divisors, $\mathrm{Div}\,(\overline{X})$ is a free Abelian group, each of whose elements has a finite orbit under $G/H = \mathrm{Gal}\,(\overline{F_0}/F_0)$. From the family of exact sequences of $G/H$-modules ($\mathrm{Pic}\,\overline{X}$ is the Picard group of $\overline{X}$)

$$0 \to {}_{\ell^n}\mathrm{Pic}\,\overline{X} \to (\overline{F_0}F)^* \otimes_{\mathbb{Z}} \mathbb{Z}/\ell^n\mathbb{Z} \to (\mathrm{Div}\,X) \otimes_{\mathbb{Z}} \mathbb{Z}/\ell^n\mathbb{Z} \to (\mathrm{Pic}\,X) \otimes_{\mathbb{Z}} \mathbb{Z}/\ell^n\mathbb{Z},$$

we obtain an inverse limit exact sequence

$$0 \to T_\ell(\mathrm{Pic}\,\overline{X}) \to \varprojlim_n (\overline{F_0}F)^* \otimes_{\mathbb{Z}} \mathbb{Z}/\ell^n\mathbb{Z} \to (\mathrm{Div}\,\overline{X}) \otimes_{\mathbb{Z}} \mathbb{Z}_\ell.$$

Since

$$H^1(\overline{F_0}F, \mathbb{Z}_\ell(2)) \cong H^1(\overline{F_0}F, \mathbb{Z}_\ell(1)) \otimes_{\mathbb{Z}_\ell} \mathbb{Z}_\ell(1) \cong \varprojlim_n (\overline{F_0}F)^* \otimes_{\mathbb{Z}} \mu_{\ell^n},$$

we obtain an exact sequence of $G/H$-modules

$$0 \to T_\ell(\mathrm{Pic}\,\overline{X}) \otimes_{\mathbb{Z}_\ell} \mathbb{Z}_\ell(1) \to H^1(\overline{F_0}F, \mathbb{Z}_\ell(2)) \to (\mathrm{Div}\,\overline{X}) \otimes_{\mathbb{Z}} \mathbb{Z}_\ell(1).$$

8. The Merkurjev–Suslin Theorem        187

Now $(\operatorname{Div}\overline{X})\otimes_{\mathbb{Z}}\mathbb{Z}_\ell(1)$ has no $G/H$-invariants, since every element of $\operatorname{Div}\overline{X}$ is fixed by a subgroup of finite index, and the cyclotomic character $G/H \to \mathbb{Z}_\ell^*$ has infinite image. Next, $T_\ell(\operatorname{Pic}\overline{X}) = T_\ell(\operatorname{Pic}{}^0\overline{X})$, where $\operatorname{Pic}{}^0(\overline{X})$, the Picard variety of $\overline{X}$, is an Abelian variety (since $\overline{X}$ is a normal projective variety) by a theorem of Chevalley. By Weil's theorem on the eigenvalues of Frobenius acting on the Tate module of an Abelian variety, we see immediately that $T_\ell(\operatorname{Pic}\overline{X}) \otimes_{\mathbb{Z}} \mathbb{Z}_\ell(1)$ has no $G/H$-invariants, since the Frobenius associated to a (suitably chosen) finite field $\mathbb{F}_q$ has eigenvalues of absolute value $q^{1/2}$ on $T_\ell(\operatorname{Pic}\overline{X})\otimes\mathbb{Q}$, while the cyclotomic character has value $q^{-1}$ on this Frobenius. This proves $H^1(\overline{F_0}F, \mathbb{Z}_\ell(2))^{G/H} = 0$, proving the lemma.

To prove (8.24)(b), we may assume without loss of generality that $F$ is finitely generated over its prime subfield. Consider the following diagram with exact rows

$$\begin{array}{ccccccc} H^1(F_0, \mathbb{Z}_p(2)) & \to & H^1(F_0, \mathbb{Q}_p(2)) & \to & H^1(F_0, \mathbb{Q}_p/\mathbb{Z}_p(2)) & \xrightarrow{\partial} & H^2(F_0, \mathbb{Z}_p(2)) \\ \beta \downarrow \cong & & \downarrow \cong & & \downarrow & & \downarrow \alpha \\ H^1(F, \mathbb{Z}_p(2)) & \to & H^1(F, \mathbb{Q}_p(2)) & \to & H^1(F, \mathbb{Q}_p/\mathbb{Z}_p(2)) & \xrightarrow{\partial} & H^2(F, \mathbb{Z}_p(2)) \end{array}$$

Here $\alpha$ is injective, and $\beta$ is an isomorphism, from the previous lemma, with $\ell = p$; also $H^i(F_0, \mathbb{Q}_p(2)) = H^i(F_0, \mathbb{Z}_p(2))\otimes_{\mathbb{Z}}\mathbb{Q}$, and similarly for $F$. Suppose $c \in F^*$ such that $\{\zeta, c\} = 0$ in $K_2(F)$. If

$$\partial' : \varinjlim_n \mu_{p^n}(F) = H^0(F, \mathbb{Q}_p/\mathbb{Z}_p(1)) \to H^1(F, \mathbb{Z}_p(1))$$

is the boundary map in the exact cohomology sequence associated to

$$0 \to \mathbb{Z}_p(1) \to \mathbb{Q}_p(1) \to \mathbb{Q}_p/\mathbb{Z}_p(1) \to 0,$$

then one checks that $\partial'(\zeta) = \chi_{F,\mathbb{Z}_p}(\zeta)$. Hence the cup product

$$\gamma = (\zeta) \cup \chi_{F,\mathbb{Z}_p}(c) \in H^1(F, \mathbb{Q}_p/\mathbb{Z}_p(1))$$

satisfies

$$\partial(\gamma) = \partial'(\zeta) \cup \chi_{F,\mathbb{Z}_p}(c) = \chi_{F,\mathbb{Z}_p}(\zeta) \cup \chi_{F,\mathbb{Z}_p}(c) = R_{F,\mathbb{Z}_p}\{\zeta, c\} = 0.$$

Hence $\gamma$ lies in the image of $H^1(F, \mathbb{Q}_p(2))$. By a diagram chase, this implies that $\gamma$ lies in the image of

$$_pH^1(F_0, \mathbb{Q}_p/\mathbb{Z}_p(2)) \to {}_pH^1(F, \mathbb{Q}_p/\mathbb{Z}_p(2))$$

(note that $\gamma$ has order dividing $p$). We have a commutative diagram with exact rows

$$\begin{array}{ccccccc} H^0(F_0, \mathbb{Q}_p/\mathbb{Z}_p(2)) & \to & H^1(F_0, \mu_p^{\otimes 2}) & \to & {}_pH^1(F_0, \mathbb{Q}_p/\mathbb{Z}_p(2)) & \to & 0 \\ \cong \downarrow & & \downarrow & & \downarrow & & \\ H^0(F, \mathbb{Q}_p/\mathbb{Z}_p(2)) & \to & H^1(F, \mu_p^{\otimes 2}) & \to & {}_pH^1(F, \mathbb{Q}_p/\mathbb{Z}_p(2)) & \to & 0 \end{array}$$

Since $\gamma$ is the image of $(\zeta) \cup \chi_{p,F}(c) \in H^1(F, \mu_p^{\otimes 2})$, we see that $(\zeta) \cup \chi_{p,F}(c)$ is in the image of $H^1(F_0, \mu_p^{\otimes 2}) \to H^1(F, \mu_p^{\otimes 2})$. But the cup product with $(\zeta)$ gives an isomorphism $H^*(F_0, \mu_p) \cong H^*(F_0, \mu_p^{\otimes 2})$, and similarly for $F$; hence $\chi_{p,F}(c)$ lies in the image of $H^1(F_0, \mu_p) \to H^1(F, \mu_p)$, i.e., we can write $c = d^p e$ with $d \in F^*$, $e \in F_0^*$, from the Kummer isomorphisms

$$H^1(F_0, \mu_p) \cong F_0^* \otimes_{\mathbb{Z}} \mathbb{Z}/p\mathbb{Z}, \quad H^1(F, \mu_p) \cong F^* \otimes_{\mathbb{Z}} \mathbb{Z}/p\mathbb{Z}.$$

Hence $\{\zeta, c\} = \{\zeta, e\}$ lies in the image of ${}_p K_2(F_0) \to {}_p K_2(F)$. If $F$ has positive characteristic, then $K_2(F_0) = 0$. If $F_0$ is a number field, Tate's theorem (8.27) and the injectivity of $H^2(F_0, \mathbb{Z}_p(2)) \to H^2(F, \mathbb{Z}_p(2))$ imply that $\{\zeta, e\} = 0$ in $K_2(F_0)$. This proves (8.24)(b).

## (8.4) Torsion in $CH^2$

We now discuss applications of the results of Merkurjev and Suslin to the study of torsion in the Chow group of codimension 2 cycles. We restrict ourselves to a typical situation, though greater generality is possible.

Let $X$ be an irreducible smooth quasi-projective variety over an algebraically closed field $k$, and $n$ an integer prime to the characteristic. We will show that ${}_n CH^2(X)$, the $n$-torsion subgroup of the Chow group of cycles of codimension 2 modulo rational equivalence, is a finite group.

Our starting point is the Gersten resolution (5.27)

$$(1) \quad \cdots \; 0 \to \mathcal{K}_{2,X} \to i_* K_2(k(X)) \xrightarrow{T} \oplus_{x \in X^1} (i_x)_* k(x)^* \xrightarrow{\partial} \oplus_{y \in X^2} (i_y)_* \mathbb{Z} \to 0$$

which gives $H^2(X, \mathcal{K}_{2,X}) = CH^2(X)$; here $\partial$, $T$ are the divisor map on rational functions, and the sum of tame symbols, respectively (since the Gersten resolution has these maps, up to sign). We have the complex of $n$-torsion subsheaves

$$(2) \quad \cdots \quad 0 \to {}_n\mathcal{K}_{2,X} \to i_* {}_n K_2(k(X)) \xrightarrow{\alpha} \oplus_{x \in X^1}(i_x)_* \mu_n \to 0,$$

where $\alpha$ is induced by $T$; the exactness of (1) implies that this complex is exact, except possibly that $\alpha$ may not be surjective. But by considering the commutative diagram

$$\begin{array}{ccc} i_* k(X)^* \otimes_{\mathbb{Z}} \mu_n & \xrightarrow{\partial \otimes 1} & (\oplus_{x \in X^1}(i_x)_* \mathbb{Z}) \otimes_{\mathbb{Z}} \mu_n \\ \theta \downarrow & & \downarrow \cong \\ i_* {}_n K_2(k(X)) & \xrightarrow{\alpha} & \oplus_{x \in X^1}(i_x)_* \mu_n \end{array}$$

(where $\theta(f \otimes \zeta) = \{f, \zeta\}$), we see that $\alpha$ is surjective. From (1) and (2) we deduce the exactness of

$$(3) \quad \cdots \; 0 \to \mathcal{K}_{2,X} \otimes_{\mathbb{Z}} \mathbb{Z}/n\mathbb{Z} \to i_* K_2(k(X)) \otimes_{\mathbb{Z}} \mathbb{Z}/n\mathbb{Z} \xrightarrow{T}$$
$$\oplus_{x \in X^1}(i_x)_* k(x)^* \otimes_{\mathbb{Z}} \mathbb{Z}/n\mathbb{Z} \xrightarrow{\partial} \oplus_{y \in X^2}(i_y)_* \mathbb{Z}/n\mathbb{Z} \to 0.$$

Consider the exact sheaf sequences (which define $\mathcal{F}$)

$$(4) \quad \cdots \quad 0 \to {}_n\mathcal{K}_{2,X} \to \mathcal{K}_{2,X} \to \mathcal{F} \to 0,$$

$$(5) \quad \cdots \quad 0 \to \mathcal{F} \to \mathcal{K}_{2,X} \to \mathcal{K}_{2,X} \otimes_{\mathbb{Z}} \mathbb{Z}/n\mathbb{Z} \to 0.$$

From (2), we have $H^i(X, {}_n\mathcal{K}_{2,X}) = 0$ for $i \geq 2$, so that from (4),

$$H^1(X, \mathcal{K}_{2,X}) \twoheadrightarrow H^1(X, \mathcal{F}), \quad H^2(X, \mathcal{F}) \cong H^2(X, \mathcal{K}_{2,X}).$$

From (5) we thus have a diagram with an exact row ($\cdot p$ denotes multiplication by $n$)

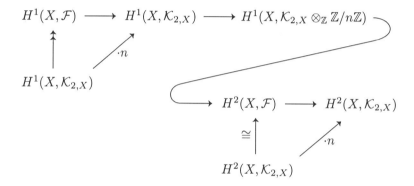

We obtain an exact sequence

$$0 \to H^1(X, \mathcal{K}_{2,X}) \otimes_{\mathbb{Z}} \mathbb{Z}/n\mathbb{Z} \to H^1(X, \mathcal{K}_{2,X} \otimes_{\mathbb{Z}} \mathbb{Z}/n\mathbb{Z}) \to {}_nCH^2(X) \to 0.$$

Let $\mathcal{H}_X^i(\mu_n^{\otimes j})$ be the sheaf on $X$ for the Zariski topology associated to the presheaf $U \mapsto H_{et}^i(U, \mu_n^{\otimes j})$. By results of S. Bloch and A. Ogus, *Gersten's conjecture and the homology of schemes*, Ann. Sci. Ecole Norm. Sup. 7 (1974) 181–201, there is a flasque resolution

$$0 \to \mathcal{H}_X^i(\mu_n^{\otimes j}) \to i_* H_{et}^i(k(X), \mu_n^{\otimes j}) \to \oplus_{x \in X^1} (i_x)_* H_{et}^{i-1}(k(x), \mu_n^{\otimes j-1}) \to \cdots$$

Further, there is a commutative diagram

$$\begin{array}{ccc} i_* K_2(k(X)) \otimes_{\mathbb{Z}} \mathbb{Z}/n\mathbb{Z} & \to & \oplus_{x \in X^1} (i_x)_* k(x)^* \otimes_{\mathbb{Z}} \mathbb{Z}/n\mathbb{Z} \\ \cong \downarrow & & \downarrow \cong \\ i_* H_{et}^2(k(X), \mu_n^{\otimes 2}) & \to & \oplus_{x \in X^1} (i_x)_* H^1(k(x), \mu_n) \end{array}$$

(which may be checked as in the proof of the commutativity of a similar diagram in the proof of lemma (8.19)). Combining this with the result of Bloch and Ogus for $\mathcal{H}_X^2(\mu_n^{\otimes 2})$, we obtain an isomorphism of sheaves $\mathcal{K}_{2,X} \otimes_{\mathbb{Z}} \mathbb{Z}/n\mathbb{Z} \cong \mathcal{H}_X^2(\mu_n^{\otimes 2})$.

In fact, the only part of the results of Bloch and Ogus which we need for this argument is the exactness of

$$0 \to \mathcal{H}^2_X(\mu_n^{\otimes 2}) \to i_* H^2_{et}(k(X), \mu_n^{\otimes 2}) \to \oplus_{x \in X^1} H^1_{et}(k(x), \mu_n),$$

which boils down to the exactness for each local ring $R$ of $X$ of

$$0 \to H^2_{et}(R, \mu_n^{\otimes 2}) \to H^2_{et}(k(X), \mu_n^{\otimes 2}) \to \oplus_{\text{ht } \mathcal{P}=1} H^1_{et}(k(\mathcal{P}), \mu_n).$$

Since $\mu_n \subset k$, this is equivalent to

$$0 \to {}_n\text{Br}(R) \to {}_n\text{Br}(k(X)) \to \oplus_{\text{ht } \mathcal{P}=1} H^1_{et}(k(\mathcal{P}), \mathbb{Z}/n\mathbb{Z}).$$

We give an *ad hoc* proof of this.

In fact, we have:

**Lemma 8.29** *Let $X$ be an irreducible smooth variety over an algebraically closed field $k$. Then we have an exact sequence*

$$0 \to {}_n\text{Br}\, X \to {}_n\text{Br}\, k(X) \to \oplus_{x \in X^1} H^1_{et}(k(x), \mathbb{Z}/n\mathbb{Z}),$$

*where* $\text{Br}\, X = H^2(X, \mathbb{G}_m)$ *is the cohomological Brauer group.*

**Proof.** $\text{Br}\, X \to \text{Br}\, k(X)$ is injective (see Milne, *Étale Cohomology*, III, (2.22)). Hence it suffices to prove that there is an exact sequence

$$H^2_{et}(X, \mu_n) \to H^2_{et}(k(X), \mu_n) \to \oplus_{x \in X^1} H^1(k(x), \mathbb{Z}/n\mathbb{Z}).$$

For any effective reduced divisor $Z \subset X$, consider the exact sequence

$$\cdots \to H^2_{et}(X, \mu_n) \to H^2_{et}(X - Z, \mu_n) \to H^3_Z(X, \mu_n) \to H^3_{et}(X, \mu_n) \to \cdots$$

We have $\varinjlim_n H^i_{et}(U, \mu_n) = H^i_{et}(k(X), \mu_n)$. So it suffices to prove that we have a natural inclusion

$$H^3_Z(X, \mu_n) \to \oplus_{x \in Z^0} H^1_{et}(k(x), \mu_n).$$

For any $W \subset Z$ of codimension 1 (so that $W$ has codimension 2 in $X$), we have an exact sequence

$$H^3_W(X, \mu_n) \to H^3_Z(X, \mu_n) \to H^3_{Z-W}(X - W, \mu_n)$$

Taking $W$ to contain the singular locus of $Z$, we have (Thom isomorphism)

$$H^3_{Z-W}(X - W, \mu_n) \cong H^1_{et}(Z - W, \mathbb{Z}/n\mathbb{Z}) \cong H^1_{et}(Z - W, \mu_n)$$

## 8. The Merkurjev–Suslin Theorem 191

(fix a choice of a primitive $n^{\text{th}}$ root of unity). Hence we have an exact sequence

$$\varinjlim_{W \subset Z} H^3_W(X, \mu_n) \to H^3_Z(X, \mu_n) \to \oplus_{x \in Z^0} H^1_{et}(k(x), \mu_n)$$

So the lemma follows from the formula $H^3_W(X, \mu_n) = 0$ for codim $W \geq 2$. We prove this formula by induction of the dimension of $W$; for dim $W = 0$, $W$ is smooth, so $H^3_W(X, \mu_n) \cong H^{3-2\dim X}_{et}(W, \mu_n^{\otimes 1 - \dim X}) = 0$, since $\dim X = \text{codim } W \geq 2$. In general, let $W_1$ be the singular locus of $W$, and consider the exact sequence

$$H^3_{W_1}(X, \mu_n) \to H^3_W(X, \mu_n) \to H^3_{W-W_1}(X - W_1, \mu_n).$$

The first term is 0 by induction, and the third term by the Thom isomorphism theorem, since $W - W_1$ is smooth of codimension $\geq 2$ in $X - W$.

From the isomorphism $\mathcal{K}_{2,X} \otimes_{\mathbb{Z}} \mathbb{Z}/n\mathbb{Z} \to \mathcal{H}^2_X(\mu_n^{\otimes 2})$, we obtain an exact sequence

(6) $\cdots 0 \to H^1(X, \mathcal{K}_{2,X}) \otimes_{\mathbb{Z}} \mathbb{Z}/n\mathbb{Z} \to H^1(X, \mathcal{H}^2_X(\mu_n^{\otimes 2})) \to {}_n CH^2(X) \to 0.$

There is a Leray spectral sequence associated to the morphism of sites $f : X_{et} \to X_{Zar}$,

$$E_2^{p,q} = H^p(X, R^q f_* \mu_n^{\otimes 2}) \Rightarrow H^{p+q}_{et}(X, \mu_n^{\otimes 2}).$$

Since $R^q f_* \mu_n^{\otimes 2}$ is the sheaf associated to the presheaf $U \mapsto H^q_{et}(U, \mu_n^{\otimes 2})$, we have $R^q f_* \mu_n^{\otimes 2} \cong \mathcal{H}^q_X(\mu_n^{\otimes 2})$, and so $E_2^{1,2} = H^1(X, \mathcal{H}^2_X(\mu_n^{\otimes 2}))$. The differentials for $E_r^{1,2}$ are

$$E_r^{1-r, 1+r} \to E_r^{1,2} \to E_r^{r+1, 3-r}$$

where $E_r^{1-r, 1+r} = 0$ as $r \geq 2$. We claim $E_2^{3,1} = E_2^{4,0} = 0$. Indeed, $R^0 f_* \mu_n^{\otimes 2}$ is the constant sheaf $\mu_n^{\otimes 2}$ for the Zariski topology on $X$, which is a flasque sheaf; hence $E_2^{p,0} = 0$ for $p > 0$. Next, from the Kummer sequence on $X_{et}$ we get an exact sequence

$$\mathcal{O}^*_X \xrightarrow{\cdot n} \mathcal{O}^*_X \to R^1 f_* \mu_n \to R^1 f_* \mathbb{G}_m$$

where $R^1 f_* \mathbb{G}_m = 0$ by Hilbert's Theorem 90 (see Milne, *Étale Cohomology*, III, (4.9)). Thus (since $\mu_p \cong \mathbb{Z}/p\mathbb{Z}$) we have $R^1 f_* \mu_p \cong \mathcal{O}^*_X \otimes_{\mathbb{Z}} \mu_p$. From the flasque resolution

$$0 \to \mathcal{O}^*_X \otimes_{\mathbb{Z}} \mu_p \to i_*(k(X)^* \otimes_{\mathbb{Z}} \mu_p) \to \oplus_{x \in X^1}(i_x)_* \mu_p \to 0,$$

we see that $R^1 f_* \mu_p$ has cohomological dimension $\leq 1$. Hence $E_2^{p,1} = 0$ for $p \geq 2$. Thus $E_2^{1,2} \cong E_\infty^{1,2}$; further $E_2^{3,0} = 0$. Since $E_\infty^{0,3} \subset E_2^{0,3}$, we obtain an exact sequence

(7) $\quad \cdots \quad 0 \to H^1(X, \mathcal{H}_X^2(\mu_n^{\otimes 2})) \to H_{et}^3(X, \mu_n^{\otimes 2}) \to H^0(X, \mathcal{H}_X^3(\mu_n^{\otimes 2})).$

Since $X$ is a smooth variety over an algebraically closed field, $H_{et}^3(X, \mu_n^{\otimes 2}) \cong H_{et}^3(X, \mathbb{Z}/n\mathbb{Z})$ is finite. Hence $_nCH^2(X)$ is finite.

If $X$ is a surface, then $\mathcal{H}_X^3(\mu_n^{\otimes 2}) = 0$, since for any affine variety of dimension $d$ over $k$, the étale cohomological dimension of an $n$-torsion sheaf is $\leq d$, by the Weak Lefschetz Theorem (Milne, *Étale Cohomology*, VI, (7.1)). Thus $H^1(X, \mathcal{H}_X^2(\mu_n^{\otimes 2})) \cong H_{et}^3(X, \mu_n^{\otimes 2})$. In particular, if $X$ is a smooth *affine* surface, then $_nCH^2(X) = 0$, since $H_{et}^3(X, \mu_n^{\otimes 2}) = 0$ by the Weak Lefschetz Theorem.

If $X$ is a smooth projective surface, Poincaré duality gives isomorphisms (Milne, *Étale Cohomology*, VI, (11.2))

$$\begin{aligned} H_{et}^3(X, \mu_n^{\otimes 2}) &\cong \mathrm{Hom}\,(H_{et}^1(X, \mathbb{Z}/n\mathbb{Z}), \mathbb{Z}/n\mathbb{Z}) \\ &\cong \mathrm{Hom}\,(H_{et}^1(X, \mu_n), \mu_n) \\ &\cong \mathrm{Hom}\,(_n\mathrm{Pic}\,X, \mu_n) \\ &\twoheadrightarrow \mathrm{Hom}\,(_n\mathrm{Pic}\,^0 X, \mu_n) \\ &\cong {}_n\mathrm{Alb}\,X. \end{aligned}$$

The last isomorphism is via the $e_n$-pairing between the $n$-torsion subgroups of $\mathrm{Pic}^0(X)$ and its dual Abelian variety $\mathrm{Alb}\,X$. Since the Néron-Severi group $NS(X) = \mathrm{Pic}\,X/\mathrm{Pic}^0 X$ is finitely generated, its torsion subgroup is finite, and so $\ker H_{et}^3(X, \mu_n) \twoheadrightarrow {}_n\mathrm{Alb}\,X$ has order bounded by a number independent of $n$.

On the other hand, the torsion subgroups of $CH^2(X)$ and $\mathrm{Alb}\,X$ are divisible (this is true for $CH^2(X)$ because the subgroup generated by cycles of degree 0 is a quotient of a direct sum of Jacobians of smooth projective curves).

Further, we claim that for any $m$, the natural map $_mCH^2(X) \to {}_m\mathrm{Alb}\,(X)$ is onto. In fact, we claim that if $C \subset X$ is a smooth hyperplane section, the map $J(C) \to \mathrm{Alb}\,X$ is surjective on $m$-torsion; this is equivalent to the claim that for any connected étale Galois $\mathbb{Z}/m\mathbb{Z}$-covering $Y \to X$, the pullback $Y \times_X C$ is connected. This is true because $Y \times_X C$ is the pullback to $Y$ of an ample divisor on $X$; hence it is an ample divisor on the non-singular projective variety $Y$ (see Hartshorne, *Algebraic Geometry*, III, Ex. 5.7). In particular, $Y \times_X C$ is connected (Hartshorne, *Algebraic Geometry*, III, Cor. 7.9).

We deduce that $\ker(CH^2(X) \to \mathrm{Alb}\,X)$ is a *divisible* group. Since

$$\ker(_nCH^2(X) \to {}_n\mathrm{Alb}\,X) = {}_n(\ker(CH^2(X) \to \mathrm{Alb}\,X))$$

## 8. The Merkurjev–Suslin Theorem

has order bounded independently of $n$, we see that $_nCH^2(X) \to {}_n\mathrm{Alb}\,X$ is in fact an *isomorphism* for each $n$ (relatively prime to char. $k$). This was first proved by A. A. Rojtman (his proof is given in "The torsion in the group of 0-cycles modulo rational equivalence," *Ann. Math.* 111 (1980) 553–569), for smooth projective varieties of any dimension. A proof along the lines given here was first obtained by S. Bloch (see his book *Lectures on Algebraic Cycles*, Duke Univ. Math. Ser. IV, Durham (1976)); he also shows that the result for smooth projective surfaces implies it for smooth projective varieties of any dimension. Milne has used an analogous argument with flat cohomology, combined with results of Kato (which describe $K_2(F)/pK_2(F)$ for function fields $F$ of characteristic $p$) to show that $_nCH^2(X) \to {}_n\mathrm{Alb}\,X$ is an isomorphism even when $n$ is a power of the characteristic; see J. S. Milne, Zero cycles on algebraic varieties in non-zero characteristic: Rojtman's theorem, *Compos. Math.* 47 (1982).

# 9. Localization for Singular Varieties

In this chapter we prove a localization theorem of Quillen for singular varieties, and a generalization of it due to Levine. These are then used to prove the so-called "Fundamental Theorem" (9.8), which computes $K_i(A[t, t^{-1}])$, and to relate the study of 0-cycles on normal surfaces to modules of finite length and finite projective dimension over the local rings at singular points. We begin with Quillen's localization theorem, proved in *Higher Algebraic K-Theory II*.

Suppose $X$ is a quasi-compact scheme which supports an ample invertible sheaf, $j : U \longrightarrow X$ the inclusion of an affine open subscheme, such that $U = X - D$ for some effective Cartier divisor $D$ in $X$. Let $\mathcal{I} = \mathcal{O}_X(-D)$ be the ideal sheaf of $D$. Let $\mathcal{H}$ be the category of quasi-coherent $\mathcal{O}_X$-modules $\mathcal{F}$ such that $j^*\mathcal{F} = 0$, and $\mathcal{F}$ has a resolution of length 1

$$0 \longrightarrow \mathcal{E}_1 \longrightarrow \mathcal{E}_0 \longrightarrow \mathcal{F} \longrightarrow 0$$

with $\mathcal{E}_0, \mathcal{E}_1 \in \mathcal{P}(X)$, the category of locally free $\mathcal{O}_X$-modules of finite rank. Since every coherent sheaf on $X$ is a quotient of a locally free sheaf, $\mathcal{H}$ is closed under extensions, and is thus an exact category.

**Theorem (9.1).** *There is a natural long exact sequence*

$$\ldots K_{i+1}(U) \to K_i(\mathcal{H}) \to K_i(X) \to K_i(U) \to \cdots \to K_0(X) \to K_0(U).$$

**Proof.** Let $\mathcal{V}$ be the category of vector bundles on $U$ which extend to vector bundles on $X$, and let $\mathcal{P}_1$ denote the category of quasi-coherent $\mathcal{O}_X$-modules $\mathcal{F}$ which have a resolution of length 1

$$0 \longrightarrow \mathcal{E}_1 \longrightarrow \mathcal{E}_0 \longrightarrow \mathcal{F} \longrightarrow 0$$

with $\mathcal{E}_i \in \mathcal{P}(X)$. Thus $\mathcal{H} \subset \mathcal{P}_1$; further $\mathcal{P} \subset \mathcal{P}_1$ is a homotopy equivalence, by the resolution theorem.

Since $U$ is affine, every exact sequence in $\mathcal{V}$ splits. Let $\mathcal{E} \longrightarrow Q\mathcal{V}$ be the extension construction over $Q\mathcal{V}$ ($\mathcal{E}$ is the category constructed in the proof of Theorem (7.7), associated to the exact category $\mathcal{V}$). Thus by (7.7), if $\mathcal{T} = \text{Iso}(\mathcal{V})$, then $\mathcal{T}^{-1}\mathcal{E} \longrightarrow Q\mathcal{V}$ is fibered with fibers homotopy

## 9. Localization for Singular Varieties

equivalent to $\mathcal{T}^{-1}\mathcal{T}$, and $\mathcal{T}^{-1}\mathcal{E}$ is contractible. Let $\mathcal{P} = \mathcal{P}(X)$, $\mathcal{S} = \operatorname{Iso}\mathcal{P}$. Then the restriction functor $j^* : \mathcal{S} \longrightarrow \mathcal{T}$ is a monoidal functor, making $\mathcal{S}$ a subcategory of $\mathcal{T}$.

**Lemma (9.2).** *Let $\mathcal{T}$ act on $\mathcal{X}$, and let $\mathcal{S}$ act on $\mathcal{X}$ by the induced action. Then $\mathcal{S}^{-1}\mathcal{X} \longrightarrow \mathcal{T}^{-1}\mathcal{X}$ is a homotopy equivalence.*

**Proof.** For any object $A \in \mathcal{T}$, there exists $B \in \mathcal{T}$ such that $A \oplus B \cong j^*C$ for some $C \in \mathcal{S}$. Hence if $\mathcal{T}$ acts on $\mathcal{X}$, and $\mathcal{S}$ acts by the induced action, translation by $C$ is a homotopy equivalence if and only if the translations by $A, B$ are homotopy equivalences. Hence $\mathcal{T}$ acts invertibly on $\mathcal{X} \Longleftrightarrow \mathcal{S}$ acts invertibly on $\mathcal{X}$. Now for any category $\mathcal{X}$ with a $\mathcal{T}$-action, we have homotopy equivalences by (7.1)

$$\mathcal{S}^{-1}\mathcal{X} \longrightarrow \mathcal{T}^{-1}\mathcal{S}^{-1}\mathcal{X}, \quad \mathcal{T}^{-1}\mathcal{X} \longrightarrow \mathcal{S}^{-1}\mathcal{T}^{-1}\mathcal{X}.$$

But $\mathcal{S} \times \mathcal{T} \times \mathcal{X} \xrightarrow{\cong} \mathcal{T} \times \mathcal{S} \times \mathcal{X}$, $(A, B, F) \mapsto (B, A, F)$ induces an isomorphism $\mathcal{S}^{-1}\mathcal{T}^{-1}\mathcal{X} \cong \mathcal{T}^{-1}\mathcal{S}^{-1}\mathcal{X}$. Now the composite $\mathcal{T}^{-1}\mathcal{X} \longrightarrow \mathcal{S}^{-1}\mathcal{T}^{-1}\mathcal{X} \cong \mathcal{T}^{-1}\mathcal{S}^{-1}\mathcal{X}$ is just the functor obtained by localizing $\mathcal{X} \longrightarrow \mathcal{S}^{-1}\mathcal{X}$ with respect to $\mathcal{T}$.

We claim the triangle

$$\begin{array}{ccc} \mathcal{S}^{-1}\mathcal{X} & \longrightarrow & \mathcal{T}^{-1}\mathcal{X} \\ & \searrow \quad \swarrow & \\ & \mathcal{T}^{-1}\mathcal{S}^{-1}\mathcal{X} & \end{array}$$

commutes up to homotopy, where $\mathcal{S}^{-1}\mathcal{X} \longrightarrow \mathcal{T}^{-1}\mathcal{X}$ is induced by the inclusion $\mathcal{S} \subset \mathcal{T}$; this will prove $\mathcal{S}^{-1}\mathcal{X} \longrightarrow \mathcal{T}^{-1}\mathcal{X}$ is a homotopy equivalence. Now $\mathcal{S}^{-1}\mathcal{X} \longrightarrow \mathcal{T}^{-1}\mathcal{S}^{-1}\mathcal{X}$ is given by $(A, F) \mapsto (0, (A, F))$, while the composite $\mathcal{S}^{-1}\mathcal{X} \longrightarrow \mathcal{T}^{-1}\mathcal{X} \longrightarrow \mathcal{T}^{-1}\mathcal{S}^{-1}\mathcal{X}$ is given by $(A, F) \mapsto (A, (0, F))$. There is a chain of natural transformations, giving a homotopy

$$(0, (A, F)) \xrightarrow{u} (A, (A, A \oplus F)) \xleftarrow{v} (A, (0, F)),$$

where $u$ is the arrow in $\mathcal{T}^{-1}(\mathcal{S}^{-1}\mathcal{X})$ given by

$$(A, A + (0, (A, F))) \xrightarrow{\cong} (A, (A, A \oplus F))),$$

and $v$ is induced by the arrow in $\mathcal{S}^{-1}\mathcal{X}$ given by

$$(A, A + (0, F)) \xrightarrow{\cong} (A, A \oplus F)).$$

This proves the lemma.

We construct below a diagram of categories with $\mathcal{S}$-action (where $\mathcal{S}$ acts trivially on $Q\mathcal{P}$, $Q\mathcal{V}$, $Q\mathcal{H}$)

$$\begin{array}{ccccc} \mathcal{C} & \xrightarrow{f} & \mathcal{D} & \xrightarrow{p} & Q\mathcal{P} \\ {\scriptstyle h}\downarrow & & \downarrow & & \downarrow \\ Q\mathcal{H} & & \mathcal{E} & \longrightarrow & Q\mathcal{V} \end{array}$$

and show that $f, h$ are homotopy equivalences. Localization then yields a diagram

$$\begin{array}{ccccc} \mathcal{S}^{-1}\mathcal{C} & \xrightarrow{\cong} & \mathcal{S}^{-1}\mathcal{D} & \longrightarrow & Q\mathcal{P} \\ {\scriptstyle \cong}\downarrow & & \downarrow & & \downarrow \\ Q\mathcal{H} & & \mathcal{S}^{-1}\mathcal{E} & \longrightarrow & Q\mathcal{V} \end{array}$$

where $\mathcal{S}^{-1}\mathcal{E} \cong \mathcal{T}^{-1}\mathcal{E}$ is contractible. This yields a long exact sequence

$$K_{i+1}(\mathcal{V}) \longrightarrow K_i(\mathcal{H}) \longrightarrow K_i(\mathcal{P}) \longrightarrow K_i(\mathcal{V}) \longrightarrow \cdots$$

where $K_i(\mathcal{P}) = K_i(X)$, and $\mathcal{V} \subset \mathcal{P}(U)$ induces maps $K_i(\mathcal{V}) \longrightarrow K_i(U)$.

**Lemma (9.3).** $K_i(\mathcal{V}) \longrightarrow K_i(U)$ *is an isomorphism for* $i > 0$, *and a monomorphism for* $i = 0$.

**Proof.** Let $\mathcal{S}' = \text{Iso}\,\mathcal{P}(U)$, where $U = \text{Spec}\,R$ is given to be affine. The inclusion $\mathcal{V} \subset \mathcal{P}(U)$ induces $\mathcal{T} \subset \mathcal{S}'$. If $\mathcal{S}_n \subset \mathcal{S}'$ is the connected component of the free module $R^{\oplus n}$, then since $R^{\oplus n} \in \mathcal{V}$, $\mathcal{S}_n \subset \mathcal{T}$. Thus the functor $\mathcal{L} \longrightarrow (\mathcal{S}'^{-1}\mathcal{S}')_0$, used to give the homotopy equivalence $BGL(R)^+ \longrightarrow B(\mathcal{S}'^{-1}\mathcal{S}')_0$ of (7.3), factors through $(\mathcal{T}^{-1}\mathcal{T})_0 \longrightarrow (\mathcal{S}'^{-1}\mathcal{S}')_0$. Hence the proof of (7.3) yields homotopy equivalences

$$BGL(R)^+ \cong B(\mathcal{T}^{-1}\mathcal{T})_0 \cong B(\mathcal{S}'^{-1}\mathcal{S}')_0.$$

Thus $K_i(\mathcal{V}) \longrightarrow K_i(U)$ is an isomorphism for $i \geq 1$. Since all exact sequences in $\mathcal{P}(U)$ are split, and $R^{\oplus n} \in \mathcal{V}$ for all $n$, we see that for $A, B \in \mathcal{V}$,

$$[A] = [B] \text{ in } K_0(U) \iff A \oplus R^{\oplus n} \cong B \oplus R^{\oplus n} \text{ for some } n$$
$$\iff [A] = [B] \text{ in } K_0(\mathcal{V}).$$

Hence $K_0(\mathcal{V}) \longrightarrow K_0(U)$ is injective.

**Remark.** Gersten has shown that if $\mathcal{M} \subset \mathcal{P}$ are exact categories, such that every exact sequence in $\mathcal{P}$ splits, and $\mathcal{M}$ is *cofinal* in $\mathcal{P}$ (i.e., $\forall A \in \mathcal{P}$, there exists $B \in \mathcal{P}$ with $A \oplus B \in \mathcal{M}$), then $BQ\mathcal{M} \longrightarrow BQ\mathcal{P}$ is homotopy equivalent to a covering space. Hence $K_i(\mathcal{M}) \longrightarrow K_i(\mathcal{P})$ is an isomorphism for $i > 0$ and a monomorphism for $i = 0$, from the corresponding properties of $\pi_{i+1}$.

## 9. Localization for Singular Varieties

Returning to the proof of (9.1), we note that $\mathcal{E}$ is equivalent to the category (again denoted $\mathcal{E}$, by abuse of notation) whose objects are admissible epimorphisms $B \twoheadrightarrow C$ with $B, C \in \mathcal{V}$, where a morphism $(B \twoheadrightarrow C) \longrightarrow (B' \twoheadrightarrow C')$ is defined to be an equivalence class of diagrams (in $\mathcal{V}$)

$$\begin{array}{ccc} B & \twoheadrightarrow & C \\ \| & & \uparrow \\ B & \twoheadrightarrow & C_1 \\ \downarrow & \square & \downarrow \\ B' & \twoheadrightarrow & C' \end{array}$$

where the bottom square is a pullback. Define the category $\mathcal{D}$ to be the pullback

$$\begin{array}{ccc} \mathcal{D} & \xrightarrow{p} & Q\mathcal{P} \\ \downarrow & & \downarrow \\ \mathcal{E} & \longrightarrow & Q\mathcal{V} \end{array}$$

Thus, an object of $\mathcal{D}$ is a pair $(B, Z \twoheadrightarrow j^*B)$ with $B \in \mathcal{P}$, $Z \in \mathcal{V}$ (where $j : U \hookrightarrow X$, $\mathcal{P} = \mathcal{P}(X)$). An arrow in $\mathcal{D}$ is represented by an equivalence class of pairs of diagrams

$$\begin{array}{ccccc} B & & Z & \twoheadrightarrow & j^*B \\ \uparrow & & \| & & \uparrow \\ B_1 & & Z & \twoheadrightarrow & j^*B_1 \\ \downarrow & & \downarrow & & \downarrow \\ B & & Z' & \twoheadrightarrow & j^*B \end{array}$$

(where the right hand column is obtained by applying $j^*$ to the left hand column). Since $\mathcal{E} \longrightarrow Q\mathcal{V}$ is fibered, so is $p : \mathcal{D} \longrightarrow Q\mathcal{P}$.

Next, let $\mathcal{C}$ be the category whose objects are admissible epimorphisms (in $\mathcal{P}_1$) $L \twoheadrightarrow M \oplus B$, with $L, B \in \mathcal{P}$, $M \in \mathcal{H}$, where an arrow $(L \twoheadrightarrow M \oplus B) \longrightarrow (L' \twoheadrightarrow M' \oplus B')$ is an equivalence class of diagrams

$$\begin{array}{ccc} L & \twoheadrightarrow & M \oplus B \\ \| & & \uparrow \\ L & \twoheadrightarrow & M_1 \oplus B_1 \\ \downarrow & & \downarrow \\ L' & \twoheadrightarrow & M' \oplus B' \end{array}$$

where the right hand column is the direct sum of columns representing arrows $M \longrightarrow M'$, $B \longrightarrow B'$ in $Q\mathcal{H}$, $Q\mathcal{P}$ respectively.

There is a functor $\mathcal{C} \longrightarrow Q(\mathcal{H} \times \mathcal{P})$ given by

$$(L \twoheadrightarrow M \oplus B) \mapsto (M, B);$$

this makes $\mathcal{C}$ fibered over $Q(\mathcal{H} \times \mathcal{P})$. Let $g : \mathcal{C} \longrightarrow Q\mathcal{P}$, $h : \mathcal{C} \longrightarrow Q\mathcal{H}$ be obtained by projection; then $g$, $h$ are also fibered functors. Let $f : \mathcal{C} \longrightarrow \mathcal{D}$ be the functor

$$(L \twoheadrightarrow M \oplus B) \mapsto (B, j^*L \twoheadrightarrow j^*B)$$

(where we note that $j^*M = 0$).

$\mathcal{S} = \operatorname{Iso} \mathcal{P}$ acts on $\mathcal{C}$ by

$$A + (L \xrightarrow{\phi} M \oplus B) = (A \oplus L \xrightarrow{(0,\phi)} M \oplus B),$$

and on $\mathcal{E}$ (through the usual action of $\mathcal{T}$) by

$$A + (B \xrightarrow{\phi} C) = (A \oplus B \xrightarrow{(0,\phi)} C).$$

$\mathcal{S}$ acts on $\mathcal{D}$ by the induced action, where we take the trivial $\mathcal{S}$-action on $Q\mathcal{P}$, $Q\mathcal{V}$, $Q\mathcal{H}$.

**Lemma (9.4).** $h : \mathcal{C} \longrightarrow Q\mathcal{H}$ *is a homotopy equivalence.*

**Proof.** Since $h$ is fibered, as remarked above, it suffices to prove the fibers of $h$ are contractible. For $M \in \mathcal{H}$, let $\mathcal{R}_M$ be the category whose objects are admissible epimorphisms (in $\mathcal{P}_1$) $L \twoheadrightarrow M$ with $L \in \mathcal{P}$, where an arrow is a diagram

where $L \rightarrowtail L'$ is admissible mono in $\mathcal{P}$. Then the category $\operatorname{Sub} \mathcal{R}_M$ (see the proof of (7.10)) is equivalent to $h^{-1}(M)$, by associating to the above diagram (an object of $\operatorname{Sub} \mathcal{R}_M$) the object $L' \twoheadrightarrow M \oplus L'/L$ in $h^{-1}(M)$. Since $\operatorname{Sub} \mathcal{R}_M$ is equivalent to $\mathcal{R}_M$, it suffices to prove $\mathcal{R}_M$ is contractible $\forall M \in \mathcal{H}$. But if $L \twoheadrightarrow M$ is a fixed object of $\mathcal{R}_M$, we have natural transformations of functors $\mathcal{R}_M \longrightarrow \mathcal{R}_M$

$$(L \mapsto M) \mapsto (L \oplus L' \twoheadrightarrow M) \leftarrow (L' \twoheadrightarrow M)$$

which connects the constant functor $\mathcal{R}_M \longrightarrow \{L \twoheadrightarrow M\}$ to the identity functor.

## 9. Localization for Singular Varieties

**Lemma (9.5).** *If $E \in \mathcal{P}$ is a vector bundle on $X$, then $E \subset j_*j^*E$, and*
$$j_*j^*E = \bigcup_{n \geq 0} \mathcal{I}^{-n}E = \bigcup_{n \geq 0} E \otimes \mathcal{O}_X(nD).$$

**Proof.** Clear.

**Lemma (9.6).** $f : \mathcal{C} \longrightarrow \mathcal{D}$ *is a homotopy equivalence.*

**Proof.** We have the triangle of categories and functors $(g = p \circ f)$

where $g, p$ are fibered functors.

**Claim.** For each $B \in Q\mathcal{P}$, the functor (induced by $f$) $g^{-1}(B) \longrightarrow p^{-1}(B)$ is a homotopy equivalence.

To prove this claim, let $\mathcal{R}_B$ be the category whose objects are admissible epimorphisms (in $\mathcal{P}$) $L \twoheadrightarrow B$ with $L \in \mathcal{P}$, and whose arrows are diagrams

$$\begin{array}{ccc} L' & \twoheadrightarrow & B \\ \downarrow & & \| \\ L & \twoheadrightarrow & B \end{array}$$

where $L' \rightarrowtail L$ is an admissible mono in $\mathcal{P}_1$ whose cokernel lies in $\mathcal{H}$, i.e., $j^*L' \longrightarrow j^*L$ is an isomorphism. Then the functor $\operatorname{Sub}\mathcal{R}_B \longrightarrow g^{-1}(B)$ given by

$$\begin{pmatrix} L' & \twoheadrightarrow & B \\ i\downarrow & & \| \\ L & \twoheadrightarrow & B \end{pmatrix} \mapsto (L \twoheadrightarrow (\operatorname{coker} i) \oplus B)$$

is clearly a homotopy equivalence.

Let $W = j^*B$, so that $p^{-1}(B) = \mathcal{E}_W$. To prove this claim for $B$, we must show that the functor $\operatorname{Sub}\mathcal{R}_B \longrightarrow \mathcal{E}_W$ given by

$$\begin{pmatrix} L' & \twoheadrightarrow & B \\ i\downarrow & & \| \\ L & \twoheadrightarrow & B \end{pmatrix} \mapsto (j^*L \twoheadrightarrow j^*B = W)$$

is a homotopy equivalence. This functor factors through the range functor $\text{Sub}\,\mathcal{R}_B \longrightarrow \mathcal{R}_B$,

$$\begin{pmatrix} L' & \twoheadrightarrow & B \\ \downarrow & & \| \\ L & \twoheadrightarrow & B \end{pmatrix} \mapsto (L \twoheadrightarrow B)$$

which is a homotopy equivalence (see (7.10)). Hence it suffices to prove that $w : \mathcal{R}_B \longrightarrow \mathcal{E}_W$, $(L \twoheadrightarrow B) \mapsto (j^*L \twoheadrightarrow j^*B = W)$ is a homotopy equivalence. To do this, by Theorem A, it suffices to prove that $w/(Z \twoheadrightarrow W)$ is contractible for any $Z \twoheadrightarrow W$ in $\mathcal{E}_W$.

An object of $w/(Z \twoheadrightarrow W)$ is a pair, consisting of an object $L \twoheadrightarrow B$ of $\mathcal{R}_B$, together with an arrow (i.e., an isomorphism)

$$(j^*L \twoheadrightarrow W) \longrightarrow (Z \twoheadrightarrow W),$$

given by an isomorphism $j^*L \cong Z$ making the triangle

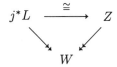

commute. Let $\underline{\mathcal{L}at}$ be the partially ordered set of vector bundles $L \in \mathcal{P}$ such that i) $L \subset j_*Z$, ii) $j^*L = Z$, iii) the image of the composite map $L \longrightarrow j_*Z \twoheadrightarrow j_*W = j_*j^*B$ is $B \subset j_*j^*B$. The partial ordering is given by inclusion. From iii), every $L \in \underline{\mathcal{L}at}$ has a given admissible epimorphism (in $\mathcal{P}_1$) $L \twoheadrightarrow B$, such that $j^*L \twoheadrightarrow j^*B = W$ is just $Z \twoheadrightarrow W$. Thus $\underline{\mathcal{L}at}$ is identified with a full subcategory of $w/(Z \twoheadrightarrow W)$, such that every object of $w/(Z \twoheadrightarrow W)$ is isomorphic to an object of $\underline{\mathcal{L}at}$, i.e., the inclusion $\underline{\mathcal{L}at} \longrightarrow w/(Z \twoheadrightarrow W)$ is an equivalence. So it suffices to prove that $\underline{\mathcal{L}at}$ is filtering, hence contractible.

To prove that $\underline{\mathcal{L}at}$ is filtering, since $\underline{\mathcal{L}at}$ is clearly nonempty (since $Z \in Q\mathcal{V}$, $Z$ extends to a vector bundle on $X$), it suffices to check that if $L_1, L_2 \in \underline{\mathcal{L}at}$, there exists $L_3 \in \underline{\mathcal{L}at}$ with $L_1 \subset L_3$, $L_2 \subset L_3$. We have an exact sequence in $\mathcal{V}$

$$0 \longrightarrow Y \longrightarrow Z \longrightarrow W \longrightarrow 0$$

where since $Y \in \mathcal{V}$, $Y = j^*C$ for some $C \in \mathcal{P}$. Then $Z \cong j^*(C \oplus B)$. Consider the corresponding lattice $C \oplus B \subset j_*Z$. If $L \longrightarrow j_*Z$ is any lattice, then since $L \longrightarrow j_*W = j_*j^*B$ has image $B$, by definition of $\underline{\mathcal{L}at}$, we must have

$$L \subset (j_*j^*(C) \oplus B) = \left(\left(\bigcup_{n \geq 1} \mathcal{I}^{-n}C\right) \oplus B\right) \subset j_*Z.$$

Since $L$ is locally finitely generated and $X$ is quasi-compact, $L \subset \mathcal{I}^{-n}C \oplus B$ for some $n \geq 1$. Take $L_3 = (\mathcal{I}^{-n}C \oplus B) \subset j_*Z$, for some sufficiently large $n$; then clearly $L_1 \subset L_3$, $L_2 \subset L_3$. This completes the proof of the claim, that $g^{-1}(B) \longrightarrow p^{-1}(B)$ is a homotopy equivalence for each $B \in \mathcal{P}$.

The lemma now follows from:

**Lemma (9.7).** *Let*

$$\begin{array}{ccc} \mathcal{A} & \xrightarrow{f} & \mathcal{B} \\ {\scriptstyle g} \searrow & & \swarrow {\scriptstyle h} \\ & \mathcal{C} & \end{array}$$

*be a commutative triangle of categories and functors, such that $g$, $h$ are fibered. Suppose that for each $B \in \mathcal{C}$, $g^{-1}(B) \longrightarrow h^{-1}(B)$ (induced by $f$) is a homotopy equivalence. Then $f$ is a homotopy equivalence.*

We postpone the proof of this lemma, and first finish the proof of Theorem (9.1). We now have the diagram ($\cong$ denotes a homotopy equivalence) on which $\mathcal{S} = \mathrm{Iso}\,\mathcal{P}$ acts.

$$\begin{array}{ccccc} \mathcal{C} & \xrightarrow{\cong} & \mathcal{D} & \longrightarrow & \mathcal{QP} \\ {\scriptstyle \cong}\downarrow & & \downarrow & & \downarrow \\ \mathcal{QH} & & \mathcal{E} & \longrightarrow & \mathcal{QV} \end{array}$$

Since $\mathcal{S}$ acts trivially on $\mathcal{QH}$, and $\mathcal{QH} \cong \mathcal{C} \cong \mathcal{D}$, $\mathcal{S}$ acts invertibly on $\mathcal{C}$, $\mathcal{D}$ so that we have a localized diagram

$$\begin{array}{ccccc} \mathcal{S}^{-1}\mathcal{C} & \xrightarrow{\cong} & \mathcal{S}^{-1}\mathcal{D} & \longrightarrow & \mathcal{QP} \\ {\scriptstyle \cong}\downarrow & & \downarrow & & \downarrow \\ \mathcal{QH} & & \mathcal{S}^{-1}\mathcal{E} & \longrightarrow & \mathcal{QV} \end{array}$$

where the arrows marked $\cong$ are homotopy equivalences, from (7.1). Further, $\mathcal{S}^{-1}\mathcal{E}$ is contractible, by (7.10). The base change arrows for the fibered functor $\mathcal{S}^{-1}\mathcal{E} \longrightarrow \mathcal{QV}$ are homotopy equivalences, from the proof of (7.9), so that the same is true for the pullback functor $\mathcal{S}^{-1}\mathcal{D} \longrightarrow \mathcal{QP}$. Thus, in the square

$$\begin{array}{ccc} \mathcal{S}^{-1}\mathcal{D} & \longrightarrow & \mathcal{QP} \\ \downarrow & & \downarrow \\ \mathcal{S}^{-1}\mathcal{E} & \longrightarrow & \mathcal{QV} \end{array}$$

the horizontal maps are quasi-fibrations, by Theorem B, and the maps from the homotopy fibers of the top row to those of the bottom row are homotopy equivalences (since the maps on the actual fibers are isomorphisms). Thus the square is homotopy Cartesian; since $\mathcal{S}^{-1}\mathcal{E}$ is contractible,

$$B\mathcal{S}^{-1}\mathcal{D} \longrightarrow BQ\mathcal{P} \longrightarrow BQ\mathcal{V}$$

has the homotopy type of a fibration. Hence,

$$BQ\mathcal{H} \longrightarrow BQ\mathcal{P} \longrightarrow BQ\mathcal{V}$$

has the homotopy of a fibration, for a certain homotopy class of maps $BQ\mathcal{H} \longrightarrow F$, where $F$ is the homotopy fiber of $BQ\mathcal{P} \longrightarrow BQ\mathcal{V}$ over a 0-object $0 \in \mathcal{V}$.

This suffices to give a long exact sequence

$$\ldots K_{i+1}(U) \longrightarrow K_i(\mathcal{H}) \longrightarrow K_i(X) \longrightarrow K_i(U) \rightarrow \cdots \rightarrow K_0(X)$$
$$\longrightarrow K_0(U).$$

However, in applications, we often need a stronger statement. Let $\mathcal{P}_1$ be the exact category of coherent $\mathcal{O}_X$-modules of homological dimension $\leq 1$, and let $\mathcal{H}(U)$ denote the category of coherent $\mathcal{O}_U$-modules of homological dimension $\leq 1$. Then $Q\mathcal{P} \subset Q\mathcal{P}_1$ is a homotopy equivalence, by the resolution theorem, while $Q\mathcal{V} \subset Q\mathcal{H}(U)$ is homotopy equivalent to a covering space of $Q\mathcal{H}(U)$, from (9.3) and the resolution theorem. Thus we have a diagram

$$\begin{array}{ccccc} F' & \longrightarrow & BQ\mathcal{P}_1 & \longrightarrow & BQ\mathcal{H}(U) \\ \uparrow \cong & & \uparrow \cong & & \uparrow \\ F & \longrightarrow & BQ\mathcal{P} & \longrightarrow & BQ\mathcal{V} \end{array}$$

where $F, F'$ are the respective homotopy fibers over a given 0-object in $\mathcal{V} \subset \mathcal{H}(U)$, and the arrows marked $\cong$ are homotopy equivalences. We have an inclusion $Q\mathcal{H} \subset Q\mathcal{P}_1$ such that the composite functor $Q\mathcal{H} \longrightarrow Q\mathcal{H}(U)$ maps every object to a 0-object. Since any two 0-objects of $\mathcal{H}(U)$ are canonically isomorphic in $Q\mathcal{H}(U)$, there is a natural nullhomotopy of the composite $BQ\mathcal{H} \longrightarrow BQ\mathcal{H}(U)$, giving a map $\phi: BQ\mathcal{H} \longrightarrow F'$. On the other hand, we have already constructed a homotopy equivalence $BQ\mathcal{H} \longrightarrow F$ (which is well defined up to homotopy); composing this with $F \xrightarrow{\cong} F'$ yield a homotopy equivalence $\psi: BQ\mathcal{H} \longrightarrow F'$. Using $\psi$, we want to prove $\phi$ is also a homotopy equivalence; the long exact sequence constructed using $\phi$ is the one with good naturality properties (see the discussion of naturality, at the end of Chapter 6, for the localization sequence associated to a Serre subcategory of an abelian category).

Our proof that $\phi$ is a homotopy equivalence is based on unpublished notes of Richard Swan. (Swan has now published his notes as an appendix to his paper "$K$-Theory of Quadratic Hypersurfaces," *Ann. Math.*

## 9. Localization for Singular Varieties 203

122 (1985), 113–153.) We first consider the commutative triangles ($\cong$ denotes a homotopy equivalence)

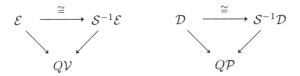

This yields the commutative diagram of classifying spaces

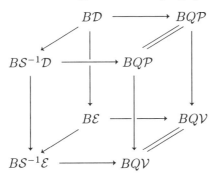

which gives a commutative square of homotopy fibers

$$\begin{array}{ccc} F_1 & \longrightarrow & F_2 \\ \downarrow & & \downarrow \\ F_3 & \longrightarrow & F_4 \end{array}$$

where $F_1 \longrightarrow B\mathcal{D} \longrightarrow B\mathcal{E}$, $F_2 \longrightarrow BQ\mathcal{P} \longrightarrow BQ\mathcal{V}$, $F_3 \longrightarrow B\mathcal{S}^{-1}\mathcal{D} \longrightarrow B\mathcal{S}^{-1}\mathcal{E}$, $F_4 \longrightarrow BQ\mathcal{P} \longrightarrow BQ\mathcal{V}$ are the homotopy fibers over corresponding points (i.e., choose any point in $B\mathcal{E}$, and take its image in $BQ\mathcal{V}$ and $B\mathcal{S}^{-1}\mathcal{E}$). Then in fact $F_2 = F_4$, and $F_3 \longrightarrow F_4$ is a homotopy equivalence since the front face is homotopy cartesian. Since $\mathcal{D} \xrightarrow{\cong} \mathcal{S}^{-1}\mathcal{D}$, $\mathcal{E} \longrightarrow \mathcal{S}^{-1}\mathcal{E}$ are homotopy equivalences, $F_1 \longrightarrow F_3$ is a homotopy equivalence (for example, compare the long exact homotopy sequences to see that $F_1 \longrightarrow F_3$ induces isomorphisms on $\pi_i$). Hence $F_1 \longrightarrow F_2$ is a homotopy equivalence. Since the $F_i$ were constructed by starting with an arbitrary point of $B\mathcal{E}$, we see that

$$\begin{array}{ccc} \mathcal{D} & \longrightarrow & Q\mathcal{P} \\ \downarrow & \text{①} & \downarrow \\ \mathcal{E} & \longrightarrow & Q\mathcal{V} \end{array}$$

is homotopy Cartesian.

Now $\mathcal{E}$ is contractible by a chain of natural transformations linking the identity functor to the constant functor with value $(0 \to 0 \to 0 \to 0 \to$

$0 \to 0) \in \mathcal{E}$; this is represented by the diagram

$$\begin{array}{ccccccccc}
0 & \to & A & \to & B & \to & C & \to & 0 \\
& & \uparrow & & \| & & \uparrow & & \\
0 & \to & 0 & \to & B & \to & B & \to & 0 \\
& & \| & & \uparrow & & \uparrow & & \\
0 & \to & 0 & \to & 0 & \to & 0 & \to & 0
\end{array}$$

(giving arrows in $\mathcal{E}$ from top and bottom sequences to the middle one). Since $\mathcal{E}$ is contractible, the homotopy Cartesian square ① yields a homotopy equivalence $B\mathcal{D} \longrightarrow F$, where $F$ is the homotopy fiber (over 0) of $Q\mathcal{P} \longrightarrow Q\mathcal{V}$. This homotopy equivalence depends on the choice of a deformation contracting $\mathcal{E}$ to a point (see the discussion in Chapter 6 of homotopy Cartesian squares, preceding the statement of Theorem B). The above chain of natural transformations thus gives one choice of a homotopy equivalence $u : B\mathcal{D} \longrightarrow F$. Regarding a map to $F$ as a pair, consisting of a map to $BQ\mathcal{P}$, and a nullhomotopy of the composite to $BQ\mathcal{V}$, we see easily that $u : B\mathcal{D} \longrightarrow F$ is given by the pair (consisting of a functor, and a chain of natural transformations)

$$\underbrace{(P, 0 \to M \to L \to j^*P \to 0)}_{\in \mathcal{D}} \mapsto (P, 0 \rightarrowtail L \twoheadrightarrow j^*P)$$

where $L$ denotes the functor $(0 \to M \to L \to j^*P \to 0) \mapsto L$, and $j^*P$ has a similar meaning (so that $(0 \to M \to L \to j^*P \to 0) \mapsto j^*P$) is the composite $\mathcal{D} \longrightarrow Q\mathcal{V}$); $0 \rightarrowtail L$, $L \twoheadrightarrow j^*P$ give the chain of natural transformations of functors $\mathcal{D} \longrightarrow Q\mathcal{V}$

$$0 \longrightarrow L \longleftarrow j^*P.$$

Next, let $\mathcal{P}' \subset \mathcal{P}_1$ be the smallest full subcategory containing $\mathcal{P}$ and $\mathcal{H}$, which is closed under extensions ($\mathcal{P}'$ is easily seen to be the full subcategory of $\mathcal{P}_1$ consisting of sheaves $M$ such that there is an admissible monomorphism $N \rightarrowtail M$ with $j^*N \cong j^*M$, and $N \in \mathcal{P}$). Let $\mathcal{D}'$ be the fiber product category

$$\begin{array}{ccc}
\mathcal{D}' & \longrightarrow & Q\mathcal{P}' \\
\downarrow & & \downarrow \\
\mathcal{E} & \longrightarrow & Q\mathcal{V}.
\end{array}$$

As above, if $\bar{F}$ is the homotopy fiber of $Q\mathcal{P}' \longrightarrow Q\mathcal{V}$ over 0, we get a map $B\mathcal{D}' \longrightarrow \bar{F}$ induced by the above contraction of $\mathcal{E}$ (we do not need to know if this is a homotopy equivalence). Since $\mathcal{P} \subset \mathcal{P}'$ is a homotopy equivalence, $F \longrightarrow \bar{F}$ is a homotopy equivalence; since $F \longrightarrow F'$ factors through $\bar{F}$, clearly $\bar{F} \longrightarrow F'$ is a homotopy equivalence.

Let $\lambda : Q\mathcal{H} \longrightarrow \mathcal{D}'$ be given by $H \mapsto (H, 0 \to 0 \to 0 \to j^*H \to 0)$. Then clearly the composite $BQ\mathcal{H} \longrightarrow BQ\mathcal{D}' \longrightarrow \bar{F} \longrightarrow F'$ is just $\phi : BQ\mathcal{H} \longrightarrow F'$, since the nullhomotopy $Q\mathcal{H} \longrightarrow Q\mathcal{V}$, as well as that of $Q\mathcal{H} \longrightarrow Q\mathcal{H}(U)$, are given by the canonical isomorphism of $j^*H$ with the given 0-object. Hence it suffices to check that the composite $BQ\mathcal{H} \longrightarrow \bar{F}$, induced by $\lambda$, is a homotopy equivalence, in order to prove $\phi$ is a homotopy equivalence.

Let $i : \mathcal{D} \longrightarrow \mathcal{D}'$ be the inclusion, $f : \mathcal{C} \longrightarrow \mathcal{D}$, $h : \mathcal{C} \longrightarrow Q\mathcal{H}$ the functors constructed earlier; we have a diagram of categories

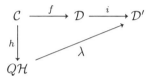

where $f, h$ are homotopy equivalences; this diagram does not commute, but does so "up to sign", as we explain below. We have a related diagram of spaces

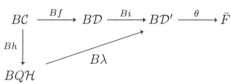

where the composite $\theta \circ Bi : B\mathcal{D} \longrightarrow \bar{F}$ is the homotopy equivalence induced by $u : B\mathcal{D} \longrightarrow F$. Thus $\theta \circ B(i \circ f) : B\mathcal{C} \longrightarrow \bar{F}$ is a homotopy equivalence.

Now $B\mathcal{D}'$ is an $H$-space under the direct sum $\oplus$; since $j^* : \mathcal{P} \longrightarrow \mathcal{V}$ is surjective on objects, one sees easily that $B\mathcal{D}'$ is connected. Hence $B\mathcal{D}'$ is an $H$-group. If we show that $(\lambda \circ h) \oplus (i \circ f) : \mathcal{C} \longrightarrow \mathcal{D}'$ is nullhomotopic, then if $\tau : B\mathcal{D}' \longrightarrow B\mathcal{D}'$ is the inverse map for the $H$-group structure, we see that $B(\lambda \circ h) = B\lambda \circ Bh$ is homotopic to $\tau \circ B(i \circ f)$. Hence $\theta \circ \tau \circ B(i \circ f)$ is homotopic to $\theta \circ B\lambda \circ h$. But $\tau$ induces multiplication by $-1$ on $\pi_n(B\mathcal{D}')$ $\forall n$; since $\theta \circ B(i \circ f)$ induces isomorphisms on homotopy groups, so does $\theta \circ \tau \circ B(i \circ f)$. Hence $\theta \circ \tau \circ B(i \circ f)$, and thus also $\theta \circ B\lambda \circ Bh$, are homotopy equivalences. Since $h$ is a homotopy equivalence, $\theta \circ B\lambda$ is a homotopy equivalence, as desired.

So we have reduced the desired naturality statement to the claim that $\ell = (i \circ f) \oplus (\lambda \circ h) : \mathcal{C} \longrightarrow \mathcal{D}'$ is nullhomotopic. Now $\mathcal{D}'$ has objects $(P', 0 \to M \to N \to j^*P' \to 0)$ with $P' \in \mathcal{P}'$, and $(0 \to M \to N \to j^*P' \to 0) \in \mathcal{E}$. The functor $\ell : \mathcal{C} \longrightarrow \mathcal{D}'$ is given by

$$(0 \to M \to L \to H \oplus P \to 0) \mapsto$$
$$(H \oplus P, 0 \to j^*M \to j^*L \to j^*(H \oplus P) \longrightarrow 0).$$

The diagrams below give a chain of natural transformations to a constant functor:

$$\begin{array}{ccccccccc}
H \oplus P & 0 & \to & j^*M & \to & j^*L & \to & j^*(H \oplus P) & \to & 0. \\
\uparrow & & & \uparrow & & \| & & \uparrow & & \\
L & 0 & \to & 0 & \to & j^*L & \to & j^*L & \to & 0. \\
\uparrow & & & \| & & \uparrow & & \uparrow & & \\
0 & 0 & \to & 0 & \to & 0 & \to & 0 & \to & 0.
\end{array}$$

(We remark here that though $L \twoheadrightarrow H \oplus P$ is a natural transformation, $L \longrightarrow P$ is not one, from the definition of morphisms in $\mathcal{C}$; hence the above does not give a nullhomotopy of $\mathcal{C} \longrightarrow \mathcal{D}$).

Thus, we have completed the proof of the localization theorem, modulo the topological lemma (9.7).

**Proof of (9.7).** Given the triangle of categories

$$\begin{array}{ccc} \mathcal{A} & \xrightarrow{f} & \mathcal{B} \\ {}_g \searrow & & \swarrow_h \\ & \mathcal{C} & \end{array}$$

we have a diagram (as in the proofs of Theorems A and B)

$$\begin{array}{ccccc}
\mathcal{A} & \xleftarrow{p_1} & S(g) & \xrightarrow{p_2} & \mathcal{C}^{\mathrm{op}} \\
{}_f \downarrow & & {}_{\tilde{f}} \downarrow & & \| \\
\mathcal{B} & \xleftarrow{p_1} & S(h) & \xrightarrow{p_2} & \mathcal{C}^{\mathrm{op}}
\end{array}$$

Here $S(g)$ is the category of triples $(X, Y, v)$ with $X \in \mathcal{A}$, $Y \in \mathcal{C}$, $v : Y \longrightarrow g(X)$ an arrow in $\mathcal{C}$, where a morphism $(X, Y, v) \longrightarrow (X', Y', v')$ is a pair of morphisms $u : X \longrightarrow X'$, $w : Y' \longrightarrow Y$ such that

$$\begin{array}{ccc}
Y & \xrightarrow{v} & g(X) \\
{}_w \uparrow & & \downarrow {}_{g(u)} \\
Y' & \xrightarrow{v} & g(X')
\end{array}$$

commutes; $S(h)$ is defined similarly, and $\tilde{f} : S(g) \longrightarrow S(h)$ is given by $\tilde{f}(X, Y, v) = (f(X), Y, v)$. As in the proof of Theorem A, $p_1 : S(g) \longrightarrow \mathcal{A}$,

## 9. Localization for Singular Varieties

$p_1 : S(h) \longrightarrow \mathcal{B}$ are homotopy equivalences. As in the proof of Theorem B, since $g$, $h$ are fibered, $Bp_2 : BS(g) \longrightarrow B\mathcal{C}^{\mathrm{op}}$, $Bp_2 : BS(h) \longrightarrow B\mathcal{C}^{\mathrm{op}}$ are quasi-fibrations, where the fiber of $Bp_2 : BS(g) \longrightarrow B\mathcal{C}^{\mathrm{op}}$ over $Y \in \mathcal{C}$ is $B(Y \setminus g)$, and the fiber of $Bp_2 : BS(h) \longrightarrow B\mathcal{C}^{\mathrm{op}}$ over $Y$ is $B(Y \setminus h)$.

By hypothesis, $B(Y \setminus g) \longrightarrow B(Y \setminus h)$ (induced by $f$) is a homotopy equivalence, since $Bg^{-1}(Y) \longrightarrow Bh^{-1}(Y)$ is one, and $g$, $h$ are fibered. Since the maps $Bp_2$ are quasi-fibrations, we deduce that in the diagram

$$\begin{array}{ccccc} F(g,Y) & \longrightarrow & BS(g) & \longrightarrow & B\mathcal{C}^{\mathrm{op}} \\ \downarrow & & \downarrow & & \| \\ F(h,Y) & \longrightarrow & BS(h) & \longrightarrow & B\mathcal{C}^{\mathrm{op}} \end{array}$$

where $F(g,Y)$, $F(h,Y)$ are the respective homotopy fibers over $Y$, the map $F(g,Y) \longrightarrow F(h,Y)$ is a homotopy equivalence. Hence, by the 5-lemma, $B\tilde{f} : BS(g) \longrightarrow BS(h)$ induces isomorphisms on homotopy groups, and hence is a homotopy equivalence. This proves (9.7).

Our next goal is to prove the so-called Fundamental Theorem. We need some notation: if $X$ is a scheme, let

$$NK_i(X) = \mathrm{coker}(K_i(X) \longrightarrow K_i(X \times_{\mathbb{Z}} \mathbb{A}^1_{\mathbb{Z}}));$$

if $A$ is a ring, let

$$NK_i(A) = NK_i(\mathrm{Spec}\, A) = \mathrm{coker}(K_i(A) \longrightarrow K_i(A[t])).$$

Next, for any scheme $X$, let $Nil(X)$ denote the exact category of pairs $(V, \alpha)$ where $V \in \mathcal{P}(X)$ is a vector bundle on $X$, and $\alpha$ is a nilpotent endomorphism of $V$; an exact sequence

$$0 \longrightarrow (V_1, \alpha_1) \longrightarrow (V_2, \alpha_2) \longrightarrow (V_3, \alpha_3) \longrightarrow 0$$

is an exact sequence

$$0 \longrightarrow V_1 \xrightarrow{i} V_2 \xrightarrow{q} V_3 \longrightarrow 0$$

in $\mathcal{P}(X)$ such that $i \circ \alpha_1 = \alpha_2 \circ i$, $q \circ \alpha_2 = \alpha_3 \circ q$. Let

$$\mathrm{Nil}_i(X) = \ker K_i(Nil(X)) \longrightarrow K_i(\mathcal{P}(X)).$$

Similarly define $\mathrm{Nil}_i(A)$ for any ring $A$.

**Theorem (9.8)** (Fundamental Theorem). *There are natural isomorphisms*

a) $NK_i(A) \cong \mathrm{Nil}_{i-1}(A)$, *for* $i \geq 1$

b) $K_i(A[t, t^{-1}]) \cong K_i(A) \oplus NK_i(A)^{\oplus 2} \oplus K_{i-1}(A)$, *for* $i \geq 1$.

Below, we only prove a), and the following weaker form of b). We show that there is an exact sequence

$$0 \longrightarrow K_i(A) \longrightarrow K_i(A[t]) \oplus K_i(A[t^{-1}]) \longrightarrow K_i(A[t,t^{-1}])$$
$$\longrightarrow K_{i-1}(A) \longrightarrow 0$$

for $i \geq 1$, which can be rewritten as a short exact sequence

$$0 \longrightarrow K_i(A) \oplus NK_i(A)^{\oplus 2} \longrightarrow K_i(A[t,t^{-1}]) \longrightarrow K_{i-1}(A) \longrightarrow 0$$

from a). The map $K_i(A[t,t^{-1}]) \longrightarrow K_{i-1}(A)$ can be shown to be a split surjection, where the splitting is given by the composite of the natural map $K_{i-1}(A) \longrightarrow K_{i-1}(A[t,t^{-1}])$ with the map $K_{i-1}(A[t,t^{-1}]) \longrightarrow K_i(A[t,t^{-1}])$ given by the product with $t \in A[t,t^{-1}]^* = K_1(A[t,t^{-1}])$ (see *Higher Algebraic K-Theory II* for details). This yields the stronger form of b).

**Proof of (9.8).** We will assume $A$ is commutative. Let $X = \mathbb{P}^1_A$, considered as the union of the open subschemes $\operatorname{Spec} A[t]$ and $\operatorname{Spec} A[t^{-1}]$; the inclusion of the open subscheme $U = \operatorname{Spec} A[t^{-1}]$ in $X$ satisfies the hypothesis of the localization theorem (9.1), so that there is an exact sequence

$$\cdots \longrightarrow K_i(\mathcal{H}) \longrightarrow K_i(X) \longrightarrow K_i(A[t^{-1}]) \longrightarrow K_{i-1}(\mathcal{H}) \longrightarrow \cdots,$$

where $\mathcal{H}$ is the category of coherent $\mathcal{O}_X$-modules which have homological dimension $\leq 1$ and vanish on $U$. Clearly $\mathcal{H}$ is isomorphic to the category of finite $A[t]$-modules of projective dimension $\leq 1$ whose localization to $A[t,t^{-1}]$ is zero. Since $\operatorname{Spec} A[t] \longrightarrow X$ is flat, we have a diagram of localization sequences

$$\begin{array}{ccccccc}
K_i(\mathcal{H}) & \to & K_i(X) & \to & K_i(A[t^{-1}]) & \to & K_{i-1}(\mathcal{H}) & \to \cdots \\
\parallel & & \downarrow & & \downarrow & & \parallel \\
K_i(\mathcal{H}) & \to & K_i(A[t]) & \to & K_i(A[t,t^{-1}]) & \to & K_{i-1}(\mathcal{H}) & \to \cdots
\end{array}$$

We first observe that $\mathcal{H} \cong Nil(A)$, as follows. If $(V,\alpha) \in Nil(A)$, regard $V$ (which is a finitely generated projective $A$-module) as an $A[t]$-module where multiplication by $t$ is the $A$-linear endomorphism $\alpha$. Then $A[t,t^{-1}] \otimes_{A[t]} V = 0$, and we have the characteristic exact sequence

$$0 \longrightarrow A[t] \otimes_A V \xrightarrow{t-\alpha} A[t] \otimes_A V \longrightarrow V \longrightarrow 0.$$

Thus $(V,\alpha)$ gives an object of $\mathcal{H}$ in a natural way, giving a functor $Nil(A) \to \mathcal{H}$. Conversely, if $M$ is an $A[t]$-module of projective dimension $\leq 1$ killed by a power of $t$, then we claim the underlying $A$-module $M$ is a projective $A$-module. Indeed, if

$$0 \longrightarrow Q \longrightarrow P \longrightarrow M \longrightarrow 0$$

is a projective resolution over $A[t]$, and $t^n M = 0$, then multiplication by $t^n$ on the terms of the above resolution yields (snake lemma) an exact sequence
$$0 \longrightarrow M \longrightarrow Q/t^n Q \longrightarrow P/t^n P,$$
and hence an exact sequence
$$0 \longrightarrow M \longrightarrow Q/t^n Q \longrightarrow Q/t^n P \longrightarrow 0$$
($t^n P \subset Q$ as $t^n M = 0$). Now $Q/t^n Q$ is a projective $A$-module (as $Q/tQ$ is one), and $Q/t^n P$ has projective dimension $\leq 1$; hence $M$ is a projective $A$-module. If $\alpha(t)$ is the $A$-linear endomorphism given by multiplication by $t$, then $M \longrightarrow (M, \alpha(t))$ gives a functor $\mathcal{H} \longrightarrow Nil(A)$ inverse to the above functor.

By Theorem (5.29), $K_i(\mathbb{P}^1_A) \cong K_i(A)^{\oplus 2}$, where $K_i(A)^{\oplus 2} \longrightarrow K_i(\mathbb{P}^1_A)$ is given by
$$(x, y) \mapsto x + \zeta y, \quad \zeta = [\mathcal{O}_{\mathbb{P}^1}(-1)] \in K_0(\mathbb{P}^1_A).$$
Thus $K_i(\mathbb{P}^1_A) \cong K_i(A) + \zeta \cdot K_i(A)$. This may be rewritten as $K_i(\mathbb{P}^1_A) \cong K_i(A) + (1 - \zeta) K_i(A)$, where
$$\mathrm{Ker}(K_i(\mathbb{P}^1_A) \longrightarrow K_i(A[t])) = \mathrm{Ker}(K_i(\mathbb{P}^1_A) \longrightarrow K_i(A[t, t^{-1}]))$$
$$= (1 - \zeta) \cdot K_i(A),$$
since $\mathcal{O}_{\mathbb{P}^1}(1)$ restricts to the trivial bundle on $\mathrm{Spec}\, A[t]$ and on $\mathrm{Spec}\, A[t^{-1}]$, while the natural maps
$$K_i(A) \longrightarrow K_i(A[t]), K_i(A) \longrightarrow K_i(A[t^{-1}]),$$
given by change of rings, are split inclusions. Hence the localization sequence for $X = \mathbb{P}^1_A$ breaks up into exact sequences (for $i \geq 1$),
$$0 \longrightarrow K_i(A) \longrightarrow K_i(A[t^{-1}]) \longrightarrow K_{i-1}(Nil(A)) \longrightarrow K_{i-1}(A) \longrightarrow 0,$$
where we have used $\mathcal{H} \cong Nil(A)$.

If $(P, \alpha) \in Nil(A)$, and we again denote the associated element of $\mathcal{H}$ by $P$, the characteristic exact sequence
$$0 \longrightarrow \mathcal{O}_{\mathbb{P}^1}(-1) \otimes_A P \xrightarrow{t - \alpha} \mathcal{O}_{\mathbb{P}^1} \otimes_A P \longrightarrow P \longrightarrow 0$$
(where $t$ is now regarded as a global section of $\mathcal{O}_{\mathbb{P}^1}(1)$) gives an exact sequence of functors $Nil(A) \longrightarrow \mathcal{P}_1(\mathbb{P}^1_A)$, where $\mathcal{P}_1(\mathbb{P}^1_A)$ is the category of coherent $\mathcal{O}_{\mathbb{P}^1}$-modules of homological dimension $\leq 1$. Hence the square below commutes:

$$\begin{array}{ccc} K_{i-1}(Nil(A)) & \xrightarrow{\cong} & K_{i-1}(\mathcal{H}) \\ f \downarrow & & \downarrow g \\ K_{i-1}(A) & \xrightarrow{1-\zeta} & K_{i-1}(X) \end{array}$$

where the map $f$ is induced by $(P, \alpha) \mapsto P \in \mathcal{P}(A)$, and $g$ is induced by $\mathcal{H} \subset \mathcal{P}_1(\mathbb{P}_A^1)$. Hence

$$\ker(K_{i-1}(\mathcal{H}) \longrightarrow K_{i-1}(X)) \cong Nil_{i-1}(A),$$

and the resulting exact sequence

$$0 \longrightarrow K_i(A) \longrightarrow K_i(A[t^{-1}]) \longrightarrow Nil_{i-1}(A) \longrightarrow 0$$

(which is obviously split) proves (9.8)(a).

Next, the diagram comparing the localization sequences of $X$ and Spec $A[t]$ gives the Mayer–Victoris sequence

$$\cdots \longrightarrow K_i(X) \longrightarrow K_i(A[t]) \oplus K_i(A[t^{-1}]) \longrightarrow K_i(A[t, t^{-1}])$$
$$\longrightarrow K_{i-1}(X) \longrightarrow \cdots.$$

As above, this breaks up into sequences

$$0 \longrightarrow K_i(A) \longrightarrow K_i(A[t]) \oplus K_i(A[t^{-1}]) \longrightarrow K_i(A[t, t^{-1}])$$
$$\longrightarrow K_{i-1}(A) \longrightarrow 0$$

which is the weak form of b). Finally, the case when $A$ is noncommutative is dealt with using arguments similar to the proof of (5.39).

**M. Levine's Localization Theorem.** We now indicate how to modify the proof of (9.1) to yield a generalization due to Levine, and apply it to the study of modules of finite length and finite projective dimension over two-dimensional normal local rings (see M. Levine, "Modules of finite length and $K$-groups of surface singularities", Compos. Math. 59 (1986) 21–40). Let $X$ be a Noetherian scheme which supports an ample line bundle, and let $Y \subset X$ be a closed subscheme of pure codimension $d$, such that for each $y \in Y$, $\mathcal{O}_{y,Y}$ has projective dimension $\leq d$ over $\mathcal{O}_{y,X}$. Fix an affine open subscheme $U \subset Y$ such that for some effective Cartier divisor $Z$ on $Y$, $Y - Z = U$. Thus the ideal sheaf $\mathcal{I}_Z$ of $Z \subset Y$ is an invertible $\mathcal{O}_Y$-module. Let $i : Y \longrightarrow X$, $j : U \longrightarrow Y$ be the inclusions. Let $C = Y - U = Z_{\text{red}}$.

We now describe some full additive subcategories of $\mathcal{M}(X)^*$, the category of quasi-coherent $\mathcal{O}_X$-modules; it is easy to see that they are closed under extensions in $\mathcal{M}(X)^*$, and hence form exact categories:

$\mathcal{P}(Y) =$ locally free $\mathcal{O}_Y$-modules of finite rank

(regarded as a subcategory of $\mathcal{M}(X)^*$ via $i_*$)

(for $r \geq d$) $\mathcal{P}^r(Y) =$ coherent $\mathcal{O}_Y$-modules of homological dimension $\leq r$ over $\mathcal{O}_X$.

$\mathcal{P}^r(Y, U) =$ full subcategory of $\mathcal{F} \in \mathcal{P}^r(Y)$ such that $\mathcal{F}$ has no associated points in $C$ (i.e., $\mathcal{F} \longrightarrow j_*j^*\mathcal{F}$ is injective), and $j^*\mathcal{F} \in \mathcal{P}(U)$.

## 9. Localization for Singular Varieties

$\mathcal{P}_C^r(Y) =$ full subcategory of $\mathcal{F} \in \mathcal{P}^r(Y)$ such that $\mathcal{F}$ is supported on $C$ (i.e., $j^*\mathcal{F} = 0$).

**Theorem (9.9).** *There is a natural long exact sequence* (for $i \geq 0$)

$$\cdots \longrightarrow K_{i+1}(U) \longrightarrow K_i(\mathcal{P}_C^{d+1}(Y)) \longrightarrow K_i(\mathcal{P}^d(Y,U)) \longrightarrow K_i(U)$$

*which is compatible with the localization sequence* (9.1) *for* $U \subset Y$, *i.e., the inclusions* $\mathcal{P}(Y) \subset \mathcal{P}^d(Y,U)$, $\mathcal{H}_C \subset \mathcal{P}_C^{d+1}(Y)$ (*where* $\mathcal{H}_C =$ *coherent* $\mathcal{O}_Y$-*modules of homological dimension* $\leq 1$ *over* $\mathcal{O}_Y$ *and supported on* $C$) *induce a commutative diagram*

$$\begin{array}{ccccccc}
\cdots \to & K_{i+1}(U) & \to & K_i(\mathcal{P}_C^{d+1}(Y)) & \to & K_i(\mathcal{P}^d(Y,U)) & \to & K_i(U) \\
& \parallel & & \uparrow & & \uparrow & & \parallel \\
\cdots \to & K_{i+1}(U) & \to & K_i(\mathcal{H}_C) & \to & K_i(Y) & \to & K_i(U)
\end{array}$$

**Remark.** Here, the naturality of the sequence means the following—let $Y' \subset X$ be a subscheme of pure codimension $d$ with $\text{proj.dim.}_{\mathcal{O}_{y,X}} \mathcal{O}_{y,Y'} \leq d$ for all $y \in Y'$, such that $Y \subset Y'$. Let $C' \subset Y'$ be a closed subset supporting an effective Cartier divisor, such that $U' = Y' - C'$ is affine, $C \subset C'$, and $U \cap U'$ is open and closed in $U'$. Let $i_1 : Y \longrightarrow Y'$, $j_1 : U \cap U' \longrightarrow U$, $h : U \cap U' \longrightarrow U'$ be the inclusions. Then we have exact functors

$$i_{1*} : \mathcal{P}^d(Y,U) \longrightarrow \mathcal{P}^d(Y',U')$$

$$i_{1*} : \mathcal{P}_C^{d+1}(Y) \longrightarrow \mathcal{P}_{C'}^{d+1}(Y')$$

$$h_* \circ j_1^* : \mathcal{P}(U) \longrightarrow \mathcal{P}(U').$$

The naturality statement is that the induced diagram below commutes:

$$\begin{array}{ccccccc}
\to & K_{i+1}(U) & \to & K_i(\mathcal{P}_C^{d+1}(Y)) & \to & K_i(\mathcal{P}^d(Y,U)) & \to & K_i(U)) \\
& \downarrow & & \downarrow & & \downarrow & & \downarrow \\
\to & K_{i+1}(U') & \to & K_i(\mathcal{P}_{C'}^{d+1}(Y)) & \to & K_i(\mathcal{P}^d(Y',U')) & \to & K_i(U')
\end{array}$$

**Proof of (9.9) (Sketch).** The proof closely follows that of (9.1). Let $\mathcal{V} \subset \mathcal{P}(U)$ be the full subcategory of sheaves of the form $j^*M$, for some $M \in \mathcal{P}^d(Y,U)$. Let $\mathcal{T} = \text{Iso}\,\mathcal{V}$, $\mathcal{S} = \text{Iso}\,\mathcal{P}^d(Y,U)$. In the following, we write $\mathcal{P}^d$ for $\mathcal{P}^d(Y,U)$, and $\mathcal{P}_C^{d+1}$ for $\mathcal{P}_C^{d+1}(Y)$.

Let $\mathcal{E}$ be the extension construction over $Q\mathcal{V}$, so that $\mathcal{E}$ has objects given by admissible epimorphisms $M \twoheadrightarrow N$ in $\mathcal{V}$, where a morphism

$(M \twoheadrightarrow N) \longrightarrow (M' \twoheadrightarrow N')$ is an equivalence class of diagrams

$$\begin{array}{ccc} M & \longrightarrow & N \\ \| & & \uparrow \\ M & \longrightarrow & N_1 \\ \downarrow & \square & \downarrow \\ M' & \longrightarrow & N' \end{array}$$

where the square marked $\square$ is a pullback. $(M \twoheadrightarrow N) \mapsto N$ gives a functor $\mathcal{E} \longrightarrow Q\mathcal{V}$; let $\mathcal{D}$ be the pullback

$$\begin{array}{ccc} \mathcal{D} & \longrightarrow & \mathcal{P}^d \\ \downarrow & & \downarrow \\ \mathcal{E} & \longrightarrow & Q\mathcal{V} \end{array}$$

so that the objects of $\mathcal{D}$ are pairs $(P, M \twoheadrightarrow j^*P)$ with $P \in \mathcal{P}^d$, $(M \twoheadrightarrow j^*P) \in \mathcal{E}$. Let $\mathcal{C}$ be the category whose objects are admissible epimorphisms (in $\mathcal{P}^{d+1}$) $M \twoheadrightarrow (P \oplus B)$ with $M, P \in \mathcal{P}^d$, $B \in \mathcal{P}_C^{d+1}$. An arrow $(M' \twoheadrightarrow P' \oplus B') \longrightarrow (M \twoheadrightarrow P \oplus B)$ is an arrow $(P', B') \leftarrow (P'', B'') \rightarrowtail (P, B)$ in $Q\mathcal{P}^d \times Q\mathcal{P}_C^{d+1}$ such that there is an induced equivalence class of diagrams

$$\begin{array}{ccc} L' & \longrightarrow & P' \oplus B' \\ \| & & \uparrow \\ L' & \longrightarrow & P \oplus B'' \\ \downarrow & & \downarrow \\ L & \longrightarrow & P \oplus B \end{array}$$

We obtain a diagram of categories and functors

$$\begin{array}{ccccc} \mathcal{C} & \xrightarrow{f} & \mathcal{D} & \xrightarrow{p} & Q\mathcal{P}^d \\ h\downarrow & & s\downarrow & & \downarrow j^* \\ Q\mathcal{C}_C^{d+1} & & \mathcal{E} & \xrightarrow{q} & Q\mathcal{V} & \longrightarrow & Q\mathcal{P}(U) \end{array}$$

where

$$q(M \twoheadrightarrow N) = N$$
$$f(M \twoheadrightarrow P \oplus B) = (P, j^*M \twoheadrightarrow j^*P)$$
$$h(M \twoheadrightarrow P \oplus B) = B,$$

and let $g = p \circ f : \mathcal{C} \longrightarrow Q\mathcal{P}^d$, so that

$$g(M \twoheadrightarrow P \oplus B) = P.$$

Let $s, p$ be the natural projections. One verifies easily (as in (9.1)) that $g$, $h$, $p$, $q$ are fibered functors.

Clearly $\mathcal{S} \subset \mathcal{T}$ is cofinal; hence $\mathcal{S}^{-1}\mathcal{S} \cong \mathcal{T}^{-1}\mathcal{T}$, and $\mathcal{S}^{-1}\mathcal{E} \cong \mathcal{T}^{-1}\mathcal{E}$. As in (9.1), we see that $\mathcal{T}^{-1}\mathcal{E}$ is contractible, and

$$\begin{array}{ccc} \mathcal{T}^{-1}\mathcal{T} & \longrightarrow & \mathcal{T}^{-1}\mathcal{E} \\ \downarrow & & \downarrow \\ \text{(point)} & \longrightarrow & Q\mathcal{V} \end{array}$$

is homotopy Cartesian, since all exact sequences in $Q\mathcal{V}$ are split. Hence $\mathcal{S}^{-1}\mathcal{E}$ is contractible, and

$$\begin{array}{ccc} \mathcal{S}^{-1}\mathcal{S} & \longrightarrow & \mathcal{S}^{-1}\mathcal{E} \\ \downarrow & & \downarrow \\ \text{(point)} & \longrightarrow & Q\mathcal{V} \end{array}$$

is homotopy Cartesian.

Next, we claim:

(i) $h : \mathcal{C} \longrightarrow Q\mathcal{P}_\mathcal{C}^{d+1}$ is a homotopy equivalence (c.f. Lemma (9.4))

(ii) if $P \in \mathcal{P}^d$, then $P \longrightarrow j_*j^*P$ is injective, and $j_*j^*P = \bigcup_n \mathcal{I}_Z^{-n}P$ (c.f. Lemma (9.5))

(iii) $f : \mathcal{C} \longrightarrow \mathcal{D}$ is a homotopy equivalence (c.f. Lemma (9.6))

(iv)
$$\begin{array}{ccc} \mathcal{S}^{-1}\mathcal{D} & \longrightarrow & Q\mathcal{P}^d \\ \downarrow & & \downarrow \\ \text{(point)} \cong \mathcal{S}^{-1}\mathcal{E} & \longrightarrow & Q\mathcal{V} \end{array}$$

is homotopy Cartesian

(v) $BQ\mathcal{P}_\mathcal{C}^{d+1} \longrightarrow BQ\mathcal{P}^d \longrightarrow BQ\mathcal{V}$ is homotopy equivalent to a fibration (this follows from (iv) and the homotopy equivalences $\mathcal{S}^{-1}\mathcal{D} \cong \mathcal{S}^{-1}\mathcal{C} \cong Q\mathcal{P}_\mathcal{C}^{d+1}$)

(vi) $K_i(\mathcal{V}) \longrightarrow K_i(\mathcal{P}(U)) = K_i(U)$ is an isomorphism for $i > 0$ and a monomorphism for $i = 0$ (c.f. Lemma (9.3)).

In fact, the proofs of the above statements closely follow those of the corresponding lemmas cited from the proof of Theorem (9.1). We leave the details to the reader as a (fairly straightforward) exercise. Clearly (v), (vi) give the desired long exact sequence. The compatibility of the above constructions with those in the proof of (9.1), gives the diagram comparing the sequence of (9.1) for $U \subset Y$ with the above new sequence.

Finally, naturality (see the remark after the statement of the theorem) is proved as follows: let $\mathcal{E}'$, $\mathcal{C}'$, $\mathcal{D}'$, $\mathcal{V}'$, $\mathcal{S}'$ be the analogous categories associated to $Y'$, $U'$. Then one checks that $i_1$, $j_1$, $h$ define a map from

$$\begin{array}{ccccc} \mathcal{S}^{-1}\mathcal{C} & \longrightarrow & \mathcal{S}^{-1}\mathcal{D} & \longrightarrow & Q\mathcal{P}^d(Y,U) & \longrightarrow & Q\mathcal{P}(U) \\ \downarrow & & \downarrow & & \downarrow & & \\ Q\mathcal{P}_C^{d+1}(Y) & & \mathcal{S}^{-1}\mathcal{E} & \longrightarrow & Q\mathcal{V} & & \end{array}$$

to the diagram

$$\begin{array}{ccccc} \mathcal{S}'^{-1}\mathcal{C}' & \longrightarrow & \mathcal{S}'^{-1}\mathcal{D}' & \longrightarrow & Q\mathcal{P}^d(Y',U') & \longrightarrow & Q\mathcal{P}(U') \\ \downarrow & & \downarrow & & \downarrow & & \\ Q\mathcal{P}_C^{d+1}(Y') & & \mathcal{S}'^{-1}\mathcal{E}' & \longrightarrow & Q\mathcal{V}' & & \end{array}$$

This gives the desired map between the two localization sequences.

We apply Theorem (9.9) to the following situation. Let $X = \operatorname{Spec} R$, where $R$ is the local ring of a normal point on a surface over an infinite field $k$, and let $Y$ be a reduced effective Cartier divisor on $X$, so that $Y = \operatorname{Spec} R/fR$ for some non-zero divisor $f \in R$ such that $R/fR$ is reduced. If $g \in R$ such that $(f,g)$ form a regular sequence, i.e., $(f,g)R$ is primary to the maximal ideal $\mathfrak{M}$ of $R$, then the image $\bar{g} \in R/fR$ of $g$ generates the ideal of an effective Cartier divisor $Z \subset Y$, such that $Y - C = U = \operatorname{Spec}((R/fR)[\bar{g}^{-1}])$ is affine. Then $\mathcal{P}_C^2(Y)$ is the category of $R$-modules $M$ of finite length and finite projective dimension which are annihilated by $f$, and $\mathcal{P}^1(Y,U)$ is the category of finite $R$-modules $P$ of projective dimension 1 and annihilated by $f$; the condition that $P$ restricts to a vector bundle on $U$ is automatically true, since $U \cong \operatorname{Spec} k(f)$, where $k(f)$ is the total quotient ring of $R/fR$ (if $fR = \mathcal{P}_1 \cap \cdots \cap \mathcal{P}_n$ is a primary decomposition, then $k(f) = k(\mathcal{P}_1) \times \cdots \times k(\mathcal{P}_n)$, where $k(\mathcal{P}) =$ residue field of $\mathcal{P}$). Thus we obtain an exact sequence

$$\longrightarrow K_{i+1}(k(f)) \longrightarrow K_i(\mathcal{P}_C^2(Y)) \longrightarrow K_i(\mathcal{P}^1(Y,U)) \longrightarrow K_i(U).$$

The set of reduced Cartier divisors in $X$ form a directed set, and the naturality statement in (9.9) yields a directed family of long exact sequences. Since $K_i$ commutes with filtered direct limits of categories, we obtain an exact sequence in the direct limit

$$\longrightarrow \bigoplus_{\mathrm{ht}\,\mathcal{P}=1} K_{i+1}(k(\mathcal{P})) \longrightarrow K_i(\mathcal{C}_R) \longrightarrow K_i(\mathcal{P}^1(R)) \longrightarrow \bigoplus_{\mathrm{ht}\,\mathcal{P}=1} K_i(k(\mathcal{P}))$$

where $\mathcal{P}$ runs over the primes of height 1 in $R$, $\mathcal{C}_R$ denotes the category of modules of finite length and finite projective dimension, and $\mathcal{P}^1(R)$ denotes the category of finite $R$-modules $M$ of projective dimension 1, such that for some $f \in \operatorname{Ann}_R M$, $R/fR$ is reduced (to see that $\varinjlim \mathcal{P}_C^2(Y) = \mathcal{C}_R$, we

## 9. Localization for Singular Varieties

need only observe that any $M \in \mathcal{C}_R$ is annihilated by some $f \in R$ such that $R/fR$ is reduced). In particular, we have an exact sequence

$$K_1(\mathcal{P}^1(R)) \longrightarrow \bigoplus_{ht\,\mathcal{P}=1} k(\mathcal{P})^* \xrightarrow{\partial} K_0(\mathcal{C}_R).$$

We wish to compute the map $\partial$ explicitly.

By the compatibility of (9.1) and (9.9), if $f \in R$ is such that $R/fR$ is reduced, and $fR = \mathcal{P}_1 \cap \cdots \cap \mathcal{P}_n$ is the primary decomposition then we have a diagram

$$\begin{array}{ccccc}
\longrightarrow & K_1(\mathcal{P}^1(Y,U)) & \longrightarrow & \bigoplus_{i=1}^n k(\mathcal{P}_i)^* & \longrightarrow & K_0(\mathcal{P}_C^2(Y)) \\
& \uparrow & & \| & & \uparrow \\
\longrightarrow & K_1(Y) & \longrightarrow & \bigoplus_{i=1}^n k(\mathcal{P}_i)^* & \xrightarrow{\partial_f} & K_0(\mathcal{H}_C)
\end{array}$$

where $Y = \operatorname{Spec} R/fR$, $C = \{\mathcal{M}\}$, and $\mathcal{H}_C$ is the category of $(R/fR)$-modules of finite length and finite projective dimension. Any element of $\oplus k(\mathcal{P}_i)^*$ is of the form $\bar{g}_1/\bar{g}_2$, where $g_1, g_2 \in R$ such that $f, g_1 g_2$ form a regular sequence, and $g_i \mapsto \bar{g}_i \in R/fR$. We claim that $\partial_f(\bar{g}_i) = \pm[R/(f,g_i)] \in K_0(\mathcal{H}_C)$ (for some universal choice of sign). If $\bar{g}_i \in (R/fR)^*$, both sides are trivial. If $\bar{g}_i$ is a non-unit, there is a local homomorphism

$$k[\bar{g}_i]_{(\bar{g}_i)} \longrightarrow R/fR$$

which is flat, since $\bar{g}_i$ is a nonzero divisor. Hence we have a diagram of localization sequences (by the naturality of the sequence of (9.1))

$$\begin{array}{ccccccc}
K_1(\mathcal{H}_C) & \longrightarrow & K_1(R/fR) & \longrightarrow & \bigoplus_j k(\mathcal{P}_j)^* & \longrightarrow & K_0(\mathcal{H}_C) \\
\uparrow & & \uparrow & & \uparrow & & \uparrow \\
K_1(k) & \longrightarrow & K_1(k[\bar{g}_i]_{(\bar{g}_i)}) & \longrightarrow & k(\bar{g}_i)^* & \xrightarrow{\phi} & K_0(k) \\
& & & & & & \| \\
& & & & & & \mathbb{Z}
\end{array}$$

where $k(\bar{g}_i)$ is the quotient field of the polynomial ring $k[\bar{g}_i]$, and $\phi(\bar{g}_i) = \pm 1$ for some universal choice of sign, by (5.28). The map $K_0(k) \longrightarrow K_0(\mathcal{H}_C)$ is given by $1 \mapsto [R/(f,g_i)]$.

This yields the following explicit formula for

$$\partial : \bigoplus_{ht\,\mathcal{P}=1} k(\mathcal{P})^* \longrightarrow K_0(\mathcal{C}_R).$$

Given a finite set $\mathcal{P}_1, \ldots, \mathcal{P}_r$ of height 1 primes, and $\alpha_i \in k(\mathcal{P}_i)^*$ for all $i$, choose $f \in \mathcal{P}_1 \cap \cdots \cap \mathcal{P}_r$ such that $R/fR$ is reduced. Let $\mathcal{P}_1, \ldots, \mathcal{P}_r, \ldots, \mathcal{P}_n$

be the set of height 1 primes containing $f$, and set $\alpha_i = 1$ for $r+1 \leq i \leq n$. Choose $g_1, g_2 \in R$ such that $(f, g_1 g_2)$ form a regular sequence, and $g_1 g_2^{-1} \mapsto \alpha_i$, $1 \leq i \leq n$. If $\alpha \in \oplus k(\mathcal{P})^*$ is the element given by $\alpha_i \in k(\mathcal{P}_i)$, $1 \leq i \leq n$, then
$$\partial(\alpha) = [R/(f, g_1)] - [R/(f, g_2)].$$

This explicit description of $\partial$ allows us to prove two statements, contained in the following two lemmas.

**Lemma (9.10)** (Hochster). *$\partial$ is surjective, i.e., $K_0(\mathcal{C}_R)$ is generated by the classes $[R/(f,g)]$ for regular sequences $(f,g)$ in $R$ such that $R/fR$ is reduced.*

**Proof.** (This proof is due to Mohan Kumar). If $(f, g)$ is any regular sequence, then a "general" $k$-linear combination $f' = af + bg$, $a, b \in k$ will have the property that $R/f'R$ is reduced, since $k$ is infinite. Since $R/(f,g) = R/(f',g)$, we are reduced to proving that $K_0(\mathcal{C}_R)$ is generated by $R/(f,g)$ for all regular sequences $(f,g)$ in $R$.

Let $M \in \mathcal{C}_R$, $f, g \in \mathrm{Ann}_R M$ a regular sequence in $R$. If $M$ is not cyclic (generated by one element, as an $R$-module), let $x, y \in M$ be part of a minimal set of generators, and let $i : fR + gR \longrightarrow M$ be defined by $i(f) = x$, $i(g) = y$. Since $f, g$ form a regular sequence and $gx = fy = 0$, $i$ is well defined. Consider the diagram obtained by pushout along $i$,

$$\begin{array}{ccccccccc}
0 & \longrightarrow & (f,g) & \longrightarrow & R & \longrightarrow & R/(f,g) & \longrightarrow & 0 \\
& & \downarrow & & \downarrow & & \parallel & & \\
0 & \longrightarrow & M & \longrightarrow & N & \longrightarrow & R/(f,g) & \longrightarrow & 0 \\
& & \downarrow & & \downarrow & & & & \\
& & (M/i(f,g)) & = & (M/i(f,g)). & & & &
\end{array}$$

If $M$ is minimally generated by $n$ elements, $M/i(f,g)$ requires $n-2$ generators, so that $N$ requires at most $n-1$ generators. Also, $[M] = [N] - [R/(f,g)]$ in $K_0(\mathcal{C}_R)$. Hence, by induction on the number of generators of $M$, we see that $K_0(\mathcal{C}_R)$ is generated by classes of cyclic modules $R/I$ of finite length and finite projective dimension.

If $R/I \in \mathcal{C}_R$, choose a regular sequence $f, g$ in $I$ which is part of a minimal set of generators, so that we have an exact sequence

$$0 \longrightarrow M \longrightarrow R/(f,g) \longrightarrow R/I \longrightarrow 0$$

where $M = I/(f,g)$ requires $n-2$ generators, if $I$ requires $n$ generators. Clearly $M \in \mathcal{C}_R$ from the above sequence. We have an exact sequence

$$0 \longrightarrow \mathrm{Tor}_2^R(M, k) \longrightarrow \mathrm{Tor}_2^R(R/(f,g), k) \longrightarrow \mathrm{Tor}_2^R(R/I, k) \longrightarrow \ldots.$$

9. Localization for Singular Varieties     217

Since proj. $\dim_R M = 2$, $\operatorname{Tor}_2^R(M, k) \neq 0$; also $\operatorname{Tor}_2^R(R/(f,g), k) \cong k$. Hence $\operatorname{Tor}_2^R(M, k) \cong k$, and the minimal resolution of $M$ has the form

$$0 \longrightarrow R \longrightarrow R^{\oplus n-1} \longrightarrow R^{\oplus n-2} \longrightarrow M \longrightarrow 0.$$

The functor $\operatorname{Ext}_R^2(-, R) : \mathcal{C}_R \longrightarrow \mathcal{C}_R$ is an exact involution which fixes the isomorphism class of any module $R/(f,g)$; since $\operatorname{Ext}_R^i(-, R) = 0$ on $\mathcal{C}_R$ for $i \leq 1$, we see from the above resolution of $M$ that $M^* = \operatorname{Ext}_R^2(M, R)$ has a minimal resolution

$$0 \longrightarrow R^{\oplus n-2} \longrightarrow R^{\oplus n-1} \longrightarrow R \longrightarrow M^* \longrightarrow 0$$

i.e., $M^* \cong R/J \in \mathcal{C}_R$ where $J$ requires only $n - 1$ generators. By induction on the number of generators, we may assume $[M^*] \in K_0(\mathcal{C}_R)$ lies in the subgroup generated by all classes $[R/(f,g)]$. Since the involution is trivial on this subgroup, we deduce that $[M] \in K_0(\mathcal{C}_R)$ lies in this subgroup. This proves (9.10).

Next, each prime $\mathcal{P} \subset R$ of height 1 gives a discrete valuation on $K$, the quotient field of $R$, and hence defines a tame symbol $T_\mathcal{P} : K_2(K) \to k(\mathcal{P})^*$ (see Example (1.15)) defined by

$$T_\mathcal{P}\{a, b\} = \phi((-1)^{v(a)v(b)} a^{v(b)} / b^{v(a)})$$

where $\phi : R_\mathcal{P}^* \longrightarrow k(\mathcal{P})^*$, and $v$ is the valuation. Let

$$T : K_2(K) \longrightarrow \bigoplus_{\operatorname{ht} \mathcal{P} = 1} k(\mathcal{P})^*$$

be the sum of the maps $T_\mathcal{P}$ (note that for any $\{a, b\} \in K_2(K)$, $a, b \in R_\mathcal{P}^*$ for all but a finite number of primes $\mathcal{P}$, so that $T_\mathcal{P}\{a, b\} = 0$ for all but a finite set of $\mathcal{P}$).

**Lemma (9.11).** $\partial \circ T : K_2(K) \longrightarrow K_0(\mathcal{C}_R)$ *is zero.*

**Proof.** We must show that $\partial \circ T\{a, b\} = 0$ for every $a, b \in K^*$. If $a, b \in R$ such that $R/ab\,R$ is reduced, then $a, b$ form a regular sequence. We compute that for any prime $\mathcal{P} \subset R$ of height 1, if $T_\mathcal{P}$ is the associated tame symbol,

$$T_\mathcal{P}\{a, b\} = \begin{cases} 1 \in k(\mathcal{P})^* & \text{if } ab \notin \mathcal{P} \\ \bar{b}^{-1} \in k(\mathcal{P})^* & \text{if } a \in \mathcal{P},\, b \notin \mathcal{P} \\ \bar{a} \in k(\mathcal{P})^* & \text{if } a \notin \mathcal{P},\, b \in \mathcal{P} \end{cases}$$

(where $\bar{a}, \bar{b}$ denote the images of $a, b$ in appropriate residue fields). Hence $\partial \circ T\{a, b\} = [R/(a, b)R] - [R/(a, b)R] = 0$.

We reduce the general case to this case, by showing that $K_2(K)$ is generated by symbols $\{a, b\}$ with $a, b \in R$ such that $R/ab\,R$ is reduced. We use the following terminology: if $f, g \in K^*$, we say $f, g$ are "relatively prime" if for any discrete valuation $v$ associated to a height 1 prime $\mathcal{P}$, then

$v(f) \cdot v(g) = 0$, i.e., at least one of $f, g$ is a unit in $R_\mathcal{P}$. If $v(f)v(g) \neq 0$, we say that $\mathcal{P}$ is a common divisor of $f, g$.

Clearly $K_2(K)$ is generated by symbols $\{a, b\}$ with $a, b \in R - \{0\}$. Fix $\{a, b\}$, and let $aR = \mathcal{P}_1^{(r_1)} \cap \cdots \cap \mathcal{P}_k^{(r_k)}$ be its primary decomposition ($\mathcal{P}^{(n)} = R \cap \mathcal{P}^n R_\mathcal{P}$). Thus if $v_i$ is the discrete valuation associated to $\mathcal{P}_i$, then $v_i(a) = r_i$. Let $s_i = \sum_{j \leq i} r_j$. Choose $a_1, \ldots, a_{s_1}, a_{s_1+1}, \ldots, a_{s_2}, \ldots, a_{s_k} \in R$ such that (i) $a_{s_{i-1}+1}, \ldots, a_{s_i} \in \mathcal{P}_i - \bigcup_{j \neq i} \mathcal{P}_j$, (ii) $a_{s_{i-1}+1}, \ldots, a_{s_i}$ each has no common divisor with $b$, except possibly $\mathcal{P}_i$, (iii) $\mathcal{P}_i$ is the only common divisor of any two of $a_{s_{i-1}+1}, \ldots, a_{s_i}$, and $a_{j_1}, a_{j_2}$ are relatively prime unless $s_{i-1} + 1 \leq j_1, j_2 \leq s_i$ for some $i$, (iv) $R/a_j R$ is reduced, for all $j$.

Then if $c = (a_1 \ldots a_{s_k})/a$, we see that $b, c$ are relatively prime, and $a_j, b$ have at most one common divisor (which must then equal $\mathcal{P}_i$ for some $i$); further $R/a_j R$, $R/cR$ are reduced. Then $\{a, b\} = (\Pi\{a_j, b\}) \cdot \{c, b\}^{-1}$. Applying a similar argument to $b$ (with respect to each of the symbols $\{a_j, b\}$, $\{c, b\}$) we see that $K_2(K)$ is generated by symbols $\{a, b\}$ with $a, b \in R - (0)$, such that $R/aR$, $R/bR$ are reduced, and $a, b$ have at most one common divisor. Suppose $\mathcal{P}$ is such a common divisor. For sufficiently general $\lambda \in k$ (since $k$ is infinite), $R/(b - \lambda a)R$ is reduced, and the only common divisor of $ab$ and $b - \lambda a$ is $\mathcal{P}$. Then

$$\left\{\frac{a}{b}, b - \lambda a\right\} = \left\{\frac{a}{b}, b\left(1 - \frac{\lambda a}{b}\right)\right\} = \{a, b\} \cdot \{b, b\}^{-1} \cdot \left\{\frac{a}{b}, 1 - \frac{\lambda a}{b}\right\}$$

$$= \{a, b\} \cdot \{-1, b\}^{-1} \cdot \left\{\lambda, 1 - \frac{\lambda a}{b}\right\}^{-1}$$

$$= \{a, b\} \cdot \{-1, b\}^{-1} \{\lambda, b - \lambda a\} \cdot \{\lambda, b\}.$$

Thus it suffices to prove that $\{\frac{a}{b}, b - \lambda a\}$ is a product of symbols of the specified type. If $aR = \mathcal{P} \cap \mathcal{Q}_1$, $bR = \mathcal{P} \cap \mathcal{Q}_2$, then choose $c \in \mathcal{Q}_2$ such that $R/cR$ is reduced, and $a, c$ as well as $b - \lambda a, c$ are relatively prime (this can be done since $\mathcal{P}$ is the only common divisor of $b, b - \lambda a$, so that the primes occuring in the primary decomposition of $\mathcal{Q}_2$ do not divide $b - \lambda a$). If $cR = \mathcal{Q}_2 \cap \mathcal{Q}_3$ then $d = \frac{ac}{b} \in R$, and $acR = \mathcal{P} \cap \mathcal{Q}_1 \cap \mathcal{Q}_2 \cap \mathcal{Q}_3$, so that $dR = \mathcal{Q}_1 \cap \mathcal{Q}_3$. Thus $d, b - \lambda a$ are relatively prime, $R/dR$ is reduced, and

$$\left\{\frac{a}{b}, b - \lambda a\right\} = \{d, b - \lambda a\} \cdot \{c, b - \lambda a\}^{-1}$$

is an expression of the desired form.

**Lemma (9.12).** *Let $S$ be a discrete valuation ring with quotient field $K$ and residue field $k$. Then in the localization sequence*

$$K_2(S) \longrightarrow K_2(K) \xrightarrow{\partial_S} K_1(k)$$
$$\parallel$$
$$k^*$$

the boundary map is given by $\partial_S(\alpha) = T_S^{\pm 1}$ (for universal choices of the signs), where $T_S$ is the tame symbol. (Here the group law in $k^*$ is written multiplicatively.)

**Proof.** Let $\Pi \in S$ be the local parameter. Using bilinearity of the Steinberg symbol and the identity $\{\Pi, -\Pi\} = 1$ in $K_2(K)$, we reduce to checking that
$$\partial_S\{\Pi, a\} = \bar{a}^{\pm 1}$$
for any $a \in R^*$, where $a \mapsto \bar{a} \in k^*$, and the sign of the exponent is universal.

We have a diagram of rings

$$\begin{array}{ccc} S & \longrightarrow & K \\ \uparrow & & \uparrow \\ \mathbb{Z}[u, u^{-1}, t] & \longrightarrow & \mathbb{Z}[u, u^{-1}, t, t^{-1}] \end{array}$$

where the vertical maps are given by $u \mapsto a$, $t \mapsto \Pi$. Let $\mathcal{M}(S)$ be the abelian category of finite $S$-modules, $\mathcal{M}^1(S)$ the Serre subcategory of torsion modules, and $\mathcal{M}(K)$ the category of finite dimensional $K$-vector spaces, so that there is a natural equivalence of categories $\mathcal{M}(S)/\mathcal{M}^1(S) \to \mathcal{M}(K)$. The map $\partial_S$ is the boundary map in the resulting localization sequence.

Let $\mathcal{M}(\mathbb{Z}[u, u^{-1}, t])$ be the abelian category of finite $\mathbb{Z}[u, u^{-1}, t]$-modules, $\mathcal{M}^1(\mathbb{Z}[u, u^{-1}, t])$ the Serre subcategory of modules on which $t$ acts nilpotently; then the quotient is naturally equivalent to $\mathcal{M}(\mathbb{Z}[u, u^{-1}, t, t^{-1}])$, the abelian category of finite $\mathbb{Z}[u, u^{-1}, t, r^{-1}]$-modules. Let
$$\mathcal{M}_*(\mathbb{Z}[u, u^{-1}, t]) \subset \mathcal{M}(\mathbb{Z}[u, u^{-1}, t])$$
be the full subcategory of modules $M$ satisfying $\mathrm{Tor}_i(M, S) = 0$ for $i > 0$; similarly define full subcategories
$$\mathcal{M}_*^1(\mathbb{Z}[u, u^{-1}, t]) \subset \mathcal{M}^1(\mathbb{Z}[u, u^{-1}, t])$$
and
$$\mathcal{M}_* = (\mathbb{Z}[u, u^{-1}t, t^{-1}]) \subset \mathcal{M}(\mathbb{Z}[u, u^{-1}, t, t^{-1}]).$$
Then we have diagrams (where arrows marked $\cong$ are homotopy equivalences, by the resolution theorem)

$$\begin{array}{ccccc} \mathcal{M}^1(S) & \longrightarrow & \mathcal{M}(S) & \longrightarrow & \mathcal{M}(K) \\ \uparrow & & \uparrow & & \uparrow \\ \mathcal{M}_*^1(\mathbb{Z}[u, u^{-1}, t]) & \longrightarrow & \mathcal{M}_*(\mathbb{Z}[u, u^{-1}, t]) & \longrightarrow & \mathcal{M}_*(\mathbb{Z}[u, u^{-1}, t, t^{-1}]) \\ \cong \downarrow & & \cong \downarrow & & \cong \downarrow \\ \mathcal{M}^1(\mathbb{Z}[u, u^{-1}, t]) & \longrightarrow & \mathcal{M}(\mathbb{Z}[u, u^{-1}, t]) & \longrightarrow & \mathcal{M}(\mathbb{Z}[u, u^{-1}, t, t^{-1}]) \end{array}$$

$$\begin{array}{ccccccc}
\longrightarrow & K_2(S) & \longrightarrow & K_2(K) & \longrightarrow & K_1(k) & \longrightarrow \\
& \uparrow & & \phi\uparrow & & \psi\uparrow & \\
& K_2(\mathbb{Z}[u,u^{-1},t]) & \xrightarrow{\alpha} & K_2(\mathbb{Z}[u,u^{-1},t,t^{-1}]) & \xrightarrow{\beta} & K_1(\mathbb{Z}[u,u^{-1}]) & \longrightarrow
\end{array}$$

$$\begin{array}{ccc}
\longrightarrow & K_1(s) & \longrightarrow \\
& \uparrow & \\
\longrightarrow & K_1(\mathbb{Z}[t,u,u^{-1}]) &
\end{array}$$

where $\phi(\{t,u\}) = \{\Pi, a\}$, $\psi(u) = \bar{a}$. Hence it suffices to prove that $\beta(\{t,u\}) = u^{\pm 1}$. Now by Theorem (5.2),

$$K_2(\mathbb{Z}[u,u^{-1},t]) \cong K_2(\mathbb{Z}[u,u^{-1}])$$
$$K_2(\mathbb{Z}[u,u^{-1},t,t^{-1}]) \cong K_2(\mathbb{Z}[u,u^{-1}]) \oplus K_1(\mathbb{Z}[u,u^{-1}]),$$

and the second isomorphism is obtained from the facts that $\alpha$ is a split inclusion, and $\beta$ a surjection. Now

$$K_1(\mathbb{Z}[u,u^{-1}]) \cong \mathbb{Z} \oplus \mathbb{Z}/2\mathbb{Z},$$

where a typical element has the form $\pm u^{\pm n}$, $n \in \mathbb{Z}$. Further, by considering $\mathbb{Z}[v,v^{-1},t] \longrightarrow \mathbb{Z}[u,u^{-1},t]$, $v \mapsto u^n$, we see that $\beta\{t,u\} = h(u) \in \mathbb{Z}[u,u^{-1}]^*$ satisfies $h(u^n) = h(u)^n$, for any $n \geq 1$; hence $h(u) = u^m$ for some $m \in \mathbb{Z}$. We claim $m = \pm 1$, which will prove the lemma. If not, let $p$ be a prime divisor of $m$, and let $\ell$ be a prime such that $\ell \equiv 1 \pmod{p}$. Then the finite field $\mathbb{F}_\ell$ contains a primitive $p$th-root of unity $\zeta \in \mathbb{F}_\ell^*$; consider the homomorphism of rings $\mathbb{Z}[u,u^{-1}] \longrightarrow \mathbb{F}_\ell$ given by $u \mapsto \zeta$. We have a diagram of localization sequences

$$\begin{array}{ccccccc}
K_2(\mathbb{F}_\ell[t]) & \longrightarrow & K_2(\mathbb{F}_\ell[t,t^{-1}]) & \longrightarrow & K_1(\mathbb{F}_\ell) & \longrightarrow \\
\uparrow & & \phi'\uparrow & & \psi'\uparrow & \\
K_2(\mathbb{Z}[u,u^{-1},t]) & \xrightarrow{\alpha} & K_2(\mathbb{Z}[u,u^{-1},t,t^{-1}]) & \xrightarrow{\beta} & K_1(\mathbb{Z}[u,u^{-1}]) & \longrightarrow
\end{array}$$

where $\phi'(\{t,u\}) = \{t,\zeta\}$. Now $K_2(\mathbb{F}_\ell[t]) \cong K_2(\mathbb{F}_\ell) = 0$ (see Example (1.19)). Also $\psi' \circ \beta\{t,u\} = \psi'(u^m) = \zeta^m = 1$. Hence $\{t,\zeta\} \in K_2(\mathbb{F}_\ell[t,t^{-1}])$ must vanish. But if

$$T: K_2(\mathbb{F}_\ell(t)) \longrightarrow \mathbb{F}_\ell^*$$

is the tame symbol associated to the valuation corresponding to the discrete valuation ring $\mathbb{F}_\ell[t]_{(t)}$, then $T(\{t,\zeta\}) = \zeta \in \mathbb{F}_\ell^*$ is non-trivial. This contradiction proves the lemma.

**Remark.** One can show that $\partial_S = T_S$ by a more natural, but somewhat less elementary argument; see Grayson, "Localization for Flat Modules in Algebraic $K$-Theory," *J. Alg.* 61 (1979), 463–496.

**Theorem (9.13).** *There is a presentation*
$$K_2(K) \xrightarrow{T} \bigoplus_{ht\,\mathcal{P}=1} k(\mathcal{P})^* \xrightarrow{\partial} K_0(\mathcal{C}_R) \longrightarrow 0.$$
*Equivalently, if $U = \operatorname{Spec} R - \{\mathcal{M}\}$ is the punctured spectrum, then $\partial$ gives an isomorphism $H^1(U, \mathcal{K}_{2,U}) \cong K_0(\mathcal{C}_R)$.*

Since $U$ is smooth over $k$, Gersten's conjecture holds for the local rings of $U$, and so (from (9.12)).
$$H^1(U, \mathcal{K}_{2,U}) \cong \operatorname{coker} T.$$
Hence the two formulations are equivalent. We already have one presentation
$$K_1(\mathcal{P}^1(R)) \xrightarrow{\eta} \bigoplus_{ht\,\mathcal{P}=1} k(\mathcal{P})^* \xrightarrow{\partial} K_0(\mathcal{C}_R) \longrightarrow 0$$
where $\mathcal{P}^1(R)$ is the category of (torsion) $R$-modules of projective dimension 1 with reduced support. We have also verified that $\partial \circ T = 0$. So it suffices to prove that image $\eta \subset$ image $T$. We have a localization sequence ($\dim U = 1$)
$$K_2(U) \longrightarrow K_2(K) \xrightarrow{T} \bigoplus_{ht\,\mathcal{P}=1} k(\mathcal{P})^* \longrightarrow K_1(U) \longrightarrow K_1(K)$$
(where the boundary map can be identified with $T$ by (9.12)). Hence it suffices to prove that the composite
$$K_1(\mathcal{P}^1(R)) \longrightarrow \bigoplus_{ht\,\mathcal{P}=1} k(\mathcal{P})^* \longrightarrow K_1(U)$$
is 0. Now $\mathcal{P}^1(R) \subset \mathcal{H}(R)$, the category of all finite $R$-modules of finite projective dimension, and if $\mathcal{P}(R)$ is the category of finite projective $R$-modules, then $\mathcal{P}(R) \longrightarrow \mathcal{H}(R)$ is a homotopy equivalence by the resolution theorem. The map
$$K_1(\mathcal{P}^1(R)) \longrightarrow K_1(U)$$
is clearly given by the restriction functor $\mathcal{P}^1(R) \longrightarrow \mathcal{M}(U)$, and hence factors through $\mathcal{H}(R)$; thus we have a triangle

$$\begin{array}{ccc} K_1(\mathcal{P}^1(R)) & \longrightarrow & K_1(U) \\ & \searrow \quad \nearrow & \\ & K_1(R) & \end{array}$$

The composite $K_1(\mathcal{P}^1(R)) \longrightarrow K_1(U) \longrightarrow K_1(K)$ is clearly 0, since the composite restriction functor $\mathcal{P}^1(R) \longrightarrow \mathcal{P}(K)$ maps every object to a

0-object. Since
$$K_1(U) \longrightarrow K_1(K)$$
$$\searrow \quad \nearrow$$
$$K_1(R)$$
commutes, and $K_1(R) \cong R^* \hookrightarrow K^* \cong K_1(K)$, we see that $K_1(\mathcal{P}^1(R)) \longrightarrow K_1(R)$ is 0, and hence $K_1(\mathcal{P}^1(R)) \longrightarrow K_1(U)$ is 0, as desired. This proves (9.13).

Let $f : Z \longrightarrow \operatorname{Spec} R$ be a resolution of singularities, $E$ the reduced exceptional divisor. Then $Z - E \cong U$; hence we have a localization sequence
$$G_1(E) \longrightarrow K_1(Z) \longrightarrow K_1(U) \longrightarrow G_0(E) \longrightarrow K_0(Z).$$

Since $R$ is the local ring of a point on an algebraic surface $X/k$, the local rings of $Z$ are regular local rings which are essentially of finite type over $k$, and hence satisfy Gersten's conjecture. Since $\dim Z = 2$, if we set $SK_1(Z) = F^1 K_1(Z)$, then
$$K_1(Z) \cong \Gamma(Z, \mathcal{O}_Z^*) \oplus SK_1(Z) \cong R^* \oplus SK_1(Z),$$
and we have an exact sequence
$$H^2(Z, \mathcal{K}_{3,Z}) \longrightarrow SK_1(Z) \longrightarrow H^1(Z, \mathcal{K}_{2,Z}) \longrightarrow 0,$$
from the $BGQ$-spectral sequence. Here
$$\operatorname{image}(H^2(Z, \mathcal{K}_{3,Z}) \longrightarrow SK_1(Z)) = F^2 K_1(Z)$$
$$= \operatorname{image}\Big(\bigoplus_{x \in Z^2} k(x)^* \longrightarrow K_1(Z)\Big) \subset \operatorname{image}(G_1(E) \longrightarrow K_1(Z)).$$
Since all closed points of $Z$ lie on $E$, this subgroup lies in the image of $G_1(E) \longrightarrow K_1(Z)$. Thus, if $F_0 G_0(E) \subset G_0(E)$ is the subgroup generated by the classes of closed points, we have an exact sequence
$$0 \longrightarrow H^1(Z, \mathcal{K}_{2,Z})/N \longrightarrow SK_1(U) \longrightarrow F_0 G_0(E) \longrightarrow 0,$$
where $N = \operatorname{image}(G_1(E) \longrightarrow SK_1(Z) \twoheadrightarrow H^1(Z, \mathcal{K}_{2,Z}))$. More explicitly, $H^1(Z, \mathcal{K}_{2,Z})$ is $H^1$ of the Gersten complex
$$0 \longrightarrow K_2(K) \longrightarrow \bigoplus_{x \in Z^1} k(x)^* \xrightarrow{\partial} \bigoplus_{x \in Z^2} \mathbb{Z} \longrightarrow 0;$$
then $N$ is the subgroup generated by
$$\ker\Big(\oplus k(E_i)^* \xrightarrow{\partial} \bigoplus_{x \in Z^2} \mathbb{Z}\Big)$$
where $E_i$ runs over the components of the exceptional divisor $E$. Note that $SK_1(U) \cong H^1(U, \mathcal{K}_{2,U})$ since $\dim U = 1$.

## 9. Localization for Singular Varieties

We now relate these remarks to the Chow group of 0-cycles. We give an ad hoc treatment for the case when the surface has one singular point; a more systematic approach can be found in Levine's paper (loc. cit.), and the sources cited there.

Let $X$ be a normal quasi-projective surface. The Chow group of 0-cycles $CH^2(X)$ is defined by

$$CH^2(X) = \frac{\text{Free abelian group on smooth point of } X}{\langle (f)_C \mid C \subset X \text{ a curve, } C \cap X_{\text{sing}} = \emptyset, f \in k(C)^* \rangle}.$$

Suppose $X_{\text{sing}} = \{P\}$, and let $R = \mathcal{O}_{P,X}$. Let $\pi : Y \longrightarrow X$ be a resolution of singularities, $\pi^{-1}(P) = E$ the reduced exceptional divisor. Then $Z = Y \times_X \operatorname{Spec} R$ is a resolution of singularities of $\operatorname{Spec} R$. Let $K = k(X)$ be the function field of $X$.

Let $(\alpha)$, $(\beta)$, $(\gamma)$ denote the following complexes:

$(\alpha):$ $\quad 0 \longrightarrow 0 \longrightarrow \bigoplus_{\substack{C \subset X \\ P \notin C}} k(C)^* \xrightarrow{\partial} \bigoplus_{Q \in X - \{P\}} \mathbb{Z} \longrightarrow 0$

$(\beta):$ $\quad 0 \longrightarrow K_2(K) \xrightarrow{T} \bigoplus_{D \subset Y} k(D)^* \xrightarrow{\partial} \bigoplus_{Q \in Y} \mathbb{Z} \longrightarrow 0$

$(\gamma):$ $\quad 0 \longrightarrow K_2(K) \xrightarrow{T} \bigoplus_{\substack{D \subset Y \\ D \cap E \neq \emptyset}} k(D)^* \xrightarrow{\partial} \bigoplus_{Q \in E} \mathbb{Z} \longrightarrow 0$

where in $(\alpha)$, $C$ runs over the irreducible curves in $X$ which do not pass through $P$, and in $(\beta)$, $(\gamma)$ $D$ runs over appropriate sets of irreducible curves in $Y$ (namely, all curves, in $(\beta)$, and curves meeting $E$, in $(\gamma)$). The map $T$ is the sum of tame symbol maps in $(\beta)$, $(\gamma)$, and $\partial$ is the divisor map on rational functions in all three complexes. We have an exact sequence of complexes

$$0 \longrightarrow (\alpha) \longrightarrow (\beta) \longrightarrow (\gamma) \longrightarrow 0.$$

Since $Y, Z$ are regular $k$-schemes whose local rings satisfy Gersten's conjecture, $(\beta)$, $(\gamma)$ compute the cohomology groups $H^i(Y, \mathcal{K}_{2,Y})$ and $H^i(Z, \mathcal{K}_{2,Z})$ respectively. We have a long exact cohomology sequence

$$H^1(Y, \mathcal{K}_{2,Y}) \to H^1(Z, \mathcal{K}_{2,Z}) \to CH^2(X) \to CH^2(Y) \to 0$$
$$\parallel$$
$$H^2(Y, \mathcal{K}_{2,Y}).$$

If $N \subset H^1(Z, \mathcal{K}_{2,Z})$ is the subgroup defined earlier, then clearly $N$ lies in the image of $H^1(Y, \mathcal{K}_{2,Y})$, since we have a commutative diagram

$$\bigoplus k(E_i)^* \xrightarrow{\partial} \bigoplus_{Q \in E} \mathbb{Z}$$

$$\bigoplus_{D \subset Y} k(D)^* \xrightarrow{\partial} \bigoplus_{Q \in Y} \mathbb{Z}$$

so that the kernel of the top row maps to $H^1((\beta)) = H^1(Y, \mathcal{K}_{2,Y})$ such that the composite map to $H^1(Z, \mathcal{K}_{2,Z})$ has image $N$. Let $N' \subset H^1(Y, \mathcal{K}_{2,Y})$ be the subgroup defined by the kernel of the top row; then we have an exact sequence

$$H^1(Y, \mathcal{K}_{2,Y})/N' \longrightarrow H^1(Z, \mathcal{K}_{2,Z})/N \longrightarrow CH^2(X) \longrightarrow CH^2(Y) \longrightarrow 0.$$

Let

$$SK_0(\mathcal{C}_R) = \ker(K_0(\mathcal{C}_R) \longrightarrow F_0 K_0(E))$$
$$\cong H^1(Z, \mathcal{K}_{2,Z})/N.$$

Then we can rewrite the above sequence as

$$H^1(Y, \mathcal{K}_{2,Y})/N' \longrightarrow SK_0(\mathcal{C}_R) \longrightarrow CH^2(X) \longrightarrow CH^2(Y) \longrightarrow 0$$

where $SK_0(\mathcal{C}_R)$ is an *analytic invariant*, i.e., depends only on the completion of $R$, and $H^1(Y, \mathcal{K}_{2,Y})/N'$ depends on the given algebraic local ring.

**Lemma (9.14).** *Let $Y$ be a smooth, quasi-projective surface over a field $k$, $x \in Y$ a $k$-rational point, $\tilde{Y} \longrightarrow Y$ the blow up at $x$, $E \subset Y$ the exceptional divisor. Then $H^1(\tilde{Y}, \mathcal{K}_{2,\tilde{Y}}) \cong H^1(Y, \mathcal{K}_{2,Y}) \oplus k^*$, where the summand $k^*$ is the image of the natural covariant map $H^0(E, \mathcal{K}_{1,E}) \cong k^* \longrightarrow H^1(Y, \mathcal{K}_{2,\tilde{Y}})$.*

**Proof.** Let $K = k(Y)$ be the function field of $Y$. Consider the complexes

$$(\alpha'): \quad 0 \longrightarrow K_2(K) \xrightarrow{T} \bigoplus_{C \subset Y} k(C)^* \longrightarrow \bigoplus_{Q \in Y} \mathbb{Z} \longrightarrow 0$$

$$(\beta'): \quad 0 \longrightarrow K_2(K) \xrightarrow{T} \bigoplus_{D \subset \tilde{Y}} k(D)^* \longrightarrow \bigoplus_{Q \in \tilde{Y}} \mathbb{Z} \longrightarrow 0$$

$$(\gamma'): \quad 0 \longrightarrow 0 \longrightarrow k(E)^* \longrightarrow \bigoplus_{Q \in E}^{0} \mathbb{Z} \longrightarrow 0$$

where $\bigoplus_{Q \in E}^{0} \mathbb{Z}$ is the group of 0-cycles of degree 0 on $E \cong \mathbb{P}^1$. We can identify $(\gamma')$ with a subcomplex of $(\beta')$, giving an exact sequence of complexes

$$0 \longrightarrow (\gamma') \longrightarrow (\beta') \longrightarrow (\alpha') \longrightarrow 0,$$

which yields an exact sequence (since $H^1((\gamma')) \cong k^* \cong H^0(E, \mathcal{K}_{1,E})$)

$$0 \longrightarrow H^0(\tilde{Y}, \mathcal{K}_{2,\tilde{Y}}) \xrightarrow{\pi_*} H^0(Y, \mathcal{K}_{2,Y}) \longrightarrow k^*$$
$$\longrightarrow H^1(\tilde{Y}, \mathcal{K}_{2,\tilde{Y}}) \xrightarrow{\pi_*} H^1(Y, \mathcal{K}_{2,Y}) \longrightarrow 0.$$

One checks immediately that the natural maps $\pi^* : H^i(Y, \mathcal{K}_{2,Y}) \longrightarrow H^i(\tilde{Y}, \mathcal{K}_{2,\tilde{Y}})$ give right inverses to the maps $\pi_*$. Hence we have a split exact sequence

$$0 \longrightarrow k^* \longrightarrow H^1(\tilde{Y}, \mathcal{K}_{2,\tilde{Y}}) \longrightarrow H^1(Y, \mathcal{K}_{2,Y}) \longrightarrow 0.$$

We note that the map $k^* \longrightarrow H^1(\tilde{Y}, \mathcal{K}_{2,\tilde{Y}})$ is induced by

$$k^* \subset k(E)^* \subset \bigoplus_{D \subset \tilde{Y}} k(D)^*.$$

This proves (9.14).

For any smooth variety $Y/k$, there is a natural map $(\operatorname{Pic} Y) \otimes_\mathbb{Z} k^* \longrightarrow H^1(Y, \mathcal{K}_{2,Y})$, given as follows: $H^1(Y, \mathcal{K}_{2,Y})$ is $H^1$ of the complex

$$0 \longrightarrow K_2(k(Y)) \longrightarrow \bigoplus_{x \in Y^1} k(x)^* \longrightarrow \bigoplus_{x \in Y^2} \mathbb{Z} \longrightarrow 0.$$

Let $\operatorname{Div} Y$ be the group of divisors on $Y$. Then

$$\bigoplus_{x \in Y^1} k^* \cong (\operatorname{Div} Y) \otimes_\mathbb{Z} k^* \subset \bigoplus_{x \in y^1} k(x)^*,$$

and clearly $\partial((\operatorname{Div} Y) \otimes_\mathbb{Z} k^*) = 0$. Hence we have an induced map $(\operatorname{Div} Y) \otimes_\mathbb{Z} k^* \longrightarrow H^1(Y, \mathcal{K}_{2,Y})$. We claim this factors through the quotient map $(\operatorname{Div} Y) \otimes_\mathbb{Z} k^* \twoheadrightarrow (\operatorname{Pic} Y) \otimes_\mathbb{Z} k^*$. To see this, consider the diagram

$$\begin{array}{ccc} K_2(k(Y)) & \xrightarrow{T} & \bigoplus_{x \in Y^1} k(x)^* \\ \{,\}\uparrow & & \uparrow \\ k(Y)^* \otimes_\mathbb{Z} k^* & \xrightarrow{\partial_Y \otimes 1} & \bigoplus_{x \in Y^1} k^* \cong (\operatorname{Div} Y) \otimes_\mathbb{Z} k^* \end{array}$$

where $\partial_Y : k(Y)^* \longrightarrow \operatorname{Div} Y$ is the divisor map on rational functions. Since $(\operatorname{coker} \partial_Y) = \operatorname{Pic} Y$, this proves the claim.

**Corollary (9.15).** *Let $Y$ be a smooth, projective surface over an algebraically closed field which is birational to a ruled surface. Then the natural map $(\operatorname{Pic} Y) \otimes_\mathbb{Z} k^* \longrightarrow H^1(Y, \mathcal{K}_{2,Y})$ is onto.*

**Proof.** This is true if $Y = C \times \mathbb{P}^1$, where $C$ is a smooth curve, since $K_1(C) \otimes K_0(Y) \longrightarrow K_1(Y)$ is onto, from (5.18), and since
$$(\text{Pic } C) \otimes_{\mathbb{Z}} k^* \twoheadrightarrow H^1(C, \mathcal{K}_{2,C}) \cong SK_1(C)$$
(as $\bigoplus_{x \in C^1} k(x)^* \cong \bigoplus_{x \in C^1} k^*$). But by (9.14),
$$\text{coker}((\text{Pic } Y) \otimes_{\mathbb{Z}} k^* \longrightarrow H^1(Y, \mathcal{K}_{2,Y}))$$
is unchanged under the blow up of a point on a smooth surface; hence it is a birational invariant. This proves the corollary.

We now return to our situation of a normal surface singularity. Let $X/k$ be a normal projective surface with a unique singular point $P \in X$, and let $\pi : Y \longrightarrow X$ be a resolution of singularities, $\pi^{-1}(P) = E$ the reduced exceptional divisor, $R = \mathcal{O}_{P,X}$, $Z = Y \times_X \text{Spec } R$. We assume below that $k$ is algebraically closed.

**Proposition (9.16).** *Under the above conditions, let $Y$ be birationally ruled, and suppose that $\pi^* : CH^2(X) \longrightarrow CH^2(Y)$ is an isomorphism. Then we have an exact sequence*
$$C\ell(R) \otimes_{\mathbb{Z}} k^* \longrightarrow K_0(\mathcal{C}_R) \longrightarrow F_0 G_0(E) \longrightarrow 0$$
*where $C\ell(R)$ is the ideal class group.*

**Proof.** The exact sequence (constructed earlier)
$$H^1(Y, \mathcal{K}_{2,Y})/N' \longrightarrow SK_0(\mathcal{C}_R) \longrightarrow CH^2(X) \xrightarrow{\pi^*} CH^2(Y) \longrightarrow 0$$
and the surjection (9.15)
$$\text{Pic } Y \otimes_{\mathbb{Z}} k^* \twoheadrightarrow H^1(Y, \mathcal{K}_{2,Y})$$
give a surjection
$$\text{Pic } Y \otimes_{\mathbb{Z}} k^* \twoheadrightarrow SK_0(\mathcal{C}_R).$$
This clearly factors as

$$\begin{array}{ccc} \text{Pic } Y \otimes_{\mathbb{Z}} k^* & \longrightarrow & \text{Pic } Z \otimes_{\mathbb{Z}} k^* \\ \downarrow & & \downarrow \\ H^1(Y, \mathcal{K}_{2,Y})/N' & \twoheadrightarrow & H^1(Z, \mathcal{K}_{2,Z})/N \cong SK_0(\mathcal{C}_R) \end{array}$$

(where $\text{Pic } Z \otimes_{\mathbb{Z}} k^* \longrightarrow H^1(Z, \mathcal{K}_{2,Z})$ is analogous to the left hand vertical arrow). One sees immediately that if $E_1, \ldots, E_n$ are the irreducible components of $E$, then
$$\left( \sum_{1 \leq i \leq n} \mathbb{Z}[E_i] \right) \otimes_{\mathbb{Z}} k^* \subset (\text{Pic } Z) \otimes_{\mathbb{Z}} k^*$$

## 9. Localization for Singular Varieties

maps to $N \subset H^1(Z, \mathcal{K}_{2,Z})$, so that we have a surjection

$$\left(\operatorname{Pic} Z / \left(\sum_{1 \leq i \leq n} \mathbb{Z}[E_i]\right)\right) \otimes_{\mathbb{Z}} k^* \twoheadrightarrow SK_0(\mathcal{C}_R).$$

But $(\operatorname{Pic} Z)/(\sum_{1 \leq i \leq n} \mathbb{Z}[E_i]) \cong \operatorname{Pic}(Z - E) \cong C\ell(R)$. This proves (9.16).

**Theorem (9.17).** *Let $G$ be a finite group acting as a group of linear transformations on a two-dimensional $k$-vector space. Let $R = k[[x,y]]^G$ be the ring of invariants for the induced action on $k[[x,y]]$ via linear substitutions. Then $K_0(\mathcal{C}_R) = \mathbb{Z}$. (In short, $K_0(\mathcal{C}_R) = \mathbb{Z}$ if $R$ has a two-dimensional quotient singularity).*

**Proof.** Let $X = \operatorname{Spec}(k[x,y]^G)$ for the action of $G$ on $\mathbb{A}^2$ induced by the given two-dimensional linear representation, and let $P \in X$ be the image of the origin under the natural quotient map $f : \mathbb{A}^2 \longrightarrow X$. Then $X$ is a normal affine surface, and $P \in X$ is the unique singular point (since $k[x,y]^G$ is a two-dimensional normal graded ring). Let $\bar{X} \supset X$ be a projective surface with $\bar{X}_{\text{sing}} = \{P\}$, and let $\pi : \bar{Y} \longrightarrow \bar{X}$ be a resolution of singularities. We then have a diagram

$$\begin{array}{ccc} V & \xrightarrow{h} & W \\ {\scriptstyle g}\downarrow & & \downarrow{\scriptstyle f} \\ \bar{Y} & \xrightarrow{\pi} & \bar{X} \end{array}$$

where $V, W$ are smooth, projective rational surfaces, $h$ is birational, $W \times_{\bar{X}} X = \mathbb{A}^2$ (i.e., $f^{-1}(X) = \mathbb{A}^2$). Thus $W$ is a smooth, projective surface containing $\mathbb{A}^2$ as an open set, and $V$ is a resolution of singularities of $\bar{Y} \times_{\bar{X}} W$. We have an induced diagram of maps of Chow groups of 0-cycles

$$\begin{array}{ccc} \mathbb{Z} \cong CH^2(V) & \xleftarrow[\cong]{h^*} & CH^2(W) \cong \mathbb{Z} \\ {\scriptstyle g^*}\uparrow & & \uparrow{\scriptstyle f^*} \\ CH^2(\bar{Y}) & \xleftarrow{\pi^*} & CH^2(\bar{X}) \end{array}$$

There are also natural transfer maps $f_* : CH^2(W) \longrightarrow CH^2(\bar{X})$, $g_* : CH^2(V) \longrightarrow CH^2(\bar{Y})$ such that $f_* \circ f^*$, $g_* \circ g^*$ both equal multiplication by $d = |G| = \deg f = \deg g$. This is clear for $g$ as $V, \bar{Y}$ are smooth, and is also true for $f$ because $f^{-1}(P) = \{0\}$ consists of only the origin in $\mathbb{A}^2 \subset W$; now to construct $f_*$ we use the isomorphism

$$CH^2(W) \cong \frac{\text{Free abelian group on points of } W - \{0\}}{\langle (f_1)_C \mid C \subset W \text{ curve}, 0 \notin C, \text{ and } f_1 \in k(C)^* \rangle}.$$

This isomorphism follows from the diagram below, whose rows are complexes, and columns are exact,

$$
\begin{array}{ccccccccc}
0 & \to & 0 & \to & \bigoplus_{\substack{x \in W^1 \\ 0 \notin \{\bar{x}\}}} k(x)^* & \to & \bigoplus_{x \in W^2 - \{0\}} \mathbb{Z} & \to & 0 \\
& & \downarrow & & \downarrow & & \downarrow & & \\
0 & \to & K_2(k(W)) & \to & \bigoplus_{x \in W^1} k(x)^* & \to & \bigoplus_{x \in W^2} \mathbb{Z} & \to & 0 \\
& & \downarrow & & \downarrow & & \downarrow & & \\
0 & \to & K_2(k(W)) & \to & \bigoplus_{x \in (\text{Spec } S)^1} k(x)^* & \to & \mathbb{Z} & \to & 0
\end{array}
$$

where $S = \mathcal{O}_{0,W}$, and the facts that $H^i(\text{Spec } S, \mathcal{K}_2) = 0$ for $i = 1, 2$ (the middle row computes $H^i(W, \mathcal{K}_2)$, and the bottom row $H^i(\text{Spec } S, \mathcal{K}_2)$). Thus the kernel of the degree map $\deg : CH^2(\bar{X}) \longrightarrow \mathbb{Z}$ is annihilated by $d$. But $\ker(CH^2(\bar{X}) \longrightarrow \mathbb{Z})$ is divisible, since it is a quotient of a direct sum of jacobians of smooth curves. Hence $CH^2(\bar{X}) \cong \mathbb{Z} \cong CH^2(\bar{Y})$. Now the function field $k(X) = k(x,y)^G$ so that $k(x,y)/k(X)$ is separable. Hence $X$ is a rational surface (see Hartshorne, *Algebraic Geometry*, Ch. V, Remark (6.2.1)). Finally, $C\ell(R)$ is finite, since it has exponent $d$, and is finitely generated (being a quotient of Pic $\bar{Y}$, where $\bar{Y}$ is a smooth rational surface). Thus $C\ell(R) \otimes_\mathbb{Z} k^* = 0$, since $k$ is algebraically closed, and hence $k^*$ is divisible. Hence $K_0(\mathcal{C}_R) \cong F_0 G_0(E) \cong \mathbb{Z}$ since $E$ is a connected tree of rational curves. It remains to note that the composite

$$
\begin{array}{ccccccc}
K_0(\mathcal{C}_R) & \longrightarrow & F_0 G_0(E) & \longrightarrow & F_0 K_0(\bar{Y}) & \xrightarrow{\deg} & \mathbb{Z} \\
& & \| & & \| & & \\
& & \mathbb{Z} & & CH^2(\bar{Y}) & &
\end{array}
$$

is just the length map $K_0(\mathcal{C}_R) \longrightarrow \mathbb{Z}$, and $F_0 G_0(E) \longrightarrow F_0 K_0(\bar{Y}) \longrightarrow \mathbb{Z}$ is an isomorphism; hence $K_0(\mathcal{C}_R) \cong \mathbb{Z}$ given by length.

**Corollary (9.8).** *Let $X$ be a normal quasi-projective surface with only quotient singularities, i.e., for each $P \in X_{\text{sing}}$, $\hat{\mathcal{O}}_{P,X} = R$ is as in (9.17). Let $\pi : Y \longrightarrow X$ be a resolution of singularities. Then $\pi^* : CH^2(X) \longrightarrow CH^2(Y)$ is an isomorphism.*

**Proof.** If $\pi : X_P \longrightarrow X$ is a resolution of the singularity at a given point $P \in X_{\text{sing}}$ (so that $X_P - \pi^{-1}(P) \cong X - \{P\}$, and $\pi^{-1}(P) \cap (X_P)_{\text{sing}} = \emptyset$), then one shows that there is an exact sequence $SK_0(\mathcal{C}_R) \longrightarrow CH^2(X) \longrightarrow CH^2(X_p) \longrightarrow 0$. (See Levine's paper, cited earlier, for details; see also

## 9. Localization for Singular Varieties

V. Srinivas, "Zero Cycles on a Singular Surface I," *Crelle's J.* 359 (1985), pg. 97, Prop. 4). This immediately implies the corollary, from (9.17).

There has been further recent progress on the topic of localization for singular varieties. Marc Levine has generalized his results on surfaces, proved above, to arbitrary dimensional isolated Cohen–Macaulay singularities (see M. Levine, "Localization on Singular Varieties, *Invent. Math.* 91 (1988), 423–464). Using the results in this paper, Levine gives a construction of two modules over the local ring $k[[X, Y, Z, W]]/(XY - ZW)$ with negative "intersection multiplicity." This was originally proved by S. Dutta, M. Hochster and J. E. McLaughlin in their paper "Modules of Finite Projective Dimension with Negative Intersection Multiplicity," *Invent. Math.* 79 (1985), 253–291. Levine's results depend on a new (equivalent) definition of $K_i$ given by H. Gillet and D. Grayson, and stronger forms of Theorems A and B proved by them, leading to a generalization of Theorem (9.1).

R. Thomason and T. Trobaugh have recently obtained the "most general" localization theorem. Their results were announced in "Le Theoreme de Localization en $K$-Theorie Algebrique," *C. R. Acad. Sci.* Paris **307** (1988), 829–831. The detailed proofs are in their paper "Higher Algebraic $K$-Theory of Schemes and of Derived Categories," in *The Grothendieck Festschrift* Vol. III, Prog. in Math. Vol. 88 (1990), pp. 247–435.

# Appendix A

# Results from Topology

A general reference (cited below as [W]) is: G.W. Whitehead *Elements of Homotopy Theory*, Grad. Texts, No. 61, Springer-Verlag (1978). Our discussion of homotopy groups, etc. is based on: P.A. Griffiths, J.W. Morgan, *Rational Homotopy Theory and Differential Forms*, Prog. Math. 16, Birkhäuser (1981).

**(A.1) Compactly Generated Spaces.** A topological space $X$ is *compactly generated* if $X$ is Hausdorff, and a subset $A \subset X$ is closed if (and only if) $A \cap K$ is closed, for every compact subset $K \subset X$. The category of compactly generated spaces is a convenient one in which to do algebraic topology (see [W], I, Chapter 4). We will assume that all spaces under consideration are compactly generated, and all categorical constructions of spaces (e.g., products, quotients) are made within this category, unless specified otherwise (however, we do not impose any such conditions on the underlying topological space of a scheme; this exception should cause no confusion).

Clearly any locally compact Hausdorff space is compactly generated. Given any Hausdorff space $X$, let $k(X)$ be the topological space with the same underlying set $X$, and the following topology—a subset $A \subset X$ is closed in $k(X)$ if and only if $A \cap K$ is closed in $K$ for every compact subset $K \subset X$. The identity map of sets gives a continuous map $k(X) \longrightarrow X$, and one verifies the following properties: (i) $k(X)$ is compactly generated, and $k(X) = X$ if and only if $X$ is compactly generated; (ii) $k(X)$ and $X$ have the same compact subsets; more generally, if $Y$ is compactly generated, then composition with the natural map $k(X) \longrightarrow X$ gives a bijection between the continuous maps $Y \longrightarrow k(X)$ and the continuous maps $Y \longrightarrow X$; (iii) as a particular case of (ii), $k(X) \longrightarrow X$ induces isomorphisms on homotopy groups, and on the complexes of singular chairs, hence on homology and cohomology with arbitrary coefficients.

The product in the category of compactly generated spaces is given by $k(X \times Y)$, where $X \times Y$ is the usual product (i.e., has the standard product topology). However, if either $X$ or $Y$ is locally compact then

$X \times Y$ is compactly generated; hence the notion of homotopy is unchanged within this category.

Next, if $X$ is compactly generated and $Y$ is a Hausdorff quotient (with the quotient topology), then $Y$ is compactly generated. If $X$ is a Hausdorff space with an increasing sequence of closed subsets $X_n$ which are compactly generated, and if $X = \varinjlim X_n$ (i.e., a subset $A \subset X$ is closed if and only if $A \cap X_n$ is closed for all $n$; we also say $X$ has the *weak topology* with respect to the collection $\{X_n\}$), then $X$ is compactly generated. In particular, any $CW$-complex (see (A.8) below) is compactly generated.

Any closed subset or open subset of a compactly generated space is compactly generated. If $f : X \longrightarrow Y$ is a function such that $f|_K$ is continuous for every compact set $K \subset X$, and if $X$ is compactly generated, then clearly $f$ is continuous.

Let $X, Y$ be compactly generated spaces, and let $C(X, Y)$ denote the space of continuous maps $X \longrightarrow Y$ with the compact-open topology (a basis of open sets for this topology is given by

$$U(K, V) = \{f \in C(X, Y) \mid f(K) \subset V\},$$

for $K \subset X$ compact, $V \subset Y$ open). Let $\mathbb{F}(X, Y) = k(C(X, Y))$. Then the *evaluation map* $e : k(X \times \mathbb{F}(X, Y)) \longrightarrow Y$ is continuous. This is one pleasant consequence of forming the function space and product in this category.

In the sequel, "space" will mean "compactly generated space" unless specified otherwise; similarly "$X \times Y$" stands for the compactly generated product, etc.

**(A.2) Homotopy Groups.** Let $X$ be a space, with a base point $x \in X$. The *n*th *homotopy group* $\pi_n(X, x)$ is defined to be the set of homotopy classes of maps $f : (I^n, \partial I^n) \longrightarrow (X, x)$ (i.e., maps $f : I^n \longrightarrow X$ with $f(\partial I^n) = x$), where $I^n$ is the unit $n$-cube in $\mathbb{R}^n$, and $\partial I^n$ is its boundary, (for $n = 0$, $I^0$ is a point, and $\partial I^0 = \emptyset$). In fact for $n = 0$, $\pi_0(X, x)$ is only a set with a distinguished point; the elements of $\pi_0(X, x)$ are in bijection with path components of $X$, and the distinguished element corresponds to the path component of $x$.

For $n = 1$, we obtain the fundamental group of homotopy classes of loops based at $x \in X$. For $n \geq 2$, $\pi_n(X, x)$ is an Abelian group, with the group operation given by juxtaposition of $n$-cubes—if $f, g$ are maps $(I^n, \partial I^n) \longrightarrow (X, x)$, and $s_1, \ldots, s_n$ are coordinates on $I^n$, then $[f] + [g] = [h]$ in $\pi_n(X, x)$ where $h : (I^n, \partial I^n) \longrightarrow (X, x)$ is the continuous map defined by

$$h(s_1, \ldots, s_n) = \begin{cases} f(2s_1, s_2, \ldots, s_n) & \text{if } 0 \leq s_1 \leq \tfrac{1}{2} \\ g(2s_1 - 1, s_2, \ldots, s_n) & \text{if } \tfrac{1}{2} \leq s_1 \leq 1. \end{cases}$$

Pictorially, this may be described by

$$\boxed{f} + \boxed{g} = \boxed{f \ g}$$

Commutativity follows immediately (for $n \geq 2$) from the sequence of diagrams ($\sim$ denotes a homotopic map, and $x$ denotes the constant map)

$$\boxed{f \ g} \sim \boxed{\begin{array}{cc} x & g \\ f & x \end{array}} \sim \boxed{\begin{array}{cc} g & x \\ x & f \end{array}} \sim \boxed{g \ f}$$

The fundamental group $\pi_1(X, x)$ acts on the right on $\pi_n(X, x)$, $n \geq 1$, as follows: let $f : (I^n, \partial I^n) \longrightarrow (X, x)$ represent a class in $\pi_n(X, x)$, and let $g : (I, \partial I) \longrightarrow (X, x)$ represent a class in $\pi_1(X, x)$. Define a map $h$ on the subset $(I^n \times \{0\}) \cup ((\partial I^n) \times I)$ of $I^{n+1}$, by

$$\begin{cases} h(s, 0) = f(s), & \text{for } (s, 0) \in I^n \times \{0\} \\ h(s, t) = g(t), & \text{for } (s, t) \in \partial I^n \times I. \end{cases}$$

Since $(I^n \times \{0\}) \cup ((\partial I^n) \times I)$ is a deformation retract of $I^{n+1}$, (as it consists of all faces of $\partial I^{n+1}$ except $I^n \times \{1\}$), we can extend $h$ to a map $h : I^{n+1} \longrightarrow X$. Let $[f]^{[g]} \in \pi_n(X, x)$ denote the class given by $h\big|_{I^n \times \{1\}}$. One verifies that this class is independent of the particular extension $h$, and depends only on the homotopy classes $[f]$, $[g]$; this gives the action of $\pi_1$ on $\pi_n$. If $n = 1$, we can represent this by the picture:

Here, the arrows represent orientations of the intervals; the fourth side represents $[f]^{[g]}$. But the square gives a homotopy of this loop to the loop $g^{-1} \cdot f \cdot g$ (where $\cdot$ is composition of loops). Hence the action of $\pi_1$ on itself is by conjugation.

We can generalize the above to the case of pairs. Let $X$ be a topological space, $A$ a subspace, $x \in A$ a base point. The $n$th *relative homotopy group* (for $n \geq 1$) $\pi_n(X, A; x)$ is defined to be the set of homotopy classes of maps $f : (I^n, \partial I^n, \partial_1 I^n) \longrightarrow (X, A, x)$, where $\partial_1 I^n = (\partial I^{n-1} \times I) \cup (I^{n-1} \times \{0\})$ consists of all of the faces of the $n$-cube $I^n$ except one (thus, $f : I^n \longrightarrow X$ is a map such that $f(\partial I^n) \subset A$, and $f(\partial_1 I^n) = \{x\}$, and the homotopy classes are with respect to homotopies through maps $I^n \longrightarrow X$ of this type). In fact, $\pi_1(X, A, x)$ is only a set with a distinguished point (but we may abuse terminology and call it the first relative homotopy group). If $n \geq 2$, we can again define a group law on $\pi_n(X, A, x)$ by "juxtaposition" of maps, i.e., if

$f, g$ are maps $(I^n, \partial I^n, \partial_1 I^n) \longrightarrow (X, A, x)$ we define $[f] + [g] \in \pi_n(X, A, x)$ to be the class of $h : (I^n, \partial I^n, \partial_1 I^n) \longrightarrow (X, A, x)$ given by

$$h(s_1, \ldots, s_n) = \begin{cases} f(2s_1, \ldots, s_n) & \text{if } 0 \leq s_1 \leq 1/2 \\ g(2s_1 - 1, \ldots, s_n) & \text{if } 1/2 \leq s_1 \leq 1. \end{cases}$$

One easily sees, by an argument analogous to that used for absolute homotopy groups, that $\pi_n(X, A, x)$ is Abelian for $n \geq 3$.

The fundamental group $\pi_1(A, x)$ acts on $\pi_n(X, A, x)$, for $n \geq 1$, as follows—let $f : (I^n, \partial I^n, \partial_1 I^n) \longrightarrow (X, A, x)$ represent a class in $\pi_n(X, A, x)$, and let $g : (I, \partial I) \longrightarrow (A, x)$ represent a class in $\pi_1(A, x)$. We can define a map

$$h : ((\partial_1 I^n) \times I) \cup (I^n \times \{0\}) \longrightarrow X$$

by $h|_{I^n \times \{0\}} = f$, and $h|_{(\partial_1 I^n) \times I} = g \circ p_2$ (where $p_2$ is the second projection). Now $((\partial_1 I^n) \times I) \cup (I^n \times \{0\})$ is a deformation retract of $I^{n+1}$, since it consists of all faces of $I^{n+1}$ except two adjacent ones; further, we can choose such a deformation retraction which maps the face $I^{n-1} \times \{1\} \times I$ into itself at all times, so that the final retraction maps this face into the "top" face $I^{n-1} \times \{1\}$ of $I^n$ (this face is precisely the one omitted in $\partial_1 I^n$). Using this retraction, we can extend $h$ to a map $\tilde{h} : I^{n+1} \longrightarrow X$, which will then satisfy $\tilde{h}(\partial I^n \times \{1\}) \subset A$, $\tilde{h}(\partial_1 I^n \times \{1\}) = \{x\}$. Thus $\tilde{h}|_{I^n \times \{1\}}$ represents a class in $\pi_n(X, A, x)$, which we define to be $[f]^{[g]}$, the result of $[g]$ acting on $[f]$. We leave it to the reader to verify (or see [W]) that this does give a well defined action.

There are natural maps (homomorphisms, if $n \geq 2$) $i : \pi_n(X, x) \longrightarrow \pi_n(X, A, x)$ and $\delta : \pi_n(X, A, x) \longrightarrow \pi_{n-1}(A, x)$ (the latter is obtained by restriction to the top face $I^{n-1} \times \{1\}$ of $I^n$). Since we have a natural homomorphism $\pi_1(A, x) \longrightarrow \pi_1(X, x)$, we have a $\pi_1(A, x)$-action on $\pi_n(X, x)$ for $n \geq 1$. One can show that the maps $i, \delta$ are $\pi_1(A)$-equivariant.

We use the following terminology—given sets $S_1, S_2, S_3$ each of which has a distinguished base point, and maps $f : S_1 \longrightarrow S_2$, $g : S_2 \longrightarrow S_3$ preserving the base points, we say that $S_1 \longrightarrow S_2 \longrightarrow S_3$ is *exact* if $g^{-1}(s_3) = f(S_1)$, where $s_3 \in S_3$ is the base point. If the $S_i$ are groups and the base points are the respective identity elements, and $f, g$ are homomorphisms, this agrees with standard algebraic terminology.

**Theorem (A.3).** *There is a long exact sequence, natural for maps of triples $(X, A, x)$,*

$$\to \pi_{n+1}(X, A, x) \to \pi_n(A, x) \to \pi_n(X, x) \to \pi_n(X, A, x) \to \ldots$$
$$\ldots \pi_1(A, x) \to \pi_1(X, x) \to \pi_1(X, A, x) \to \pi_0(A, x) \to \pi_0(X, x).$$

*The maps in the sequence are compatible with the $\pi_1(A, x)$-actions.*

The proof of this result is left as an exercise to the reader (compare [W], IV Chapter 2). We define the $\pi_1(A,x)$-actions on $\pi_0(A,x)$ and $\pi_0(X,x)$ to be trivial. The sequence may be completed to the right

$$\to \pi_0(A,x) \to \pi_0(X,x) \to \pi_0(X,A,x) \to 0$$

if we define $\pi_0(X,A,x)$ to be the pointed set obtained by identifying all the points of $\pi_0(X,x)$ which correspond to the path components of $X$ which meet $A$.

Given a pair $(X,A)$ and a map $f_* : (I^n, \partial I^n, \partial_1 I^n) \longrightarrow (X,A,x)$, representing a class in $\pi_n(X,A,x)$, we have an induced map

$$f : H_n(I^n, \partial I^n; \mathbb{Z}) \longrightarrow H_n(X,A; \mathbb{Z}).$$

Since $H_n(I^n, \partial I^n; \mathbb{Z}) \cong \mathbb{Z}$, and the homomorphism $f_*$ depends only on the homotopy class of $f$, we have a map $\pi_n(X,A,x) \longrightarrow H_n(X,A;\mathbb{Z})$ given by $[f] \mapsto f_*(\delta_n)$, where $\delta_n \in H_n(I^n, \partial I^n; \mathbb{Z})$ is the standard generator (corresponding to the standard choice of orientation—see [W], IV, pg. 169). This is called the *Hurewicz map*, and is a homomorphism for $n \geq 2$ (for $n \geq 1$, if $A = \{x\}$). Since the base point plays no role in the definition of the homology groups $H_n(X,A;\mathbb{Z})$, the Hurewicz map kills the $\pi_1(A,x)$-action (the $\pi_1(X,x)$-action, if $A = \{x\}$), i.e., it factors through the coinvariants for this action (see [W], IV(4.9), (4.10)).

The first computation of homotopy groups is given by the result of Brouwer, that homotopy classes of maps from an $n$-sphere into itself are classified by their degree, if $n \geq 1$. This can be restated as

**Theorem (A.4).** *The Hurewicz maps*

$$\pi_n(S^n, x) \longrightarrow H_n(S^n, \mathbb{Z}) \quad (n \geq 1),$$
$$\pi_n(I^n, \partial I^n) \longrightarrow H_n(I^n, \partial I^n; \mathbb{Z}) \quad (n \geq 2)$$

*are isomorphisms of groups.*

(For the proofs, see [W], IV(4.5), (4.6)).

Next, we state the Hurewicz theorems. Recall that a space $X$ is *n-connected* if $\pi_i(X,x) = 0$ for $i \leq n$ (in particular, 0-connected is the same as path connected). A pair $(X,A)$ is called *n-connected* if $\pi_i(X,A,x) = 0$ for $i \leq n$.

**Theorem (A.5).** (i) *For any 0-connected space $X$, the Hurewicz map gives an exact sequence*

$$1 \longrightarrow [\pi_1(X,x), \pi_1(X,x)] \longrightarrow \pi_1(X,x) \longrightarrow H_1(X, \mathbb{Z}) \longrightarrow 1.$$

(ii) *For any n-connected space $X$, $n \geq 1$, the Hurewicz map*

$$\pi_{n+1}(X,x) \longrightarrow H_{n+1}(X, \mathbb{Z})$$

*is an isomorphism.*

(iii) *For any n-connected pair* $(X, A)$ *such that* $A$ *is 1-connected, and* $n \geq 1$, *the Hurewicz map*

$$\pi_{n+1}(X, A, x) \longrightarrow H_{n+1}(X, A; \mathbb{Z})$$

*is an isomorphism.*

(For the proofs, see [W], IV, Chapter 7).

**Corollary (A.6).** *Let* $f : X \longrightarrow Y$ *be a map between simply connected spaces which induces isomorphisms on integral homology groups. Then* $f$ *induces isomorphisms on homotopy groups.*

**Proof.** We first replace $f$ by an inclusion, using the mapping cylinder $M_f$. Regarding $f$ as a map $X \times \{1\} \longrightarrow Y$, we define $M_f$ to be the pushout

$$\begin{array}{ccc} X \times \{1\} & \xrightarrow{f} & Y \\ \downarrow & & \downarrow \\ X \times I & \longrightarrow & M_f. \end{array}$$

Clearly, a deformation retraction of $X \times I$ onto $X \times \{1\}$ induces one of $M_f$ onto $Y$, and the composite $X \times \{1\} \longrightarrow Y \longrightarrow M_f$ is homotopic to the inclusion $X \times \{0\} \subset M_f$. Hence, replacing $Y$ by $M_f$, we are reduced to the case when $f$ is an inclusion. Now the hypothesis on $f$ and the exact homology sequence for $(X, Y)$ give $H_n(Y, X; \mathbb{Z}) = 0$ for all $n$. Hence, by induction, (A.5)(iii) gives $\pi_n(Y, X, x) = 0$ for all $n$ (and any base point $x \in X$) (to start the induction, note that $\pi_1(Y, X, x) = 0$ since $X, Y$ are simply connected). Hence, by (A.3), $f$ induces isomorphisms on homotopy groups.

**(A.7) Products.** If $(X, x)$, $(Y, y)$ are spaces with base points, and $X \vee Y = (X \times \{y\}) \cup (\{x\} \times Y) \subset X \times Y$, the smash product $X \wedge Y$ is defined to be the quotient $(X \times Y)/(X \wedge Y)$. Let $x \wedge y \in X \wedge Y$ be the image of $(x, y)$, which we take to be the base point. Given maps $f : (I^n, \partial I^n) \longrightarrow (X, x)$; $g : (I^m, \partial I^m) \longrightarrow (Y, y)$ the product map $f \times g : I^{m+n} \longrightarrow X \times Y$ maps $\partial I^{m+n} = (\partial I^n \times I^m) \cup (I^n \times \partial I^m)$ into $X \vee Y$. Hence the induced map to the smash product $f \wedge g : (I^{m+n}, \partial I^{m+n}) \longrightarrow (X \wedge Y, x \wedge y)$ represents a class in the $(m+n)$th-homotopy group of $X \wedge Y$. One checks that this gives a product (which is bilinear, for $m, n \geq 1$)

$$\pi_n(X, x) \times \pi_m(Y, y) \longrightarrow \pi_{m+n}(X \wedge Y, x \wedge y).$$

Thus, given a space $(X, x)$ and a map $(X \wedge X, x \wedge x) \longrightarrow (X, x)$ we have an induced product

$$\pi_n(X, x) \times \pi_m(X, x) \longrightarrow \pi_{m+n}(X, x).$$

This product satisfies $x \cdot y = (-1)^{mn} y \cdot x$ for $x \in \pi_n(X,x)$, $y \in \pi_m(X,x)$ (this follows from the fact that the map $I^{m+n} \longrightarrow I^{m+n}$, induced by the switch map $I^m \times I^n \longrightarrow I^n \times I^m$, induces multiplication by $(-1)^{mn}$ on $H^{m+n}(I^{m+n}, \partial I^{m+n}; \mathbb{Z})$, i.e., has degree $(-1)^{mn}$, and from (A.4)).

**(A.8) CW-Complexes.** (See [W], II): Let $A$ be a compactly generated topological space, $X$ a topological space containing $A$ as a closed subset. A *CW-decomposition* of the pair $(X, A)$ is a nested sequence of closed subsets $X_n \subset X$, $n \geq 0$, such that

(i) $A \subset X_0$, and $X_0 - A$ is a discrete closed subspace of $X_0$.

(ii) $X = \cup X_n$ and $X$ has the weak topology with respect to $\{X_n\}$ (i.e., $B \subset X$ is closed $\iff B \cap X_n$ is closed $\forall n$).

(iii) $X_n$ is an $n$-cellular extension of $X_{n-1}$ (for $n > 0$), i.e., for some index set $S_n$ (possibly empty), which we regard as a discrete space, there is a continuous map $f_n : (\Delta_n \times S_n, \partial \Delta_n \times S_n) \longrightarrow (X_n, X_{n-1})$ such that the diagram below is a pushout:

$$\begin{array}{ccc} \partial \Delta_n \times S_n & \longrightarrow & \Delta_n \times S_n \\ f_n \downarrow & & \downarrow f_n \\ X_{n-1} & \longrightarrow & X_n \end{array}$$

(here $\Delta_n = \{(x_0, \ldots, x_n) \in \mathbb{R}^{n+1} \mid x_i \geq 0, \sum x_i = 1\}$ is the standard $n$-simplex, and $\partial \Delta_n$ is the subset of points with at least one vanishing coordinate, i.e., $\partial \Delta_n = \Delta_n \cap (x_0 x_1 \ldots x_n = 0)$). Since $X_n$ is a Hausdorff quotient of $X_{n-1} \coprod (\Delta_n \times S_n)$, it follows by induction that $X_n$ is compactly generated for all $n$. Hence $X = \varinjlim X_n$ is compactly generated. The image of $\Delta_n \times \{s\}$, for $s \in S_n$, is called a (closed) $n$-cell of $(X, A)$ (for the given $CW$-decomposition).

A pair $(X, A)$ consisting of a compactly generated space and a closed subspace is called a *relative CW-complex* if it has a $CW$-decomposition; $X$ is a $CW$-complex if $(X, \emptyset)$ is a relative $CW$-complex. A closed subset $A$ of a $CW$-complex $X$ is called a *subcomplex* of $X$ if $A$ is also a $CW$-complex, and $X, A$ have $CW$-decomposition such that every $n$-cell of $A$ is also an $n$-cell of $X$. Then $(X, A)$ is also a relative $CW$-complex.

A space $X$ is said to have the *homotopy type of a CW-complex* if there is a $CW$-complex $Y$ together with a homotopy equivalence $f : Y \longrightarrow X$. Two useful facts about spaces with the homotopy type of a $CW$-complex are given by the following results, due to Whitehead and Milnor, respectively.

Appendix A: Results from Topology    237

**Theorem (A.9)** (Whitehead theorem). *Let $f : X \longrightarrow Y$ be a continuous map between connected spaces each of which has the homotopy type of a CW-complex. Suppose $f$ induces isomorphisms on homotopy groups. Then $f$ is a homotopy equivalence.*

(For the proof, see [W], V(3.8)).

**Corollary (A.10).** *Let $X, Y$ be simply connected spaces, each of which has the homotopy type of a CW-complex, and let $f : X \longrightarrow Y$ be a map which induces isomorphisms on integral homology. Then $f$ is a homotopy equivalence.*

**Proof.** This follows at once from (A.6) and (A.9).

If $X, Y$ are spaces, let $\mathbb{F}(X, Y)$ denote the function space of all (continuous) maps from $X$ to $Y$, topologized as in (A.1).

**Theorem (A.11)** (Milnor). *Let $X$ be a compact space, $Y$ a space with the homotopy type of a CW-complex. Then $\mathbb{F}(X, Y)$ has the homotopy type of a CW-complex.*

(This is proved in Milnor's article, J. Milnor, On spaces having the homotopy type of a $CW$-complex, *Trans. A.M.S.* 90 (1959), 272–280).

We recall the standard method used to compute the homology of a $CW$-complex, using the complex of cellular chains. If $(X, A)$ is a relative $CW$-complex, and $S_n$ denotes the set of $n$-cells, let $C_n(X, A)$ be the free Abelian group on $S_n$; then

$$H_i(X_n, X_{n-1}; \mathbb{Z}) = \begin{cases} C_n(X, A) & \text{if } i = n \\ 0 & \text{otherwise.} \end{cases}$$

The homology sequence of the triple $(X_{n+1}, X_n, X_{n-1})$ (where we define $X_{-1} = A$) gives an exact sequence

$$\ldots \longrightarrow H_i(X_n, X_{n-1}; \mathbb{Z}) \longrightarrow H_i(X_{n+1}, X_{n-1}; \mathbb{Z}) \longrightarrow H_i(X_{n+1}, X_n; \mathbb{Z})$$
$$\xrightarrow{\partial} H_{i-1}(X_n, X_{n-1}; \mathbb{Z}) \longrightarrow \ldots .$$

In particular, taking $i = n + 1$, we have a map

$$d_{n+1} : C_{n+1}(X, A) \longrightarrow C_n(X, A),$$

for each $n \geq 0$; one sees easily that $d_n \circ d_{n+1} = 0$, where we define $d_0 = 0$ (this follows from the construction of the exact sequence of the triple from the three exact sequence of the pairs $(X_n, X_{n-1}) \subset (X_{n+1}, X_{n-1}) \subset (X_{n+1}, X_n)$; in particular the map $\partial$ above is defined to be the composite of the boundary map

$$H_i(X_{n+1}, X_n; \mathbb{Z}) \longrightarrow H_{i-1}(X_n, \mathbb{Z})$$

and the natural map $H_{i-1}(X_n, \mathbb{Z}) \longrightarrow H_{i-1}(X_n, X_{n-1}; \mathbb{Z}))$.

**Theorem (A.12).** *For any Abelian group $G$, the homology groups $H_i(X, A; G)$ of a relative CW-complex $(X, A)$ are given by*

$$H_i(X, A; G) = H_i(C_*(X, A) \otimes_{\mathbb{Z}} G, d_* \otimes 1),$$

*the homology groups of the relative cellular chain complex with coefficients in $G$.*

**Proof.** We have

$$H_i(X_n, X_{n-1}; G) = \begin{cases} C_n(X, A) \otimes_{\mathbb{Z}} G & \text{if } i = n \\ 0 & \text{otherwise} \end{cases}$$

(this follows from the case $G = \mathbb{Z}$ by the universal coefficient theorem). Hence, by induction on $p$, and the long exact homology sequence of a triple, we obtain

$$H_i(X_{n+p}, X_n; G) = 0 \text{ if } i > n + p \text{ or } i \leq n.$$

Since any compact set in $X$ lies in some $X_m$, the complex of singular chains with coefficients in $G$ is the direct limit of the corresponding complexes for the $X_m$. Hence

$$H_n(X, A; G) = \varinjlim H_n(X_m, A; G) \text{ for any } n \geq 0.$$

We deduce that the natural maps

$$H_n(X, A; G) \longrightarrow H_n(X, X_{n-2}; G),$$
$$H_n(X_{n+1}, X_{n-2}; G) \longrightarrow H_n(X, X_{n-2}; G)$$

are isomorphisms for all $n \geq 0$ (here $X_i = A$ for $i < 0$).

Next, from the exact sequence of the triple $X_{n-2} \subset X_{n-1} \subset X_n$, we have an exact sequence

$$0 \longrightarrow H_n(X_n, X_{n-2}; G) \xrightarrow{i} H_n(X_n, X_{n-1}; G) \xrightarrow{\partial} H_{n-1}(X_{n-1}, X_{n-2}; G)$$

where $\partial = d_n \otimes 1 : C_n(X, A) \otimes G \longrightarrow C_{n-1}(X, A) \otimes G$. This computes

$$\ker(d_n \otimes 1) \cong H_n(X_n, X_{n-2}; G).$$

Finally, from the exact sequence of the triple $X_{n-2} \subset X_n \subset X_{n+1}$, we have an exact sequence

$$H_{n+1}(X_{n+1}, X_n; G) \xrightarrow{\partial} H_n(X_n, X_{n-2}; G) \longrightarrow H_n(X_{n+1}, X_{n-2}; G) \longrightarrow 0.$$

But the composite

$$H_{n+1}(X_{n+1}, X_n; G) \xrightarrow{\partial} H_n(X_n, X_{n-2}; G) \xrightarrow{i} H_n(X_n, X_{n-1}; G)$$

is just $d_{n+1} \otimes 1$, by the naturality of the exact sequence of a triple, for the map of triples $(X_{n+1}, X_n, X_{n-2}) \longrightarrow (X_{n+1}, X_n, X_{n-1})$. Thus image $\partial \cong$ image $d_{n+1} \otimes 1$.

Appendix A: Results from Topology                239

**(A.13) Local Coefficients.** (See [W], VI). Let $X$ be a topological space. A *local system* on $X$ consists of the following data:

(i) for each $x \in X$ we are given a group $G_x$

(ii) for each path $\gamma$ from $x$ to $y$, we have an isomorphism $\gamma^* : G_x \longrightarrow G_y$, which depends only on the homotopy class of $\gamma$ (among paths joining $x$ to $y$).

(iii) if $\gamma_1, \gamma_2$ are paths joining $x$ to $y$ and $y$ to $z$, respectively, and $\gamma_1, \gamma_2$ is the composite path joining $x$ to $z$, then
$$(\gamma_1, \gamma_2)^* = (\gamma_2)^* \circ (\gamma_1)^* : G_x \longrightarrow G_z.$$

These three conditions are equivalent to the condition that $x \mapsto G_x$ is a group valued contravariant functor on the fundamental groupoid of $X$. Clearly, a local system on a path connected space $X$ is determined by the following data: if $x \in X$ is a base point, it suffices to be given the group $G_x$, and an action of $\pi_1(X, x)$ on $G_x$. For any space $X$, and any integer $i \geq 1$, $x \mapsto \pi_i(X, x)$ is a local system on $X$.

We can define homology and cohomology groups $H_*(X, A; G)$, $H^*(X, A; G)$ for a pair $(X, A)$ and a local system $G$ on $X$. For our purposes, it suffices to do this under the hypothesis that each path component of $X$ and of $A$ is open, and has a universal covering space; if $\{X_\alpha \mid \alpha \in J\}$ are the path components, we take

$$H_*(X, A; G) = \bigoplus_{\alpha \in J} H_*(X_\alpha, A \cap X_\alpha; G|_{X_\alpha}),$$

and

$$H^*(X, A; G) = \prod_{\alpha \in J} (X_\alpha, A \cap X_\alpha; G|_{X_\alpha}).$$

Hence it suffices to define homology and cohomology with local coefficients of pairs $(X, A)$ where $X$ is path connected and $X, A$ have universal covering spaces.

Let $p : \tilde{X} \longrightarrow X$ be the universal covering space. If we fix a base point $x \in X$, we have a natural free action of $\Pi = \pi_1(X, x)$ on $\tilde{X}$, with quotient $X$. If $\sigma : \Delta_n \longrightarrow X$ is a singular $n$-simplex, then $\tilde{X} \times_X \Delta_n \longrightarrow \Delta_n$ is a covering space with group $\Pi$, i.e., is a disjoint union of copies of $\Delta_n$ permuted freely by $\Pi$. Thus, the complex $S_*(\tilde{X})$ of singular chains is a complex of free $\mathbb{Z}[\Pi]$-modules, which in degree $n$ has a basis in bijection with the set of $n$-simplices in $X$. If $A \subset X$, and $\tilde{A} = p^{-1}(A)$, the singular complex $S_*(\tilde{A})$ is a subcomplex of free $\mathbb{Z}[\Pi]$-modules of $S_*(\tilde{X})$, which in degree $n$ has a basis in bijection with the set of $n$-simplices in $A$. Hence $\mathbb{Z}[\Pi]$-bases for $S_n(\tilde{A})$, $S_n(\tilde{X})$ can be chosen so that the basis for $S_n(\tilde{A})$ is a subset of that for $S_n(\tilde{X})$. Thus, the complex of relative singular chains

$S_*(\tilde{X}, \tilde{A}) = S_*(\tilde{X})/S_*(\tilde{A})$ is a complex of free $\mathbb{Z}[\Pi]$-modules, such that for each $n$, the short exact sequence

$$0 \longrightarrow S_n(\tilde{A}) \longrightarrow S_n(\tilde{X}) \longrightarrow S_n(\tilde{X}, \tilde{A}) \longrightarrow 0$$

is a split exact sequence of $\mathbb{Z}[\Pi]$-modules.

If $G$ is a local coefficient system on $X$, let $G_x \cong G_0$, which is a (right) $\Pi$-module in a natural way; we can also regard $G_0$ as a left $\Pi$-module (using the anti-automorphism $g \mapsto g^{-1}$). We define

$$H_n(X, A; G) = H_n(G_0 \times_{\mathbb{Z}[\Pi]} S_*(\tilde{X}, \tilde{A}))$$
$$H^n(X, A; G) = H^n(\mathrm{Hom}_{\mathbb{Z}[\Pi]}(S_*(\tilde{X}, \tilde{A}), G_0)).$$

If $A_0$ is a path component of $A$, let $\tilde{A}_0 = \tilde{X} \times_X A_0$ be the induced covering (possibly not connected) of $A_0$, and let $\hat{A}_0$ denote the universal covering of $A_0$. Let $\Pi' = \pi_1(A_0, x_0)$, and choose a path joining $x_0$ and $x$ (in $X$), to give a homomorphism $\Pi' \longrightarrow \Pi$. Then $S_*(\hat{A}_0)$ is a complex of free $\Pi'$-modules. The universal covering $\hat{A}_0 \longrightarrow A_0$ factors through (some path component of) $\tilde{A}_0$. This gives a map $S_*(\hat{A}_0) \longrightarrow S(\tilde{A}_0)$. One sees easily that under this homomorphism, $S_*(\hat{A}_0) \cong \mathbb{Z}[\Pi] \otimes_{\mathbb{Z}[\Pi']} S_*(\hat{A}_0)$. If $G$ is a local coefficient system, corresponding to the $\Pi$-module $G_0$, then $G|_{A_0}$ is the local coefficient system corresponding to the $\Pi'$-module structure on $G_0$ induced by the map $\Pi' \longrightarrow \Pi$, and so

$$G_0 \otimes_{\mathbb{Z}[\Pi]} S_*(\tilde{A}_0) \cong G_0 \otimes_{\mathbb{Z}[\Pi']} S_*(\hat{A}_0).$$

From this, one sees at once that the homology groups of $G_0 \otimes_{\mathbb{Z}[\Pi]} S_*(\tilde{A})$ are precisely the groups $H_*(A, G|_A)$. A similar argument works for cohomology. Since the exact sequence of complexes of $\Pi$-modules

$$0 \longrightarrow S_*(\tilde{A}) \longrightarrow S_*(\tilde{X}) \longrightarrow S_*(\tilde{X}, \tilde{A}) \longrightarrow 0$$

consists of a split exact sequence of free $\Pi$-modules in each degree,

$$0 \longrightarrow G_0 \otimes_{\mathbb{Z}[\Pi]} S_*(A) \longrightarrow G_0 \otimes_{\mathbb{Z}[\Pi]} S_*(\tilde{X}) \longrightarrow G_0 \otimes_{\mathbb{Z}[\Pi]} S_*(\tilde{X}, \tilde{A}) \longrightarrow 0$$

is an exact sequence of complexes, which yields the long exact sequence of the pair $(X, A)$

$$\longrightarrow H_{i+1}(X, A; G) \longrightarrow H_i(A, G) \longrightarrow H_i(X, G) \longrightarrow H_i(X, A; G)$$
$$\longrightarrow H_{i-1}(A, G) \longrightarrow \ldots$$

for any local coefficient system $G$ on $X$ (where $H_i(A, G)$ denotes $H_i(A, G|_A)$). A similar argument yields the exact cohomology sequence for a pair $(X, A)$ with coefficients in $G$.

Suitable reformulations of the Eilenberg–Steenrod axioms hold for homology and cohomology with local coefficients (see [W], VI, Chapter 2). In particular excision holds. Hence, we can compute homology with local coefficients using a $CW$-decomposition. Let $(X, A)$ be a relative $CW$-complex such that $X$ is path connected; let $p : \tilde{X} \longrightarrow X$ be the universal covering, $\tilde{A} = p^{-1}(A)$ the induced covering space of $A$. Then we have isomorphisms (from excision) for $n \geq 0$

$$H_i(X_n, X_{n-1}; G|_{X_n}) = \begin{cases} \cong G_0 \otimes_{\mathbb{Z}[\Pi]} H_n(\tilde{X}_n, \tilde{X}_{n-1}; \mathbb{Z}), & \text{if } i = n \\ = 0, & \text{otherwise} \end{cases}$$

(where $X_n = A$ if $n < 0$). Here $\Pi = \pi_1(X, x)$ for a base point $x \in A$, and $G_0$ is the $\Pi$-module associated to $G$. Thus, the complex of relative cellular chains $C_n(\tilde{X}, \tilde{A})$ (see (A.12)) is a complex of free $\mathbb{Z}[\Pi]$-modules, and we can form the complex of relative cellular chains on $(X, A)$ with coefficients in $G$, denoted $C_*(X, A; G)$, by

$$C_n(X, A; G) = H_n(X_n, X_{n-1}; G) \cong G_0 \otimes_{\mathbb{Z}[\Pi]} C_n(\tilde{X}, \tilde{A}).$$

The proof of (A.12) goes through in this context to prove

**Theorem (A.14).** *We have natural isomorphisms*

$$H_n(X, A; G) \cong H_n(C_*(X, A; G))$$
$$\cong H_n(G_0 \otimes_{\mathbb{Z}[\Pi]} C_*(\tilde{X}, \tilde{A})).$$

*for all $n \geq 0$.*

An analogous result is valid for cohomology with local coefficients. Let $C^n(X, A; G) = \mathrm{Hom}_{\mathbb{Z}[\Pi]}(C_n(\tilde{X}, \tilde{A}), G_0) \cong H^n(X_n, X_{n-1}; G)$. As in the case of homology, we also have the vanishing result

$$H^i(X_n, X_{n-1}; G) = 0 \text{ for } i \neq n.$$

However, the proof of

$$H^n(X, A; G) = \varprojlim_m H^n(X_m, A; G)$$

requires some care (see [W], VI, (4.3)). This yields, by the standard argument, the result

**Theorem (A.15).** *We have natural isomorphisms*

$$H^n(X, A; G) \cong H^n(C^*(X, A; G))$$
$$\cong H^n(\mathrm{Hom}_{\mathbb{Z}[\Pi]}(C_*(\tilde{X}, \tilde{A}), G_0))$$

*for all $n \geq 0$.*

**(A.16) Obstruction Theory.** (*Reference*: P. Olum, Obstructions to extensions and homotopies, *Ann. Math.* 52 (1950), 1–50).

Let $(X, A)$ be a relative $CW$-complex, where $A$ is non-empty, and fix a base point $x_0 \in A$. We will assume that $X, A$ are path connected. Consider the following two extension problems:

(i) Given a continuous map $f : (A, x_0) \longrightarrow (Y, y_0)$ for a path connected space $Y$, when does $f$ extend to a map $F : (X, x_0) \longrightarrow (Y, y_0)$ (with $F|_A = f$)?

(ii) Given two continuous maps $F_0, F_1 : (X, x_0) \longrightarrow (Y, y_0)$ such that $F_0|_A = F_1|_A = f$, when is $F_0$ homotopic to $F_1$ relative to $A$, i.e., when does there exist a map $H : X \times I \longrightarrow Y$ such that $H(x, 0) = F_0(x)$, $H(x, 1) = F_1(x)$, $H(x, t) = f(x)$ for $(x, t) \in A \times I$?

Obstruction theory gives a set of necessary conditions for the solutions of these problems. First, consider the question (i). If $f : A \longrightarrow Y$ is given we can always find extensions of $f$ to the 1-skeleton $X_1$ of $(X, A)$. Indeed, we can assign arbitrary values to the points of $X_0 - A$, and then extend to $X_1$ using the fact that $Y$ is path connected. If it is possible to extend $f$ to the 2-skeleton $X_2$, then since $\pi_1(X_2, x_0) \cong \pi_1(X, x_0)$, we would be able to find a homomorphism $\theta : \pi_1(X, x_0) \longrightarrow \pi_1(Y, f(x_0))$ such that

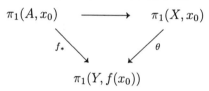

commutes. It turns out that this is the only obstruction to finding some extension of $f$ to the 2-skeleton $X_2$.

Next, fix a homomorphism $\theta : \pi_1(X, x_0) \longrightarrow \pi_1(Y, f(x_0))$ as above, which is compatible with $f_*$, and consider the problem of extending $f$ to a map $F : (X, x_0) \longrightarrow (Y, f(x_0))$ such that the homomorphism on $\pi_1$ induced by $F$ is $\theta$. Now $\pi_1(Y, f(x_0))$ acts on $\pi_i(Y, f(x_0))$, $i \geq 1$, giving a local system $\pi_i(Y)$ on $Y$ with local group $\pi_i(Y, f(x_0))$ at $f(x_0)$. Given the homomorphism $\theta$, we can form the pullback local system $\theta^*\pi_i(Y)$ on $X$ with local group $\pi_i(Y, f(x_0))$ at $x_0$, and $\pi_1(X, x_0)$-action induced by $\theta$.

Suppose that one can find an extension $f_n : X_n \longrightarrow Y$ of $f$ to the $n$-skeleton $X_n$ of $(X, A)$, for some $n \geq 2$, such that the induced homomorphism on $\pi_1$ is $\theta$. Then one can define an *obstruction cocycle*, giving an *obstruction class*

$$O^{n+1}(\theta, f_n) \in H^{n+1}(X, A; \theta^*\pi_n(Y))$$

in the cohomology group with coefficients in the local system $\theta^*\pi_n(Y)$, whose vanishing is a necessary and sufficient condition that $f_n|_{X_{n-1}}$ extends to a map $f_{n+1} : X_{n+1} \longrightarrow Y$ where $X_{n+1}$ is the $(n + 1)$-skeleton (i.e.,

$O^{n+1}(\theta, f_n) = 0 \Leftrightarrow$ after possibly modifying $f_n$ on $X_n - X_{n-1}$, we can extend $f_n$ to $f_{n+1}$ defined on $X_{n+1}$). More generally, consider all possible extensions $f_n$ of $f$ to the $n$-skeleton which induce $\theta$ on fundamental groups, and let the *obstruction set*

$$O^{n+1}(\theta, f) \subset H^{n+1}(X, A; \theta^*\pi_n(Y))$$

be the subset consisting of all the obstruction classes. Then for any $n \geq 2$, $f$ extends to $X_{n+1}$ (compatibly with $\theta$) $\Leftrightarrow 0 \in O^{n+1}(\theta, f) \Leftrightarrow O^{n+2}(\theta, f)$ is non-empty. This gives a set of necessary conditions for the solution of (i). In the special case when $H^n(X, A; \theta^*\pi_{n-1}(Y)) = 0$ for $n \geq 3$, $f$ extends to $X$.

The question (ii) can be regarded as a special case of (i), with $X$ replaced by $X \times I$, and $A$ replaced by $(X \times \{0,1\} \cup A \times I) \subset X \times I$. By van Kampen's theorem, $\pi_1(X \times \{0,1\} \cup A \times I)$ is the amalgamated free product $\pi_1(X) \times_{\pi_1(A)} \pi_1(X)$, and the map

$$\pi_1(X \times \{0,1\} \cup A \times I) \longrightarrow \pi_1(X \times I) \cong \pi_1(X)$$

is the natural quotient. Let $\tilde{f} : X \times \{0,1\} \cup A \times I \longrightarrow Y$ be given by $\tilde{f}|_{X \times \{0\}} = F_0$, $\tilde{f}|_{X \times \{1\}} = F_1$, $\tilde{f}(x,t) = f(x)$ for $(x,t) \in A \times I$. Then $F_0$ is homotopic to $F_1$ relative to $A$ if and only if $\tilde{f}$ extends to a map $X \times I \longrightarrow Y$. Clearly the condition that $\tilde{f}$ extends to the 2-skeleton of $(X \times I, X \times \{0,1\} \cup A \times I)$ is just that $(F_0)_* = (F_1)_* : \pi_1(X, x_0) \longrightarrow \pi_1(Y, f(x_0))$, from the above calculation of fundamental groups.

Hence, we need only consider the following special case of (ii). Let $f : (A, x_0) \longrightarrow (Y, f(x_0))$ be given, and let $\theta : \pi_1(X, x_0) \longrightarrow \pi_1(Y, f(x_0))$ be a homomorphism compatible with $f$. We consider maps $F_0, F_1 : X \longrightarrow Y$ extending $f$ and inducing $\theta$ on $\pi_1$, and ask for conditions that $F_0$ be homotopic to $F_1$. Then for each choice of a homotopy between $F_0|_{X_n}$ and $F_1|_{X_n}$, where $n \geq 1$, which we regard as an extension of $\tilde{f}$ to the $(n+1)$-skeleton of $(X \times I, X \times \{0,1\} \cup A \times I)$, there is an obstruction class in $H^{n+2}(X \times I, X \times \{0,1\} \cup A \times I; \tilde{\theta}^*\pi_{n+1}(Y))$, where $\tilde{\theta} : \pi_1(X \times I, (x_0, 0)) \longrightarrow \pi_1(Y, f(x_0))$ is induced by $\theta$ and the natural isomorphism $\pi_1(X, x_0) \cong \pi_1(X, \times I, (x_0, 0))$ (given by $X \times \{0\} \subset X \times I$). There is a suspension isomorphism

$$H^{n+1}(X, A; \theta^*\pi_{n+1}(Y)) \cong H^{n+2}(X \times I, X \times \{0,1\} \cup A \times I; \tilde{\theta}^*\pi_{n+1}(Y)),$$

so that we can regard the obstruction class as an element of the former group. Thus, for any $n \geq 1$, we have an *obstruction set* $O^{n+1}(\theta; F_0, F_1) \subset H^{n+1}(X, A; \theta^*\pi_{n+1}(Y))$, which is non-empty $\Longleftrightarrow F_0|_{X_n}$ is homotopic (relative to $A$) to $F_1|_{X_n}$, and $0 \in O^{n+1}(\theta; F_0, F_1) \Longleftrightarrow F_0|_{X_{n+1}}$ and $F_1|_{X_{n+1}}$ are homotopic relative to $A$. If $H^n(X, A; \theta^*\pi_n(Y)) = 0$ for $n \geq 2$, then $F_0, F_1$ are homotopic relative to $A$.

**(A.17) Fibrations.** A surjective map $p : E \longrightarrow B$ is called a *fibration* if it has the *homotopy lifting property*, i.e., for any diagram

$$\begin{array}{ccc} Z \times \{0\} & \xrightarrow{f} & E \\ \downarrow & & \downarrow p \\ Z \times I & \xrightarrow{\bar{H}} & B \end{array}$$

(where $f : Z \longrightarrow E$ is a map, $\bar{H} : Z \times I \longrightarrow B$ a homotopy of $p \circ f : Z \longrightarrow B$), there exists a map $H : Z \times I \longrightarrow E$ such that $p \circ H = \bar{H}$, with $H|_{Z \times \{0\}} = f$.

Clearly if $p$ is a fibration, then for any diagram

$$\begin{array}{ccc} Z \times \{\underline{0}\} & \xrightarrow{f} & E \\ \downarrow & & \downarrow p \\ Z \times I^n & \xrightarrow{\bar{H}} & B \end{array}$$

where $\underline{0} = (0, \ldots, 0) \in I^n$, there is a map $H : Z \times I^n \longrightarrow E$ with $H|_{Z \times \{\underline{0}\}} = f$, such that $p \circ H = \bar{H}$. This implies that if $x \in B$, $F_x = p^{-1}(x)$, $y \in F_x$, then for any $n \geq 1$ the natural maps $\pi_n(E, F_x; y) \longrightarrow \pi_n(B, x)$ are bijective (see [W], IV, (8.5)). The long exact homotopy sequence for the pair $(E, F_x)$ yields:

**Theorem (A.18).** *Let $p : E \longrightarrow B$ be a fibration, $x \in B$, $F_x = p^{-1}(x)$, $y \in F_x$.*

*Then there is a long exact homotopy sequence*

$$\to \pi_{n+1}(B, x) \to \pi_n(F_x, y) \to \pi_n(E, y) \to \pi_n(B, x) \to \pi_{n-1}(F_x, y) \to \cdots$$
$$\cdots \to \pi_0(F_x, y) \to \pi_0(E, y).$$

*If $p : E \longrightarrow B$, $p' : E' \longrightarrow B'$ are two fibrations such that there is a fiber preserving map $E \longrightarrow E'$, then the sequence for $p$ maps to that for $p'$.*

Next, we remark that if $p : E \longrightarrow B$ is a fibration, $f : B' \longrightarrow B$ any map, $E' = B' \times_B E$, $p' : E' \longrightarrow B'$ the natural map, then $p'$ is a fibration, since the homotopy lifting property for $p$ immediately yields the homotopy lifting property for $p'$.

The homotopy lifting property implies the following more general lifting property:

**Lemma (A.19).** *Let $Y$ be a space, $A$ a closed subspace such that $(Y, A)$ is a DR-pair (i.e., $A$ is a deformation retract of $Y$, and there is a map*

Appendix A: Results from Topology    245

$u : Y \longrightarrow I$ with $u^{-1}(0) = A$). Let $p : E \longrightarrow B$ be a fibration, $f : Y \longrightarrow B$ a given map and $g : A \longrightarrow E$ a lifting of $f|_A$ (i.e., $p \circ g = f|_A$). Then there exists a lifting $h : Y \longrightarrow E$ of $f$ which extends $g$ (i.e., $p \circ h = f$ and $h|_A = g$).

**Proof.** Let $\Phi : Y \times I \longrightarrow Y$ be a deformation retraction of $Y$ onto $A$, i.e., $\Phi(Y \times \{0\}) = A$, $\Phi(y,t) = y \; \forall (y,t) \in A \times I$, $\Phi(y,1) = y \; \forall y \in Y$. Let $\Psi : Y \times I \longrightarrow Y$ be defined by

$$\Psi(y,t) = \begin{cases} \Phi(y, t/u(y)) & \text{if } t < u(y) \\ \Phi(y, 1) & \text{if } t \geq u(y). \end{cases}$$

One easily checks that $\Psi$ is continuous. Then $f \circ \Psi : Y \times I \longrightarrow B$ and $k : Y \longrightarrow E$, given by $k(y) = g(\Phi(y,0))$, are the data for a homotopy lifting problem, since $p \circ k = f \circ \Psi|_{Y \times \{0\}}$. If $H : Y \times I \longrightarrow E$ is a lifting of $f \circ \Psi$ which extends the lifting $k$ of $f \circ \Psi|_{Y \times \{0\}}$, then $h : Y \longrightarrow E$ given by $h(y) = H(y, u(y))$ is the desired lifting of $f$ extending $g$.

We can use this lemma to prove a certain local triviality property of a fibration. If $p : E \longrightarrow B$, $p' : E' \longrightarrow B$ are fibrations with the same base $B$, we say that $p$ and $p'$ are *fiber homotopy equivalent* if there exist maps $f : E \longrightarrow E'$, $g : E' \longrightarrow E$ and homotopies $H : E \times I \longrightarrow E$, $H' : E' \times I \longrightarrow E'$ such that (i) $p \circ H(x,t) = p(x)$ for all $(x,t) \in E \times I$, and $H(x,0) = x$, $H(x,1) = g \circ f(x) \; \forall x \in E$; and

(ii) $p' \circ H'(x',t') = p'(x') \; \forall \, (x',t') \in E' \times I$, $H'(x',0) = x'$, $H'(x',1) = f \circ g(x') \; \forall x' \in E'$. In particular, $f, g$ are homotopy equivalences; we call such a map $f : E \longrightarrow E'$ a fiber homotopy equivalence from $E$ to $E'$ (and similarly for $g : E' \longrightarrow E$). We also say that $E$ and $E'$ have the same fiber homotopy type.

**Theorem (A.20).** *Let $p : E \longrightarrow B$ be a fibration over a contractible base $B$. Let $b_0 \in B$, $F_0 = f^{-1}(b_0)$, and let $p' : F_0 \times B \longrightarrow B$ be the projection. Then there is a fiber homotopy equivalence $f : E \longrightarrow F_0 \times B$ such that $f|_{F_0} : F_0 \longrightarrow F_0 \times \{b\}$ is homotopic to the identity.*

**Proof.** Let $h : B \times I \longrightarrow B$ be a deformation retraction from $B$ to $\{b_0\}$, so that $h(b,0) = b_0$, $h(b,1) = b$, $h(b_0,t) = b_0 \; \forall b \in B, t \in I$. Let $f_0 : F_0 \times B \longrightarrow E$ be the composite of projection $F_0 \times B \longrightarrow F_0$ with the inclusion $F_0 \subset E$, so that $p \circ f_0 : F_0 \times B \longrightarrow \{b_0\}$, and let $h_0 : F_0 \times B \times I \longrightarrow B$ be the composite of projection $F_0 \times B \times I \longrightarrow B \times I$ with $h : B \times I \longrightarrow B$. Thus $h_0(x,b,t) = h(b,t)$, $\forall x \in F_0, b \in B, t \in I$, and $f_0, h_0$ are data for a homotopy lifting problem. Let $H_0 : F_0 \times B \times I \longrightarrow E$ be a lifting of $h_0$, i.e., $p \circ H_0 = h_0$, such that $H_0|_{F_0 \times B \times \{0\}} = f_0$. Let $g : F_0 \times B \longrightarrow E$ be given by $g(x,b) = H_0(x,b,1)$, so that $p \circ g = p'$, i.e., $g$ is a fiber preserving map $E' \longrightarrow E$, where $E' = F_0 \times B$.

Similarly the identity map $1_E : E \longrightarrow E$ and the map $h_1 : E \times I \longrightarrow B$, given by the composition $h_1 = h \circ (p \times 1_I)$ (where $p \times 1_I : E \times I \longrightarrow B \times I$ is the obvious map), are data for a homotopy lifting problem. Hence there is a map $H_1 : E \times I \longrightarrow E$ with $p \circ H_1 = h_1$, and $H_1\big|_{E \times \{1\}} = 1_E$. Let $f : E \to E' (= F_0 \times B)$ be defined by $f(x) = (H_1(x,0), p(x)) \; \forall x \in E$; then $p' \circ f = p$, so that $f$ is fiber preserving.

We claim that $f, g$ give a fiber homotopy equivalence between $E$ and $E'$. Indeed, let

$$h_2 : F_0 \times B \times I \times I \longrightarrow B$$
$$m : F_0 \times B \times \{1\} \times I \cup F_0 \times B \times I \times \{0,1\} \longrightarrow E$$

be defined by $h_2(x,b,s,t) = h_0(x,b,s) = h(b,s)$, and

$$m(x,b,s,t) = \begin{cases} H_0(x,b,s) & \text{if } t = 0,\; s \in I \\ H_0(x,b,1) & \text{if } s = 1,\; t \in I \\ H_1(g(x,b),s) & \text{if } t = 1,\; s \in I. \end{cases}$$

To see that $m$ is well defined and continuous, we need only verify that $m(x,b,1,1) = H_0(x,b,1) = H_1(g(x,b),1)$; but $H_1(x,1) = x \; \forall x \in E$, and $g(x,b) = H_0(x,b,1)$ by definition. We claim that there is a diagram

$$\begin{array}{ccc} F_0 \times B \times \{1\} \times I \cup F_0 \times B \times I \times \{0,1\} & \xrightarrow{\;m\;} & E \\ \downarrow & & \downarrow p \\ F_0 \times B \times I \times I & \longrightarrow & B \end{array}$$

i.e., for $(x,b,s,t) \in F_0 \times B \times \{1\} \times I \cup F_0 \times B \times I \times \{0,1\}$, we must verify that $h_2(x,b,s,t) = p \circ m(x,b,s,t)$. Now $h_2(x,b,s,t) = h(b,s)$. On the other hand, $p \circ m(x,b,s,0) = p \circ H_0(x,b,s) = h_0(x,b,s) = h(b,s)$, and a similar calculation works for points $(x,b,1,t)$; finally $p \circ m(x,b,s,1) = p \circ H_1(g(x,b),s) = h_1(g(x,b),s) = h(p \circ g(x,b),s) = h(b,s)$. Hence the above diagram commutes. Since $(I \times I, \{1\} \times I \cup I \times \{0,1\})$ is a $DR$-pair, so is $(E' \times I \times I, E' \times \{1\} \times I \cup E' \times I \times \{0,1\})$. Hence by (A.19), $b_2$ can be lifted to $\tilde{H}_2 : F_0 \times B \times I \times I \longrightarrow E$ extending $m$. Let $H_2 : F_0 \times B \times I \longrightarrow F_0 \times B$ be given by

$$H_2(x,b,t) = (\tilde{H}_2(x,b,0,t), b).$$

Then $H_2(x,b,0) = (\tilde{H}_2(x,b,0,0), b) = (m(x,b,0,0), b) = (H_0(x,b,0), b) = (f_0(x,b), b) = (x,b)$ for $(x,b) \in F_0 \times B$, and

$$H_2(x,b,1) = (\tilde{H}_2(x,b,0,1), b) = (m(x,b,0,1), b)$$
$$= (H_1(g(x,b),0), b) = f \circ g(x,b).$$

Finally $p' \circ H_2(x,b,t) = b \; \forall (x,b,t) \in F_0 \times B \times I$.

An analogous argument works for the other composite $g \circ f$. We define
$$h_3 : E \times I \times I \longrightarrow B$$
$$n : E \times \{0\} \times I \cup E \times I \times \{0,1\} \longrightarrow E$$
by $h_3(x,s,t) = h_1(x,s) = h(p(x),s)$,
$$n(x,s,t) = \begin{cases} H_1(x,s) & \text{if } t=0,\ s \in I \\ H_1(x,0) & \text{if } s=0,\ t \in I \\ H_0(f(x),s) & \text{if } t=1,\ s \in I. \end{cases}$$
Then $n$ is well defined, since
$$n(x,0,1) = H_0(f(x),0) = H_0(H_1(x,0),p(x),0)$$
$$= f_0(H_1(x,0),p(x)) = H_1(x,0).$$
Further we have a commutative diagram

$$\begin{array}{ccc} E \times \{0\} \times I \cup E \times I \times \{0,1\} & \xrightarrow{n} & E \\ \downarrow & & \downarrow p \\ E \times I \times I & \xrightarrow{h_3} & B \end{array}$$

(as before, we immediately reduce to checking $p \circ H_0(f(x),s) = h(p(x),s)$; but $p \circ H_0(f(x),s) = h_0(f(x),s) = h_0(H_1(x,0),p(x),s) = h(p(x),s)$ by definition of $h_0$). As before, a lifting $H_3 : E \times I \times I \longrightarrow E$ extending $n$ determines a fiber-preserving homotopy between the identity map of $E$ and $g \circ f$, by restricting $H_3$ to $E \times \{1\} \times I$. This completes the proof of (A.20).

**Corollary (A.21).** *If $p : E \longrightarrow B$ is a fibration over a path connected base $B$, then any two fibers of $p$ are homotopy equivalent.*

**Proof.** If $b_0, b_1 \in B$, let $\gamma : I \longrightarrow B$ be a path joining $b_0$ to $b_1$. Then $\gamma^* E = I \times_B E \longrightarrow I$ is fiber homotopy equivalent to the product fibrations $p^{-1}(b_0) \times I \longrightarrow I$, $p^{-1}(b_1) \times I \longrightarrow I$. Hence these two fibrations are fiber homotopy equivalent, and in particular $p^{-1}(b_0)$ is homotopy equivalent to $p^{-1}(b_1)$.

**Corollary (A.22).** *Let $p : E \longrightarrow B$ be a fibration, $\gamma : I \longrightarrow B$ a path with $\gamma(0) = b_0$, $\gamma(1) = b_1$. Let $F_i = p^{-1}(b_i)$, $i = 0,1$. Then there are canonical isomorphisms*
$$\gamma^* : H_n(F_0, G) \longrightarrow H_n(F_1, G), \quad \gamma_* : H^n(F_1, G) \longrightarrow H^n(F_0, F)$$
*for any coefficient group $G$ and $n \geq 0$. Further, $\gamma^*$, $\gamma_*$ depend only on the homotopy class of $\gamma$ (among paths joining $b_0$ to $b_1$), and are compatible with composition of paths. $((\gamma \circ \delta)^* = \delta^* \circ \gamma^*, (\gamma \circ \delta)_* = \gamma_* \circ \delta_*)$. In particular, $b \mapsto H_n(p^{-1}(b), G)$, $b \mapsto H^n(p^{-1}(b), G)$ give local systems on $B$.*

**Proof.** Let $\gamma^* E = I \times_B E \longrightarrow I$ be the pull-back fibration. Then by (A.20) there is a fiber homotopy equivalence $f : F_0 \times I \longrightarrow \gamma^* E$, where $f$ is a homotopy equivalence on each fiber, such that $f|_{F_0 \times \{0\}} : F_0 \longrightarrow F_0$ is homotopic to the identity map of $F_0$. The restriction of $f$ to $F_0 \times \{1\}$ is a map $\bar{f} : F_0 \longrightarrow F_1$ which is a homotopy equivalence. Let $\gamma^*$, $\gamma_*$ be the maps on homology and cohomology induced by $\bar{f}$. To check that they depend only on $\gamma$, it suffices to show that if $f_1, f_2$ are two fiber homotopy equivalences $F_0 \times I \longrightarrow \gamma^* E$ as above, then $\bar{f}_1, \bar{f}_2 : F_0 \longrightarrow F_1$ are homotopic.

Let $g_1, g_2 : \gamma^* E \longrightarrow F_0 \times I$ be the "inverse" maps to $f_1, f_2$ respectively, so that the $g_i$ are fiber preserving, and the composites $f_i \circ g_i$, $g_i \circ f_i$ are homotopic to the respective identity maps through fiber preserving homotopies. Then we have maps $g_2 \circ f_1$, $g_1 \circ f_2 : F_0 \times I \longrightarrow F_0 \times I$ such that the two composites $g_2 \circ f_1 \circ g_1 \circ f_2$ and $g_1 \circ f_2 \circ g_1 \circ f_1$ are each homotopic to the identity by fiber preserving homotopies, and such that the maps $F_0 \times \{0\} \longrightarrow F_0 \times \{0\}$ induced by restricting $g_2 \circ f_1$ and $g_1 \circ f_2$ are each homotopic to the identity. If we prove $g_2 \circ f_1$, $g_1 \circ f_2$ are each homotopic to the identity by fiber preserving homotopies, then we have fiber preserving homotopies

$$f_1 \sim (f_2 \circ g_2) \circ f_1 = f_2 \circ (g_2 \circ f_1) \sim f_2,$$

and

$$g_1 \sim g_1 \circ (f_2 \circ g_2) = (g_1 \circ f_2) \circ g_2 \sim g_2.$$

Hence it suffices to remark that if $f : F \times I \longrightarrow F \times I$ is a self fiber-homotopy equivalence of the product fibration $p_1 : F \times I \longrightarrow F$, such that $f_0 = f|_{F \times \{0\}} : F \longrightarrow F$ is homotopic to the identity, then $f$ is homotopic to the identity, by a fiber preserving homotopy. Let $H : F \times I \times I \longrightarrow F \times I$ be defined by $H(x, s, t) = (p_1 \circ f(x, st), s)$; then $H(x, s, 1) = (p_1 \circ f(x, s), s) = f(x, s)$ since $f$ is fiber preserving, while $H(x, s, 0) = (p_1 \circ f(x, 0), s) = (f_0 \times 1_I)(x, s)$, and clearly $H$ is a fiber preserving homotopy. But since $f_0 : F \longrightarrow F$ is homotopic to the identity, $f_0 \times 1_I : F \times I \longrightarrow F \times I$ is homotopic to the identity through a fiber preserving homotopy. This completes the proof that $\gamma_*, \gamma^*$ are well defined.

If $\gamma_1, \gamma_2$ are homotopic paths in $B$ joining $b_0$ to $b_1$, and $H : I \times I \longrightarrow B$ is a homotopy from $\gamma_1$ to $\gamma_2$ preserving end points, then $H^* E = (I \times I) \times_B E \longrightarrow I \times I$ is also homotopy equivalent to a product, so that we obtain fiber homotopy equivalences of $F_0 \times I$ with $\gamma_1^* E$ and $\gamma_2^* E$ such that the induced maps $F_0 \longrightarrow F_1$ are homotopic. Hence $(\gamma_1)_* = (\gamma_2)_*$, $(\gamma_1)^* = (\gamma_2)^*$. We leave to the reader the verification that $\gamma \mapsto \gamma_*, \gamma \mapsto \gamma^*$ are compatible with composition of paths (the latter assignment is, of course, contravariant).

**(A.23) The Leray–Serre Spectral Sequence.** Let $p : E \longrightarrow B$ be a fibration, where $B$ is a connected $CW$-complex. If $\pi : \tilde{B} \longrightarrow B$ is the universal covering, $\tilde{p}: \tilde{E} \longrightarrow \tilde{B}$ the pullback fibration, then by (A.22), for any two points $\tilde{b}_1$, $\tilde{b}_2$ of $\tilde{B}$ there are canonical isomorphisms $H^i(F_1, G) \cong H^i(F_2, G)$ where $F_j = \tilde{p}^{-1}(\tilde{b}_j)$, $j = 1, 2$, for any $i \geq 0$ and any coefficient group $G$. Let $B_p$ be the $p$-skeleton of $B$ for a fixed $CW$-decomposition, $\tilde{B}_p = \pi^{-1}(B_p)$ the $p$-skeleton of $\tilde{B}$ for the induced $CW$-decomposition, and let $E_p = p^{-1}(B_p)$, $\tilde{E}_p = \tilde{p}^{-1}(\tilde{B}_p)$.

If $S_p$ denotes the set of $p$-cells of $B_p$, the set of $p$-cells of $\tilde{B}_p$ can be identified with $S_p \times \Pi$, where $\Pi = \pi_1(B, b_0)$ for some fixed base point $b_0 \in B$, such that the natural action of $\Pi$ permuting the $p$-cells of $\tilde{B}_p$ becomes the action of $\Pi$ on $S_p \times \Pi$ induced by the trivial action of $\Pi$ on $S_p$ and the left regular representation of $\Pi$ on itself.

We have a pull-back fibration

$$p_1 : (\Delta_p \times S_p \times \Pi) \times_{\tilde{B}} \tilde{E} \longrightarrow (\Delta_p \times S_p \times \Pi)$$

induced by the natural map $\Delta_p \times S_p \times \Pi \longrightarrow \tilde{B}_p$ associated with the $CW$-decomposition of $\tilde{B}_p$. Fix a base point $\underline{0} \in \Delta_p$. Then by (A.20) $p_1$ is fiber homotopy equivalent to

$$\coprod_{(s,g) \in S_p \times \Pi} p_1^{-1}(\underline{0}, s, g) \times \Delta_p \longrightarrow \coprod_{(s,g) \in S_p \times \Pi} \Delta_p,$$

since $\Delta_p$ is contractible. Since

$$(((\Delta_p \times S_p \times \Pi) \times_{\tilde{B}} \tilde{E}) \coprod \tilde{E}_{p-1}, ((\partial \Delta_p \times S_p \times \Pi) \times_{\tilde{B}} \tilde{E}) \coprod \tilde{E}_{p-1})$$
$$\longrightarrow (\tilde{E}_p, \tilde{E}_{p-1})$$

is a relative homeomorphism of $NDR$-pairs (see (A.42); $f : (X, A) \longrightarrow (Y, B)$ is a relative homeomorphism if $f$ is onto, $Y$ has the quotient topology from $f$, and $f : X - A \longrightarrow Y - B$ is a homeomorphism), the excision theorem gives isomorphisms for each $n \geq 0$

$$H_n(\tilde{E}_p, \tilde{E}_{p-1}; G) \cong H_n((\Delta_p \times S_p \times \Pi) \times_{\tilde{B}} \tilde{E}, (\partial \Delta_p \times S_p \times \Pi) \times_{\tilde{B}} \tilde{E}; G)$$
$$\cong \bigoplus_{(s,g) \in S_p \times \Pi} H_n(p_1^{-1}(\underline{0}, s, g) \times \Delta_p, p_1^{-1}(\underline{0}, s, g) \times \partial \Delta_p; G)$$
$$\cong \bigoplus_{(s,g) \in S_p \times \Pi} H_{n-p}(p_1^{-1}(\underline{0}, s, g), G) \text{ (Kunneth theorem)}$$
$$\cong \bigoplus_{(s,g) \in S_p \times \Pi} H_{n-p}(F_0, G) \cong H_{n-p}(F_0, G) \otimes_{\mathbb{Z}} C_p(\tilde{B})$$

where $F_0 = p^{-1}(b_0) = \tilde{p}^{-1}(\tilde{b}_0)$ for some fixed base point $\tilde{b}_0 \in \tilde{p}^{-1}(b_0)$, and we have used the canonical isomorphism

$$H_{n-p}(p_1^{-1}(\underline{0}, s, g), G) \cong H_{n-p}(F_0, G)$$

given by (A.22). Here $C_p(\tilde{B})$ is the group of cellular $p$-chains on $\tilde{B}$ (see (A.12)).

Now $\Pi$ acts on $(\tilde{B}_p, \tilde{B}_{p-1})$, hence on $C_p(\tilde{B}) = H_p(\tilde{B}_p, \tilde{B}_{p-1}; \mathbb{Z})$ and on $H_p(\tilde{B}_p, \tilde{B}_{p-1}; G)$. If we let $\Pi$ act on itself through the left regular representation and act trivially on $\Delta_p \times S_p$, the induced action on $\Delta_p \times S_p \times \Pi$ makes

$$((\Delta_p \times S_p \times \Pi) \coprod \tilde{B}_{p-1}, (\partial \Delta_p \times S_p \times \Pi) \coprod \tilde{B}_{p-1}) \longrightarrow (\tilde{B}_p, \tilde{B}_{p-1})$$

a $\Pi$-equivariant relative homeomorphism. Also, $\Pi$ acts on $(\tilde{E}_p, \tilde{E}_{p-1})$ so that $(\tilde{E}_p, \tilde{E}_{p-1}) \longrightarrow (\tilde{B}_p, \tilde{B}_{p-1})$ is $\Pi$-equivariant. Hence $\Pi$ acts on $H_n(\tilde{E}_p, \tilde{E}_{p-1}; G)$. Under the isomorphism constructed above, we claim the induced $\Pi$-action on $H_{n-p}(F_0, G) \otimes_\mathbb{Z} C_p(\tilde{B})$ is the diagonal $\Pi$-action on the tensor product corresponding to the natural action on $C_p(\tilde{B})$ described above, and the $\Pi$-action on $H_{n-p}(F_0, G)$ corresponding to the local system $b \mapsto H_n(p^{-1}(b), G)$ on $B$ (from (A.22)). To verify the claim, we observe that if $g \in \Pi$, $\tilde{b} \in \tilde{B}$, $b = \pi(\tilde{b}) \in B$, then $\tilde{p}^{-1}(\tilde{b}) = \tilde{p}^{-1}(g \cdot \tilde{b}) = p^{-1}(b)$, but the two homotopy equivalences of $F_b = p^{-1}(b)$ with $F_0 = p^{-1}(b_0) = \tilde{p}^{-1}(\tilde{b}_0)$ (determined by paths joining $\tilde{b}$, $g\tilde{b}$ respectively to $\tilde{b}_0$) differ by a self homotopy equivalence of $F_0$ which induces the action of $g \in \pi$ on $H_*(F_0, G)$ given by (A.22). In fact the union of the images in $B$ of the paths joining $\tilde{b}$, $g\tilde{b}$ to $\tilde{b}_0$ give a loop at $b_0$ representing $g \in \Pi = \pi_1(B, b_0)$.

Thus for $x \in H_{n-p}(F_0, G)$, $y \in C_p(\tilde{B})$, $g \in \Pi$ we have $g \cdot (x \otimes y) = (g \cdot x) \otimes (g \cdot y)$ where $\cdot$ denotes the $\Pi$-action). Regarding $H_{n-p}(F_0, G)$ as a right $\Pi$-module by $x \cdot g = g^{-1} \cdot x$ for $x \in H_{n-p}(F_0, G)$, we can rewrite the formula for the action as $g(x \cdot g \otimes y) = (x \otimes g \cdot y)$.

Next, consider the long exact homology sequence for the triple $(\tilde{E}_p, \tilde{E}_{p-1}, \tilde{E}_{p-2})$ which yields the ($\Pi$-equivariant) boundary map

$$H_n(\tilde{E}_p, \tilde{E}_{p-1}; G) \longrightarrow H_{n-1}(\tilde{E}_{p-1}, \tilde{E}_{p-2}; G).$$

By our earlier calculations this can be viewed as a map

$$H_{n-p}(F_0, G) \otimes_\mathbb{Z} C_p(\tilde{B}) \longrightarrow H_{n-p}(F_0, G) \otimes_\mathbb{Z} C_{p-1}(\tilde{B}).$$

It can be shown (see [W], XIII, (4.8)) that this boundary map is of the form $1 \otimes d_p$, where $d_p : C_p(\tilde{B}) \longrightarrow C_{p-1}(\tilde{B})$ is the cellular boundary map (see (A.12)).

Finally, the natural map $(\tilde{E}_p, \tilde{E}_{p-1}) \longrightarrow (E_p, E_{p-1})$ is just the quotient map modulo the free $\Pi$-action on the former, so that $H_n(\tilde{E}_p, \tilde{E}_{p-1}; G) \longrightarrow H_n(E_p, E_{p-1}; G)$ factors through the $\Pi$-coinvariants

$$H_n(\tilde{E}_p, \tilde{E}_{p-1}; G) \, / \, \{x - g^{-1}x \mid x \in H_n(\tilde{E}_p, \tilde{E}_{p-1}; G), g \in \Pi\}.$$

Appendix A: Results from Topology 251

We have a diagram of pairs whose horizontal maps are relative homeomorphisms, and vertical maps are quotients modulo a free $\Pi$-action,

$$(((\Delta_p \times S_p \times \Pi) \times_{\tilde{B}} \tilde{E}) \coprod \tilde{E}_{p-1}, ((\partial \Delta_p \times S_p \times \Pi) \times_{\tilde{B}} E) \coprod \tilde{E}_{p-1})) \longrightarrow (\tilde{E}_p, \tilde{E}_{p-1})$$

$$\downarrow \qquad\qquad\qquad\qquad\qquad\qquad\qquad\qquad\qquad\qquad\qquad \downarrow$$

$$(((\Delta_p \times S_p) \times_B E) \coprod E_{p-1}, ((\partial \Delta_p \times S_p) \times_B E) \coprod E_{p-1}) \longrightarrow (E_p, E_{p-1}).$$

Computing with the left-hand vertical arrow, one sees that in fact $H_n(E_p, E_{p-1}; G)$ is identified with the $\Pi$-coinvariants of $H_n(\tilde{E}_p, \tilde{E}_{p-1}; G)$, i.e., we have an isomorphism

$$H_n(E_p, E_{p-1}; G) \cong \Pi\text{-coinvariants of } H_{n-p}(F_0, G) \otimes_{\mathbb{Z}} C_p(\tilde{B})$$
$$= H_{n-p}(F_0, G) \otimes_{\mathbb{Z}[\Pi]} C_p(\tilde{B}),$$

since the kernel of the surjection onto the coinvariants

$$H_{n-p}(F_0, G) \otimes_{\mathbb{Z}} C_p(\tilde{B}) \longrightarrow H_n(E_p, E_{p-1}, G)$$

is precisely the subgroup generated by all elements of the form

$$g \cdot (x \otimes y) - x \otimes y = xg^{-1} \otimes g \cdot y - x \otimes y.$$

Thus, the long exact homology sequence for the pair $(E_p, E_{p-1})$ can be written as

$$\cdots \longrightarrow H_n(E_{p-1}, G) \longrightarrow H_n(E_p, G) \longrightarrow H_{n-p}(F_0, G) \otimes_{\mathbb{Z}[\Pi]} C_p(\tilde{B})$$
$$\longrightarrow H_{n-1}(E_{p-1}, G) \longrightarrow \cdots$$

and the composite

$$H_n(E_p, E_{p-1}; G) \xrightarrow{\partial} H_{n-1}(E_{p-1}, G) \longrightarrow H_{n-1}(E_{p-1}, E_{p-2}; G)$$

(the boundary map in the exact sequence of the triple $(E_p, E_{p-1}, E_{p-2})$) is identified with

$$1 \otimes d_p : H_{n-p}(F_0, G) \otimes_{\mathbb{Z}[\Pi]} C_p(\tilde{B}) \longrightarrow H_{n-p}(F_0, G) \otimes_{\mathbb{Z}[\Pi]} C_{p-1}(\tilde{B}).$$

Since $\varinjlim_p H_n(E_p, G) = H_n(E, G)$, the method of exact couples (see Appendix C) yields a spectral sequence (of homological type) whose complexes of $E^1$ terms are identified (by (A.14)) with the complexes of cellular chains used to compute the homology groups with local coefficients $H_*(B, H_*(F_0, G))$. Thus we have shown:

**Theorem (A.24).** *Let $p : E \longrightarrow B$ be a fibration, where $B$ is a connected CW-complex. Then there is a first quadrant spectral sequence of homological type*

$$E^2_{p,q} = H_p(B, H_q(F, G)) \Longrightarrow H_{p+q}(E, G)$$

for any coefficient group $G$, where $H_q(F,G)$ denotes the local coefficient system on $B$ given by $b \mapsto H_q(p^{-1}(b), G)$ (and (A.22)).

An analogous argument (but some care is required in passing from $E_p$, for large $p$, to $E = \varinjlim E_p$) yields

**Theorem (A.25).** *Let $p : E \longrightarrow B$ be a fibration, where $B$ is a connected CW-complex. Then there is a first quadrant spectral sequence of cohomological type*

$$E^2_{p,q} = H^p(B, H^q(F,G)) \Longrightarrow H^{p+q}(E, G)$$

*for any coefficient group $G$, where $H^q(F,G)$ is the local coefficient system on $B$ given by $b \mapsto H^q(p^{-1}(b), G)$ and (A.22)).*

We state a lemma about fibrations needed below.

**Lemma (A.26).** *Let $p : E \longrightarrow B$ be a fibration, $b \in B$, $F = p^{-1}(b)$, $x \in F$. Then with the natural action of $\pi_1(F,x)$ on $\pi_i(F,x)$, $i \geq 1$, $\pi_1(F,x)$ acts trivially on $\ker(\pi_i(f,x) \longrightarrow \pi_i(E,x))$.*

**Proof.** The exact homotopy sequence for the pair $(E, F)$ shows that we have a $\pi_1(F,x)$-equivariant isomorphism

$$\ker(\pi_i(F,x) \longrightarrow \pi_i(E,x)) \cong \operatorname{image}(\partial : \pi_{i+1}(E,F;x) \longrightarrow \pi_i(F,x)).$$

So it suffices to prove that under the above hypothesis, $\pi_1(F,x)$ acts trivially on $\pi_n(E,F;x)$ for $n \geq 2$. But the map of pairs $(E, F) \longrightarrow (B, \{b\})$ yields an isomorphism $p_* : \pi_n(E,F;x) \cong \pi_n(B,x)$, (since $p$ is a fibration), which is $\pi_1(F,x)$-equivariant, where $\pi_1(F,x)$ acts on the homotopy sequence for $(B, \{b\})$ through the map $\pi_1(F,x) \longrightarrow \pi_1(\{b\}, b) = 0$. This follows from the naturality of the homotopy sequence for a pair and the naturality of the $\pi_1$-action of the subspace on the sequence.

**(A.27) Homotopy Fibers.** Let $f : X \longrightarrow Y$ be an arbitrary (continuous) map of spaces. Let $P(Y)$ denote the free path space of $Y$, i.e., $P(Y) = \mathbb{F}(I, Y)$ is the space of maps $I \longrightarrow Y$ (topologized as in (A.1)). There are maps $f_i : P(Y) \longrightarrow Y$, $i = 0, 1$ given by associating to any path its initial and final point, respectively. The evaluation map $e : P(Y) \times I \longrightarrow Y$ gives a homotopy between the $f_i$. Let $M(f)$ be defined as the pullback

$$\begin{array}{ccc} M(f) & \xrightarrow{f'} & P(Y) \\ {\scriptstyle p_0}\downarrow & & \downarrow{\scriptstyle f_0} \\ X & \xrightarrow{f} & Y \end{array}$$

Then the composite $p : M(f) \xrightarrow{f'} P(Y) \xrightarrow{f_1} Y$ is called the *mapping path fibration* of $f$. There is a section $\lambda : X \longrightarrow M(f)$ of $p_0$ given by $\lambda(x) = (x, \omega_{f(x)})$, where for any $y \in Y$, $\omega_y$ denotes the constant path $I \longrightarrow \{y\}$.

**Lemma.** *With the above notation*

(i) $p_0, \lambda$ *are homotopy inverses*

(ii) $p \circ \lambda : X \longrightarrow M(f) \longrightarrow Y$ *is just* $f$

(iii) $p = f_1 \circ f' \cong f_0 \circ f' = f \circ p_0$ ($\cong$ *denotes a homotopy*)

(iv) $p : M(f) \longrightarrow Y$ *is a fibration*.

**Proof.** (i), (ii), (iii) are left to the reader as easy exercises ([W], I, Chapter 7). To prove (iv), let $g : Z \longrightarrow M(f)$ be a map, $H : Z \times I \longrightarrow Y$ a homotopy beginning with $p \circ g$ i.e., $H(z, 0) = p \circ g(z) \; \forall z \in Z$. Let $H' : Z \times I \longrightarrow M(f)$ be defined as follows: if $g(z) = (x, \gamma)$ for some path $\gamma : I \longrightarrow Y$ with $\gamma(0) = f(x)$, then $p \circ g(z) = \gamma(1)$; let $H'(z,t) = (x, \gamma_t)$ where

$$\gamma_t(s) = \begin{cases} \gamma(2s/2 - t) & \text{if } 0 \leq s \leq 1 - t/2 \\ H(z, (2s + t - 2)) & \text{if } 1 - t/2 \leq s \leq 1. \end{cases}$$

(i.e., juxtapose $\gamma$ and the path beginning at $x = p \circ g(z)$ given by $H$). Clearly $p \circ H'(z, t) = \gamma_t(1) = H(z, t)$. We leave it to the reader to verify that $H'$ is continuous.

**Definition.** The fiber $p^{-1}(y)$ of $p : M(f) \longrightarrow Y$ is called the *homotopy fiber* of $f$ at $y$, denoted by $F(f, y)$ (or just $F(f)$).

By (A.21), we see that if $Y$ is path connected, the homotopy type of $F(f, y)$ is independent of $Y$. Explicitly, $F(f, y)$ is the set of pairs $(x, \gamma) \in X \times P(Y)$ with $\gamma(0) = f(x)$, $\gamma(1) = y$. Since $p$ is a fibration, and $p_0 : M(f) \longrightarrow X$ is a homotopy equivalence such that $f \circ p_0 : M(f) \longrightarrow Y$ is homotopic to $p$, we have a long exact homotopy sequence

$$\ldots \to \pi_i(F(f, y), (x, \omega_y)) \xrightarrow{(p_0)_*} \pi_i(X, x) \xrightarrow{f_*} \pi_i(Y, y)$$
$$\longrightarrow \pi_{i-1}(F(f, y), (x, \omega_y)) \ldots \to$$

where $x \in f^{-1}(y)$ is a base point of $X$ (and $\omega_y : I \longrightarrow \{y\}$). Taking $X = \{y\}$, $F(f, y)$ is just $\Omega(Y)$, the space of loops based in $Y$ based at $y$; we obtain isomorphisms $\pi_{i+1}(Y, y) \cong \pi_i(\Omega(Y), \omega_y)$.

If $Y$ is a connected $CW$-complex, we have Leray–Serre spectral sequences ((A.24), (A.25)) for homology and cohomology

$$E^2_{p,q} = H_p(Y, H_q(F(f), G)) \Longrightarrow H_{p+q}(X, G)$$
$$E_2^{p,q} = H^p(Y, H^q(F(f), G)) \Longrightarrow H^{p+q}(X, G)$$

for any coefficient group $G$, where $H_q(F(f),G)$, $H^q(F(f),G)$ are the local coefficient systems on $Y$ associated to $p$ (by A.22). In this case, the homotopy class of the self homotopy equivalence of $F(f,y)$, induced by the class of a loop $\sigma: I \longrightarrow Y$ based at $y$, has the explicit representative $\tilde{\sigma}: F(f,y) \longrightarrow F(f,y)$ given by $\tilde{\sigma}(x,\gamma) = (x, \sigma \cdot \gamma)$ where

$$\sigma \cdot \gamma(t) = \begin{cases} \gamma(2t) & \text{if } 0 \leq t \leq 1/2 \\ \sigma(2t-1) & \text{if } 1/2 \leq t \leq 1. \end{cases}$$

The homotopy fiber $F(f,y)$ has the following property which characterizes it: to give a map $Z \longrightarrow F(f,y)$ is equivalent to giving (i) a map $g: Z \longrightarrow X$ (ii) a homotopy of $f \circ g: Z \longrightarrow Y$ to the constant map $Z \longrightarrow \{y\}$. (A.26) immediately yields the fact that $\pi_1(F(f,y),(x,\omega_y))$ acts trivially on

$$\ker(\pi_n(F(f,y),(x,\omega_y)) \xrightarrow{p_*} \pi_n(X,x))$$

for any $n \geq 1$.

Finally if $f: X \longrightarrow Y$ is a map to a path connected space $Y$, and if $g: \tilde{Y} \longrightarrow Y$ is a covering space, $\tilde{f}: \tilde{X} = X \times_Y \tilde{Y} \longrightarrow \tilde{Y}$ the induced map, and $y \in Y$, then the natural map,

$$F(\tilde{f}, \tilde{y}) \longrightarrow F(f, y)$$

(where $y = g(\tilde{y})$) is a homeomorphism. Indeed, if $(x, \gamma) \in F(f,y)$ where $\gamma$ is a path in $Y$ with $\gamma(0) = f(x)$, $\gamma(1) = y$, then there is a unique path $\tilde{\gamma}$ in $\tilde{Y}$ lifting $\gamma$ such that $\tilde{\gamma}(1) = \tilde{y}$; there is a unique point $\tilde{x} \in \tilde{X}$ with $\tilde{x} \mapsto x$ under $\tilde{X} \longrightarrow X$, and $\tilde{f}(\tilde{x}) = \tilde{y}$. Then $(x, \gamma) \mapsto (\tilde{x}, \tilde{\gamma})$ gives the inverse homeomorphism $F(f,y) \longrightarrow F(\tilde{f}, \tilde{y})$.

**(A.28) Spectral Sequence for a Covering.** Let $p: Y \longrightarrow X$ be a Galois covering space with (discrete) covering group $G$. If $f: \Delta_p \longrightarrow X$ is a singular $p$-simplex, then $\Delta_p \times_X Y \longrightarrow \Delta_p$ is a trivial covering, since $\Delta_p$ is simply connected. Hence the group of singular $p$-chains $S_p(Y)$ is in a natural way a free $\mathbb{Z}[G]$-module, such that $S_p(X) = \mathbb{Z} \otimes_{\mathbb{Z}[G]} S_p(Y)$ where $\mathbb{Z}$ is regarded as a $G$-module with trivial $G$-action.

Let

$$\cdots \longrightarrow P_n \longrightarrow P_{n-1} \longrightarrow \cdots \longrightarrow P_0 \longrightarrow \mathbb{Z} \longrightarrow 0$$

be a projective $\mathbb{Z}[G]$-resolution of $\mathbb{Z}$, and $A_{pq} = P_p \times_{\mathbb{Z}[G]} S_q(Y)$ so that $\{A_{pq}\}$ form the terms of the double complex associated to the tensor product over $\mathbb{Z}[G]$ of the singular complex of $Y$ and the resolution of $\mathbb{Z}$. Then the total complex $\text{Tot}(A_{pq})$ has homology groups $H_n(\text{Tot}(A_{pq})) = H_n(X, \mathbb{Z})$, since the spectral sequence with $E^1_{p,q}$ obtained by taking the homology groups in the $p$-direction has $E^2_{p,q} = 0$ for $p \neq 0$, and $E^2_{0,n} = H_n(\mathbb{Z} \otimes_{\mathbb{Z}[G]}$

$S_*(Y)) = H_n(S_*(X)) = H_n(X, \mathbb{Z})$. The second spectral sequence for the double complex has
$$E^1_{p,q} = P_p \otimes_{\mathbb{Z}[G]} H_q(Y, \mathbb{Z}),$$
so that $E^2_{p,q} = \operatorname{Tor}^{\mathbb{Z}[G]}_p(\mathbb{Z}, H_q(Y, \mathbb{Z})) = H_p(G, H_q(Y, \mathbb{Z}))$. Hence we obtain a spectral sequence of homological type
$$E^2_{p,q} = H_p(G, H_q(Y, \mathbb{Z})) \Longrightarrow H_{p+q}(X, \mathbb{Z}).$$
Similarly, one has a spectral sequence of cohomological type
$$E_2^{p,q} = H^p(G, H^q(Y, \mathbb{Z})) \Longrightarrow H^{p+q}(X, \mathbb{Z}).$$

**(A.29) Quasi-Fibrations.** A continuous surjective map $p : E \longrightarrow B$ is called a *quasi fibration* if for each $x \in B$, $y \in p^{-1}(x)$ and $i \geq 0$ the natural map $p_* : \pi_i(E, p^{-1}(x); y) \longrightarrow \pi_i(B, x)$ is a bijection. (See A. Dold, R. Thom, Quasifaserungen und unendlische symmetrische produkte, *Ann. Math.* 67 (1958)).

Note that $p_*$ is an isomorphism of groups for $i \geq 2$, and a bijection of sets with a distinguished point for $i = 0, 1$. We can use $p_*$ to give $\pi_1(E, p^{-1}(x); y)$ the structure of a group, so that $p_*$ becomes an isomorphism of groups even for $i = 1$. The long exact homotopy sequence for a pair yields a long exact sequence
$$\longrightarrow \pi_{i+1}(B, x) \longrightarrow \pi_i(p^{-1}(x), y) \longrightarrow \pi_i(E, y) \longrightarrow \pi_i(B, x)$$
$$\longrightarrow \cdots \longrightarrow \pi_0(B, x) \longrightarrow 0.$$

Thus for each $x \in B$, the natural map $p^{-1}(x) \longrightarrow F(p, x)$ to the homotopy fiber at $x$ (see (A.27)) induces a bijection on homotopy groups, i.e., is a weak homotopy equivalence. Thus if $B$ is path connected, so that the homotopy fibers of $p$ at any two points of $B$ are homotopy equivalent, we see that for any $x_1, x_2 \in B$ the fibers $p^{-1}(x_1)$, $p^{-1}(x_2)$ are weakly homotopy equivalent.

Given two quasi fibrations $p : E \longrightarrow B$, $p' : E' \longrightarrow B'$, a map $f : E \longrightarrow E'$ is called fiber preserving if there is a map of sets $g : B \longrightarrow B'$ making

$$\begin{array}{ccc} E & \xrightarrow{f} & E' \\ p \downarrow & & \downarrow p' \\ B & \xrightarrow{g} & B' \end{array}$$

commute (note that $g$ need not be continuous). If $f : E \longrightarrow E'$ is fiber preserving, $x \in B$, $x' = g(x)$, $y \in p^{-1}(x)$, $y' = f(x') \in p'^{-1}(x')$ then we have a map of pairs $(E, p^{-1}(x)) \longrightarrow (E', p'^{-1}(x'))$ which induces a map between their exact homotopy sequences; hence we have a commutative diagram

$$\begin{array}{ccccccc}
\to \pi_{i+1}(B,x) & \to & \pi_i(p^{-1}(x),y) & \to & \pi_i(E,y) & \to & \pi_i(B,x) & \to \\
\downarrow g_* & & \downarrow f_* & & \downarrow f_* & & \downarrow g_* & \\
\to \pi_{i+1}(B',x') & \to & \pi_i(p'^{-1}(x'),y') & \to & \pi_i(E',y') & \to & \pi_i(B',x') & \to \\
& \cdots \to & \pi_0(B,x) & \to & 0 & & & \\
& & \downarrow g_* & & & & & \\
& \cdots \to & \pi_0(B',x') & \to 0 & & & &
\end{array}$$

where $g_*$ is the unique arrow (for each $i$) making the diagram below commute:

$$\begin{array}{ccc}
\pi_i(E, p^{-1}(x); y) & \xrightarrow{p_*} & \pi_i(B, x) \\
\downarrow f_* & & \downarrow g_* \\
\pi_i(E', p'^{-1}(x'); y') & \xrightarrow{p'_*} & \pi_i(B', x)
\end{array}$$

If $p : E \longrightarrow B$ is any (continuous) map, a subset $U \subset B$ is called *distinguished* for $p$ if $p^{-1}(U) \longrightarrow U$ is a quasi fibration.

**Theorem (A.30).** *Let $p : E \longrightarrow B$ be a map, $\mathcal{U} = \{U_i\}$ an open cover of $B$. Assume that*

(i) *each $U_i$ is distinguished for $p$*

(ii) *for each $x \in U_i \cap U_j$ there exists $U_k \in \mathcal{U}$ with $x \in U_k \subset U_i \cap U_j$.*

*Then $B$ is distinguished for $p$, i.e., $p$ is a quasi fibration.*

The proof of this theorem requires the following results.

**Lemma (A.31).** *Let $p : E \longrightarrow B$ be a map, $U \subset B$ a distinguished subset for $p$. Then the following two statements are equivalent:*

(a) $p_* : \pi_i(E, p^{-1}(x); y) \cong \pi_1(B, x)$ *for all $x \in U$, $y \in p^{-1}(x)$, $i \geq 0$.*

(b) $p_* : \pi_i(E, p^{-1}(U); y) \cong \pi_i(B, U; x)$ *for all $x \in U$, $y \in p^{-1}(x)$, $i \geq 0$.*

**Proof.** We have a diagram whose rows are the exact homotopy sequences of the triples $(E, p^{-1}(U), p^{-1}(x))$ and $(B, U, \{x\})$

$$\begin{array}{ccccc}
\pi_i(p^{-1}(U), p^{-1}(x); y) & \to & \pi_i(E, p^{-1}(x); y) & \to & \pi_i(E, p^{-1}(U); y) \to \cdots \\
\downarrow & & \downarrow & & \downarrow \\
\pi_i(U, x) & \to & \pi_i(B, x) & \to & \pi_i(B, U; x) \to \cdots
\end{array}$$

By hypothesis, $\pi_i(p^{-1}(U), p^{-1}(x); y) \cong \pi_i(U, x)$ for all $i \geq 0$. Hence the equivalence of (a) and (b) follows from the 5-lemma, except that a little care is required for $i = 0, 1$ (for $i = 2$ we are dealing with possibly non-Abelian groups, but the proof of the 5-lemma goes through).

In case $i = 0$, recall that $\pi_0(X, Y)$ is obtained from the set $\pi_0(X)$ of path components of $X$ by identifying the points corresponding to path components meeting $Y$. From this, (a) $\Rightarrow$ (b) (for $\pi_0$) and the surjectivity of $p_*$ on $\pi_0$ in (b) $\Rightarrow$ (a) follow easily. To prove injectivity of $p_* : \pi_0(E, p^{-1}(x); y) \longrightarrow \pi_0(B, x)$ in (b) $\Rightarrow$ (a), we have to prove that if $E_1$, $E_2$ are path components of $E$ lying over the same path component of $B$, then either $E_1 = E_2$ or $E_1$, $E_2$ both meet $p^{-1}(x)$. Since $[E_1] = [E_2]$ in $\pi_0(E, p^{-1}(U); y)$; let $V_1$, $V_2$ be two path components of $p^{-1}(U)$ with $V_i \cap E_i \neq \emptyset$ (so that $V_i \subset E_i$). Then $p(V_1)$, $p(V_2)$ lie in the same path component of $B$, namely that which contains $p(E_1)$ and $p(E_2)$. Hence for some $z \in p(V_1)$ we can find a path in $B$ beginning at $z$ and ending in $p(V_2)$, giving a class $\alpha \in \pi_1(B, U; z)$. Since by (b)

$$p_* : \pi_1(E, p^{-1}(U); w) \cong \pi_1(B, U; z)$$

for any $z \in U$, $w \in p^{-1}(z)$, we may choose $z$ as above in $p(V_1)$ and $w \in V_1$, and find a path in $E$ joining $V_1$ to some path component $V_3$ of $p^{-1}(U)$, such that $p(V_2)$, $p(V_3)$ lie in the same path component of $U$ (since the image of this path under $p$ must be homotopic to the original one used to define $\alpha$). Since we have an isomorphism

$$p_* : \pi_0(p^{-1}(U), p^{-1}(x)) \cong \pi_0(U, x),$$

$V_2$ and $V_3$ give the same class in $\pi_0(p^{-1}(U), p^{-1}(x))$, i.e., either $V_2 = V_3$, or $V_2$, $V_3$ meet $p^{-1}(x)$. Since $V_1$ and $V_3$ are joined by a path in $E$, and $V_1 \subset E_1$, we also have $V_3 \subset E_1$. Hence either $V_2 \subset E_1$, i.e., $E_1 = E_2$, or $E_1$, $E_2$ meet $p^{-1}(x)$.

Similarly to prove the statement about $\pi_1$ in (a) $\Rightarrow$ (b) or (b) $\Rightarrow$ (a), we again use the trick of shifting the base point. For example, if $\alpha_1, \alpha_2 \in \pi_1(E, p^{-1}(U); y)$ have the same image in $\pi_1(B, U; x)$, let $w_1, w_2$ be paths in $E$ beginning at $y$ and ending in $p^{-1}(U)$ such that $[w_i] = \alpha_i \in \pi_1(E, p^{-1}(U); y)$. Then $p \circ w_i(0) = x$ and there is a homotopy between $p \circ w_1$ and $p \circ w_2$, fixing $p \circ w_i(0) = x$, and moving $p \circ w_1(1)$ to $p \circ w_2(1)$ along a path in $U$. We can regard $w_1^{-1} \cdot w_2$ (the composite path) as a path in $E$ beginning at $y' = w_1(1)$ and ending in $p^{-1}(U)$, i.e., as giving an element of $\pi_1(E, p^{-1}(U); y')$, and the homotopy between $p \circ w_1$ and $p \circ w_2$ can be regarded a nullhomotopy of $p \circ (w_1^{-1} \cdot w_2)$. Thus to prove injectivity of $p_*$ on $\pi_1$ in (a) $\Rightarrow$ (b) or (b) $\Rightarrow$ (a), one reduces to proving that the inverse image of the class of the constant path (loop) is the class of the constant path, for all possible choices of the base points. This follows from a straightforward diagram chase.

We will prove the surjectivity of $p_*$ on $\pi_1$ in (a) $\Rightarrow$ (b), and leave (b) $\Rightarrow$ (a) to the reader. Suppose (a) holds; let $\alpha \in \pi_1(B, U; x)$. Then by a diagram chase, one can find $\beta \in \pi_1(E, p^{-1}(U); y)$ (for any given $y \in p^{-1}(x)$) such that $p_*\beta, \alpha$ have the same image in $\pi_0(U, x)$. Choose paths $w_1 : I \longrightarrow B$, $w_2 : I \longrightarrow E$ representing $\alpha, \beta$ respectively. Then $w_1(0) = x$, $w_1(1) \in U$, $w_2(0) = y$, $w_2(1) \in p^{-1}(U)$, and $p \circ w_2(1)$ and $w_1(1)$ lie in the same path component of $U$ (this expresses the fact that $\alpha$, $p_*\beta$ have the same image in $\pi_0(U, x)$). Let $y' = w_2(1)$, $x' = p(y')$, $x'' = w_1(1)$. Let $w_3 : I \longrightarrow U$ be a path with $w_3(0) = x''$, $w_3(1) = x'$. Then $\gamma = ((p \circ w_2)^{-1} \cdot w_1) \cdot w_3$ is a loop in $B$ based at $x'$ (here $\cdot$ denotes composition of paths). By (a), $p_* : \pi_1(E, p^{-1}(x'); y') \cong \pi_1(B, x')$ so that we can find a path $w_4 : I \longrightarrow E$ with $w_4(0) = y'$, $w_4(1) \in p^{-1}(x')$, such that $p \circ w_4$ is a loop in $B$ based at $x'$ which is homotopic to $\gamma$ through a homotopy fixing $x'$. Now $w_2 \cdot w_4$ is a path in $E$ beginning at $y$ and ending at $w_4(1) \in p^{-1}(x') \subset p^{-1}(U)$, so that it represents a class in $\pi_1(E, p^{-1}(U); x)$. Further, the homotopy of $p \circ w_4$ with $\gamma$ can be regarded a homotopy between $w_1$ and $p \circ (w_2 \cdot w_4)$ keeping the first end point fixed at $x$, and moving the second end point $w_1(1)$ to $p \circ (w_2 \cdot w_4)(1) = x'$ along the path $w_3$, i.e., $p_*[(w_2 \cdot w_4)] = \alpha$ (a picture of the homotopy is given below; the top and bottom sides are mapped to $x'$)

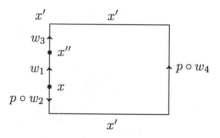

The other result needed to prove (A.30) is a weak form of the homotopy lifting property.

**Theorem (A.32).** *Let $p : E \longrightarrow B$ be a map, $\mathcal{U} = \{U_i\}$ an open cover of $B$ satisfying the hypothesis of (A.30). For some $n > 0$, assume given a diagram*

$$\begin{array}{ccc} I^{n-1} \times \{0\} & \xrightarrow{h} & E \\ \downarrow & & \downarrow \\ I^{n-1} \times I & \xrightarrow{\tilde{H}} & B \end{array}$$

Appendix A: Results from Topology

and an open set $U \in \mathcal{U}$ such that $\bar{H}(\partial I^{n-1} \times I \cup I^{n-1} \times \{1\}) \subset U$. Then there exists a map $H : I^{n-1} \times I \longrightarrow E$ with $H|_{I^{n-1} \times \{0\}} = h$, and a deformation $D : I^{n-1} \times I \times I \longrightarrow B$ such that

i) $D(z, s, 0) = \bar{H}(z, s)$, $D(z, s, 1) = p \circ H(z, s)$, $D(z, 0, t) = p \circ h(z)$ for all $z \in I^{n-1}$, $s, t \in I$.

ii) $D(\partial I^{n-1} \times I \times I \cup I^{n-1} \times \{1\} \times I) \subset U$.

We first prove (A.30) using (A.31) and (A.32).

**Proof of (A.30).** From (A.31) (since the $U_i$ cover $B$) it suffices to show that for $x \in U \in \mathcal{U}$, $y \in p^{-1}(x)$, $i \geq 0$ the map

$$p_* : \pi_i(E, p^{-1}(U); y) \cong \pi_i(B, U; x).$$

(a) $p_*$ is onto. let $\alpha \in \pi_i(B, U; x)$, $i > 0$ ($i = 0$ is trivial since $p$ is clearly onto). Let $H_0 : (I^{i-1} \times I, I^{i-1} \times \{1\}) \longrightarrow (B, U)$ be a map representing $\alpha$, such that $H_0(I^{i-1} \times \{0\} \cup \partial I^{i-1} \times I) = \{x\}$. We can find a homeomorphism $\theta : I^{i-1} \times I \longrightarrow I^{i-1} \times I$ such that

$$\theta(I^{i-1} \times \{0\}) = I^{i-1} \times \{0\} \cup \partial I^{i-1} \times I,$$

$$\theta(I^{i-1} \times \{1\} \cup \partial I^{i-1} \times I) = I^{i-1} \times \{1\}$$

(see the figure below for $\theta^{-1}$; the lines are $\theta(\{s\} \times I)$ for $s \in I^{i-1}$)

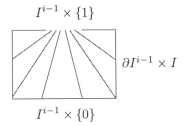

We apply (A.32) with $n = i$, $\bar{H} = H_0 \circ \theta$, $h(I^{i-1} \times \{0\}) = \{y\}$; for the given $U \in \mathcal{U}$, $\bar{H}(I^{i-1} \times \{1\} \cup \partial I^{i-1} \times I) = H_0(I^{i-1} \times \{i\}) \subset U$. Then if $H, D$ are as given by (A.32), $H \circ \theta^{-1} : (I^{i-1} \times I, I^{i-1} \times \{1\}) \longrightarrow (E, p^{-1}(U))$ gives a class $\beta \in \pi_i(E, p^{-1}(U); y)$, and $D$ gives a homotopy between $H_0$ and $p \circ H \circ \theta^{-1}$, so that $p_*(\beta) = \alpha$.

(b) $p_*$ is one-one. Let $\alpha \in \pi_i(E, p^{-1}(U); y)$ such that $p_*(\alpha) = 0$ (it suffices to consider this situation, and show that $\alpha = 0$, to prove the result for $i \geq 1$—see the proof of (A.31); the case $i = 0$ is left to the reader). Let $h : (I^i, \partial I^i) \longrightarrow (E, p^{-1}(U))$, $\bar{H} : (I^i \times I, \partial I^i \times I) \longrightarrow (B, U)$ be maps such that $h$ represents $\alpha$, and $\bar{H}$ is a null homotopy of $p \circ h$ to the constant map $I^i \longrightarrow \{x\}$. By (A.32), with $n = i + 1$, we can find

$$H : (I^i \times I, \partial I^i \times I) \longrightarrow (E, p^{-1}(U)) \text{ with } H|_{I^i \times \{0\}} = h,$$

and $H(I^i \times \{1\}) \subset p^{-1}(U)$. Then $H|_{I^i \times \{1\}}$ is also a representative for $\alpha$, i.e., $\alpha = 0$. This completes the proof of (A.30).

This proof of (A.32) uses the following lemma.

**Lemma (A.33).** *Let $p : F \longrightarrow U$ be a map, $V \subset U$, $G = p^{-1}(V)$. Let $m \geq 0$ be an integer such that for each $x \in V$, $y \in p^{-1}(x)$, the map*

$$p_* : \pi_i(F, G; y) \longrightarrow \pi_i(U, V; x)$$

*is injective for $i = m$ and surjective for $i = m + 1$. Assume given maps*

(i) $\bar{H} : (I^m \times I, I^m \times \{1\}) \longrightarrow (U, V)$

(ii) $h : (I^m \times \{0\} \cup \partial I^m \times I, \partial I^m \times \{1\}) \longrightarrow (F, G)$

(iii) $d : (I^m \times \{0\} \times I \cup \partial I^m \times I \times I, \partial I^m \times \{1\} \times I) \longrightarrow (U, V)$ *with*

$$d(z, s, 0) = \bar{H}(z, s), d(z, s, 1) = p \circ h(z, s) \quad \forall z \in Z, s \in I$$

*(such that $(z, s) \in I^m \times \{0\} \cup \partial I^m \times I$).*

*Then we can find maps*

(a) $H : (I^m \times I, I^m \times \{1\}) \longrightarrow (F, G)$ *which extends $h$*

(b) $D : (I^m \times I, I^m \times \{1\} \times I) \longrightarrow (U, V)$ *which extends $d$, and satisfies*

$$D(z, s, 0) = \bar{H}(z, s), D(z, s, 1) = p \circ H(z, s) \quad \forall (z, s) \in I^m \times I.$$

**Proof.** $h$ determines a class $\alpha \in \pi_m(F, G)$ and $d, \bar{H}$ give a homotopy of $p \circ h$ to a map whose image lies in $V$, i.e., $p_*(\alpha) = 0$ in $\pi_m(U, V)$. Since $p_*$ is injective on $\pi_m$, $\alpha = 0$, so that we can find a map $H' : (I^m \times I, I^m \times \{1\}) \longrightarrow (F, G)$ which extends $h$; we may assume without loss of generality that $H'$ is constant on the subset $K \times [3/4, 1] \subset I^m \times I$, where $K = [1/4, 3/4]^m \subset I^m$ (we can deform the identity map of $I^m \times I$ to a map shrinking $K \times [3/4, 1]$ to a point but which is the identity on $I^m \times \{0\} \cup \partial I^m \times I$, and maps $I^m \times \{1\}$ into itself). Let $y = H'(K \times [3/4, 1])$ so that $y \in G$ (as $H'(K \times \{1\}) \subset H'(I^m \times \{1\}) \subset G$).

Let $\bar{K} = I^m \times I \times \{0, 1\} \cup I^m \times \{0\} \times I \cup \partial I^m \times I \times I =$ closure of $\partial(I^m \times I \times I) - I^m \times \{1\} \times I$, so that

$$\partial \bar{K} = I^m \times \{(1, 0), (1, 1)\} \cup \partial I^m \times \{1\} \times I.$$

Define $D' : (\bar{K}, \partial \bar{K}) \longrightarrow (U, V)$ by $D'(z, s, 0) = \bar{H}(z, s)$, $D'(z, s, 1) = p \circ H'(z, s)$, for $(z, s) \in I^m \times I$, and $D'(z, s, t) = d(z, s, t)$ for $(z, s, t) \in I^m \times \{0\} \times I \cup \partial I^m \times I \times I$. Further $D'$ contracts $K \times [3/4, 1] \times \{1\}$ to the point $x = p(y)$. Now $\bar{K}$ is homeomorphic to $I^{m+1}$, so that $D'$ determines a class $\beta \in \pi_{m+1}(U, V)$. Since $p_*$ is onto on $\pi_{m+1}$, we can find a map

Appendix A: Results from Topology    261

$f : (I^{m+1}, \partial I^{m+1}) \longrightarrow (F, G)$ with $f(I^m \times \{0\} \cup \partial I^m \times I) = \{y\}$, such that $p_*[f] = -\beta$ in $\pi_{m+1}(U, V, x)$. Define $H : (I^m \times I, I^m \times \{1\}) \longrightarrow (F, G)$ by

$$H(z, t) = \begin{cases} f \circ \psi(z, t) & \text{for } (z, t) \in K \times [3/4, 1] \\ H'(z, t) & \text{otherwise}, \end{cases}$$

where $\psi : K \times [3/4, 1] \longrightarrow I^m \times I$ is the homeomorphism obtained from the obvious linear homeomorphisms $x \mapsto 2(x - 1/4)$ of $[1/4, 3/4]$ with $I$, and $x \mapsto 4(x - 3/4)$ of $[3/4, 1]$ with $I$. We note that

$$f \circ \psi(K \times \{3/4\} \cup \partial K \times [3/4, 1]) = \{y\}$$

so that $H$ is continuous. Next, let $D_1 : (\bar{K}, \partial \bar{K}) \longrightarrow (U, V)$ be defined by

$$D_1(z, s, 1) = p \circ H(z, s) \text{ for } (z, s) \in K \times [3/4, 1] \text{ and}$$

$$D_1(z, s, t) = D'(z, s, t) \text{ elsewhere on } \bar{K}.$$

Then by construction, $D_1$ represents $(-\beta) + \beta$ in $\pi_{m+1}(U, V; x)$ so that $D_1$ extends to a map $D : (I^m \times I \times I, I^m \times \{1\} \times I) \longrightarrow (U, V)$. The pair $H, D$ have all the desired properties.

**Proof of (A.32).** We can find an integer $N > 0$ which is so large that if we divide $I = [0, 1]$ into $N$ equal subintervals $I_j = \left[\frac{j-1}{N}, \frac{j}{N}\right]$, $1 \leq j \leq N$, then for any $m \geq 0$, and any $m$-cube $\sigma$ which is a face of the resulting subdivision of $I^{n-1} \times I$ into $n$-cubes of size $1/N$, we can choose an open set $U^\sigma \in \mathcal{U}$ such that

(i) $\bar{H}(\sigma) \subset U^\sigma$ for all such $\sigma$

(ii) if $\sigma$ is a face of $\sigma'$, then $U^\sigma \subset U^{\sigma'}$

(iii) if $\sigma$ meets $\partial I^{n-1} \times I \cup I^{n-1} \times \{1\}$, then $U^\sigma \subset U$.

The existence of such an assignment follows from a standard argument, whose details are left to the reader (work by descending induction on $m$, using successively finer subdivisions; this uses hypothesis (ii) of (A.30) about the cover $\mathcal{U}$).

Now each $m$-cube $\sigma$ of the subdivision has the form $\sigma_1 \times \sigma_2$ where $\sigma_1$ is a cube in the subdivision of $I^{n-1}$, and $\sigma_2 \subset I$ is either a point or a subinterval $I_j$. Clearly $I^n$ is the union of all $m$-cubes $\sigma = \sigma_1 \times \sigma_2$ with $\sigma_2 = I_j$ for some $j$. We'll order the cubes of this type, so that $\sigma = \sigma_1 \times \sigma_2$ precedes $\tau = \tau_1 \times \tau_2$ if $\sigma_2$ precedes $\tau_2$ in the obvious ordering of the $I_j$, or else $\sigma_2 = \tau_2$ and $\dim \sigma_1 < \dim \tau_1$; we choose an arbitrary ordering of the $m$-cubes $\sigma_1 \times \sigma_2$ with a fixed interval $\sigma_2$ and $\sigma_1$ ranging over all $(m-1)$-cubes in $I^{n-1}$.

We will successively construct the desired maps $H, D$ on the above cubes $\sigma_1 \times \sigma_2$ (with $\sigma_2$ an interval) taken in order. If $\sigma$ is one such $m$-cube, so that $\sigma = \sigma_1 \times \sigma_2$ is homeomorphic to $I^{m-1} \times I$, then from $H, D$ on the earlier cubes we are given maps

(i) $\bar{H}_\sigma : (I^{m-1} \times I, I^{m-1} \times \{1\}) \longrightarrow (U^\sigma, U^{\sigma_1 \times \{1\}})$ (where $\sigma_1 \times \{1\}$ denotes the product of $\sigma_1$ with the second end-point of $\sigma_2 \subset I$; $\sigma_1 \times \{0\}$ has a similar meaning)

(ii) $H_\sigma : (I^{m-1} \times \{0\} \cup \partial I^{m-1} \times I, \partial I^{m-1} \times \{1\}) \longrightarrow (p^{-1}(U^\sigma), p^{-1}(U^{\sigma_1 \times \{1\}}))$

(iii) $D_\sigma : (I^{m-1} \times \{0\} \times I \cup \partial I^{m-1} \times I \times I, \partial I^{m-1} \times \{1\} \times I) \longrightarrow (U^\sigma, U^{\sigma_1 \times \{1\}})$

satisfying the hypothesis of (A.33) with $V = U^{\sigma_1 \times \{1\}}$, $U = U^\sigma$. Hence we can extend $H, D$ continuously over $\sigma$, as desired, proving (A.32) by induction on the position of $\sigma$ in the above ordering.

We give two more results about quasi fibrations which are needed in the main text.

**Lemma (A.34).** *Let $p : E \longrightarrow B$ be a surjective map, $B' \subset B$ a subset which is distinguished for $p$, and $E' = p^{-1}(B')$. Assume given deformations $D_t : E \longrightarrow E$, $d_t : B \longrightarrow B$, $t \in I$ such that $D_0 = 1_E$, $d_0 = 1_B$, $D_t(E') \subset E'$, $d_t(B') \subset B'$, $D_1(E) \subset E'$, $d_1(B) \subset B'$, and $p \circ D_1 = d_1 \circ p$. Suppose that for each $x \in B$, $y \in p^{-1}(x)$, and $i \geq 0$*

$$(D_1)_* : \pi_i(p^{-1}(x), y) \cong \pi_i(p^{-1}(d_1(x)), D_1(y)).$$

*Then $p$ is a quasi fibration.*

**Proof.** Since $d_t, D_t$ are deformations of the respective identity maps,

$$(d_1)_* : \pi_i(B, x) \cong \pi_i(B', x'), \quad x' = d_1(x)$$

$$(D_1)_* : \pi_i(E, y) \cong \pi_i(E', y'), \quad y' = D_1(y).$$

Since $p \circ D_1 = d_1 \circ p$, $D_1(p^{-1}(x)) \subset p^{-1}(x')$, and we have a map from the exact homotopy sequence of $(E, p^{-1}(x))$ to that of $(E', p^{-1}(x'))$. Since $(D_1)_* : \pi_i(p^{-1}(x), y) \cong \pi_i(p^{-1}(x'), y')$ by hypothesis, the 5-lemma gives isomorphisms $(D_1)_* : \pi_i(E, p^{-1}(x); y) \cong \pi_i(E', p^{-1}(x'), y')$. Hence in the diagram below, the horizontal and right-hand vertical maps are isomorphisms, so the left-hand vertical map is one too.

$$\begin{array}{ccc}
\pi_i(E, p^{-1}(x); y) & \xrightarrow{(D_1)_*} & \pi_i(E', p^{-1}(x'), y') \\
{\scriptstyle p_*}\downarrow & & \downarrow{\scriptstyle p_*} \\
\pi_i(B, x) & \xrightarrow[(d_1)_*]{} & \pi_i(B', x')
\end{array}$$

**Lemma (A.35).** *Let $p: E \longrightarrow B$ be a map, $B_1 \subset B_2 \subset \cdots \subset B$ an increasing sequence of subsets that* (i) *each $B_i$ is distinguished for $p$* (ii) *$B$ has the weak topology relative to the $B_i$ (i.e., a subset of $B$ is closed if and only if its intersection with each $B_i$ is closed in $B_i$)* (iii) *points are closed in $B$. Then $p$ is a quasi fibration.*

**Proof.** From (ii) and (iii) one sees easily that any compact subset of $B$ lies in some $B_i$. Hence any compact subset of $E$ lies in some $p^{-1}(B_i)$. If $x \in B_i$, $y \in p^{-1}(x)$, then

$$\pi_n(E, p^{-1}(x); y) \cong \varinjlim_{j \geq 1} \pi_n(p^{-1}(B_j), p^{-1}(x); y)$$
$$\cong \varinjlim_{j \geq 1} \pi_n(B_j, x) \quad \text{(by (i))}$$
$$\cong \pi_n(B, x).$$

**(A.36) Cofibrations and NDR-Pairs.** Let $i: A \longrightarrow X$ be the inclusion of a closed subspace. Then $i$ is called a *cofibration* if $i$ has the homotopy extension property, i.e., any map $F: X \times \{0\} \cup A \times I \longrightarrow Y$ extends to a map $G: X \times I \longrightarrow Y$. Taking $Y = X \times \{0\} \cup A \times I$, $F$ = identity, we see that $i$ is a cofibration $\iff X \times \{0\} \cup A \times I$ is a retract of $X \times I$.

Let $i: A \longrightarrow X$ be the inclusion of a closed subspace. Following N. Steenrod (A convenient category of topological spaces, *Mich. Math. J.* 14 (1967), 133–152), we call $(X, A)$ an $NDR$-pair ("neighborhood deformation retract pair") if there exist maps $u: X \longrightarrow I$, $h: X \times I \longrightarrow X$ such that $u^{-1}(0) = A$, and $h(x, 0) = x$, $h(x, 1) \in A$ if $u(x) < 1$ and $h(x, t) = x$ if $x \in A$, $t \in I$. Thus, $A$ is a deformation retract of the open set $u^{-1}([0, 1])$. If $h(x, 1) \in A$ $\forall x \in X$, we say that $(X, A)$ is a $DR$-pair.

**Lemma (A.37).** *If $(X, A)$, $(Y, B)$ are $NDR$-pairs, so is $(X \times Y, X \times B \cup A \times Y)$. If one of them is a $DR$-pair (and the other an $NDR$-pair) then the product pair is a $DR$-pair.*

**Theorem (A.38).** *If $i: A \longrightarrow X$ is the inclusion of a closed subspace, the following are equivalent*

(i) *$(X, A)$ is an $NDR$-pair*

(ii) *$X \times \{0\} \cup A \times I$ is a deformation retract of $X \times I$*

(iii) *$X \times \{0\} \cup A \times I$ is a retract of $X \times I$*

(iv) *$i$ is a cofibration.*

**Proof.** Since $(I, \{0\})$ is a $DR$-pair, (i) $\Rightarrow$ (ii) follows from (A.37); clearly (ii) $\Rightarrow$ (iii), and (iii) $\iff$ (iv) as remarked above. So it suffices to show

(iii) $\Rightarrow$ (i). Let $R : X \times I \longrightarrow X \times \{0\} \cup A \times I$ be a retraction, $p_1 : X \times I \longrightarrow X$, $p_2 : X \times I \longrightarrow I$ the projections, $h = \bar{p}_1 \circ R$, $k = \bar{p}_2 \circ R$, where $\bar{p}_i$ is the restriction of $p_i$ to $X \times \{0\} \cup A \times I$. Then $h(x,0) = p_1 \circ R(x,0) = p_1(x,0) = x$, and if $x \in A$, $h(x,t) = \bar{p}_1 \circ R(x,t) = p_1(x,t) = x$. Let $v_m : X \to [0, 1/2^m]$ be the map $v_m(x) = \min(1/2^m, k(x, 1/2^m))$ and let $u : X \to I$ be given by the (uniformly convergent) series

$$u(x) = 1 - \sum_{m=1}^{\infty} v_0(x)\, v_m(x) = \sum_{m=1}^{\infty} ((1/2^m) - v_0(x)\, v_m(x)).$$

Then $u(x) = 0 \iff v_0(x) v_m(x) = 1/2^m\ \forall m \geq 1$. If $x \in A$, then $k(x, 1/2^m) = \bar{p}_2 \circ R(x, 1/2^m) = p_2(x, 1/2^m) = 1/2^m$, so that $v_m(x) = 1/2^m$ $\forall m \geq 0$, and $u(x) = 0$. If $x \notin A$, then $R(x, 0) = (x, 0) \in (X - A) \times \{0\}$ which is open in $X \times \{0\} \cup A \times I$, so that $R(V) \subset (X - A) \times \{0\}$ for some open set $V \subset X \times I$ with $(x, 0) \in V$. In particular $k(V) = \{0\}$, so that for all large $m$, $k(x, 1/2^m) = 0$, and $v_m(x) = 0$; thus $u(x) > 0$. Finally $u(x) < 1 \iff v_0(x) v_m(x) > 0$ for some $m > 0$; in particular $v_0(x) = k(x, 1) > 0$, i.e., $R(x, 1) \in A \times I \subset X \times \{0\} \cup A \times I$. Hence $h(x, 1) \in A$ if $u(x) < 1$. The pair $(h, u)$ represent $(X, A)$ as an $NDR$-pair.

**Proof of (A.37).** Let $u : X \longrightarrow I$, $h : X \times I \longrightarrow I$ represent $(X, A)$ as an $NDR$-pair, and let $v : Y \longrightarrow I$, $j : Y \times I \longrightarrow Y$ represent $(Y, B)$ as an $NDR$-pair. Let $w : X \times Y \longrightarrow I$ be given by $w(x, y) = u(x) \cdot v(y)$ (pointwise product) so that $w^{-1}(0) = X \times B \cup A \times Y$. Define $q : X \times Y \times I \longrightarrow X \times Y$ by

$$q(x, y, t) = \begin{cases} (x, y) & \text{if } (x, y) \in A \times B \\ \left(h(x, t), j\left(y, t\tfrac{u(x)}{v(y)}\right)\right) & \text{if } v(y) \geq u(x) \text{ and } v(y) \neq 0 \\ \left(h\left(x, t\tfrac{v(y)}{u(x)}\right), j(y, t)\right) & \text{if } u(x) \geq v(y) \text{ and } u(x) \neq 0. \end{cases}$$

One checks that $q$ is continuous, and $w, q$ represent $(X \times Y, X \times B \cup A \times Y)$ as an $NDR$-pair. If $u, h$ represent $(X, A)$ as a $DR$-pair, and $u' = \tfrac{1}{2}u$, then $u', h$ also represent $(X, A)$ as a $DR$-pair; if we carry out the above construction with $u', h$ then $w' = u' \cdot v < 1$ on all of $X \times Y$, so $w', q$ represent $(X \times Y, X \times B \cup A \times Y)$ as a $DR$-pair.

**Lemma (A.39).** *Let $f : (X, A) \longrightarrow (Y, B)$ be a relative homeomorphism (i.e., $f$ induces a homeomorphism $X - A \cong Y - B$, $f$ is onto and $Y$ has the quotient topology from $X$). Then if $(X, A)$ is an $NDR$-pair, so is $(Y, B)$.*

**Proof.** Let $u : X \longrightarrow I$, $h : X \times I \longrightarrow X$ represent $(X, A)$ as an $NDR$-pair. Then $v : Y \longrightarrow I$, $j : Y \times I \longrightarrow Y$ given by $v(y) = u(f^{-1}(y))$,

$$j(y, t) = \begin{cases} f \circ h(f^{-1}(y), t) & \text{if } y \notin B,\ t \in I \\ y & \text{if } y \in B,\ t \in I \end{cases}$$

are continuous, since $f$ is a relative homeomorphism, and represent $(Y, B)$ as an $NDR$-pair.

**Corollary (A.40).** *If $(X, A)$ is a relative $CW$-complex, then $(X, A)$ is an $NDR$-pair, i.e., $A \hookrightarrow X$ is a cofibration.*

**Proof.** If $A \subset B \subset X$ are inclusions of closed subspaces such that $(B, A)$ and $(X, B)$ are $NDR$-pairs, then from (A.37) (iii), $(X, A)$ is an $NDR$-pair. If $(X, A)$ is an $NDR$-pair, $(X \coprod Y, A \coprod Y)$ is an $NDR$-pair for any $Y$, and $(\Delta_n, \partial \Delta_n)$ is an $NDR$-pair for any $n \geq 0$. Hence $(X_n, A)$ is an $NDR$-pair for any $n \geq 0$, where $X_n$ is the $n$-skeleton of $(X, A)$, since by the earlier remarks $(X_n, X_{n-1})$ is an $NDR$-pair $\forall n \geq 0$ (where $X_{-1} = A$). There is a retraction $R_n : X_n \times I \longrightarrow X_n \times \{0\} \cup X_{n-1} \times I$; extend this to retraction
$$S_n : X \times \{0\} \cup X_n \times I \longrightarrow X \times \{0\} \cup X_{n-1} \times I$$
by letting $S_n(x, 0) = (x, 0) \; \forall x \in X$. Let $T_n : X \times \{0\} \cup X_n \times I \longrightarrow X \times \{0\} \cup A \times I$ be the composite retraction $T_n = S_0 \circ \cdots \circ S_n$. Then $T_{n+1}$ restricts to $T_n$ on $X \times \{0\} \cup X_n \times I$; since $X \times I = \underrightarrow{\lim}\, (X \times \{0\} \cup X_n \times I)$ has the weak topology, the set theoretic retraction $X \times I \longrightarrow X \times \{0\} \cup A \times I$, which restricts to $T_n$ on $X \times \{0\} \cup X_n \times I$, is continuous. Hence $(X, A)$ is an $NDR$-pair.

As a simple example of the use of the above notions, let $(X, x_0)$ be a space with a base point. The base point is called *non-degenerate* if $(X, \{x_0\})$ is an $NDR$-pair. Given any pointed space $(X, x_0)$ let $\tilde{X}$ be the quotient space of $X \coprod I$ obtained by identifying $x_0$ with $0 \in I$; let $\tilde{x} \in \tilde{X}$ be the image of $1 \in I$. Then $(\tilde{X}, \tilde{x})$ has a non-degenerate base point, and $(\tilde{X}, \tilde{x}) \longrightarrow (X, x)$, given by collapsing the image of $I$ in $\tilde{X}$ to a point, is a homotopy equivalence. The process of forming $\tilde{X}$ is sometimes called "adding a whisker."

**Lemma (A.41).** *Let $f, g : (X, x_0) \longrightarrow (Y, y_0)$ be two maps which are freely homotopic, i.e., there is a homotopy $\bar{H}; X \times I \longrightarrow Y$ with $\bar{H}(x, 0) = f(x)$, $\bar{H}(x, 1) = g(x) \; \forall x \in X$ (but which may not satisfy $\bar{H}(x_0, t) = y_0 \; \forall t \in I$). Let $\omega : I \to Y$ be the loop at $y_0$ given by $\omega(t) = \bar{H}(x_0, t)$, and assume $[\omega] = 0$ in $\pi_1(Y, y_0)$. Then there exists a homotopy $H : X \times I \longrightarrow Y$ with $H(x, 0) = f(x)$, $H(x, 1) = g(x)$ and $H(x_0, t) = y_0 \; \forall x \in X, t \in I$, provided $x_0 \in X$ is non-degenerate.*

**Proof.** Since $X \times \{0\} \cup \{x_0\} \times I$ is a retract of $X \times I$, $X \times \{0\} \times I \cup \{x_0\} \times I \times I$ is a retract of $X \times I \times I$, so that any map $F : X \times \{0\} \times I \cup \{x_0\} \times I \times I \longrightarrow Y$ extends to a map $G : X \times I \times I \longrightarrow Y$. Let $F$ be given by
$$F(x, s, t) = \begin{cases} \bar{H}(x, t) & \text{if } s = 0, \, (x, t) \in X \times I \\ H_1(s, t) & \text{if } x = x_0, \, (s, t) \in I \times I \end{cases}$$

where $H_1 : I \times I \longrightarrow Y$ is a null homotopy of $\omega$, i.e., $H_1(0,t) = \omega(t)$, $H_1(s,0) = H_1(s,1) = H_1(1,t) = y_0$ for all $s,t \in I$. Let $G : X \times I \times I \to Y$ extend $F$, and let $H : X \times I \to Y$ be given by

$$H(x,s) = \begin{cases} G(x,3s,0) & \text{if } 0 \le s \le 1/3 \\ G(x,1,3s-1) & \text{if } 1/3 \le s \le 2/3 \\ G(x,3-3s,1) & \text{if } 2/3 \le s \le 1. \end{cases}$$

One sees that $H$ has the desired properties.

**Lemma (A.42).** *Let $f : (X,A) \longrightarrow (Y,B)$ be a relative homeomorphism, where $(X,A)$ is an NDR-pair. Then $f_* : H_n(X,A;G) \longrightarrow H_n(Y,B;G)$, $f^* : H^n(Y,B;G) \longrightarrow H^n(X,A;G)$ are isomorphisms for any coefficient group $G$, $\forall n \ge 0$.*

**Proof.** There is a neighborhood $U$ of $A$ in $X$ such that $A$ is a deformation retract of $U$, and $f(U) = V$ is a similar neighborhood of $B$ in $Y$. Then we have isomorphisms (for any coefficient group)

$$\begin{aligned} H_n(X,A) &\cong H_n(X,U) \cong H_n(X-A, U-A) \quad \text{(excision)} \\ &\cong H_n(Y-B, V-B) \\ &\cong H_n(Y,V) \quad \text{(excision)} \\ &\cong H_n(Y,B) \end{aligned}$$

and a similar argument works for cohomology.

**(A.43) H Spaces.** A space $(X,x_0)$ with a non-degenerate base point is called an *H-space* if there is a map $\mu : (X \times X, (x_0,x_0)) \longrightarrow (X,x_0)$ such that $\mu_1(x) = \mu(x,x_0)$, $\mu_2(x) = \mu(x_0,x)$ are each homotopic to the identity map of $X$ (through homotopies fixing $x_0$).

**Lemma (A.44).** *Let $(X,x_0)$ be an H-space. Then there is a map $\mu : X \times X \longrightarrow X$ with $\mu(x,x_0) = \mu(x_0,x) = x \ \forall x \in X$.*

**Proof.** Let $\mu' : X \times X \longrightarrow X$ be a map giving an $H$-space structure on $X$, i.e., $\mu'_1(x) = \mu'(x,x_0)$, $\mu'_2(x) = \mu'(x_0,x)$ are each homotopic to the identity map of $X$. Equivalently, if $X \vee X = X \times \{x_0\} \cup \{x_0\} \times X \subset X \times X$, $k : X \vee X \hookrightarrow X \times X$, $\nabla : X \vee X \longrightarrow X$ the folding map (i.e., $\nabla(x,x_0) = \nabla(x_0,x) = x \ \forall x \in X$), then $\mu' \circ k \cong \nabla$ (where $\cong$ denotes a homotopy preserving the base points). By (A.37), $(X \times X, X \vee X)$ is an $NDR$-pair, and so by (A.38) (iv), $X \vee X \longrightarrow X \times X$ has the homotopy extension property. Hence, given $\mu' : X \times X \longrightarrow X$ and the homotopy $\mu' \circ k \cong \nabla$ of $\mu'|_{X \vee X}$, we can find a homotopy $\mu' \cong \mu$, where $\mu : X \times X \longrightarrow X$ satisfies $\mu \circ k = \nabla$, i.e., $\mu(x,x_0) = \mu(x_0,x) = x \ \forall x \in X$.

**Lemma (A.45).** *If $(X, x_0)$ is an $H$-space, then $(X, x_0)$ is simple, i.e., $\pi_1(X, x_0)$ acts trivially on $\pi_n(X, x_0)$ for $n \geq 1$.*

**Proof.** Let $\mu : X \times X \longrightarrow X$ be a map giving an $H$-space structure such that $\mu(x, x_0) = \mu(x_0, x) = x \ \forall x \in X$, as in (A.44). Then if $f : (I^n, \partial I^n) \longrightarrow (X, x_0)$, $g : (I, \partial I) \longrightarrow (X, x_0)$ represent given classes $\alpha \in \pi_n(X, x_0)$, $\beta \in \pi_1(X, x_0)$ then $h(s, t) = \mu(f(s), g(t))$ gives a map $h : I^n \times I \longrightarrow X$ satisfying $h(s, 0) = f(s) \ \forall s \in I^n$, $h(s, t) = g(t)$ for all $s \in \partial I^n$ (so that $f(s) = x_0$) and all $t \in I$. Hence by definition of the action of $\pi_1$ on $\pi_n$, the map $h(s, 1) : (I^n, \partial I^n) \longrightarrow (X, x_0)$ represents the result of $\beta \in \pi_1$ acting on $\alpha \in \pi_n$; but clearly $h(s, 1) = f(s)$ since $g(1) = x_0$. This proves the lemma.

An $H$-space is *homotopy associative* if the maps $\mu \circ (\mu \times 1)$, $\mu \circ (1 \times \mu) : X \times X \times X \to X$ are homotopic (1 denotes the identity map of $X$, $\mu \circ (\mu \times 1)(x, y, z) = \mu(\mu(x, y), z)$, etc.). It is *homotopy commutative* if $\mu \circ s$ is homotopic to $\mu$, where $s : X \times X \longrightarrow X \times X$ is the switch map $s(x, y) = (y, x)$. Finally, $(X, x_0)$ is an *$H$-group* if it is homotopy associative, and there is a map $j : (X, x_0) \longrightarrow (X, x_0)$ (an *inverse up to homotopy* for the product) such that $\mu \circ (1 \times j) \circ \Delta$, $\mu \circ (j \times 1) \circ \Delta : X \longrightarrow X$ are both homotopic to the constant map $X \longrightarrow \{x_0\}$, where $\Delta(x) = (x, x)$ is the diagonal.

**Lemma (A.46).** *A homotopy associative $H$-space is an $H$-group if and only if the shear map $\phi : X \times X \longrightarrow X \times X$, $\phi(x, y) = (x, \mu(x, y))$ is a homotopy equivalence.*

**Proof.** If $(X, x_0)$ is an $H$-space, with product $\mu : X \times X \longrightarrow X$, $j : X \longrightarrow X$ an inverse up to homtopy for $\mu$, then $\mu(x, y) = (x, \mu(j(x), y))$ is clearly a homotopy inverse for $\phi$. Conversely if $\phi$ is a homotopy equivalence, and $\psi : X \times X \longrightarrow X \times X$ is a homotopy inverse for $\phi$, and $p_i : X \times X \longrightarrow X$, $i = 1, 2$ are the projections, then let $j$ be defined by

$$j(x) = p_2 \circ \psi(x, x_0) = p_2 \circ \psi \circ i_1, \text{ where } i_1(x) = (x, x_0), i_2(x) = (x_0, x).$$

We have homotopies

$$p_1 \cong p_1 \circ \phi \circ \psi = p_1 \circ \psi, \quad p_2 \cong p_2 \circ \phi \circ \psi = \mu \circ \psi,$$

so that $p_1 \circ \psi \circ i_1 \cong p_1 \circ i_1 = 1$, and so

$$\mu \circ (1 \times j) \circ \Delta \cong \mu \circ ((p_1 \circ \psi \circ i_1) \times (p_2 \circ \psi \circ i_1)) \circ \Delta$$
$$= \mu \circ (p_1 \times p_2) \circ ((\psi \circ i_1) \times (\psi \circ i_1)) \circ \Delta$$
$$= \mu \circ (p_1 \times p_2) \circ \Delta \circ \psi \circ i_1$$
$$= \mu \circ \psi \circ i_1 \cong p_2 \circ i_1 = *,$$

the constant map (with value $x_0$). Since $j$ is a right inverse up to homotopy for $\mu$, and $\mu$ is homotopy associative, $j$ is also a left inverse up to homotopy for $\mu$.

**Corollary (A.47).** *Let $(X, x_0)$ be a path connected, homotopy associative H-space with the homotopy type of a CW-complex. Then $X$ is an H-group.*

**Proof.** Let $\mu : X \times X \longrightarrow X$ be a map giving an H-space structure such that $\mu(x, x_0) = \mu(x_0, x) = x$, and let $\phi : X \times X \longrightarrow X \times X$ be the shear map (of (A.46)). Since $\phi(x, x_0) = (x, x)$ and $\phi(x_0, x) = (x_0, x)$, one sees that under the natural direct product decomposition

$$\pi_n(X \times X, (x_0, x_0)) \cong \pi_n(X, x_0) \times \pi_n(X, x_0)$$

(induced by the inclusions $X \to X \times X$, $x \mapsto (x, x_0)$ and $x \mapsto (x_0, x)$) $\phi_* : \pi_n(X \times X, (x_0, x_0)) \longrightarrow \pi_n(X \times X, (x_0, x_0))$ becomes the group theoretic shear map $(\alpha, \beta) \longrightarrow (\alpha, \alpha+\beta)$ on $\pi_n(X, x_0) \times \pi_n(X, x_0)$. Hence $\phi_*$ induces isomorphisms on homotopy groups, so by the Whitehead theorem (A.9) $\phi$ is a homotopy equivalence.

**Covering Spaces of Simplicial Sets.** In this section we follow the treatment given in: P. Gabriel, M. Zisman, *Calculus of Fractions and Homotopy Theory*, Ergeb. Math. 35, Springer-Verlag, 1967).

We will use the notation and terminology of Chapter 3 of the main text. A map $p : E \longrightarrow B$ of simplicial sets is said to be *locally trivial* with fiber $F$ (also a simplicial set) if for each simplex $\sigma \in B(\underline{n})$ (for any $n$), if $\Delta(n) \longrightarrow E$ is the associated map of simplicial sets, and $\Delta(n) \times_B E$ is the fiber product (so that $(\Delta(n) \times_B E)(\underline{p}) = \Delta(n)(\underline{p}) \times_{B(\underline{p})} E(\underline{p})$ for all $p \geq 0$), there is an isomorphism $g$ of simplicial sets making the diagram below commute:

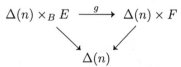

If $f : Y \longrightarrow X$ is a map of topological spaces, we say that $f$ is *locally trivial* with fiber $Z$ (a topological space) if $X$ has a covering by open subsets $U$ with the property that there exists a homeomorphism $g$ making the diagram below commute:

$$\begin{array}{ccc} U \times_X Y & \xrightarrow{g} & U \times Z \\ & \searrow \swarrow & \\ & U & \end{array}$$

Note that by convention of (A.1) all the spaces are compactly generated, and products, fiber products are formed in the category of compactly generated spaces, so that the above notion of local triviality differs from the "standard" one. However, the two notions do agree if $Z$ is locally compact, and in particular if $Z$ is discrete.

**Theorem (A.48).** *Let $p : E \longrightarrow B$ be a locally trivial morphism of simplicial sets with fiber $F$, a simplicial set. Then the geometric realization $|p| : |E| \to |B|$ is a locally trivial map of topological spaces with fiber $|F|$.*

Before proving this result, we describe an open neighborhood $U_x$ of a given point $x \in |B|$, for any simplicial set $B$ (and any $x \in |B|$). These neighborhoods will be used to prove (A.48). First, if $m \geq 1$ and $V \subset \partial \Delta_m$ is any (relatively) open subset, let $V^* \subset \Delta_m$ denote the open subset (the "open cone over $V$")

$$V^* = \{tx + (1-t)x_0 \mid x \in V, 0 < t \leq 1\},$$

where we regard $\Delta_m$ as the subset of $\mathbb{R}^{m+1}$

$$\Delta_m = \{(s_0, \ldots, s_m) \in \mathbb{R}^{m+1} \mid s_i \geq 0, \Sigma s_i = 1\},$$

and $x_0 = \left(\frac{1}{m+1}, \ldots, \frac{1}{m+1}\right)$ is its barycenter (the sum in the definition of $V^*$ is as vectors in $\mathbb{R}^{m+1}$ where $t, 1-t$ are scalars). Clearly $V$ is a deformation retract of $V^*$. Next, given a $CW$-complex $X$, with $n$-skeleton $X_n$ for each $n \geq 0$, and a point $x \in X$, then we can find a unique integer $n \geq 0$ such that $x \in X_n - X_{n-1}$ (where $X_{-1}$ is empty), and hence we can find a unique $n$-cell $\sigma : \Delta_n \longrightarrow X$ such that $x \in \sigma\,(\Delta_n - \partial\Delta_n)$ (where $\partial \Delta_0$ is empty). Let $U_n = \sigma(\Delta_n - \partial\Delta_n))$ so that $U_n \subset X_n$ is open. For $m > n$, we inductively define open subsets $U_m \subset X_m$ by

$$U_m = U_{m-1} \cup \left( \bigcup_\sigma \sigma(\sigma^{-1}(U_{m-1})^*) \right)$$

where $\sigma$ runs over the $m$-cells of $X$. Then $U_m$ is indeed open in $X_m$, and $U_m \cap X_{m-1} = U_{m-1}$. Let $U_x = \bigcup_{m \geq n} U_m$. Since $X$ has the weak topology with respect to the $X_m$, and $U_x \cap X_m = U_m$ is open in $X_m$ for all $m \geq n$, $U_x$ is open in $X$. This works, in particular, for $X = |B|$ where $B$ is any simplicial set. Note that $U_x$ has the weak topology relative to the $U_m$.

The above construction has the following property. Let $f : Y \longrightarrow X$ be a map of topological spaces where $X$ is a $CW$-complex and let $x \in X_n - X_{n-1}$. Assume that there exists a space $Z$ and homeomorphisms $g_m : U_m \times_X Y \longrightarrow U_m \times Z$, for $m \geq n$, such that $g_{m+1}|_{U_m \times_X Y} = g_m$, and

$$\begin{CD} U_m \times_X Y @>{g_m}>> U_m \times Z \\ @VVV @VVV \\ & U_m & \end{CD}$$

commutes. Then the resulting set theoretic map $g : U_x \times_X Y \longrightarrow U_x \times Y$ (given by $g|_{U_m \times_X Y} = g_m$) is a homeomorphism making the diagram below commute:

$$\begin{array}{ccc} U_m \times_X Y & \xrightarrow{g} & U_x \times Z \\ & \searrow \quad \swarrow & \\ & U_x & \end{array}$$

This is one advantage of working with compactly generated spaces.

**Proof of (A.48).** From the discussion above, if $x \in |B|$ lies in the interior of an $n$-cell of the $CW$-complex $|B|$, it suffices to construct a compatible family of homeomorphisms

$$g_m : U_m \times_{|B|} |E| \longrightarrow U_m \times |F|$$

for $m \geq n$, such that the diagram below commute:

$$\begin{array}{ccc} U_m \times_{|B|} |E| & \xrightarrow{g_m} & U_m \times |F| \\ & \searrow \quad \swarrow & \\ & U_m & \end{array}$$

The existence of such a map $g_n$ follows from the fact that for any $m$-cell $\sigma : \Delta_m \longrightarrow |B|$ we are given a homeomorphism $g(\sigma)$ (compatible with projection to $\Delta_m$),

$$g(\sigma) : \Delta_m \times_{|B|} |E| \longrightarrow \Delta_m \times |F|,$$

which is obtained by realizing a corresponding isomorphism of simplicial sets; in particular this is valid for the $n$-cell whose interior contains $x$. Given $g_m : U_m \times_{|B|} |E| \longrightarrow U_m \times |F|$, which we wish to extend to $g_{m+1} : U_{m+1} \times_{|B|} |E| \longrightarrow U_{m+1} \times |F|$, we see that for any $(m+1)$-cell $\sigma : \Delta_{m+1} \longrightarrow |B|$, $g_m$ determines the restriction of $g_{m+1}$ to $(\sigma(\Delta_{m+1}) \cap U_{m+1}) \times_{|B|} |E|$. The maps $g(\sigma)$, $g_m$ thus give two homeomorphisms $g(\sigma)'$, $g_m(\sigma)'$

$$\sigma^{-1}(U_m) \times_{|B|} |E| \longrightarrow \sigma^{-1}(U_m) \times |F|$$

compatible with projection on the first factor. Thus, if $\mathrm{Aut}(|F|)$ is the subspace of $\mathbb{F}(|F|, |F|)$ consisting of homeomorphisms, we can find a map $\theta : \sigma^{-1}(U_m) \longrightarrow \mathrm{Aut}(|F|)$, such that

$$g_m(\sigma)' = \tilde{\theta} \circ g(\sigma)', \text{ where } \tilde{\theta} : \sigma^{-1}(U_m) \times |F| \longrightarrow \sigma^{-1}(U_m) \times |F|$$

is given by $\tilde{\theta}(x, y) = (x, \theta(x)(y))$. Now $\sigma^{-1}(U_m)$ is a deformation retract of $\sigma^{-1}(U_m)^*$, so we can extend $\theta$ to a map $\theta^* : \sigma^{-1}(U_m)^* \longrightarrow \mathrm{Aut}(|F|)$ and so define

$$\tilde{\theta}^* : \sigma^{-1}(U_m)^* \times |F| \longrightarrow \sigma^{-1}(U_m)^* \times |F|$$

by $\tilde{\theta}^*(x,y) = (x, \theta^*(x)(y))$. Next, let
$$g_{m+1}(\sigma) : \sigma^{-1}(U_m)^* \times_{|B|} |E| \longrightarrow \sigma^{-1}(U_m)^* \times |F|$$
be the composite $g_{m+1}(\sigma) = \tilde{\theta}^* \circ \bar{g}(\sigma)$ (where $\bar{g}(\sigma)$ is the restriction of $g(\sigma)$ to $\sigma^{-1}(U_m^*)$. Since $U_{m+1} = U_m \cup (\bigcup_\sigma \sigma(\sigma^{-1}(U_m)^*))$, the reader can verify that $g_{m+1}(\sigma)$ do patch together to give an extension of $g_m$ to a map $g_{m+1} : U_{m+1} \times_{|B|} |E| \longrightarrow U_{m+1} \times |F|$, as desired. This completes the proof of (A.48).

We define a morphism $p : E \longrightarrow B$ of simplicial sets to be a *simplicial covering* if $E(\underline{0}) \longrightarrow B(\underline{0})$ is surjective, and for every diagram of maps of simplicial sets

$$\begin{array}{ccc} \Delta(0) & \longrightarrow & E \\ \downarrow & & \downarrow p \\ \Delta(n) & \longrightarrow & B \end{array}$$

there is a unique map of simplicial sets $\Delta(n) \longrightarrow E$ making the diagram commute. If $|B|$ is connected (equivalently, any two 0-simplices of $B$ are joined by an edge-path formed from 1-simplices) then one easily verifies that $p : E \longrightarrow B$ is a simplicial covering if and only if it is locally trivial with a non-empty discrete simplicial set as fiber (a discrete simplicial set is one which has no non-degenerate positive dimensional simplices). Since the realization of a discrete simplicial set is a discrete space, we deduce from (A.48):

**Corollary (A.49).** *Let $p : E \to B$ be a simplicial covering. Then the realization $|p| : |E| \longrightarrow |B|$ is a covering of topological spaces.*

In particular, the construction in Example (3.10) in the main text does give a construction of the classifying space $BG$ of a discrete group $G$.

**Lemma (A.50).** *Let $f, g : G_1 \longrightarrow G_2$ be two homomorphisms between (discrete) groups $G_1$, $G_2$. Assume that $f, g$ are conjugate, i.e., for some $\delta \in G_2$, $f(x) = \delta^{-1} g(x) \delta$ for all $x \in G_1$. Then there is a homotopy $h : BG_1 \times I \longrightarrow BG_2$ between $Bf$ and $Bg$ such that the loop in $BG_2$ traced out by the image of the base point of $BG_1$ under $h$, represents the class $\delta \in \pi_1(BG_2) \cong G_2$.*

**Proof.** The homomorphisms $f, g$ induce functors $\tilde{f}, \tilde{g} : \underline{G}_1 \longrightarrow \underline{G}_2$ (in the notation of Example (3.10), Chapter 3) such that $B\tilde{f} = Bf$, $B\tilde{g} = Bg$. Now $\underline{G}_2$ is a category with a single object $*$, such that $\mathrm{Hom}_{\underline{G}_2}(*,*) = G_2$. Hence $\delta \in G_2$ gives a morphism $\delta : * \longrightarrow *$ in $\underline{G}_2$, such that for any $x \in G_1$, $\delta \circ f(x) = g(x) \circ \delta$, where $f(x), g(x) \in G_2$ are again regarded as morphisms $* \longrightarrow *$ in $\underline{G}_2$. Thus $\delta$ defined a natural transformation of functors $\tilde{f} \longrightarrow \tilde{g}$, whose realization gives the desired homotopy (we take the 0-cells corresponding to the unique objects of $\underline{G}_1$, $\underline{G}_2$ to be the base points of $B\underline{G}_1$, $B\underline{G}_2$ respectively).

**(A.51) The Hurewicz Theorem for $H$-Spaces.** If $f : X \longrightarrow Y$ is a map of path connected topological spaces which induces isomorphisms on integral homology groups in dimensions $\leq n$, then if $X, Y$ are simply connected, $f$ induces isomorphisms on homotopy groups $\pi_i$ for $i < n$, and a surjection on $\pi_n$ (see (A.5), (A.6)). This may not be true if we drop the hypothesis of simple connectivity. However, if $X, Y$ are $H$-spaces which have universal covering spaces, then the result holds. Our proof of this in (A.53) is based on discussions with M.S. Raghunathan.

**Lemma (A.52).** *Let $X$ be a path-connected $H$-space which has a universal covering space $\pi : \tilde{X} \longrightarrow X$. Then $\pi_1(X)$ acts trivially on the integral homology groups $H_q(\tilde{X}, \mathbb{Z})$.*

**Proof.** Let $x_0 \in X$ be the base point. By (A.43) we may assume that the multiplication $\mu : X \times X \longrightarrow X$, giving the $H$-space structure on $X$, satisfies $\mu(x_0, x) = \mu(x, x_0) = x \; \forall \in X$, i.e., $x_0$ is a two-sided identity for $\mu$.

A point $y \in \tilde{X}$ is given by a pair $(x, [\gamma])$ where $x \in X$, $[\gamma]$ a homotopy class of paths in $X$ joining $x_0$ to $x$; the map $\pi : \tilde{X} \longrightarrow X$ is then $\pi(x, [\gamma]) = x$. In particular, the fundamental group $\pi_1(X, x_0)$ is identified with the fiber $\pi^{-1}(x_0)$ by $[\omega] \mapsto (x_0, [\omega])$ for any class $[\omega] \in \pi_1(X, x_0)$. The action of $\pi_1(X, x_0)$ on $\tilde{X}$ is described by

$$[\omega](x, [\gamma]) = (x, [\omega \cdot \gamma])$$

and $\omega \cdot \gamma$ is the composite path $I \longrightarrow X$,

$$(\omega \cdot \gamma)(t) = \begin{cases} \omega(2t) & \text{if } 0 \leq t \leq 1/2 \\ \gamma(2t-1) & \text{if } 1/2 \leq t \leq 1. \end{cases}$$

The identity element of $\pi_1(X, x)$ gives a natural base point $y_0 \in \tilde{X}$.

If $\omega : I \longrightarrow X$ is a loop based at $x_0$, and $\gamma : I \longrightarrow X$ a path from $x_0$ to $x$ in $X$, define $\omega_1 : I \longrightarrow X$, $\gamma_1 : I \longrightarrow X$ by

$$\omega_1(t) = \begin{cases} \omega(2t) & \text{if } 0 \leq t \leq 1/2 \\ x_0 & \text{if } 1/2 \leq t \leq 1 \end{cases}$$

$$\gamma_1(t) = \begin{cases} x_0 & \text{if } 0 \leq t \leq 1/2 \\ \gamma(2t-1) & \text{if } 1/2 \leq t \leq 1. \end{cases}$$

Then clearly $[\omega_1] = [\omega]$ in $\pi_1(X, x_0)$, and $[\gamma] = [\gamma_1]$ in the set of homotopy classes of paths joining $x_0$ to $x$ in $X$. The path $(\omega_1, \gamma_1) : I \longrightarrow X \times X$ satisfies $\mu \circ (\omega_1, \gamma_1) = \omega \cdot \gamma$, the composite path.

Let $\tilde{\mu} : \tilde{X} \times \tilde{X} \longrightarrow X$ be given by $\tilde{\mu}((x_1, [\gamma_1]), (x_2, [\gamma_2])) = (\mu(x_1, x_2), [\mu \circ (\gamma_1, \gamma_2)])$ where $(\gamma_1, \gamma_2) : I \longrightarrow X \times X$. One sees at once that $\tilde{\mu}$ is well defined and continuous, and satisfies

$$\tilde{\mu}(y_0, y) = \tilde{\mu}(y, y_0) = y \ \forall \, y \in \tilde{X},$$

i.e., $\tilde{\mu}$ has a two-sided identity $y_0$. Thus $(\tilde{X}, \tilde{\mu})$ is an $H$-space. From the above computations, if $y \in \tilde{X}$, $\alpha \in \pi_1(X, x_0)$, then $\alpha(y) = \tilde{\mu}((x_0, \alpha), y)$. Since $\tilde{X}$ is path connected, left multiplication with respect to $\tilde{\mu}$ by $(x_0, \alpha)$ is homotopic to left multiplication by $y_0$, i.e., translation by $\alpha$ on $\tilde{X}$ is homotopic to the identity map of $\tilde{X}$. This proves (A.52).

**Theorem (A.53).** *Let $f : X \longrightarrow Y$ be a map between path connected $H$-spaces which have universal covering spaces. Suppose $f$ induces isomorphisms on integral homology groups in dimensions $\leq n$. Then $f$ induces isomorphisms*

$$f_* : \pi_i(X, x_0) \longrightarrow \pi_i(Y, f(x_0)), \quad i < n,$$

*and a surjection for $i = n$.*

**Proof.** Replacing $f$ by its mapping cylinder $M_f$ (see (A.6)) we may assume $f$ is an inclusion. The theorem is then equivalent to the statement $\pi_i(Y, X, x_0) = 0$ for $i \leq n$. This is trivial for $n = 0$, so assume $n \geq 1$. Since $X, Y$ are $H$-spaces, their fundamental groups are Abelian, by (A.44); now by (A.5) (i), $\pi_1(X, x_0) \cong \pi_1(Y, x_0) \cong G$, say. Thus if $\tilde{Y} \longrightarrow Y$ is the universal covering, $\tilde{X} = \tilde{Y} \times_Y X$, then $\tilde{X} \longrightarrow X$ is the universal covering and $\tilde{X} \hookrightarrow \tilde{Y}$. As in (A.28), one has a spectral sequence of homological type

$$E^2_{p,q} = H_p(G, H_q(\tilde{Y}, \tilde{X}; \mathbb{Z})) \Longrightarrow H_{p+q}(Y, X; \mathbb{Z}).$$

Since $f_* : H_i(X, \mathbb{Z}) \longrightarrow H_i(Y, \mathbb{Z})$ is an isomorphism for $i \leq n$, $H_i(Y, X; \mathbb{Z}) = 0$ for $i \leq n$; hence $E^\infty_{p,q} = 0$ for $p + q \leq n$. Further, from (A.52),

$$E^2_{0,q} = H_0(G, H_q(\tilde{Y}, \tilde{X}; \mathbb{Z})) = H_q(\tilde{Y}, \tilde{X}; \mathbb{Z})$$

(actually, this follows from the proof of (A.52)). Let $q$ be the smallest integer such that $H_q(\tilde{Y}, \tilde{X}; \mathbb{Z}) \neq 0$ (if $q$ does not exist, then $H_q(\tilde{Y}, \tilde{X}, \mathbb{Z}) = 0$ $\forall \, q$); we claim $q \geq n + 1$. Indeed, for $r \geq 2$, we have the complex of $E^r$ terms

$$E^r_{r,q-r+1} \longrightarrow E^r_{0,q} \longrightarrow E^r_{-r,q+r-1}$$

where $E^2_{r,q-r+1} = 0$ since $q - r + 1 < q$, while $E^2_{-r,q+r-1} = 0$ since $-r < 0$. Hence $E^2_{0,q} \cong E^\infty_{0,q}$, i.e., $H_q(Y, X; \mathbb{Z}) \neq 0$. Thus $q \geq n + 1$. Hence, in any case $H_i(\tilde{Y}, \tilde{X}; \mathbb{Z}) = 0$ for $i \leq n$. Since $\tilde{X}, \tilde{Y}$ are simply connected, (A.5) (iii) implies $\pi_i(\tilde{Y}, \tilde{X}; x_0) = 0$ for $i \leq n$, as desired.

**Corollary (A.54).** *Let $f : X \longrightarrow Y$ be a map between connected H-spaces, each of which has the homotopy type of a CW-complex. Suppose $f$ induces isomorphisms on integral homology groups. Then $f$ is a homotopy equivalence.*

**Proof.** This follows at once from (A.53) and the Whitehead theorem (A.9).

Finally, we prove a result of Milnor, which is the basic homeomorphism used to show that geometric realization of simplicial sets commutes with products, provided the topological product is formed in the category of compactly generated spaces.

Consider the simplicial set $\Delta(r) \times \Delta(s)$, given by the functor $\Delta^{\mathrm{op}} \longrightarrow$ Set, $\underline{p} \longrightarrow \mathrm{Hom}_\Delta(\underline{p},\underline{r}) \times \mathrm{Hom}_\Delta(\underline{p},\underline{s})$. The projections $\Delta(r) \times \Delta(s) \longrightarrow \Delta(r)$, $\Delta(r) \times \Delta(s) \longrightarrow \Delta(s)$ induce maps on the geometric realizations $|\Delta(r) \times \Delta(s)| \longrightarrow \Delta_r$, $|\Delta(r) \times \Delta(s)| \longrightarrow \Delta_s$, giving the product map $f : |\Delta(r) \times \Delta(s)| \longrightarrow \Delta_r \times \Delta_s$.

**Theorem (A.55).** *$f$ is a homeomorphism.*

**Proof.** We begin by making some general remarks about geometric realizations. Given any simplicial set $T$, any point $x \in |T|$ lies in the interior of a unique non-degenerate simplex, since any point in a CW-complex lies in the interior of a unique cell (here a 0-simplex is its own interior). Thus, given $x \in |T|$, there exists a unique integer $n \geq 0$, a unique non-degenerate simplex $\sigma \in T(\underline{n})$, and a unique point $P \in \Delta_n^0$ (where $\Delta_n^0 = \{(t_0,\ldots,t_n) \in \mathbb{R}^{n+1} \mid t_i > 0, \Sigma t_i = 1\}$ is the interior of $\Delta_n$), such that $x$ is the image of $(\sigma, P)$ under the quotient map

$$\coprod_{n \geq 0} T(\underline{n}) \times \Delta_n \twoheadrightarrow |T|.$$

$(\sigma, P)$ is called the non-degenerate representative of $x$ (any $(\sigma', P')$ mapping to $x$ under the quotient map will be called a representative of $x$).

For any $n$-simplex $\sigma \in T(\underline{n})$, let $\lambda(\sigma)$ be the corresponding non-degenerate simplex (it is uniquely determined by $\sigma$). If $\lambda(\sigma)$ is an $(n-p)$-simplex, then there is a unique sequence $j_1, \ldots, j_p$ of integers such that

$$\lambda(\sigma) = T(S_{j_p}) \circ \cdots \circ T(S_{j_1})(\sigma), \quad 0 \leq j_1 < \cdots < j_p < n,$$

where the $S_j$ are the degeneracy morphisms in $\Delta$. Since the arrows $(S_j)_* : \Delta_m \longrightarrow \Delta_{m-1}$ (for any $m \geq 1$) induce maps $\Delta_m^0 \longrightarrow \Delta_{m-1}^0$, one sees that if $x \in |T|$, and $(\sigma, P)$ is any representative of $x$, with $\sigma \in T(n)$ and $P \in \Delta_n^0$, then the non-degenerate representative of $x$ is $(\lambda(\sigma), (S_\sigma)_*, (P))$, where

$$(S_\sigma)_* = (S_{j_1})_* \circ \cdots \circ (S_{j_p})_*$$

for the above sequence $j_1, \ldots, j_p$.

We now show that $f$ is bijective. Suppose $x \in |\Delta(r) \times \Delta(s)|$ has non-degenerate representative $((\sigma_1, \sigma_2), P)$ where $(\sigma_1, \sigma_2)$ is a non-degenerate simplex of $\Delta(r) \times \Delta(s)$. Then $f(x) = (f_1(x), f_2(x))$ where $f_1(x) \in |\Delta(r)| = \Delta_r$ has non-degenerate representative $(\lambda(\sigma_1), (S_{\sigma_1})_*(P))$, and $f_2(x) \in \Delta_s$ has non-degenerate representative $(\lambda(\sigma_2), (S_{\sigma_2})_*(P))$.

On the other hand, suppose $x_1 \in |\Delta(r)|$, $x_2 \in |\Delta(s)|$ have non-degenerate representatives $(\sigma_1, P_1)$ and $(\sigma_2, P_2)$, respectively. Let $\sigma_1$ be an $\ell$-simplex, $\sigma_2$ an $m$-simplex, $P_1 = (s_0, \ldots, s_\ell) \in \Delta_\ell^0$, $P_2 = (t_0, \ldots, t_m) \in \Delta_m^0$. Let $u_0 < \cdots < u_k$ be the sequence obtained by putting the distinct elements of

$$\{s_0 + \cdots + s_i\}_{0 \leq i \leq \ell} \cup \{t_0 + \cdots + t_j\}_{0 \leq j \leq m}$$

in increasing order. Let $v_0 = u_0$, $v_1 = u_1 - u_0, \ldots, v_k = u_k - u_{k-1}$. Then $P = (v_0, \ldots, v_k) \in \Delta_k^0$. Let $p_1 < \cdots < p_{k-\ell}$ be the integers $p$ such that $u_p \notin \{s_0 + \cdots + s_i\}_{0 \leq i \leq \ell}$, and let $q_1 < \cdots < q_{k-m}$ be the integers $q$ such that $u_q \notin \{t_0, \ldots, t_j\}_{0 \leq j \leq m}$. Let $\sigma = (\bar{\sigma}_1, \bar{\sigma}_2)$, where

$$\bar{\sigma}_1 = \Delta(r)(S_{p_{k-\ell}}) \circ \cdots \circ \Delta(r)(S_{p_1})(\sigma_1)$$
$$\bar{\sigma}_2 = \Delta(s)(S_{q_{k-m}}) \circ \cdots \circ \Delta(s)(S_{q_1})(\sigma_2).$$

By construction, the indices $p, q$ run over disjoint sets, so that $\sigma$ is a non-degenerate $k$-simplex of $\Delta(r) \times \Delta(s)$.

Define $g : |\Delta(r)| \times |\Delta(s)| \longrightarrow |\Delta(r) \times \Delta(s)|$ by

$$g((\sigma_1, P_1), (\sigma_2, P_2)) = ((\bar{\sigma}_1, \bar{\sigma}_2), P)$$

where $\bar{\sigma}_i, P$ are given by the above formulas. By definition, $\lambda(\bar{\sigma}_i) = \sigma_i$, $(S\sigma)_*(P) = P_i$, $i = 1, 2$. Hence $f \circ g : |\Delta(r)| \times |\Delta(s)| \longrightarrow |\Delta(r)| \times |\Delta(s)|$ is the identity. By a straightforward calculation, one checks (left to the reader) that $g \circ f$ is the identity on $|\Delta(r) \times \Delta(s)|$. Hence $f$ is a continuous bijection of compact, Hausdorff spaces, i.e., is a homeomorphism. This proves (A.55).

# Appendix B
# Results from Category Theory

A *small* category is a category with a set of objects. Thus, if $\mathcal{C}$ is a small category, we have a set $\operatorname{Ob}\mathcal{C}$ of objects, and a collection $\operatorname{Mor}\mathcal{C}$ of sets indexed by $\operatorname{Ob}\mathcal{C} \times \operatorname{Ob}\mathcal{C}$,

$$\operatorname{Mor}\mathcal{C} = \{\operatorname{Mor}_\mathcal{C}(A,B) \mid (A,B) \in \operatorname{Ob}\mathcal{C} \times \operatorname{Ob}\mathcal{C}\},$$

which satisfy the usual axioms for a category.

Let $\mathcal{C}$, $\mathcal{D}$ be small categories, $F$, $G$ functors from $\mathcal{C}$ to $\mathcal{D}$. We say that $F$ is *isomorphic* to $G$ if there is a natural transformation $\tau : F \longrightarrow G$ such that for each $C \in \operatorname{Ob}\mathcal{C}$, $\tau(C) : F(C) \longrightarrow G(C)$ is an isomorphism. Clearly isomorphism of functors is an equivalence relation.

A functor $F : \mathcal{C} \longrightarrow \mathcal{D}$ is called an *equivalence* of categories if there is a functor $G : \mathcal{D} \longrightarrow \mathcal{C}$ such that the two composite functors $G \circ F : \mathcal{C} \longrightarrow \mathcal{C}$, $F \circ G : \mathcal{D} \longrightarrow \mathcal{D}$ are isomorphic to the identity functors of $\mathcal{C}$, $\mathcal{D}$ respectively.

**Lemma (B.1).** *Let $F : \mathcal{C} \longrightarrow \mathcal{D}$ be a functor between small categories $\mathcal{C}$, $\mathcal{D}$. Then $F$ is an equivalence of categories $\iff$ i) every object of $\mathcal{D}$ is isomorphic to an object of the form $F(C)$, for some $C \in \operatorname{Ob}\mathcal{C}$, and ii) for any pair of objects $C_1, C_2 \in \operatorname{Ob}\mathcal{C}$,*

$$F_* : \operatorname{Hom}_\mathcal{C}(C_1, C_2) \longrightarrow \operatorname{Hom}_\mathcal{D}(F(C_1), F(C_2)) \quad \textit{is bijective.}$$

**Proof.** ($\Rightarrow$) Let $G : \mathcal{D} \longrightarrow \mathcal{C}$ be a functor such that there exist isomorphisms of functors $\eta_\mathcal{C} : 1_\mathcal{C} \longrightarrow G \circ F$, $\eta_\mathcal{D} : 1_\mathcal{D} \longrightarrow F \circ G$. Then given $D \in \operatorname{Ob}\mathcal{D}$, we have an isomorphism $\eta_\mathcal{D}(D) : D \longrightarrow F \circ G(D)$. Hence i) holds. Next, given $C_1, C_2 \in \operatorname{Ob}\mathcal{C}$, there is a diagram

$$\operatorname{Mor}_\mathcal{C}(C_1, C_2) \xrightarrow{F_*} \operatorname{Mor}_\mathcal{D}(F(C_1), F(C_2)) \xrightarrow{G_*} \operatorname{Mor}_\mathcal{C}(GF(C_1), GF(C_2)) \downarrow{\eta}$$
$$\operatorname{Mor}_\mathcal{C}(C_1, C_2)$$

with the composite being the identity.

Hence $F_*$ is injective on morphisms. A similar argument shows that $G_*$ is injective. Hence, in the above diagram, all the arrows are injective, while the composite "identity" is surjective. Hence $F_*$, $G_*$ are also surjective.

Appendix B: Results from Category Theory

($\Leftarrow$) For each $D \in \mathrm{Ob}\,\mathcal{D}$, consider the (non-empty) set ($\mathcal{C}, \mathcal{D}$ are small categories)
$$X(D) = \{(C, \phi) \mid C \in \mathrm{Ob}\,\mathcal{C}, \phi : D \longrightarrow F(C) \text{ an isomorphism}\}.$$
By the axiom of choice, we can choose one such pair $(C_D, \phi_D)$ for each $D \in \mathrm{Ob}\,\mathcal{D}$. We define a functor $G : \mathcal{D} \longrightarrow \mathcal{C}$ by letting $G(D) = C_D$; if $D_1, D_2 \in \mathrm{Ob}\,\mathcal{D}, \psi : D_1 \longrightarrow D_2$ a morphism, let $G(\psi) = F_*^{-1}(\phi_{D_2} \circ \psi \circ \phi_{D_1}^{-1})$, where
$$F_* : \mathrm{Hom}_{\mathcal{C}}(C_{D_1}, C_{D_2}) \longrightarrow \mathrm{Hom}_{\mathcal{D}}(F(C_{D_1}), F(C_{D_2}))$$
is the bijection given by ii). Clearly $G$ is a functor, and $\eta_{\mathcal{D}}(D) = \phi_D$ gives an isomorphism of functors $1_{\mathcal{D}} \longrightarrow F \circ G$. Given $C \in \mathrm{Ob}\,\mathcal{C}$, we have an isomorphism $\phi_{F(C)} : F(C) \longrightarrow F(C_{F(C)})$. Since $F_* : \mathrm{Hom}_{\mathcal{C}}(C, C_{F(C)}) \longrightarrow \mathrm{Hom}_{\mathcal{D}}(F(C), F(C_{F(C)}))$ is bijective, we have an arrow $\eta_{\mathcal{C}}(C) : C \longrightarrow C_{F(C)}$, where $C_{F(C)} = G \circ F(C)$. One verifies at once that $\eta_{\mathcal{C}} : 1_{\mathcal{C}} \longrightarrow G \circ F$ is an isomorphism of functors. This proves (B.1).

We say that a category $\mathcal{C}$ has a set of isomorphism classes of objects if there exists a set $S$, a collection of objects of $\mathcal{C}$ indexed by $S$, $\{C_i \in \mathrm{Ob}\,\mathcal{C} \mid i \in S\}$, and for each $C \in \mathrm{Ob}\,\mathcal{C}$ we are given an index $i \in S$ and an isomorphism. $\phi_C : C \longrightarrow C_i$. Let $\mathcal{D}$ be the full subcategory of $\mathcal{C}$ with the $C_i$ as objects. Then there is a functor $F : \mathcal{C} \longrightarrow \mathcal{D}$ given by $F(C) = C_i$, where $i \in S$ is associated to $C$ as above, and given an arrow $u : C \longrightarrow C'$, we let $F(u) = \phi_{C'} \circ u \circ \phi_C^{-1}$. Then if $G : \mathcal{D} \longrightarrow \mathcal{C}$ is the inclusion functor, clearly $F \circ G : \mathcal{D} \longrightarrow \mathcal{D}$, $G \circ F : \mathcal{C} \longrightarrow \mathcal{C}$ are isomorphic to the respective identity functors. Hence $G : \mathcal{D} \longrightarrow \mathcal{C}$ is an equivalence of categories. In particular, $\mathcal{C}$ is equivalent to a full small subcategory. If $\mathcal{C}$ is equivalent to two small full subcategories $\mathcal{D}_1, \mathcal{D}_2$ then clearly $\mathcal{D}_1$ and $\mathcal{D}_2$ are equivalent. Any of the categories we need to deal with (below and in the main text) are either small, or have a set of isomorphism class of objects, and hence are equivalent to small full subcategories. In the latter situation we often tacitly assume that a given category has been replaced by an equivalent small full subcategory.

**(B.2) Abelian Categories.** A category $\mathcal{C}$ is called an *additive category* if it satisfies the following conditions:

i) there exists a 0-object $0 \in \mathrm{Ob}\,\mathcal{C}$ (i.e., 0 is both an initial and a final object)

ii) for any $M, N \in \mathrm{Ob}\,\mathcal{C}$, the direct sum and direct product exist

iii) $\mathrm{Hom}_{\mathcal{C}}(M, N)$ has a structure of an Abelian group, for any $M, N \in \mathcal{C}$, such that composition of morphisms in $\mathcal{C}$ is bilinear.

Let $\mathcal{C}$ be an additive category. An arrow $g : K \longrightarrow M$ is said to be a *kernel* of an arrow $f : M \longrightarrow N$ if for any arrow $h : P \longrightarrow M$ with $f \circ h = 0$, there exists a unique arrow $k : P \longrightarrow K$ with $h = g \circ k$. Equivalently, for any object $P$ of $\mathcal{C}$, we have an exact sequence of Abelian groups

$$0 \longrightarrow \mathrm{Hom}_{\mathcal{C}}(P, K) \xrightarrow{g_*} \mathrm{Hom}_{\mathcal{C}}(P, M) \xrightarrow{f_*} \mathrm{Hom}_{\mathcal{C}}(P, N).$$

If a morphism $f$ has a kernel, it is unique up to unique isomorphism. Similarly, an arrow $g_1 : M \longrightarrow C$ is a *cokernel* of $f : M \longrightarrow N$ if for any object $P$ in $\mathcal{C}$, we have an exact sequence

$$0 \longrightarrow \mathrm{Hom}_{\mathcal{C}}(C, P) \xrightarrow{g_1^*} \mathrm{Hom}_{\mathcal{C}}(N, P) \xrightarrow{f^*} \mathrm{Hom}_{\mathcal{C}}(M, P).$$

Suppose every morphism $f$ in $\mathcal{C}$ has a kernel and a cokernel. Let $\ker f$, $\mathrm{coker}\, f$ denote the kernel and cokernel respectively (we fix one representative for each arrow within the respective isomorphism class). Let $\mathrm{im}\, f = \ker(\mathrm{coker}\, f)$, $\mathrm{coim}\, f = \mathrm{coker}(\ker f)$ be the *image* and *coimage* of $f$, respectively. By the universal properties defining kernels and cokernels, there is a canonically defined arrow $\mathrm{coim}\, f \longrightarrow \mathrm{im}\, f$.

An additive category $\mathcal{C}$ is called an *Abelian category* if $\mathcal{C}$ also satisfies

i) every morphism has a kernel and a cokernel

ii) for any morphism $f$, the canonical arrow $\mathrm{coim}\, f \longrightarrow \mathrm{im}\, f$ is an isomorphism.

Note that condition ii) does not depend on the choices of the various kernels and cokernels within their isomorphism classes.

If $\mathcal{A}$ is a small Abelian category, we will assume below and in the main text that fixed choices have been made of representatives within the isomorphism classes of arrows representing the kernel and cokernel of any given arrow $f : M \longrightarrow N$. These representatives will be called *the* kernel and cokernel of $f$, respectively. Similarly, consider monomorphisms (arrows with kernel a 0-object) $f : M \longrightarrow N$. For fixed $N$, consider the isomorphism classes of pairs $(M, f)$, where $(M, f) \cong (M', f')$ if there is an isomorphism $g : M \longrightarrow M'$ making

$$\begin{array}{c} M \\ g \downarrow \phantom{xx} \searrow^{f} \\ M' \phantom{xx} \nearrow_{f'} N \end{array}$$

commute. Within each such isomorphism class of pairs, we fix a choice of one pair, and we call the collection of chosen pairs the set of *subobjects* of $N$. Similarly we fix choices of all *quotient objects* of each object $M \in \mathcal{A}$.

The collection of subobjects of an object $N$ forms a partially ordered set in a natural way, with respect to the partial order (denoted $\subset$)

$$(N_1 \xrightarrow{i_1} N) \subset (N_2 \xrightarrow{i_2} N)$$

Appendix B: Results from Category Theory    279

if and only if there is a commutative diagram

$$\begin{array}{c} N_1 \\ j \downarrow \phantom{xxx} \searrow^{i_1} \\ \phantom{xxx} \phantom{x} N \\ \phantom{xxx} \nearrow_{i_2} \\ N_2 \end{array}$$

for a (necessarily unique) monomorphism $j$. If

$$i : N' \longrightarrow N$$

is a monomorphism, then for any subobject

$$i_1 : N_1 \longrightarrow N'$$

there is a unique subobject of $N$ isomorphic to

$$f \circ i_1 : N_1 \longrightarrow N.$$

This gives an injective order-preserving map from the subobjects of $N'$ to that of $N$. In considering the partially ordered set of subobjects of a given object $N$, we identify the partially ordered set of subobjects of any given subobject $i : N' \longrightarrow N$ with its image under the canonical order-preserving injection; loosely speaking, "a subobject of a subobject is a subobject." Similarly, when we speak of "layers" of length $r$

$$N_1 \longrightarrow N_2 \longrightarrow \cdots \longrightarrow N_r$$

of subobjects of an object $N$, we mean the corresponding linearly ordered chain in the partially ordered set of subobjects of $N$. A filtration by subobjects has a similar meaning.

Let $i_j = N_j \longrightarrow N$, $j = 1, 2$ be subobjects of $N$. We define $N_1 \cap N_2 \longrightarrow N$ and $N_1 + N_2 \longrightarrow N$ as the unique subobjects isomorphic to

$$\ker(N \longrightarrow N/N_1 \oplus N/N_2)$$

and

$$\mathrm{im}(N_1 \oplus N_2 \longrightarrow N)$$

respectively. These have the standard properties.

Finally, given any morphism $f : M \longrightarrow N$, there is an order preserving map from the partially ordered set of subobjects of $M$ to that of $N$, induced by $(M' \xrightarrow{i} M) \longmapsto \mathrm{im}(f \circ i)$.

Similar remarks apply to the partially ordered set of quotients of a given object.

**(B.3) Quotient Abelian Categories.** Let $\mathcal{B} \subset \mathcal{A}$ be an inclusion of an Abelian category $\mathcal{B}$ as a full additive subcategory of a small Abelian category $\mathcal{A}$. In particular, the inclusion functor is assumed to preserve finite direct sums, 0-objects and the Abelian group structure on Hom-sets. Then $\mathcal{B}$ is called a *Serre subcategory* (or an *épaisse* subcategory, or *thick* subcategory) if

i) any object of $\mathcal{A}$ isomorphic to an object of $\mathcal{B}$ lies in $\mathcal{B}$

ii) $\mathcal{B}$ is closed under taking subobjects, quotients and extensions in $\mathcal{A}$, i.e., if $0 \to M' \to M \to M'' \to 0$ is exact in $\mathcal{A}$, then $M \in \mathcal{B} \iff M', M'' \in \mathcal{B}$.

If $\mathcal{B} \subset \mathcal{A}$ is a Serre subcategory, we construct below the *quotient Abelian* category $\mathcal{A}/\mathcal{B}$. Our treatment follows the article by P. Gabriel, Des catégories abéliennes, *Bull. Math. Soc. France* 90 (1962), 323–448. The category $\mathcal{A}/\mathcal{B}$ has the following description:

i) $\operatorname{Ob} \mathcal{A}/\mathcal{B} = \operatorname{Ob} \mathcal{A}$

ii) let $M, N \in \operatorname{Ob} \mathcal{A} = \operatorname{Ob} \mathcal{A}/\mathcal{B}$, and let $M' \subset M$, $N' \in N$ be subobjects such that $M/M' \in \mathcal{B}$, $N' \in \mathcal{B}$. There is a natural homomorphism $\operatorname{Hom}_{\mathcal{A}}(M, N) \longrightarrow \operatorname{Hom}_{\mathcal{A}}(M', N/N')$. As $M'$, $N'$ range over such pairs of subobjects, the groups

$$\operatorname{Hom}_{\mathcal{A}}(M', N/N')$$

from a directed system of Abelian groups, and we define

$$\operatorname{Hom}_{\mathcal{A}/\mathcal{B}}(M, N) = \varinjlim_{(M', N')} \operatorname{Hom}_{\mathcal{A}}(M', N/N').$$

One easily verifies that the composition law for morphisms of $\mathcal{A}$ yields a bilinear composition law for morphisms of $\mathcal{A}/\mathcal{B}$, so that $\mathcal{A}/\mathcal{B}$ is an additive category. Let $T: \mathcal{A} \longrightarrow \mathcal{A}/\mathcal{B}$ be the canonical additive functor.

**Lemma (B.4).** *Let $u \in \operatorname{Hom}_{\mathcal{A}}(M, N)$. Then $Tu$ is null $\iff \operatorname{im} u \in \operatorname{Ob} \mathcal{B}$. Similarly $Tu$ is a monomorphism $\iff \ker u \in \operatorname{Ob} \mathcal{B}$, and $Tu$ is an epimorphism $\iff \operatorname{coker} u \in \operatorname{Ob} \mathcal{B}$.*

**Proof.** If $\operatorname{im} u \in \operatorname{Ob} \mathcal{B}$, then $u \mapsto 0$ in $\operatorname{Hom}_{\mathcal{A}}(M, N/\operatorname{im} u)$ so that $u \mapsto 0$ in $\operatorname{Hom}_{\mathcal{A}/\mathcal{B}}(M, N)$ (which is the direct limit). Hence $Tu = 0$. Conversely if $Tu = 0$, then for some pair $M'$, $N'$ of subobjects of $M, N$ respectively, with $M/M', N' \in \operatorname{Ob} \mathcal{B}$, the image of $u$ in $\operatorname{Hom}_{\mathcal{A}}(M', N/N')$ vanishes. Thus $u(M') \subset N'$ so that $u(M') \in \operatorname{Ob} \mathcal{B}$. Also $M/M' \in \operatorname{Ob} \mathcal{B}$. Hence $M/(M' + \ker u) \in \operatorname{Ob} \mathcal{B}$. We have an exact sequence

$$0 \longrightarrow u(M') \longrightarrow \operatorname{im} u \longrightarrow M/(M' + \ker u) \longrightarrow 0$$

with extreme terms in $\operatorname{Ob} \mathcal{B}$, so that $\operatorname{im} u \in \operatorname{Ob} \mathcal{B}$.

Appendix B: Results from Category Theory    281

Next, we show $Tu$ is mono $\iff \ker u \in \operatorname{Ob} \mathcal{B}$. Suppose $Tu$ is mono. Let $i : \ker u \longrightarrow M$ be the canonical arrow; then $u \circ i = 0$, so $(Tu)(Ti) = 0$. Since $Tu$ is mono, $Ti = 0$. Hence by the first part of the lemma, proved above, $\operatorname{im} i \cong \ker u$ is an object of $\mathcal{B}$. Conversely, suppose $\ker u \in \operatorname{Ob} \mathcal{B}$. To prove $Tu$ is mono, it suffices to show that if $\bar{f} : TP \to TM$ is any non-null arrow in $\mathcal{A}/\mathcal{B}$, then $Tu \circ \bar{f} : TP \longrightarrow TN$ is also non-null. Now $\bar{f}$ is induced by some arrow $f : P' \longrightarrow M/M'$ in $\mathcal{A}$, where $P/P'$, $M' \in \operatorname{Ob} \mathcal{B}$. Since $\ker u \in \operatorname{Ob} \mathcal{B}$, we may assume without loss of generality that $\ker u \subset M'$. Then $u$ induces a monomorphism $u' : M/M' \longrightarrow N/u(M')$. Since $\bar{f} \neq 0$, $\operatorname{im} f \notin \operatorname{Ob} \mathcal{B}$, hence $\operatorname{im}(u' \circ f) \notin \operatorname{Ob} \mathcal{B}$. Hence $T(u' \circ f) \neq 0$. Thus $Tu \circ \bar{f} \neq 0$.

The proof that $Tu$ is epi $\iff \operatorname{coker} u \in \operatorname{Ob} \mathcal{B}$ is similar.

**Lemma (B.5).** *Let $u \in \operatorname{Hom}_{\mathcal{A}}(M, N)$, $i : K \longrightarrow M$ the kernel, and $p : N \longrightarrow C$ the cokernel, of $u$. Then $Ti : TK \longrightarrow TM$ is a kernel, and $Tp : TN \longrightarrow TC$ a cokernel, of $Tu$.*

**Proof.** Clearly $Ti$ is mono, from (B.4). Let $\bar{f} : TX \longrightarrow TM$ such that $Tu \circ \bar{f} = 0$. Then $\bar{f}$ is induced by some $f \in \operatorname{Hom}_{\mathcal{A}}(X', M/M')$ where $X/X'$, $M' \in \operatorname{Ob} \mathcal{B}$. We have a diagram with exact rows

$$\begin{array}{ccccccc} 0 & \longrightarrow & K & \stackrel{i}{\longrightarrow} & M & \stackrel{u}{\longrightarrow} & N \\ & & \downarrow & & \downarrow & & \downarrow \\ 0 & \longrightarrow & K/K \cap M' & \stackrel{i'}{\longrightarrow} & M/M' & \stackrel{u'}{\longrightarrow} & N/u(M') \end{array}$$

where $u'$, $i'$ are induced by $u, i$ respectively. Since $Tu \circ \bar{f} = 0$, $T(u' \circ f) = 0$, and so $\operatorname{im}(u' \circ f) \in \operatorname{Ob} \mathcal{B}$. Let $X'' = f^{-1}(\operatorname{im} i')$; then $X'/X'' \cong \operatorname{im}(u' \circ f)$, so that $X'/X''$, $X/X'' \in \operatorname{Ob} \mathcal{B}$. The restriction of $f$ to $X''$ factors through a morphism $g : X'' \longrightarrow K/K \cap M'$ since $u' \circ f|_{X''} = 0$, and $i' = \ker u'$. If $\bar{g} \in \operatorname{Hom}_{\mathcal{A}/\mathcal{B}}(X, K)$ is the morphism determined by $g$, then $Ti \circ \bar{g} = \bar{f}$.

The statement about cokernels is proved similarly.

**Lemma (B.6).** *Let $u \in \operatorname{Hom}_{\mathcal{A}}(M, N)$. Then $Tu$ is an isomorphism $\iff \ker u$, $\operatorname{coker} u \in \operatorname{Ob} \mathcal{B}$.*

**Proof.** Clearly $Tu$ is an isomorphism $\Rightarrow Tu$ is mono and epi $\Rightarrow \ker u$, $\operatorname{coker} u \in \operatorname{Ob} \mathcal{B}$ by (B.4). Conversely, if $\ker u$, $\operatorname{coker} u$ lie in $\operatorname{Ob} \mathcal{B}$, then let $q : M \longrightarrow \operatorname{coim} u$, $j : \operatorname{im} u \longrightarrow N$ be the canonical epi and mono $\mathcal{A}$, respectively, and let $\theta : \operatorname{coim} u \longrightarrow \operatorname{im} u$ be the canonical isomorphism. Then we have a diagram

$$\begin{array}{ccc} M & \xrightarrow{u} & N \\ q\downarrow & & \uparrow j \\ \operatorname{coim} u & \xrightarrow[\theta]{\cong} & \operatorname{im} u \end{array}$$

So it suffices to prove $Tq$, $Tj$ are isomorphisms. Now

$$\operatorname{coim} u = \operatorname{coker}(\ker u) = M/i(K),$$

where $i : K \longrightarrow M$ is the kernel, with $K \in \operatorname{Ob} \mathcal{B}$. The identity map of $\operatorname{coim} u$ is thus an element of $\operatorname{Hom}_{\mathcal{A}}(\operatorname{coim} u, M/i(K))$, and hence gives an element $\bar{q} \in \operatorname{Hom}_{\mathcal{A}/\mathcal{B}}(\operatorname{coim} u, M)$. Clearly $\bar{q}$ is the inverse of $Tq$; hence $Tq$ is an isomorphism. The proof that $Tj$ is an isomorphism is similar.

**Proposition (B.7).** $\mathcal{A}/\mathcal{B}$ *is an Abelian category, such that* $T : \mathcal{A} \longrightarrow \mathcal{A}/\mathcal{B}$ *is an exact functor.*

**Proof.** Let $\bar{f} : TM \longrightarrow TN$ be a morphism in $\mathcal{A}/\mathcal{B}$. We must show that $\bar{f}$ has a kernel, cokernel, coimage and image, and that the natural morphism from the coimage to the image is an isomorphism. Let $\bar{f}$ be induced by $f \in \operatorname{Hom}_{\mathcal{A}}(M', N/N')$, where $M/M'$, $N' \in \operatorname{Ob} \mathcal{B}$, and let $i : M' \longrightarrow M$, $q : N \longrightarrow N/N'$ be the natural mono and epi (in $\mathcal{A}$) respectively. Then we have a diagram in $\mathcal{A}/\mathcal{B}$

$$\begin{array}{ccc} TM & \xrightarrow{\bar{f}} & TN \\ Ti\uparrow & & \downarrow \\ TM' & \xrightarrow{Tf} & T(N/N') \end{array}$$

where by (B.6), $Ti$, $Tq$ are isomorphisms. But by (B.4), (B.5) $Tf$ has a kernel and a cokernel, namely $T(\ker f)$ and $T(\operatorname{coker} f)$. Similarly $T(\operatorname{coim} f)$, $T(\operatorname{im} f)$ are a coimage and image for $Tf$, respectively. Since the natural map $T(\operatorname{coim} f) \longrightarrow T(\operatorname{im} f)$ has the form $Tg$, where $g : \operatorname{coim} f \longrightarrow \operatorname{im} f$ is the natural map, and since $g$ is an isomorphism (as $\mathcal{A}$ is an Abelian category), $Tg$ is an isomorphism. Hence $\mathcal{A}/\mathcal{B}$ is an Abelian category. The exactness of $T$ follows at once from (B.4) and (B.5).

**Example (B.8).** Let $R$ be a Noetherian commutative ring, $S \subset R$ a multiplicative set, $\mathcal{A}$ the Abelian category of finitely generated $R$-modules, $\mathcal{B}$ the full Serre-subcategory of $S$-torsion $R$-modules (thus $M \in \mathcal{B} \iff$ for some $s \in S$, $sM = 0$). Let $\mathcal{C}$ be the category of finite $S^{-1}R$-modules, $L : \mathcal{A} \longrightarrow \mathcal{C}$ the localization functor $L(M) = S^{-1}R \otimes_R M$. We claim there is an equivalence of categories $U : \mathcal{A}/\mathcal{B} \longrightarrow \mathcal{C}$ such that $U \circ T$ and $L$ are isomorphic functors $\mathcal{A} \longrightarrow \mathcal{C}$.

To see this, we first note that there is a natural isomorphism of rings $R \cong \operatorname{Hom}_{\mathcal{A}}(R, R)$ (where the multiplication on the latter is given by composition). For any $M \in \operatorname{Ob} \mathcal{A}$, $\operatorname{Hom}_{\mathcal{A}}(R, M)$ is a module over $\operatorname{Hom}_{\mathcal{A}}(R, R)$ in a canonical way, and is naturally isomorphic to $M$ as an $R$-module (under the isomorphism $R \cong \operatorname{Hom}_{\mathcal{A}}(R, R)$).

Now consider the ring $\operatorname{Hom}_{\mathcal{A}/\mathcal{B}}(TR, TR)$. There is a canonical ring homomorphism

$$R \xrightarrow{\cong} \operatorname{Hom}_{\mathcal{A}}(R, R) \xrightarrow{T_*} \operatorname{Hom}_{\mathcal{A}/\mathcal{B}}(TR, TR).$$

If $s \in S$, multiplication by $s$ gives a map $\phi(s) : R \longrightarrow R$ whose kernel and cokernel lie in $\mathcal{B}$. Hence $T_* \circ \phi(s)$ is an isomorphism. One easily verifies from the definitions that $\operatorname{Hom}_{\mathcal{A}/\mathcal{B}}(TR, TR)$ is commutative; now by the universal property of the localization homomorphism $R \longrightarrow S^{-1}R$, the map $T_* \circ \phi : R \longrightarrow \operatorname{Hom}_{\mathcal{A}/\mathcal{B}}(TR, TR)$ factors uniquely through $R \longrightarrow S^{-1}R$. One can now verify that $S^{-1}R \longrightarrow \operatorname{Hom}_{\mathcal{A}/\mathcal{B}}(TR, TR)$ is an isomorphism.

For any $M \in \operatorname{Ob} \mathcal{A}$, we have a module $\operatorname{Hom}_{\mathcal{A}/\mathcal{B}}(TR, TM)$ over the ring $\operatorname{Hom}_{\mathcal{A}/\mathcal{B}}(TR, TR)$. Using the above isomorphism of rings we can regard $\operatorname{Hom}_{\mathcal{A}/\mathcal{B}}(TR, TM)$ as an $S^{-1}R$-module. Further there is an $R$-module homomorphism

$$M \underset{\cong}{\longrightarrow} \operatorname{Hom}_{\mathcal{A}}(R, M) \xrightarrow{T_*} \operatorname{Hom}_{\mathcal{A}/\mathcal{B}}(TR, TM).$$

By the universal property of $M \longrightarrow S^{-1}M = S^{-1}R \otimes_R M$, there is a unique $S^{-1}R$-module map $S^{-1}M \xrightarrow{\psi_M} \operatorname{Hom}_{\mathcal{A}/\mathcal{B}}(TR, TM)$ compatible with the above map. One can check that this is an isomorphism. Now define $U : \mathcal{A}/\mathcal{B} \longrightarrow \mathcal{C}$ by sending $TM$ to the $S^{-1}R$-module $\operatorname{Hom}_{\mathcal{A}/\mathcal{B}}(TR, TM)$, and define

$$U_* : \operatorname{Hom}_{\mathcal{A}/\mathcal{B}}(TM, TN) \to \operatorname{Hom}_{\mathcal{C}}(\operatorname{Hom}_{\mathcal{A}/\mathcal{B}}(TR, TM), \operatorname{Hom}_{\mathcal{A}/\mathcal{B}}(TR, TN))$$

to be the map induced by the composition law for morphisms in $\mathcal{A}/\mathcal{B}$. Then the $\psi_M$ define a natural transformation $\psi : L \longrightarrow U \circ T$ which is an isomorphism of functors.

**Example (B.9).** Let $X$ be a Noetherian scheme, $Z \subset X$ a closed subscheme, $U = X - Z$, $\mathcal{A}$ = Abelian category of coherent $\mathcal{O}_X$-modules, $\mathcal{B}$ = Serre subcategory of modules supported on $Z$, $\mathcal{C}$ = Abelian category of coherent $\mathcal{O}_U$-modules. Let $j : U \longrightarrow X$ be the inclusion. Then there is an equivalence of categories $F : \mathcal{A}/\mathcal{B} \longrightarrow \mathcal{C}$ and an isomorphism of functors $j^* \cong F \circ T : \mathcal{A} \longrightarrow \mathcal{C}$. The proof is left to the reader.

**(B.10) Adjoint Functors.** Let Set denote the category of sets. Given a functor $F: \mathcal{C} \longrightarrow \mathcal{D}$ between categories,

$$(M, N) \longmapsto \mathrm{Hom}_{\mathcal{D}}(M, FN)$$

gives a functor $\mathcal{D}^{\mathrm{op}} \times \mathcal{C} \longrightarrow$ Set. Similarly given a functor $G: \mathcal{D} \longrightarrow \mathcal{C}$, we have a functor $\mathcal{D}^{\mathrm{op}} \times \mathcal{C} \longrightarrow$ Set,

$$(M, N) \longmapsto \mathrm{Hom}_{\mathcal{C}}(GM, N).$$

A pair of functors $F: \mathcal{C} \longrightarrow \mathcal{D}$, $G: \mathcal{D} \longrightarrow \mathcal{C}$ form an *adjoint pair* if the two functors above $\mathcal{D}^{\mathrm{op}} \times \mathcal{C} \longrightarrow$ Set are isomorphic, i.e, if we have natural bijections

$$\mathrm{Hom}_{\mathcal{C}}(GM, N) \cong \mathrm{Hom}_{\mathcal{D}}(M, FN).$$

In this situation we also say that $G$ is a *left adjoint* of $F$, and $F$ is a *right adjoint* of $G$.

Suppose $F: \mathcal{C} \longrightarrow \mathcal{D}$ has a left adjoint $G: \mathcal{D} \longrightarrow \mathcal{C}$. Then given $M \in \mathrm{Ob}\,\mathcal{D}$, we have a bijection

$$\mathrm{Hom}_{\mathcal{C}}(GM, GM) \cong \mathrm{Hom}_{\mathcal{D}}(M, FGM).$$

In particular the identity map of $GM$ gives an arrow $i_M: M \longrightarrow FGM$. This arrow has the following universal property: given any arrow $u: M \longrightarrow FN$ with $N \in \mathrm{Ob}\,\mathcal{C}$, there is a unique arrow $v: GM \longrightarrow N$ such that

$$\begin{array}{ccc} M & \xrightarrow{i_M} & FGM \\ & {\scriptstyle u}\searrow & \downarrow{\scriptstyle Fv} \\ & FN & \end{array}$$

commutes. Indeed we have a diagram

$$\begin{array}{ccc} \mathrm{Hom}_{\mathcal{C}}(GM, GM) & \xrightarrow{\cong} & \mathrm{Hom}_{\mathcal{D}}(M, FGM) \\ {\scriptstyle v_*}\downarrow & & \downarrow{\scriptstyle (Fv)_*} \\ \mathrm{Hom}_{\mathcal{C}}(GM, N) & \xrightarrow[\cong]{} & \mathrm{Hom}_{\mathcal{D}}(M, FN) \end{array}$$

where $v \in \mathrm{Hom}_{\mathcal{C}}(GM, N)$ corresponds (under the bijection of the bottom row) to $u \in \mathrm{Hom}_{\mathcal{D}}(M, FN)$, showing that $(Fv)_* i_M = u$, which is just the commutativity of the above triangle. Conversely, if $v': GM \longrightarrow N$ is such that $Fv'$ makes the above triangle commute, then $(Fv')_* i_M = u$, so that $v'_*(1_{GM}) \in \mathrm{Hom}_{\mathcal{C}}(GM, N)$ maps to $u$ under $\mathrm{Hom}_{\mathcal{C}}(GM, N) \xrightarrow{\cong} \mathrm{Hom}_{\mathcal{D}}(M, FN)$, i.e., $v'_*(1_{GM}) = v$, i.e., $v' = v$.

Conversely, suppose $F: \mathcal{C} \longrightarrow \mathcal{D}$ is a functor with the following universal property: given any $M \in \mathrm{Ob}\,\mathcal{D}$, there is an arrow $i_M: M \longrightarrow F\bar{M}$

Appendix B: Results from Category Theory     285

universal among arrows $u : M \longrightarrow FN$, i.e., given $u : M \longrightarrow FN$, there exists a unique $v : \bar{M} \longrightarrow N$ in $\mathcal{C}$ such that

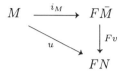

commutes. If $\mathcal{C}, \mathcal{D}$ are small categories, then we can choose a universal arrow $i_M : M \longrightarrow F\bar{M}$ for each $M \in \mathcal{D}$. Then given an arrow $w : M_1 \longrightarrow M_2$, the composite $i_{M_2} \circ w : M_1 \longrightarrow F\bar{M}_2$ factors through an arrow $F\bar{w} : F\bar{M}_1 \longrightarrow F\bar{M}_2$, for a unique arrow $\bar{w} : \bar{M}_1 \longrightarrow \bar{M}_2$ in $\mathcal{C}$. Then $G(M) = \bar{M}$, $G(w) = \bar{w}$ gives a functor $G : \mathcal{D} \longrightarrow \mathcal{C}$ which is left adjoint to $F : \mathcal{C} \longrightarrow \mathcal{D}$.

A similar characterization exists for a functor $G : \mathcal{D} \longrightarrow \mathcal{C}$ which has a right adjoint, in terms of an appropriate universal property for maps $GM \longrightarrow N$, for each $N \in \mathcal{C}$.

**(B.11) Filtering Categories.** A small category $\mathcal{I}$ is said to be *filtering* if

i) Ob $\mathcal{I}$ is non-empty

ii) given $M, M' \in \mathcal{I}$ there exists $N \in \mathcal{I}$ and a diagram

iii) given $M, M' \in \mathcal{I}$ and arrows $f, g : M \longrightarrow M'$, there is an arrow $h : M' \longrightarrow N$ in $\mathcal{I}$ such that $h \circ f = h \circ g : M \longrightarrow N$.

For example, let $(S, \leq)$ be a partially ordered set. Then we may regard $S$ as a category, with objects given by elements of $S$, and for any $x, y \in S$, $\text{Hom}_S(x, y)$ is empty unless $x \leq y$, in which case there is a unique morphism $x \to y$. Then $S$ is filtering $\iff$ $S$ is non-empty and directed.

A directed family of sets $\{A_i \mid i \in \mathcal{I}\}$ indexed by $\mathcal{I}$ is a functor $\mathcal{I} \longrightarrow \underline{\text{Set}}$ from $\mathcal{I}$ to the category of sets. Then

$$\varinjlim_{i \in \mathcal{I}} A_i$$

is defined to be the quotient of $\coprod_{i \in \mathcal{I}} A_i$ modulo the equivalence relation, that if $x \in A_i$, $y \in A_j$, then $x \sim y \iff$ there exists $k \in \mathcal{I}$ and arrows $i \longrightarrow k$, $j \longrightarrow k$ such that $(i \longrightarrow k)_*(x) = (j \longrightarrow k)_*(y)$ in $A_k$.

A direct family of small categories $\{C_i \mid i \in \mathcal{I}\}$ indexed by $\mathcal{I}$ is a functor $\mathcal{I} \longrightarrow \underline{\mathrm{Cat}}$, the category of small categories. Then $i \longmapsto \mathrm{Ob}\, C_i$ is a family of sets indexed by $\mathcal{I}$. We define $\varinjlim_{i \in \mathcal{I}} C_i$ to be a small category $C$, together with a natural transformation of functors $\mathcal{I} \longrightarrow \underline{\mathrm{Cat}}$ from the given functor $i \longmapsto C_i$ to the constant functor $i \longmapsto C$, with the following universal property: given any natural transformation between the functor $i \longmapsto C_i$ and the constant functor $i \longmapsto \mathcal{D}$, for a small category $\mathcal{D}$, there is a unique functor $C \longrightarrow \mathcal{D}$ such that

commutes for all $i \in \mathcal{I}$. One can easily check that $\varinjlim_{i \in \mathcal{I}} C_i = C$ exists, with

$$\mathrm{Ob}\, C = \varinjlim_{i \in \mathcal{I}} \mathrm{Ob}\, C_i,$$

and with the following morphisms: given $M, N \in \mathrm{Ob}\, C$, we can find an index $i \in \mathcal{I}$ and objects $M_i, N_i \in C_i$ such that $M_i \longmapsto M$, $N_i \longmapsto N$; if $i \setminus \mathcal{I}$ denotes the full subcategory of objects $j$ of $\mathcal{I}$ such that there exists an arrow $i \longrightarrow j$ in $\mathcal{I}$, then we let

$$\mathrm{Hom}_C(M, N) = \varinjlim_{j \in (i \setminus \mathcal{I})} \mathrm{Hom}_{C_j}(M_j, N_j)$$

where $M_i \longmapsto M_j$, $N_i \longmapsto N_j$ under $C_i \longrightarrow C_j$.

# Appendix C
# Exact Couples

**(C.1) The Spectral Sequence of an Exact Couple.** Let $\mathcal{A}$ be an Abelian category. An *exact couple* in $\mathcal{A}$ is an exact triangle

$$D_1 \xrightarrow{b_1} D_1$$
$$\nwarrow_{a_1} \swarrow_{c_1}$$
$$E_1$$

Thus $E_1, D_1 \in \mathrm{Ob}\,\mathcal{A}$, and we have a long exact sequence in $\mathcal{A}$

$$\ldots \to E_1 \xrightarrow{a_1} D_1 \xrightarrow{b_1} D_1 \xrightarrow{c_1} E_1 \xrightarrow{a_1} D_1 \to \ldots$$

Let $d_1 : E_1 \longrightarrow E_1$ be the composite $d_1 = c_1 \circ a_1$. Then $d_1^2 = c_1 \circ a_1 \circ c_1 \circ a_1 = 0$ as $a_1 \circ c_1 = 0$, by exactness. Thus $(E_1, d_1)$ is a differential group. Let $E_2 = \ker d_1 / \operatorname{im} d_1$, $D_2 = \operatorname{im} b_1 \cong D_1 / \ker b_1$, and define a new triangle

$$D_2 \xrightarrow{b_2} D_2$$
$$\nwarrow_{a_2} \swarrow_{c_2}$$
$$E_2$$

with maps

i) $b_2 = b_1|_{D_2}$

ii) $a_2$ is induced by $a_1|_{\ker d_1}$ (this is well defined as $a_1 \circ d_1 = 0$, so that $a_1(\operatorname{im} d_1) = 0$)

iii) $c_1(\ker b_1) = c_1(\operatorname{im} a_1) = \operatorname{im} d_1$, so that $c_1$ induces a map $D_1/\ker b_1 \longrightarrow \ker d_1/\operatorname{im} b_1$, i.e., a map $c_2 : D_2 \longrightarrow E_2$. The reader can easily verify that the new triangle is exact (c.f. [W], Ch. XIII). This exact couple is called the (first) *derived couple* of the original exact couple.

We can iterate this construction, and by induction define (for $n \geq 2$) the $(n-1)$th-derived couple

$$D_n \xrightarrow{b_n} D_n$$
$$\nwarrow_{a_n} \swarrow_{c_n}$$
$$E_n$$

to be the derived couple of the $(n-2)$th-derived couple (where the 0th-derived couple is the original one). The sequence of derived couples associated to the given exact couple is called the *spectral sequence of the couple*.

One computes by induction that for any $n \geq 1$, $D_n = \operatorname{im} b_1^{n-1} \cong D_1/\ker b_1^{n-1}$, $E_n = a_1^{-1}(\operatorname{im} b_1^{n-1})/c_1(\ker b_1^{n-1})$ and the maps $a_n$, $b_n$, $c_n$ are given by

i) $a_n(x + c_1(\ker b_1^{n-1})) = a_1(x)$ for $x \in a_1^{-1}(\operatorname{im} b_1^{n-1})$
ii) $b_n(y) = b_1(y)$ for $y \in \operatorname{im} b_1^{n-1}$
iii) $c_n(b_1^{n-1}(y)) = c_1(y) + c_1(\ker b_1^{n-1})$ for $y \in D_1$.

Here and below we follow the convention that $b_1^0$ is the identity map of $D_1$. Let $Z_n = a_1^{-1}(\operatorname{im} b_1^{n-1})$, $B_n = c_1(\ker b_1^{n-1})$. Then

$$E_1 = Z_1 \supset Z_2 \supset Z_3 \cdots \supset Z_n \supset Z_{n+1} \supset \cdots \supset B_{n+1}$$
$$\supset B_n \supset \cdots B_2 \supset B_1 = 0,$$

and $E_n = Z_n/B_n$. Define

$$Z_\infty = \bigcap_{n \geq 1} Z_n, \qquad B_\infty = \bigcup_{n \geq 1} B_n$$

(if they exist). Then $0 \subset B_\infty \subset Z_\infty \subset E_1$. Define $E_\infty = Z_\infty/B_\infty$, which we call the *limit term* of the spectral sequence.

The exact couples which we need to consider are *bigraded*, and are all obtained by the following construction (or its version for homology, see below). Assume given two collections of objects of $\mathcal{A}$ indexed by $\mathbb{Z} \times \mathbb{Z}$

$$\{A^{m,n}\}_{(m,n) \in \mathbb{Z} \times \mathbb{Z}}, \quad \{E^{m,n}\}_{(m,n) \in \mathbb{Z} \times \mathbb{Z}},$$

and suppose that we are given long exact sequences for all $p \in \mathbb{Z}$

$$\to A^{p+1,q-1} \xrightarrow{f_{p,q}} A^{p,q} \xrightarrow{g_{p,q}} E^{p,q} \xrightarrow{h_{p,q}} A^{p+1,q} \xrightarrow{f_{p,q+1}} A^{p,q+1} \to$$

We can combine all of these exact sequences into an exact couple

$$\begin{array}{ccc} D_1 & \xrightarrow{b_1} & D_1 \\ & \nwarrow_{a_1} \quad \swarrow_{c_1} & \\ & E_1 & \end{array}$$

by setting $D_1 = \bigoplus_{p,q} A^{p,q}$, $E_1 = \bigoplus_{p,q} E^{p,q}$, so that $D_1$, $E_1$ are bigraded with $D_1^{p,q} = A^{p,q}$, $E_1^{p,q} = E^{p,q}$, and defining maps

$$a_1 = \oplus h_{p,q}, \quad b_1 = \oplus f_{p-1,q+1}, \quad c_1 = \oplus g_{p,q}.$$

Thus, with the given bigradings, $a_1$ has bidegree $(1,0)$, $b_1$ has bidegree $(-1,1)$ and $c_1$ has bidegree $(0,0)$. Thus, from the formulae given above

Appendix C: Exact Couples

for the terms and maps in the $(n-1)$st-derived couple, we see that $B_n$, $Z_n \subset E_1$ and $D_n \subset D_1$ are graded submodules, so that $D_n$, $E_n = Z_n/B_n$ have natural bigradings with respect to which $a_n$ has bidegree $(1,0)$, $b_n$ has bidegree $(-1,1)$ and $c_n$ has bidegree $(n-1,1-n)$. Thus $d_n = c_n \circ a_n$ has bidegree $(n, 1-n)$, so that $d_n$ has $(p,q)$-component $d_n^{p,q} : E_n^{p,q} \longrightarrow E_n^{p+n,q-n+1}$. A bigraded exact couple with the above choices of gradings gives rise to a *spectral sequence of cohomological type*, characterized by the fact that $E_n$, $D_n$ are bigraded and the maps $a_n$, $b_n$, $c_n$, $d_n$ have the above bidegrees.

Next, we discuss convergence of the spectral sequence. Consider the directed system indexed by $q \in \mathbb{Z}$

$$\cdots \to A^{n-q,q} \xrightarrow{f_{n-q-1,q+1}} A^{n-q-1,q+1} \xrightarrow{f_{n-q-2,q+2}} A^{n-q-2,q+2} \to \cdots$$

and let $A^n = \varinjlim_q A^{n-q,q}$. There is a decreasing filtration

$$F^p A^n = \mathrm{im}(A^{p,n-p} \longrightarrow A^n), \quad p \in \mathbb{Z}, \ F^{p+1} A^n \subset F^p A^n.$$

We make two further assumptions about our data, which will be valid in all our applications:

i) for each $n$, there is an integer $q_1(n)$ such that $f_{n-q,q} : A^{n-q+1,q-1} \longrightarrow A^{n-q,q}$ is an isomorphism for $q \geq q_1(n)$

ii) for each $n$, there is an integer $q_0(n)$ such that $A^{n-q,q} = 0$ for $q < q_0(n)$.

Then the above filtration on $A^n$ is finite, for each $n$, and we claim that there are natural isomorphisms for any $p, q$ and for $n = n(p,q)$ sufficiently large

$$E_n^{p,q} \cong E_\infty^{p,q} \cong F^p A^{p+q}/F^{p+1} A^{p+q}.$$

In this situation we say that the spectral sequence *converges*; $\{A^n\}_{n \in \mathbb{Z}}$ is called the *abutment* of the spectral sequence, $\{F^p A^n\}_{p \in \mathbb{Z}}$ the *filtration induced by the spectral sequence*, and we write $E_1^{p,q} \Rightarrow A^{p+q}$ (or $E_n^{p,q} \Rightarrow A^{p+q}$) to express the fact that $\{A^n\}$ is the abutment.

To prove the claimed isomorphisms, we note that the exact triangle

$$\begin{array}{ccc} D_n & \xrightarrow{b_n} & D_n \\ & \nwarrow_{a_n} \quad \swarrow_{c_n} & \\ & E_n & \end{array}$$

yields an exact sequence (taking into account the bidegrees of $a_n$, $b_n$, $c_n$)

$$\cdots \to D_n^{p-n+2,q+n-2} \to D_n^{p-n+1,q+n-1} \to E_n^{p,q} \to D_n^{p+1,q} \to D_n^{p,q+1} \to \cdots$$

Now assume $p, q$ are fixed and $n$ is sufficiently large. Then

$$D_n^{p-n+2,q+n-2} = D_1^{p-n+2,q+n-2} \cap \operatorname{im} b_1^{n-1}$$
$$= \operatorname{im}(D_1^{p+1,q-1} \longrightarrow D_1^{p-n+2,q+n-2})$$
$$\cong \operatorname{im}(D_1^{p+1,q-1} \longrightarrow A^{p+q}) = F^{p+1} A^{p+q}$$
$$D_n^{p-n+1,q+n-1} = D_1^{p-n+1,q+n-1} \cap \operatorname{im} b_1^{n-1}$$
$$= \operatorname{im}(D_1^{p,q} \longrightarrow D_1^{p-n+1,q+n-1}) \cong F^p A^{p+q},$$

and the map $D_n^{p-n+2,q+n-2} \longrightarrow D_n^{p-n+1,q+n-1}$ is identified with the inclusion $F^{p+1} A^{p+q} \subset F^p A^{p+q}$. Further

$$D_n^{p+1,q} = D_1^{p+1,q} \cap \operatorname{im} b_1^{n-1}$$
$$= \operatorname{im}(D_1^{p+n,q-n+1} \longrightarrow D_1^{p+1,q})$$
$$= 0$$

since $D_1^{p+n,q-n+1} = 0$ if $n > q + 1 - q_0(p+q+1)$. Hence the claimed isomorphism $F^p A^{p+q} / F^{p+1} A^{p+q} \cong E_n^{p,q}$ follows from the above exact sequence, provided $n$ is sufficiently large. To see that $E_n^{p,q} \cong E_\infty^{p,q}$ we see that for sufficiently large $n$, we have a complex of $E_n$-terms, whose homology is $E_{n+1}^{p,q}$,

$$E_n^{p-n,q+n-1} \xrightarrow{d_n} E_n^{p,q} \xrightarrow{d_n} E_n^{p+n,q-n+1}.$$

But $E_1^{p+n,q-n+1} = 0 = E_1^{p-n,q+n-1}$ for sufficiently large $n$, from the exact sequences

$$D_1^{p+n,q-n+1} \longrightarrow E_1^{p+n,q-n+1} \longrightarrow D_1^{p+n+1,q-n+1}$$

$$D_1^{p-n+1,q+n} \xrightarrow{f_{p-n,q+n-1}} D_1^{p-n,q+n-1} \longrightarrow E_1^{p-n,q+n-1}$$

$$\longrightarrow D^{p-n+1,q+n-1} \xrightarrow{f_{p-n,q+n}} D_1^{p-n,q+n}$$

and the facts that $D_1^{p+n,q-n+1} = D_1^{p+n-1,q-n+1} = 0$, and $f_{p-n,q+n-1}$, $f_{p-n,q+n}$ are isomorphisms, for sufficiently large $n$. Thus

$$E_n^{p,q} \cong E_{n+1}^{p,q} \cong \cdots \cong E_\infty^{p,q} \cong F^p A^{p+q} / F^{p+1} A^{p+q}$$

for any given $p, q$ and sufficiently large $n$.

We compute explicitly, in terms of the original data of a family of long exact sequences, the differentials $d_1$ of the spectral sequence. We have $E_1^{p,q} = E^{p,q}$, and $a_1|_{E_{1\,p,q}}$ is just $h_{p,q} : E^{p,q} \longrightarrow A^{p+1,q}$. Next $c_1|_{A^{p+1,q}}$ is the map $g_{p+1,q} : A^{p+1,q} \longrightarrow E^{p+1,q}$. Thus $d_1 : E_1^{p,q} \longrightarrow E_1^{p+1,q}$ is the

Appendix C: Exact Couples 291

dashed arrow in the diagram below:

$$\cdots \to A^{p,q} \to E^{p,q} \to A^{p+1,q} \to A^{p,q+1} \to E^{p,q+1} \to \cdots$$

$$\cdots \to E^{p+1,q-1} \to A^{p+2,q-1} \to A^{p+1,q} \to E^{p+1,q} \to A^{p+2,q} \to \cdots$$

In an analogous fashion one can construct spectral sequences of *homological type* from the following data: assume given objects

$$\{A_{m,n}\}_{(m,n)\in\mathbb{Z}\times\mathbb{Z}}, \quad \{E_{m,n}\}_{(m,n)\in\mathbb{Z}\times\mathbb{Z}},$$

together with long exact sequences for each $p \in \mathbb{Z}$

$$\cdots \to A_{p-1,q+1} \to A_{p,q} \to E_{p,q} \to A_{p-1,q} \to A_{p,q-1} \to \cdots .$$

Then one can combine them into an exact couple of bigraded objects

$$\begin{array}{ccc} D_1 & \xrightarrow{b_1} & D_1 \\ & {}_{a_1}\nwarrow \swarrow_{c_1} & \\ & E_1 & \end{array}$$

with $E_1 = \bigoplus_{p,q} E_{p,q}$, $D_1 = \bigoplus_{p,q} A_{p,q}$, bigradings $E^1_{p,q} = E_{p,q}$, $D^1_{p,q} = A_{p,q}$, and maps $a_1, b_1, c_1$ constructed from the maps in the exact sequences in a natural way. Thus $a_1$ has bidegree $(-1, 0)$, $b_1$ has bidegree $(1, -1)$ and $c_1$ has bidegree $(0, 0)$. Then the $(n-1)$th-derived couple

$$\begin{array}{ccc} D_n & \xrightarrow{b_n} & D_n \\ & {}_{a_n}\nwarrow \swarrow_{c_n} & \\ & E_n & \end{array}$$

consists of bigraded objects $E_n = \oplus E^n_{p,q}$, $D_n = \oplus D^n_{p,q}$ such that $a_n, b_n, c_n$, $d_n$ respectively have bidegrees $(-1, 0)$, $(1, -1)$, $(1-n, n-1)$, $(-n, n-1)$. In particular $d_n$ has $(p, q)$-component $d^{p,q}_n : E^n_{p,q} \to E^n_{p-n,q+n-1}$. We have subgroups $B^n_{p,q} \subset Z^n_{p,q} \subset E^1_{p,q}$ such that

$$B^n_{p,q} \subset B^{n+1}_{p,q} \subset Z^{n+1}_{p,q} \subset Z^n_{p,q}$$

for all $n, p, q$. Let

$$Z^\infty_{p,q} = \bigcap_n Z^n_{p,q}, \quad B^\infty_{p,q} = \bigcup_n B^n_{p,q}, \quad E^\infty_{p,q} = Z^\infty_{p,q}/B^\infty_{p,q};$$

then $E_\infty = \oplus E^\infty_{p,q}$ is the limit term of the spectral sequence. The spectral sequence will *converge* if we assume that for each $n \in \mathbb{Z}$, there are integers $p_0(n), p_1(n)$ such that (i) $A_{p,n-p} \to A_{p+1,n-p-1}$ is isomorphism for $p > p_1(n)$, and (ii) $A_{p,n-p} = 0$ for $p < p_0(n)$. Then if

$$A_n = \varinjlim_p A_{p,n-p}, \quad F_p A_n = \mathrm{im}(A_{p,n-p} \to A_n),$$

$\{A_n\}_{n\in\mathbb{Z}}$ is the abutment of the spectral sequence, $\{F_pA_n\}_{p\in\mathbb{Z}}$ is the induced filtration on the abutment, which is a finite increasing filtration of $A_n$, for each $n$. Further, given $p, q$ we have isomorphisms for all sufficiently large $n$

$$E_{p,q}^n \cong E_{p,q}^{n+1} \cong \cdots \cong E_{p,q}^\infty \cong F_pA_{p+q}/F_{p-1}A_{p+q}.$$

We write $E_{p,q}^1 \Longrightarrow A_{p+q}$ (or $E_{p,q}^n \Longrightarrow A_{p+q}$) to denote that the abutment is $\{A_{p+q}\}$, and $E_{p,q}^\infty = F_pA_{p+q}/F_{p-1}A_{p+q}$.

We discuss two examples below.

**(C.2) The BGQ Spectral Sequence.** This is the spectral sequence of Theorem (5.20) of the main text. It is a spectral sequence of cohomological type, with

$$A^{p,q} = K_{-p-q}(\mathcal{M}^p(X)), \quad E^{p,q} = \bigoplus_{x\in X^p} K_{-p-q}(k(x)),$$

and the family of long exact sequences

$$\ldots A^{p+1,q-1} \longrightarrow A^{p,q} \longrightarrow E^{p,q} \longrightarrow A^{p+1,q} \longrightarrow \ldots$$

is taken to be the family of localization sequences

$$\ldots K_{-p-q}(\mathcal{M}^{p+1}(X)) \longrightarrow K_{-p-q}(\mathcal{M}^p(X)) \longrightarrow \bigoplus_{x\in X^p} K_{-p-q}(k(x))$$

$$\longrightarrow K_{-1-p-q}(\mathcal{M}^{p+1}(X)) \longrightarrow \ldots .$$

Here, we make the conventions that $\mathcal{M}^p(X) = \mathcal{M}(X)$, $X^p = X^0$ for $p < 0$, and $K_n(\mathcal{M}^p(X)) = 0$ for $n < 0$, $p \in \mathbb{Z}$. We have $A^{p,q} = 0$ unless $p + q < 0$, and $A^{p,q} \cong A^{p-1,q+1} \cong K_{-p-q}(\mathcal{M}(X))$ for $p \leq 0$, $q \in \mathbb{Z}$, so that the abutment terms are $\{K_{-n}(\mathcal{M}(X))\}_{n\in\mathbb{Z}}$; the filtration on $K_{-n}(\mathcal{M}(X))$ induced by the spectral sequence is

$$F^p K_{-n}(\mathcal{M}(X)) = F^p A^n = \text{im}(A^{p,n-p} \longrightarrow A)$$
$$= \text{im}(K_{-n}(\mathcal{M}^p(X)) \longrightarrow K_{-n}(\mathcal{M}(X)))$$

i.e., is the topological filtration on $K_{-n}(\mathcal{M}(X))$. Finally, if $X$ has Krull dimension $d$, then $\mathcal{M}^p(X) = 0$ for $p > d$, so $A^{p,q} = 0$ for $p > d$. Hence $A^{n-q,q} = 0$ for $q < n - d$, i.e., the spectral sequence is convergent.

**(C.3) The Spectral Sequence of a Filtered Complex.** Let $\mathcal{A}$ be an Abelian category, and let $C^\bullet$ be a cochain complex in $\mathcal{A}$;

$$C^\bullet : 0 \longrightarrow C^0 \xrightarrow{d^0} C^1 \xrightarrow{d^1} C^2 \xrightarrow{d^2} \ldots .$$

Assume given for each $n \geq 0$ a finite, decreasing filtration

$$C^n = F^0 C^n \supset F^1 C^n \supset \cdots \supset F^m C^n = (0), \quad m = m(n),$$

Appendix C: Exact Couples

such that $d^n(F^pC^n) \subset F^pC^{n+1}\ \forall p, n \geq 0$, so that
$$F^pC^\bullet : 0 \longrightarrow F^pC^0 \longrightarrow F^pC^1 \longrightarrow \ldots$$
is a subcomplex of $C^\bullet$. Let $gr_F^p C^\bullet$ be the quotient complex
$$gr_F^p C^\bullet : 0 \longrightarrow F^pC^0/F^{p+1}C^0 \longrightarrow F^pC^1/F^{p+1}C^1 \longrightarrow \ldots$$
so that we have a short exact sequence of complexes
$$0 \longrightarrow F^{p+1}C^\bullet \longrightarrow F^pC^\bullet \longrightarrow gr_F^p C^\bullet \longrightarrow 0,$$
giving rise to a long exact sequence of cohomology objects (with $H^n(\ ) = 0$ for $n < 0$)
$$\to H^n(F^{p+1}C^\bullet) \to H^n(F^pC^\bullet) \to H^n(gr_F^p C^\bullet) \to H^{n+1}(F^{p+1}C^\bullet) \to$$
Thus if we define $F^pC^\bullet = C^\bullet$, $gr_F^p C^\bullet = 0$ for $p < 0$, the above family of long exact sequences yields a bigraded exact couple

$$\begin{array}{ccc} D_1 & \longrightarrow & D_1 \\ & \nwarrow \swarrow & \\ & E_1 & \end{array}$$

with $A^{p,q} = D_1^{p,q} = H^{p+q}(F^pC^\bullet)$, $E^{p,q} = E_1^{p,q} = H^{p+q}(gr_F^p C^\bullet)$, and abutment $A^n = \varinjlim_q A^{n-q,q} = \varinjlim_q H^n(F^{n-q}C^\bullet) = H^n(C^\bullet)$, and induced filtration
$$F^pA^n = \operatorname{im}(A^{p,n-p} \longrightarrow A^n) = \operatorname{im}(H^n(F^pC^\bullet) \longrightarrow H^n(C^\bullet)) = F^pH^n(C^\bullet),$$
say. Thus we have a convergent spectral sequence of cohomological type
$$E_1^{p,q} = H^{p+q}(gr_F^p C^\bullet) \Longrightarrow H^{p+q}(C^\bullet),$$
and limit terms
$$E_\infty^{p,q} = gr_F^p H^{p+q}(C^\bullet).$$
Similarly, given a chain complex in $\mathcal{A}$
$$C_\bullet : \ldots C_2 \xrightarrow{d_2} C_1 \xrightarrow{d_1} C_0 \longrightarrow 0,$$
and finite increasing filtrations
$$0 = F_0C_n \subset F_1C_n \subset \cdots \subset F_mC_n = C_n, \quad m = m(n),$$
such that $d_n(F_pC_n) \subset F_pC_{n-1}\ \forall n, p$, we let $gr_p^F C_\bullet$ be the complex
$$gr_p^F C_\bullet : \ldots \to F_pC_2/F_{p-1}C_2 \to F_pC_1/F_{p-1}C_1 \to F_pC_0/F_{p-1}C_0 \to 0$$
for each $p \in \mathbb{Z}$, where we let $F_pC_n = 0$ for $p < 0$, $F_pC_n = C_n\ \forall p \geq m(n)$. Then we obtain a convergent spectral sequence of homological type (obtained from a suitable bigraded exact couple)
$$E^1_{p,q} = H_{p+q}(gr_p^F C_\bullet) \Rightarrow H_{p+q}(C_\bullet), \text{ with limit terms}$$
$$E^\infty_{p,q} = gr_p^F H_{p+q}(C_\bullet).$$

The spectral sequence of a filtered complex includes as special cases the usual spectral sequences encountered in homological algebra (see Cartan and Eilenberg's book *Homological Algebra*, for example), e.g., the spectral sequences of a double complex, and the Grothendieck spectral sequence for the derived functors of a composite; thus these spectral sequences can be regarded as arising from suitable exact couples.

# Appendix D
# Results from Algebraic Geometry

A general reference for algebraic geometry from the point of view most useful to us is

[H]  R. Hartshorne, *Algebraic Geometry*, Grad. Texts in Math. No. 52, Springer-Verlag, New York (1978).

We omit most of the proofs in this appendix, but they can be found in [H] (or [H] cites an appropriate reference).

## Sheaves

Let $X$ be a topological space. The open sets in $X$ are partially ordered by inclusion, hence may be regarded as a category $\mathcal{T}_X$.

(D.1)   A *presheaf of sets* on $X$ is a functor $\mathcal{T}_X^{op} \to \mathbf{Set}$, where $\mathbf{Set}$ is the category of sets.

Thus if $\mathcal{F}$ is a presheaf on $X$, then for each open set $U \subset X$, we are given a set $\mathcal{F}(U)$, and for any smaller open subset $V \subset U$, a *restriction map* $\rho_{UV} : \mathcal{F}(U) \to \mathcal{F}(V)$, such that $\rho_{VW} \circ \rho_{UV} = \rho_{UW}$ for $W \subset V \subset U$. Elements of $\mathcal{F}(U)$ are called *sections* of $\mathcal{F}$ over $U$; if $U = X$, they are called *global sections*. We sometimes also use the notation $\Gamma(U, \mathcal{F})$ instead of $\mathcal{F}(U)$.

*Morphisms* of presheaves are just natural transformations of functors. A presheaf $\mathcal{F}'$ is called a *sub-presheaf* of $\mathcal{F}$ if $\mathcal{F}'(U) \subset \mathcal{F}(U)$ for each $U$, and the restriction maps for $\mathcal{F}'$ are obtained by restricting those for $\mathcal{F}$. If $f : \mathcal{F} \to \mathcal{G}$ is a morphism of presheaves, then $U \mapsto (\text{image}\mathcal{F}(U))$ is a presheaf, which is a sub-presheaf of $\mathcal{G}$.

For example, let $A$ be a set. For any topological space $X$, let $\mathcal{F}(U) = A$ for all open sets $U$, and let $\rho_{UV}$ be the identity for all $V \subset U$. Then $\mathcal{F}$ is a presheaf on $X$ called the *constant presheaf* associated to $A$.

(D.2)   A presheaf $\mathcal{F}$ is called a *sheaf* if for any open set $U$ of $X$ and any open cover $\{U_\alpha\}_{\alpha \in \mathcal{A}}$ of $U$, the following conditions hold.

(i) For any sections $s, t \in \mathcal{F}(U)$, if $\rho_{UU_\alpha}(s) = \rho_{UU_\alpha}(t)$ for all $\alpha \in \mathcal{A}$, then $s = t$.

(ii) Let $U_{\alpha\beta} = U_\alpha \cap U_\beta$ for any $\alpha, \beta \in \mathcal{A}$; then for any family of sections $s_\alpha \in \mathcal{F}(U_\alpha)$, $\alpha \in \mathcal{A}$, such that $\rho_{U_\alpha U_{\alpha\beta}}(s_\alpha) = \rho_{U_\beta U_{\alpha\beta}}(s_\beta)$ for all $\alpha, \beta \in \mathcal{A}$, there exists a (necessarily unique, by (i)) $s \in \mathcal{F}(U)$ such that $\rho_{UU_\alpha}(s) = s_\alpha$ for all $\alpha \in \mathcal{A}$.

(iii) If $U = \phi$ is empty, then $\mathcal{F}(U)$ is a 1-point set (i.e., a *final object* in the category **Set**).

*Morphisms of sheaves* are defined to be morphisms of the underlying presheaves. Thus we can make sense of subsheaves of a sheaf. However, if $f : \mathcal{F} \to \mathcal{G}$ is a morphism of sheaves, the image presheaf is *not* a sheaf in general.

(D.3)  The *stalk* $\mathcal{F}_x$ of a presheaf $\mathcal{F}$ at $x \in X$ is defined as

$$\mathcal{F}_x = \varinjlim_{U \ni x} \mathcal{F}(U).$$

(D.4)  Define
$$\mathcal{G}(\mathcal{F})(U) = \prod_{x \in U} \mathcal{F}_x.$$

(If $U = \phi$, define $\mathcal{G}(\mathcal{F})(U)$ to be a final object in **Set**.) Then $\mathcal{G}(\mathcal{F})$ is a sheaf, such that all the restriction maps $\rho_{UV}$ are surjective; a sheaf with this property is called *flasque* (or *flabby*). For each open set $U$, there is a natural map $\mathcal{F}(U) \to \prod_{x \in U} \mathcal{F}_x$, giving a morphism of presheaves $\mathcal{F} \to \mathcal{G}(\mathcal{F})$. If $\mathcal{F}$ is a sheaf, this is injective, giving an isomorphism of $\mathcal{F}$ with its image. In general, the image of $\mathcal{F}$ is a sub-presheaf. Let $a(\mathcal{F})$ be the intersection of all the subsheaves of $\mathcal{G}(\mathcal{F})$ which contain the image of $\mathcal{F}$ (since $\mathcal{G}(\mathcal{F})$ is one such, the family of subsheaves is non-empty, and clearly any intersection of subsheaves is a subsheaf). If $f : \mathcal{F} \to \mathcal{F}'$ is a morphism of presheaves, there is an induced morphism of sheaves $\mathcal{G}(\mathcal{F}) \to \mathcal{G}(\mathcal{F}')$ compatible with $f$, and hence a morphism $a(\mathcal{F}) \to a(\mathcal{F}')$. In particular, if $\mathcal{F}'$ is a sheaf, so that $\mathcal{F}' \to a(\mathcal{F}')$ is an isomorphism, we see that $f$ factors uniquely through $\mathcal{F} \to a(\mathcal{F})$. Thus $a$ is a functor from presheaves to sheaves on $X$, which is left adjoint to the inclusion functor from sheaves to presheaves. We call $a(\mathcal{F})$ the *sheaf associated to the presheaf* $\mathcal{F}$.

(D.5)  A presheaf of Abelian groups (or rings, or modules over a ring ...) on $X$ is a functor from $T_X^{op}$ to the category **Ab** of Abelian groups (or rings, or modules over a ring, ...). It is a sheaf if the analogues of the conditions (i), (ii), (iii) above are satisfied. If $\mathcal{F}$ is a presheaf of Abelian groups, $\mathcal{G}(\mathcal{F})$,

$a(\mathcal{F})$ are sheaves of Abelian groups; a similar claim holds for sheaves of rings, modules, etc. In particular, for any Abelian group $A$, we have the *constant sheaf* $A_X$ associated to $A$, which is the sheaf associated to the constant presheaf determined by $A$ (discussed earlier in (D.1)). If $A$ is a ring, $A_X$ is a sheaf of rings. An important example is the sheaf $\mathbb{Z}_X$ of rings determined by the ring $\mathbb{Z}$ of integers.

More generally, we may consider sheaves with values in any category with arbitrary products and finite inverse limits, and which has a final object, since the sheaf conditions may be rephrased using only these notions.

(D.6)   Let $\mathcal{F}$ be a sheaf of Abelian groups on $X$, and $s \in \mathcal{F}(U)$. Then the *support* of $s$ is the set $|s| = \{x \in U \mid s_x \neq 0\}$, where $s_x$ is the image of $s$ in the stalk $\mathcal{F}_x$. One sees easily that $|s| \subset U$ is closed. We define the support of $\mathcal{F}$ to be the union of the supports of its sections, which is the set $|\mathcal{F}| = \{x \in X \mid \mathcal{F}_x \neq 0\}$. This need not be closed in general. However, we will see later that this is the case for *coherent* sheaves of $\mathcal{O}_X$-modules on a 'reasonable' scheme $X$ (see (D.45)).

(D.7)   Let $\mathcal{O}_X$ be a presheaf of rings on a topological space $X$. A *presheaf of $\mathcal{O}_X$-modules* is a presheaf $\mathcal{F}$ of Abelian groups together with an $\mathcal{O}_X(U)$-module structure on each Abelian group $\mathcal{F}(U)$, such that if $V \subset U$, then $\rho_{UV} : \mathcal{F}(U) \to \mathcal{F}(V)$ is $\mathcal{O}_X(U)$-linear, where $\mathcal{F}(V)$ is regarded as an $\mathcal{O}_X(U)$-module via the ring homomorphism $\rho_{UV} : \mathcal{O}_X(U) \to \mathcal{O}_X(V)$ and the given $\mathcal{O}_X(V)$-module structure. If $\mathcal{O}_X$ is a sheaf of rings, a *sheaf of $\mathcal{O}_X$-modules* is a sheaf of Abelian groups which has the structure of a presheaf of $\mathcal{O}_X$-modules. A sheaf of $\mathbb{Z}_X$-modules is just a sheaf of Abelian groups. In another direction, if $X = \{x\}$, then all presheaves which satisfy the sheaf condition (iii) are in fact sheaves; a sheaf of rings $\mathcal{O}_X$ is identified with a ring $R$ (the stalk of $\mathcal{O}_X$ at $x$), and the category of sheaves of $\mathcal{O}_X$-modules is identified with the category of $R$-modules.

*Convention*: whenever we consider *sheaves* of $\mathcal{O}_X$-modules, we will assume that $\mathcal{O}_X$ *is a sheaf of rings*.

The category of presheaves of $\mathcal{O}_X$-modules on a topological space $X$ forms an Abelian category in a natural way. The category of sheaves of $\mathcal{O}_X$-modules is a full additive subcategory, which is also an Abelian category; for any morphism $f : \mathcal{F} \to \mathcal{F}'$, the sheaf kernel of $f$ is the presheaf kernel, but the sheaf cokernel is defined to be $a(\operatorname{coker}_p(f))$ where '$\operatorname{coker}_p$' denotes the presheaf cokernel. In particular, one sees that a sequence

$$0 \to \mathcal{F}' \to \mathcal{F} \to \mathcal{F}'' \to 0$$

of sheaves of $\mathcal{O}_X$-modules is *exact* iff

$$0 \to \mathcal{G}(\mathcal{F}') \to \mathcal{G}(\mathcal{F}) \to \mathcal{G}(\mathcal{F}'') \to 0$$

is exact as a sequence of presheaves; this is equivalent to the exactness of

$$0 \to \mathcal{F}'_x \to \mathcal{F}_x \to \mathcal{F}''_x \to 0$$

for each $x \in X$.

The category of presheaves of $\mathcal{O}_X$-modules has direct sums, and direct and inverse limits over directed sets. A finite (presheaf) direct sum of sheaves of $\mathcal{O}_X$-modules is a sheaf. The inverse limit presheaf of a directed family of sheaves of $\mathcal{O}_X$-modules is in fact a sheaf, but the direct limit in the category of sheaves of $\mathcal{O}_X$-modules is the sheaf associated to the presheaf direct limit. However there is one case where the presheaf and sheaf direct limits coincide: when the topological space $X$ is *Noetherian*, i.e., satisfies the *descending chain condition for closed subsets*, that any strictly descending chain of closed subsets of $X$ is finite.

(D.8) We mention some other basic operations on presheaves and sheaves. If $f : X \to Y$ is a continuous map, and $\mathcal{F}$ is a presheaf on $X$, then we can define a presheaf $f_*\mathcal{F}$ on $Y$ by $f_*\mathcal{F}(U) = \mathcal{F}(f^{-1}(U))$. We call $f_*\mathcal{F}$ the *direct image* of $\mathcal{F}$. If $\mathcal{F}$ is a sheaf, so is $f_*\mathcal{F}$. If $\mathcal{O}_X$ is a sheaf of rings on $X$, then $f_*\mathcal{O}_X$ is a sheaf of rings, and for any $\mathcal{O}_X$-module $\mathcal{F}$, the direct image $f_*\mathcal{F}$ is an $f_*\mathcal{O}_X$-module in a natural way. The direct image functor is *left exact*.

The direct image functor $f_*$ from presheaves (or sheaves) of Abelian groups on $X$ to those on $Y$ has a left adjoint $f^{-1}$, called the *inverse image* functor. On presheaves, it is defined (on objects) by

$$(f^{-1}\mathcal{F})(U) = \varinjlim_{V \supset f(U)} \mathcal{F}(V).$$

This clearly defines a presheaf on $X$, and the adjointness property

$$\mathrm{Hom}\,(f^{-1}\mathcal{F}', \mathcal{F}) \cong \mathrm{Hom}\,(\mathcal{F}', f_*\mathcal{F})$$

is easily verified. The sheaf inverse image is the sheaf associated to the presheaf inverse image; the adjointness property follows from the adjointness at the level of presheaves, and the adjointness of the 'associated sheaf' functor $a$. If $f(x) = y$, then for any presheaf $\mathcal{F}$ on $Y$, we have an identification of stalks $f^{-1}(\mathcal{F})_x \cong \mathcal{F}_y$. In particular, $f^{-1}$ is an exact functor. If $\mathcal{O}_Y$ is a sheaf of rings, then so is $f^{-1}\mathcal{O}_Y$, and $f^{-1}$ takes $\mathcal{O}_Y$-modules into $f^{-1}\mathcal{O}_Y$-modules, and converts $\mathcal{O}_Y$-linear maps into $f^{-1}\mathcal{O}_Y$-linear ones.

In particular, if $j : U \hookrightarrow X$ is the inclusion of an open subset, we have $(j^{-1}\mathcal{F})(V) = \mathcal{F}(V)$ for any open set $V \subset U$. We also denote $j^{-1}\mathcal{F}$ by $\mathcal{F}|_U$. The functor $j^{-1}$ from sheaves of Abelian groups on $X$ to those on $U$ has a left adjoint $j_!$, called *extension by 0*, where $j_!\mathcal{F}$ is the sheaf associated

to the presheaf
$$V \mapsto \begin{cases} \mathcal{F}(V) & \text{if } V \subset U \\ 0 & \text{if } V \not\subset U \end{cases}$$

The sheaf $j_!\mathcal{F}$ is characterized by the properties that $j^{-1}j_!\mathcal{F} \cong \mathcal{F}$ and $(j_!\mathcal{F})_x = 0$ for $x \in X - U$. Note that there is a natural inclusion $j_!(\mathcal{F}|_U) \to \mathcal{F}$ for any sheaves $\mathcal{F}$ of Abelian groups.

If $i : Z \hookrightarrow X$ is the inclusion of a closed subset, let $\mathcal{F}|_Z = i^{-1}\mathcal{F}$. The functor $i_*$ gives an equivalence of categories between sheaves of Abelian groups on $Z$ and the full subcategory of sheaves of Abelian groups $\mathcal{F}$ on $X$ with $\mathcal{F}|_{X-Z} = 0$.

If $\mathcal{O}_X$ is a sheaf of rings on $X$, let $\mathcal{O}_U = \mathcal{O}_X|_U$. If $\mathcal{F}, \mathcal{G}$ are presheaves of $\mathcal{O}_X$-modules, define a presheaf of Abelian groups $\mathcal{H}om_{\mathcal{O}_X}(\mathcal{F}, \mathcal{G})$ by the assignment $U \mapsto \text{Hom}_{\mathcal{O}_U}(\mathcal{F}|_U, \mathcal{G}|_U)$. If $\mathcal{F}, \mathcal{G}$ are sheaves, so is $\mathcal{H}om_{\mathcal{O}_X}(\mathcal{F}, \mathcal{G})$. If $\mathcal{O}_X$ is a sheaf of commutative rings, $\mathcal{H}om_{\mathcal{O}_X}(\mathcal{F}, \mathcal{G})$ is a sheaf of $\mathcal{O}_X$-modules in a natural way. In particular, if $\mathcal{O}_X$ is commutative, we have a notion of *dual*; the dual $\mathcal{F}^*$ of a sheaf $\mathcal{F}$ of $\mathcal{O}_X$-modules is $\mathcal{H}om_{\mathcal{O}_X}(\mathcal{F}, \mathcal{O}_X)$.

If $\mathcal{O}_X$ is a sheaf of rings, not necessarily commutative, let $\mathcal{O}_X^{\text{op}}$ be the corresponding sheaf of opposite rings, so that an $\mathcal{O}_X^{\text{op}}$-module is a right $\mathcal{O}_X$-module. For any $\mathcal{O}_X^{\text{op}}$-module $\mathcal{F}$, and any sheaf $\mathcal{H}$ of Abelian groups, the sheaf $\mathcal{H}om_{\mathbb{Z}_X}(\mathcal{F}, \mathcal{H})$ is an $\mathcal{O}_X$-module in a natural way, via the action $(s \cdot \varphi)(f) = \varphi(s \cdot f)$ for sections $s \in \mathcal{O}_X^{\text{op}}(U) = \mathcal{O}_X(U)$, $f \in \mathcal{F}(U)$ and $\varphi \in \mathcal{H}om_{\mathbb{Z}_X}(\mathcal{F}, \mathcal{H})(U)$.

Let $\mathcal{O}_X$ be a sheaf of rings, $\mathcal{F}$ an $\mathcal{O}_X^{\text{op}}$-module. The functor $\mathcal{H} \mapsto \mathcal{H}om_{\mathbb{Z}_X}(\mathcal{F}, \mathcal{H})$ (from the category of $\mathbb{Z}_X$-modules to that of $\mathcal{O}_X$-modules) has a left adjoint $\mathcal{G} \mapsto \mathcal{F} \otimes_{\mathcal{O}_X} \mathcal{G}$. Thus, by definition, there are natural isomorphisms

$$\text{Hom}_{\mathcal{O}_X}(\mathcal{G}, \mathcal{H}om_{\mathbb{Z}_X}(\mathcal{F}, \mathcal{H})) \cong \text{Hom}_{\mathbb{Z}_X}(\mathcal{F} \otimes_{\mathcal{O}_X} \mathcal{G}, \mathcal{H}),$$

which characterizes $\mathcal{F} \otimes_{\mathcal{O}_X} \mathcal{G}$ in terms of the usual universal property for bilinear maps of sheaves $\mathcal{F} \times \mathcal{G} \to \mathcal{H}$. One checks that the sheaf associated to the presheaf $U \mapsto \mathcal{F}(U) \otimes_{\mathcal{O}_X(U)} \mathcal{G}(U)$ satisfies this universal property, so that this defines the sheaf $\mathcal{F} \otimes_{\mathcal{O}_X} \mathcal{G}$. When $\mathcal{O}_X$ is commutative, if $\mathcal{H}$ is also an $\mathcal{O}_X$-module, then we have a commutative diagram

$$\begin{array}{ccc} \text{Hom}_{\mathcal{O}_X}(\mathcal{G}, \mathcal{H}om_{\mathbb{Z}_X}(\mathcal{F}, \mathcal{H})) & \xrightarrow{\cong} & \text{Hom}_{\mathbb{Z}_X}(\mathcal{F} \otimes_{\mathcal{O}_X} \mathcal{G}, \mathcal{H}) \\ \uparrow & & \uparrow \\ \text{Hom}_{\mathcal{O}_X}(\mathcal{G}, \mathcal{H}om_{\mathcal{O}_X}(\mathcal{F}, \mathcal{H})) & \xrightarrow{\cong} & \text{Hom}_{\mathcal{O}_X}(\mathcal{F} \otimes_{\mathcal{O}_X} \mathcal{G}, \mathcal{H}) \end{array}$$

where the vertical arrows are each induced by the natural inclusion of the Abelian group of $\mathcal{O}_X$-linear maps into that of $\mathbb{Z}_X$-linear ones.

In a similar fashion, one may define symmetric powers, exterior powers, etc. when $\mathcal{O}_X$ is commutative.

*Convention*: From now on, we will assume $\mathcal{O}_X$ is a sheaf of *commutative* rings, unless explicitly mentioned otherwise. Some statements made below may have generalizations to the non-commutative case; we leave these to the interested reader.

(D.9) One way to define a sheaf on a space $X$ is through patching: let $\{U_i\}_{i \in I}$ be an open cover of $X$, and let $\mathcal{F}_i$ be a sheaf on $U_i$, for each $i$, such that

(i) for any pair of distinct indices $i, j$ there is an isomorphism
$$\varphi_{ij} : \mathcal{F}_i \mid_{U_i \cap U_j} \xrightarrow{\cong} \mathcal{F}_j \mid_{U_i \cap U_j}$$

(ii) $\varphi_{ji} = \varphi_{ij}^{-1}$

(iii) for any 3 distinct indices $i, j, k$ we have $\varphi_{jk} \circ \varphi_{ij} = \varphi_{ik}$ on $U_i \cap U_j \cap U_k$. Then there is a sheaf $\mathcal{F}$, such that there are isomorphisms $\varphi_i : \mathcal{F} \mid_{U_i} \to \mathcal{F}_i$ compatible with the $\varphi_{ij}$; further such an $\mathcal{F}$ is unique up to unique isomorphism compatible with the $\varphi_{ij}$. One way to construct $\mathcal{F}$ is to define a presheaf $\mathcal{F}_0$ as follows: let $\mathcal{U}$ be the collection of open subsets of $X$ which are contained in some $U_i$, and choose a function $f : \mathcal{U} \to I$ such that $V \subset U_{f(V)}$ for all $V \in \mathcal{U}$. Define $\mathcal{F}_0(V) = 0$ for $V \notin \mathcal{U}$, and $\mathcal{F}_0(V) = \mathcal{F}_{f(V)}(V)$ for $V \in \mathcal{U}$. Using the isomorphisms $\varphi_{ij}$ we see that there are natural restriction maps making $\mathcal{F}_0$ a presheaf, together with given isomorphisms $\mathcal{F}_0 \mid_{U_i} \cong \mathcal{F}_i$. Then $\mathcal{F} = a(\mathcal{F}_0)$ is the desired sheaf obtained by patching the $\mathcal{F}_i$ using the isomorphisms $\varphi_{ij}$. We leave it to the reader to check the uniqueness assertion.

(D.10) An $\mathcal{O}_X$-module is *free* of rank $n$ if it is isomorphic to $\mathcal{O}_X^{\oplus n}$. An $\mathcal{O}_X$-module $\mathcal{F}$ is called *locally free* (of finite rank) if each $x \in X$ has an open neighborhood $U$ such that $\mathcal{F} \mid_U$ is a free $\mathcal{O}_U = \mathcal{O}_X \mid_U$-module of finite rank. A locally free $\mathcal{O}_X$-module of rank 1 is called an *invertible* $\mathcal{O}_X$-module.

Locally free modules have several good properties. For example, if $\mathcal{E}$ is locally free, then the functors $\mathcal{F} \mapsto \mathcal{E} \otimes_{\mathcal{O}_X} \mathcal{F}$, $\mathcal{F} \mapsto \mathcal{H}om_{\mathcal{O}_X}(\mathcal{E}, \mathcal{F})$ are exact. We also have isomorphisms of functors (in $\mathcal{F}$) $\mathcal{H}om_{\mathcal{O}_X}(\mathcal{E}, \mathcal{F}) \cong \mathcal{E}^* \otimes_{\mathcal{O}_X} \mathcal{F}$, and $\text{Hom}_{\mathcal{O}_X}(\mathcal{E}, \mathcal{F}) \cong (cE^* \otimes_{\mathcal{O}_X} \mathcal{F})(X)$. The natural map $\mathcal{E} \to (\mathcal{E}^*)^*$ from $\mathcal{E}$ to its double dual is an isomorphism. For any locally free $\mathcal{O}_X$-module $\mathcal{E}$, there is a natural $\mathcal{O}_X$-linear surjection $\mathcal{E} \otimes_{\mathcal{O}_X} \mathcal{E}^* \to \mathcal{O}_X$, which is an isomorphism if $\mathcal{E}$ is invertible. However, note that in general, locally free $\mathcal{O}_X$-modules are *not* projective objects in the category of $\mathcal{O}_X$-modules.

(D.11) Recall that an object $I$ of an Abelian category $\mathcal{A}$ is *injective* if the functor $X \mapsto \text{Hom}_{\mathcal{A}}(X, I)$ is exact. For any (possibly non-commutative) sheaf of rings $\mathcal{O}_X$, the Abelian category of sheaves of $\mathcal{O}_X$-modules *has enough injectives*, i.e., for any sheaf $\mathcal{F}$ of $\mathcal{O}_X$-modules, there is a monomorphism $\mathcal{F} \to \mathcal{I}$, where $\mathcal{I}$ is an injective $\mathcal{O}_X$-module. To prove this, one notes that if $\mathcal{J}$ is an injective sheaf of Abelian groups, then $\mathcal{I} = \mathcal{H}om_{\mathbb{Z}_X}(\mathcal{O}_X^{\text{op}}, \mathcal{J})$, which is naturally an $\mathcal{O}_X$-module, is in fact injective; this follows from the natural isomorphism

$$\text{Hom}_{\mathcal{O}_X}(\mathcal{F}, \mathcal{H}om_{\mathbb{Z}_X}(\mathcal{O}_X^{\text{op}}, \mathcal{J})) \cong \text{Hom}_{\mathbb{Z}_X}(\mathcal{O}_X^{\text{op}} \otimes_{\mathcal{O}_X} \mathcal{F}, \mathcal{J}) = \text{Hom}_{\mathbb{Z}}(\mathcal{F}, \mathcal{J})$$

for any sheaf $\mathcal{F}$ of $\mathcal{O}_X$-modules. This reduces us to proving the result when $\mathcal{O}_X = \mathbb{Z}_X$. One sees easily that if $\{I_x\}_{x \in X}$ is a family of injective (= divisible) Abelian groups indexed by points of $X$, and $\mathcal{I}(U) = \prod_{x \in U} I_x$, then $\mathcal{I}$ is an injective sheaf of Abelian groups. Now for any sheaf $\mathcal{F}$ of Abelian groups, if we choose inclusions $\mathcal{F}_x \hookrightarrow I_x$ into injective Abelian groups, then we obtain an injection of sheaves $\mathcal{G}(\mathcal{F}) \hookrightarrow \mathcal{I}$, where $\mathcal{I}$ is defined by the chosen family $\{I_x\}_{x \in X}$; composing with the natural injection $\mathcal{F} \hookrightarrow \mathcal{G}(\mathcal{F})$ (since $\mathcal{F}$ is a sheaf, the natural map is an inclusion), we are done.

Thus any sheaf $\mathcal{F}$ of $\mathcal{O}_X$-modules has an *injective resolution*

$$0 \to \mathcal{F} \to \mathcal{I}_0 \to \mathcal{I}_1 \to \cdots \to \mathcal{I}_n \to \cdots$$

in the category of $\mathcal{O}_X$-modules, and this is unique up to chain homotopy (by standard arguments using the universal property of an injective object). Hence for any left exact functor $F$ from the category of sheaves of $\mathcal{O}_X$-modules to an Abelian category, we may define its *derived functors* $R^i F$ by

$$R^i F(\mathcal{F}) = i^{\text{th}} \text{ cohomology object of the complex } F(\mathcal{I}_\bullet).$$

If $0 \to \mathcal{F}' \to \mathcal{F} \to \mathcal{F}'' \to 0$ is an exact sequence of sheaves, we have functorial boundary maps $R^i F(\mathcal{F}'') \to R^{i+1} F(\mathcal{F}')$ giving a long exact sequence of derived functors (where we identify $R^0 F$ with $F$)

$$0 \to F(\mathcal{F}') \to F(\mathcal{F}) \to F(\mathcal{F}'') \to R^1 F(\mathcal{F}') \to \cdots \to$$
$$R^i F(\mathcal{F}') \to R^i F(\mathcal{F}) \to R^i F(\mathcal{F}'') \to R^{i+1}(\mathcal{F}') \to \cdots$$

Any natural transformation between left exact functors induces a unique natural transformation between their derived functors, compatible with boundary maps in the respective long exact sequences.

(D.12) Important examples of left exact functors on sheaves and their derived functors are as follows.

(i) Let $f : X \to Y$ be a continuous map, $\mathcal{O}_X$ a sheaf of (possibly non-commutative) rings on $X$. Then $f_*$ is a left exact functor from $\mathcal{O}_X$-modules to $f_* \mathcal{O}_X$-modules, whose derived functors $R^i f_*$ are called the *higher direct image* functors of the map $f$. In particular, if $Y = \{y\}$ is a point, then $f_*\mathcal{O}_X$ is identifed with the ring $R = \mathcal{O}_X(X)$, and $f_*\mathcal{F}$ is identified with the $R$-module $\mathcal{F}(X)$ of global sections. The sheaves $R^i f_*\mathcal{F}$ yield $R$-modules $H^i(X,\mathcal{F})$ called the *cohomology groups* (really, cohomology $R$-modules) of $\mathcal{F}$.

(ii) Let $\mathcal{G}$ be an $\mathcal{O}_X$-module. Then

$$\mathcal{F} \mapsto \mathrm{Hom}_{\mathcal{O}_X}(\mathcal{G},\mathcal{F}), \quad \mathcal{F} \mapsto \mathcal{H}om_{\mathcal{O}_X}(\mathcal{G},\mathcal{F})$$

are left exact functors. Their $i^{\text{th}}$ derived functors are denoted by $\mathrm{Ext}^i_{\mathcal{O}_X}(\mathcal{G},\mathcal{F})$ and $\mathcal{E}xt^i_{\mathcal{O}_X}(\mathcal{G},\mathcal{F})$, respectively.

We have a natural isomorphism $\mathrm{Hom}_{\mathcal{O}_X}(\mathcal{O}_X,\mathcal{F}) \cong \mathcal{F}(X) = H^0(X,\mathcal{F})$. Hence there are natural isomorphisms $\mathrm{Ext}^i_{\mathcal{O}_X}(\mathcal{O}_X,\mathcal{F}) \cong H^i(X,\mathcal{F})$. Note that if $\mathcal{O}_X$ is commutative, and $\mathcal{E}$ is a locally free $\mathcal{O}_X$-module, then

$$\mathcal{E}xt^i_{\mathcal{O}_X}(\mathcal{E},\mathcal{F}) = 0 \text{ for all } i > 0,$$

and there are natural isomorphisms

$$\mathrm{Ext}^i_{\mathcal{O}_X}(\mathcal{E},\mathcal{F}) \cong H^i(X,\mathcal{E}^* \otimes_{\mathcal{O}_X} \mathcal{F}).$$

(D.13) Derived functors may also be computed using *acyclic resolutions*, i.e., if $0 \to \mathcal{F} \to \mathcal{F}_\bullet$ is a resolution, and $F$ a left exact functor with $R^i F(\mathcal{F}_j) = 0$ for all $i > 0$, $j \geq 0$, then the $i^{\text{th}}$ cohomology object of the complex $F(\mathcal{F}_\bullet)$ is naturally isomorphic to $R^i F(\mathcal{F})$.

We claim that flasque sheaves of Abelian groups are acyclic for $f_*$ for any map $f: X \to Y$. Indeed, one shows that the following statements hold (see [H], II, Ex. 1.16).

(i) Injective sheaves of Abelian groups are flasque. Indeed, if $j: U \hookrightarrow X$ is an open set, then the map $\mathrm{Hom}_{\mathbb{Z}_X}(\mathbb{Z}_X,\mathcal{I}) \to \mathrm{Hom}_{\mathbb{Z}_X}(j_!\mathbb{Z}_U,\mathcal{I})$, induced by the inclusion of sheaves $j_!\mathbb{Z}_U \to \mathbb{Z}_X$, is *surjective* for any injective sheaf $\mathcal{I}$, i.e., $\rho_{X,U}: \mathcal{I}(X) \to \mathcal{I}(U)$ is surjective. Similarly, working with $\mathcal{O}_X$ and $j_!\mathcal{O}_U$, we see that injective $\mathcal{O}_X$-modules are flasque for any (possibly non-commutative) sheaf of rings $\mathcal{O}_X$.

(ii) If $0 \to \mathcal{F}' \to \mathcal{F} \to \mathcal{F}'' \to 0$ is exact with $\mathcal{F}'$ flasque, then $\mathcal{F}(U) \to \mathcal{F}''(U)$ is surjective for each open $U \subset X$.

(iii) If $0 \to \mathcal{F}' \to \mathcal{F} \to \mathcal{F}'' \to 0$ is exact with $\mathcal{F}',\mathcal{F}$ flasque, then $\mathcal{F}''$ is flasque.

From (i) and (iii), the quotient of an injective sheaf by a flasque subsheaf is flasque. From (ii), given a short exact sequence $0 \to \mathcal{F}' \to \mathcal{F} \to \mathcal{F}'' \to 0$ with $\mathcal{F}'$ flasque, we get that for any continuous map $f : X \to Y$, the direct image sequence $0 \to f_*\mathcal{F}' \to f_*\mathcal{F} \to f_*\mathcal{F}'' \to 0$ is exact. Hence if $\mathcal{F}$ is a flasque sheaf of $\mathcal{O}_X$-modules, $0 \to \mathcal{F} \to \mathcal{I}_\bullet$ is an injective resolution by sheaves of $\mathcal{O}_X$-modules, then $0 \to f_*\mathcal{F} \to f_*\mathcal{I}_\bullet$ is exact for any continuous map $f : X \to Y$. Hence $R^i f_*\mathcal{F} = 0$ for all $i > 0$.

Since injective $\mathcal{O}_X$-modules are flasque, we see that the cohomology (or higher direct images) of an $\mathcal{O}_X$-module $\mathcal{F}$, computed with resolutions by injective $\mathcal{O}_X$-modules, equals the cohomology (or higher direct images) of the underlying sheaf of Abelian groups $\mathcal{F}$. Another application is the following: if $f : X \to Y$ is the inclusion of a closed subset, then $f_*$ is an exact functor from sheaves of Abelian groups on $X$ to those on $Y$, which sends flasque sheaves on $X$ to flasque sheaves on $Y$. Hence $R^i f_*\mathcal{F} = 0$ for all $i > 0$, and there are natural isomorphisms $H^i(X, \mathcal{F}) \cong H^i(Y, f_*\mathcal{F})$ for all $i \geq 0$ (an injective resolution of $\mathcal{F}$ on $X$ yields a flasque resolution of $f_*\mathcal{F}$ with the same complex of global sections).

(D.14) One important tool in computing sheaf cohomology is Leray's theorem, which relates the cohomology groups defined above to Čech cohomology. We first recall the definition of Čech cohomology (in a simple context sufficient for our needs). Let $\mathcal{U} = \{U_i\}_{i \in I}$ be an open covering of a topological space $X$, where we fix a well ordering of the index set $I$, and let $\mathcal{F}$ be a sheaf of Abelian groups on $X$. Define groups

$$\check{C}^p(\mathcal{U}, \mathcal{F}) = \prod_{i_0 < i_1 < \cdots < i_p} \mathcal{F}(U_{i_0} \cap \cdots \cap U_{i_p})$$

and maps $\delta^p : \check{C}^p(\mathcal{U}, \mathcal{F}) \to \check{C}^{p+1}(\mathcal{U}, \mathcal{F})$ by

$$(\delta^p \alpha)_{i_0, \ldots, i_{p+1}} = \sum_{j=0}^{p+1} (-1)^j \alpha_{i_0, \ldots, \widehat{i_j}, \ldots, i_{p+1}} |_{U_{i_0} \cap \cdots \cap U_{i_{p+1}}},$$

where $\widehat{i_j}$ means that the index $i_j$ is omitted. Then $(\check{C}^\bullet(\mathcal{U}, \mathcal{F}), \delta^\bullet)$ is a complex, called the *Čech complex* of $\mathcal{F}$ with respect to $\mathcal{U}$, whose cohomology groups are called the *Čech cohomology* groups of $\mathcal{F}$ with respect to $\mathcal{U}$, and are denoted by $\check{H}^i(\mathcal{U}, \mathcal{F})$.

There is a natural map $\check{H}^i(\mathcal{U}, \mathcal{F}) \to H^i(X, \mathcal{F})$ for each $i$ (see [H], III, (4.4)). *Leray's theorem* asserts that if $H^j(U_{i_0} \cap \cdots \cap U_{i_p}, \mathcal{F}) = 0$ for all finite intersections of open sets in the covering, and for all $j > 0$, then these natural maps are isomorphisms (see [H], III, Ex. 4.11).

## Schemes

**(D.15)** The basic building block for schemes is the spectrum of a commutative ring. If $R$ is a commutative ring (with 1), its *spectrum* $\operatorname{Spec} R$ is the set of all prime ideals in $R$. (Recall that an ideal $P \subset R$ is *prime* if $R/P$ has no nontrivial zero-divisors; equivalently, if $a, b \in R$ with $ab \in P$, then $a \in P$ or $b \in P$.) This set is given the *Zariski topology* in which the closed sets are of the form $V(I) = \{P \in \operatorname{Spec} R \mid I \subset P\}$, where $I$ is an ideal. The complementary open set $\operatorname{Spec} R - V(I)$ is denoted by $D(I)$; if $I = (f)$ is a principal ideal, it is also denoted by $D(f)$. The sets $D(f)$ form a basis for the Zariski topology on $\operatorname{Spec} R$. The closed points of $\operatorname{Spec} R$ are precisely the maximal ideals. Note that for any ideal $I \subset R$, $V(I) = V(\sqrt{I})$, where $\sqrt{I} = \{x \mid x^n \in I \text{ for some } n\}$ is the radical of $I$. But $\sqrt{I} = \cap_{P \in V(I)} P$. Hence $V(I_1) = V(I_2) \Leftrightarrow \sqrt{I_1} = \sqrt{I_2}$. We also have $\cap_\alpha V(I_\alpha) = V(\sum_\alpha I_\alpha)$, $V(I) \cup V(J) = V(IJ)$.

If $k$ is an algebraically closed field, $X$ an affine variety over $k$ with coordinate ring $R$, then from the Nullstellensatz, evaluation at points of $X$ determines surjections $R \to k$ whose kernels are precisely the maximal ideals of $R$; the subspace topology on the set of maximal ideals from $\operatorname{Spec} R$ is precisely the classical Zariski topology on the affine variety $X$. This also motivates the following definition: if $R$ is a ring, the *affine n-space* $\mathbb{A}^n_R$ over $R$ is defined to be $\operatorname{Spec} R[X_1, \ldots, X_n]$ where $R[X_1, \ldots, X_n]$ is the polynomial ring in $n$ variables over $R$.

Let $R_P$ denote the localization of $R$ at a prime ideal $P$, i.e., with respect to the multiplicative set $R - P$. For $f \in R$, let $R_f$ denote the localization with respect to powers of $f$; thus $R_f \cong R[T]/(fT - 1)$. There is a sheaf of rings $\mathcal{O}_{\operatorname{Spec} R}$ defined on the topological space $\operatorname{Spec} R$, by

$$\mathcal{O}_{\operatorname{Spec} R}(D(I))$$
$$= \{s \in \prod_{P \in D(I)} R_P \mid \text{locally on } D(I), s \text{ equals a quotient of two elements of } R\}.$$

Here, we mean that for each $P \in D(I)$, there is

(i) a neighborhood $D(f)$ of $P$ in $D(I)$ (i.e., $f \in I$, $f \notin P$), and

(ii) an element $a/f^n \in R_f$,

such that for each $Q \in D(f)$, $s$ and $a/f^n$ have the same image in $R_Q$. Because of the 'local' nature of the condition defining the sections $s$, the above definition yields a sheaf of rings on $\operatorname{Spec} R$, such that (see [H], II, (2.2))

(i) the stalk of $\mathcal{O}_{\operatorname{Spec} R}$ at $P$ is $R_P$

Appendix D: Results from Algebraic Geometry    305

(ii) there is a natural isomorphism $\mathcal{O}_{\mathrm{Spec}\,R}(D(f)) \cong R_f$; in particular, $\mathcal{O}_{\mathrm{Spec}\,R}(\mathrm{Spec}\,R) \cong R$.

In (ii) above, one can show that in fact there is an isomorphism of sheaves of rings $\mathcal{O}_{\mathrm{Spec}\,R}|_{D(f)} \cong \mathcal{O}_{\mathrm{Spec}\,R_f}$ on $D(f)$.

From now onwards, we regard $\mathrm{Spec}\,R$ as the pair consisting of the above topological space equipped with its sheaf of rings, the *structure sheaf* $\mathcal{O}_{\mathrm{Spec}\,R}$.

(D.16)   A *ringed space* is a pair $(X, \mathcal{O}_X)$ consisting of a topological space $X$ and a sheaf of rings $\mathcal{O}_X$ on it, which we call the structure sheaf on $X$. This is called a *locally ringed space* if the stalks of $\mathcal{O}_X$ are local rings.

For example, $(\mathrm{Spec}\,R, \mathcal{O}_{\mathrm{Spec}\,R})$ is a locally ringed space.

Given a point $x$ of a locally ringed space $(X, \mathcal{O}_X)$, the *local ring of $X$ at $x$* is the stalk $\mathcal{O}_{X,x} = (\mathcal{O}_X)_x$, and the residue field of this local ring is called the *residue field at $x$*, sometimes denoted by $k(x)$.

A *morphism* $f : (X, \mathcal{O}_X) \to (Y, \mathcal{O}_Y)$ of locally ringed spaces is a pair, consisting of a continuous map $f : X \to Y$, and a map of sheaves of rings $f^\# : f^{-1}\mathcal{O}_Y \to \mathcal{O}_X$, where $f^\#$ is required to satisfy the following property: for any $x \in X$, if $y = f(x)$, then the map on stalks at $x$ induced by $f^\#$ is a homomorphism of local rings $f_x^\# : (\mathcal{O}_{Y,y}, \mathcal{M}_y) \to (\mathcal{O}_{X,x}, \mathcal{M}_x)$ where $\mathcal{M}_x \subset \mathcal{O}_{X,x}$, $\mathcal{M}_y \subset \mathcal{O}_{Y,y}$ are the respective maximal ideals; then we require that $f_x^\#(\mathcal{M}_y) \subset \mathcal{M}_x$.

If $(X, \mathcal{O}_X)$ is a locally ringed space, $U \subset X$ an open subset, $\mathcal{O}_U = \mathcal{O}_X|_U$, then a simple example of a morphism is the inclusion $j_U : (U, \mathcal{O}_U) \to (X, \mathcal{O}_X)$, where $j_U^{-1}\mathcal{O}_X = \mathcal{O}_U$ and $j_U^\#$ is the identity.

We remark that if $f : R \to S$ is a homomorphism of rings, then $P \mapsto f^{-1}(P)$ gives a continuous map $\widetilde{f} : \mathrm{Spec}\,S \to \mathrm{Spec}\,R$, and the ring homomorphism $f$ induces a homomorphism of sheaves of rings $\widetilde{f}^\# : \widetilde{f}^{-1}\mathcal{O}_{\mathrm{Spec}\,R} \to \mathcal{O}_{\mathrm{Spec}\,S}$; one checks that $(\widetilde{f}, \widetilde{f}^\#)$ is in fact a morphism of locally ringed spaces. We recover the ring homomorphism $f : R \to S$ by the map induced by $\widetilde{f}^\#$ on global sections. This construction in fact gives us an equivalence of categories between the opposite of the category of commutative rings and the full subcategory of the category of locally ringed spaces consisting of spectra of rings. This generalizes the classical equivalence between the category of affine varieties over an algebraically closed field $k$ and the opposite category of the category of finitely generated $k$-algebras which are reduced (see (D.21)). The spectrum of a ring, considered as a locally ringed space, is called an *affine scheme*.

(D.17)   A *scheme* is a locally ringed space $(X, \mathcal{O}_X)$ such that for some open cover $\{U_i\}$ of $X$, we have isomorphisms of locally ringed spaces $(U_i, \mathcal{O}_X|_{U_i}) \cong (\mathrm{Spec}\,R_i, \mathcal{O}_{\mathrm{Spec}\,R_i})$; in particular, we must have $R_i \cong$

$\mathcal{O}_X(U_i)$. A *morphism* of schemes is defined to be a morphism of locally ringed spaces; thus schemes form a full subcategory of locally ringed spaces.

For example, for any scheme $(X, \mathcal{O}_X)$, there is a unique morphism $f : (X, \mathcal{O}_X) \to \operatorname{Spec} \mathbb{Z}$, where $\mathbb{Z}$ is the ring of integers.

If in a given context, the structure sheaf $\mathcal{O}_X$ is known, we often denote the scheme $(X, \mathcal{O}_X)$ by just $X$. We have already used this convention in writing $\operatorname{Spec} R$ to mean both the topological space and the affine scheme.

If $X = \operatorname{Spec} A$, and $U \subset X$ is open, then $(U, \mathcal{O}_X |_U)$ is a scheme. Indeed, if $U = X - V(I)$, and $I$ is generated by $\{f_\alpha\}$, then $U$ is covered by the open subsets $D(f_\alpha) = \operatorname{Spec} A_{f_\alpha}$ of $X$. Hence if $(X, \mathcal{O}_X)$ is any scheme, $U \subset X$ an open subset, then $(U, \mathcal{O}_X |_U)$ is a scheme. This is called an *open subscheme* of $(X, \mathcal{O}_X)$.

(D.18) The construction of affine schemes is modeled after that of affine varieties in 'classical' algebraic geometry, through their relationship with their coordinate rings; in a similar manner, one can associate a scheme $\operatorname{Proj} R$ to a graded ring $R$ which mimics the relationship between a 'classical' projective variety and its homogeneous coordinate ring. This is done as follows.

Let $R = \oplus_{n \geq 0} R_n$ be a (non-negatively) graded ring. An ideal $I \subset R$ is called *homogeneous* if it is generated by homogeneous elements; thus an element of $R$ lies in $I$ if and only if its homogeneous components lie in $I$. Thus an ideal is homogeneous if and only if it is a graded $R$-submodule of $R$.

A homogeneous ideal $P$ is prime if and only if for any pair $a, b \in R$ of homogeneous elements with $ab \in P$, either $a \in P$ or $b \in P$. The ideal $R_+ = \oplus_{n>0} R_n$ is called the 'irrelevant homogeneous ideal'. Define $\operatorname{Proj} R$ to be the set of all homogeneous prime ideals $P$ of $R$ such that $R_+ \not\subset P$. If $I$ is a homogeneous ideal, let $V(I)$ be the subset of $\operatorname{Proj} R$ of all prime ideals containing $I$; give $\operatorname{Proj} R$ the topology such that these are the closed subsets. Let $D_+(I) = \operatorname{Proj} R - V(I)$. If $I = (f)$, we also denote $D_+(I)$ by $D_+(f)$, and these open sets give a basis for the topology of $\operatorname{Proj} R$.

Finally, we define a sheaf of rings $\mathcal{O}_{\operatorname{Proj} R}$ on $\operatorname{Proj} R$, as follows. For any homogeneous prime $P \in \operatorname{Proj} R$, let $S_{(P)}$ be the multiplicative set of homogeneous elements of $R - P$; let $R_{(P)}$ be the subring of $S_{(P)}^{-1} R$ consisting of elements $a/b$ where $a \in R$, $b \in S_{(P)}$ are homogeneous of the same degree (so that $a/b$ has degree 0). This is seen to be a local ring. Now define

$$\mathcal{O}_{\operatorname{Proj} R}(D_+(I)) = \{s \in \prod_{P \in D_+(I)} R_{(P)} \mid s \text{ is locally equal to an element}$$
$$a/b \text{ where } a, b \in R \text{ have the same degree}\}.$$

The condition on $s$ means that for each $P \in D_+(I)$, there exists a homogeneous $f \in I - P$, say of degree $d$, such that for some homogeneous element $a \in R_{nd}$ for some $n$, $s$ and $a/f^n$ have the same image in $R_{(Q)}$ for each $Q \in D_+(f)$.

One checks (see [H], II, (2.5)) that $\mathcal{O}_{\text{Proj } R}$ is a sheaf of rings, whose stalk at $P$ is $R_{(P)}$, and whose ring of sections over $D_+(f)$ is $R_{(f)}$, the subring of the $\mathbb{Z}$-graded ring $R_f$ consisting of elements of degree 0. In fact, $(D_+(f), \mathcal{O}_{\text{Proj } R} |_{D_+(f)}) \cong (\text{Spec } R_{(f)}, \mathcal{O}_{\text{Spec } R_{(f)}})$ as locally ringed spaces. Hence $(\text{Proj } R, \mathcal{O}_{\text{Proj } R})$ is a scheme.

(D.19)  In particular, if $A$ is any commutative ring, $R = A[T_0, \ldots, T_n]$ the polynomial ring, graded by defining elements of $A$ to have degree 0, and each variable to have degree 1, then $\text{Proj } R$ is called *projective n-space over* $\text{Spec } A$ (or just projective $n$-space over $A$); it is denoted by $\mathbb{P}_A^n$. This terminology is justified to some extent by the fact that if $A = k$ is an algebraically closed field, then the set of closed points of the scheme $\mathbb{P}_k^n$, with the subspace topology, is naturally identified with the 'classical' projective $n$-space of hyperplanes in an $n+1$-dimensional $k$-vector space, with its Zariski topology.

(D.20)  Unlike in the case of affine schemes, the scheme $\text{Proj } R$ does not determine the graded ring $R$. For example, if $S = \oplus_n S_n \subset R = \oplus_n R_n$ is a graded subring such that $S_n = R_n$ for all sufficiently large $n$, then $\text{Proj } R$ and $\text{Proj } S$ are naturally isomorphic (see also [H], II, 5.16.1). Further, if $R \to S$ is a homomorphism between graded rings, we do not necessarily get a corresponding morphism $\text{Proj } S \to \text{Proj } R$ by taking inverse images of homogeneous prime ideals. The problem is that even if a homogeneous prime $P$ of $S$ does not contain the irrelevant ideal $S_+$, its inverse image in $R$ may contain the irrelevant ideal $R_+$. However, if $I = R_+S$, so that $I$ is a graded ideal in $S$, then a homogenous prime ideal $P$ of $S$ which does not contain $I$ has an inverse image in $R$ which does not contain $R_+$, and conversely. Using this, one sees that there is a naturally defined morphism of schemes $\text{Proj } S - V(I) \to \text{Proj } R$, which on points is given by taking the inverse images of homogeneous prime ideals (see [H], II, Ex. 2.14).

## Some Properties of Schemes

In this section we recall the definitions of certain basic properties of schemes and morphisms.

(D.21)  A scheme $(X, \mathcal{O}_X)$ is *reduced* if for each open set $U \subset X$, the ring $\mathcal{O}_X(U)$ has no non-zero nilpotent elements (i.e., $\mathcal{O}_X(U)$ is a reduced ring).

For any commutative ring $A$, let nil $A$ denote the ideal of nilpotent elements of $A$. Then, nil $A$ is the intersection of all prime ideals of $A$. If $(X, \mathcal{O}_X)$ is an arbitrary scheme, then there is an associated reduced scheme

$(X, (\mathcal{O}_X)_{red})$, denoted for short by $X_{red}$, where

$$(\mathcal{O}_X)_{red}(U) = \mathcal{O}_X(U)/(\text{nil } \mathcal{O}_X(U)).$$

One checks that this does define a scheme structure on the topological space $X$, such that if $X = \text{Spec } A$, then $X_{red} = \text{Spec }(A/\text{nil } A)$.

(D.22) A scheme $(X, \mathcal{O}_X)$ is *integral* if for each open subset $U \subset X$, the ring $\mathcal{O}_X(U)$ is an integral domain.

A scheme $(X, \mathcal{O}_X)$ is integral $\Leftrightarrow$ $(X, \mathcal{O}_X)$ is reduced and $X$ is irreducible (i.e., $X$ is not the union of 2 proper closed subsets). If $U = \text{Spec } A$ is a non-empty affine open subscheme of an integral scheme $(X, \mathcal{O}_X)$, then $(0) \subset A$ is a prime ideal, so yields a point $\eta$ of $\text{Spec } A$. Now $\{\eta\}$ is dense in $\text{Spec } A$, hence in $X$, and so lies in any non-empty affine open subset $V = \text{Spec } B$ of $X$. We see easily that $\eta$ also corresponds to the prime ideal $(0)$ of $B$. The point $\eta$ thus lies in all non-empty open subsets of $X$, and is called the *generic point* of $X$. We may define the *function field* of an integral scheme $(X, \mathcal{O}_X)$ as (the residue field of) the local ring $\mathcal{O}_{X,\eta}$ at the generic point, which is in fact the quotient field of $A$ for any non-empty affine open subset $U = \text{Spec } A$ of $X$.

(D.23) A topological space $X$ is *quasi-compact* if every covering of $X$ by open sets has a finite subcover.

If $(X, \mathcal{O}_X) = \text{Spec } R$ is an affine scheme, then $X$ is quasi-compact. Indeed, if $U_i = X - V(I_i)$ is an open covering, then $\cap_i V(I_i) = V(\sum_i I_i)$ is empty, so that $\sum_i I_i = R$. But then $1 \in \sum_i I_i$, so $1 = a_1 + \cdots + a_r$ for some $a_j \in I_{i_j}$. Then $\sum_j I_{i_j} = R$, and $\cup_j U_{i_j} = X$.

(D.24) A topological space $X$ is called *Noetherian* if it satisfies the descending chain condition for closed subsets.

The motivating example for this definition is $X = \text{Spec } A$, where $A$ is a Noetherian ring. A space $X$ is Noetherian if and only if any open subset is quasi-compact. A Noetherian topological space $X$ has a unique decomposition $X = \cup_{i=1}^r X_i$ for closed subsets $X_i$ such that (i) each $X_i$ is irreducible, i.e., cannot be expressed as the union of 2 proper closed subsets (ii) the decomposition is irredundant, i.e., for each $i$, we have $X_i \not\subset \cup_{j \neq i} X_j$. The $X_i$ are called the *irreducible components* of the Noetherian space $X$. If $X$ is irreducible, a *generic point* of $X$ is a point $x \in X$ such that the set $\{x\}$ is dense in $X$. If $(X, \mathcal{O}_X)$ is a scheme, then every non-empty irreducible closed subset of $X$ has a unique generic point.

A scheme $(X, \mathcal{O}_X)$ is called *Noetherian* if it has a finite open cover by affine open subschemes $U_i = \text{Spec } A_i$ such that each $A_i$ is a Noetherian ring.

One can show (see [H], II, 3.2) that this implies the condition that for any affine open subscheme $U = \text{Spec } A$ of $X$, the ring $A$ is Noetherian. The topological space underlying a Noetherian scheme is Noetherian, though the converse is false.

(D.25) Recall that a Noetherian local ring $(R, \mathcal{M})$ of dimension $d$ is called *regular* if $\mathcal{M}$ is generated by $d$ elements; equivalently, $\mathcal{M}/\mathcal{M}^2$ is an $R/\mathcal{M}$-vector space of dimension $d$. The local ring $(R, \mathcal{M})$ is called *normal* if it is an integral domain which is integrally closed in its quotient field; it is called *Cohen-Macaulay* if it has depth $d$, i.e., there exist $x_1, \ldots, x_d \in \mathcal{M}$ which are non-zero-divisors in $R$, such that $x_i R \cap (\sum_{j \neq i} x_j R) = \sum_{j \neq i} x_i x_j R$ for all $1 \leq i \leq d$ (this is equivalent to the exactness of the *Koszul complex* of the $x_i$ with respect to $R$).

A scheme $(X, \mathcal{O}_X)$ is called *regular* (respectively *normal* or *Cohen-Macaulay*) if every local ring $\mathcal{O}_{X,x}$ is regular (respectively normal, or Cohen-Macaulay).

(D.26) A morphism $f : (X, \mathcal{O}_X) \to (Y, \mathcal{O}_Y)$ is of *finite type* if $Y$ has a covering by affine open subsets $U_i = \operatorname{Spec} A_i$ such that for each $i$, $f^{-1}(U_i)$ has a *finite* covering by affine open subschemes $\operatorname{Spec} B_{ij}$ of $X$, where $B_{ij}$ is a finitely generated $A_i$ algebra. We say that $(X, \mathcal{O}_X)$ is *of finite type over* $Y$ if there is a morphism $f : X \to Y$ of finite type.

One can show that $f$ is of finite type $\Leftrightarrow$ for every affine open subscheme $U = \operatorname{Spec} A$ of $Y$, $f^{-1}(U)$ has a finite open covering by the spectra of finitely generated $A$-algebras. The Hilbert basis theorem implies that if $Y$ is a Noetherian scheme and $f : X \to Y$ is a morphism of finite type, then $X$ is a Noetherian scheme. In particular, projective space $\mathbb{P}^n_A$ over a Noetherian ring $A$ is a Noetherian scheme.

(D.27) A morphism $f : (X, \mathcal{O}_X) \to (Y, \mathcal{O}_Y)$ is *finite* if there exists a covering of $Y$ by affine open subschemes $U_i = \operatorname{Spec} A_i$ such that $f^{-1}(U_i) = \operatorname{Spec} B_i$ is affine, and $B_i$ is a finite $A_i$-module, for each $i$.

This is equivalent to requiring that for every affine open subset $U = \operatorname{Spec} A$ of $Y$, $f^{-1}(U) = \operatorname{Spec} B$ is affine and $B$ is a finite $A$-module. If $(Y, \mathcal{O}_Y)$ is an integral scheme of finite type over a field $k$ (i.e., over $\operatorname{Spec} k$), then we can construct a finite morphism $f : (X, \mathcal{O}_X) \to (Y, \mathcal{O}_Y)$ called the *normalization* of $Y$, with the following property: for each affine open set $U = \operatorname{Spec} A \subset Y$, $f^{-1}(U) = \operatorname{Spec} B$, where $B$ is the integral closure of $A$ in its quotient field, which is the function field of $Y$. This is a finite morphism, because $B$ is a finite $A$-module (see [H], I, 3.9 A). A finite morphism $f : X \to Y$ is *closed*, i.e., if $Z \subset X$ is closed, then $f(Z) \subset Y$ is closed. To prove this, one reduces to the case when $Y$, and hence $X$, is affine; further one may assume $Z = X$. If $X = \operatorname{Spec} B$, $Y = \operatorname{Spec} A$, we may replace $Y$ by the closed subscheme $\operatorname{Spec} A/I$, where $I = \ker(A \to B)$; hence we may further assume the ring homomorphism $A \to B$ is injective. Now we want to prove $\operatorname{Spec} B \to \operatorname{Spec} A$ is surjective. If $P \subset A$ is prime,

$S_P = A - P$, then $S_P^{-1}B$ is a finite $A_P$-module; so it suffices to prove that when $A$ is local with maximal ideal $M$, then $B$ has a prime ideal contracting to $M$. Now $B$ is a non-zero finitely generated $A$-module; hence $MB \neq B$ by Nakayama's lemma; if $M'$ is a maximal ideal of $B$ containing $MB$, then $M' \cap A$ is an ideal $\neq A$ (since $1 \notin M'$) which contains $M$, hence equals $M$.

(D.28) A morphism $f : (X, \mathcal{O}_X) \to (Y, \mathcal{O}_Y)$ is *affine* if there exists a covering of $Y$ by affine open subschemes $U_i = \operatorname{Spec} A_i$ such that $f^{-1}(U_i)$ is affine for each $i$.

Thus finite morphisms are affine. A morphism $f : X \to Y$ is affine if and only if $f^{-1}(U)$ is affine for every affine open subscheme $U$ of $Y$.

(D.29) A morphism of schemes $f : (X, \mathcal{O}_X) \to (Y, \mathcal{O}_Y)$ is an *immersion* if $f^\# : f^{-1}\mathcal{O}_Y \to \mathcal{O}_X$ is surjective. In particular, if $f$ is the inclusion of a closed subset of $Y$, we call $(X, \mathcal{O}_X)$ a *closed subscheme* of $(Y, \mathcal{O}_Y)$. The morphism $f$ is called a *closed immersion* if it is an immersion which is an isomorphism of $(X, \mathcal{O}_X)$ with a closed subscheme of $(Y, \mathcal{O}_Y)$.

If $X = \operatorname{Spec} R$, and $I$ is an ideal of $R$, then there is a natural bijection between the prime ideals of $R/I$ and the set $V(I)$ of prime ideals of $R$ containing $I$. This bijection in fact makes $\operatorname{Spec} R/I$ a closed subscheme of $\operatorname{Spec} R$. One can show ([H], II, 5.10) that every closed subscheme of $\operatorname{Spec} R$ is obtained this way. If $R$ is a graded ring, $X = \operatorname{Proj} R$, and $I \subset R$ is a homogeneous ideal, then again the bijection $\operatorname{Proj} R/I \cong V(I)$ makes $\operatorname{Proj} R/I$ a closed subscheme of $\operatorname{Proj} R$. If $R = \oplus_{i \geq 0} R_i$ where $R_1$ is a finite $R_0$-module, and $R$ is generated by $R_1$ as an $R_0$-algebra, then any closed subscheme of $\operatorname{Proj} R$ is of this form ([H], II, 5.16).

(D.30) The *Krull dimension* $\dim X$ of a topological space $X$ is the supremum of all integers $n$ such that there exists a chain $Z_0 \subset Z_1 \subset \cdots \subset Z_n$ of distinct, irreducible closed subsets of $X$. The Krull dimension of a scheme is that of its underlying topological space.

This notion of dimension is called Krull dimension because for $X = \operatorname{Spec} A$, this agrees with the Krull dimension (as defined in commutative algebra) of $A$.

The *codimension* $\operatorname{codim}_X Z$ of an irreducible closed subset $Z \subset X$ is the supremum of all integers $n$ such that there is a chain $Z = Z_0 \subset Z_1 \subset \cdots \subset Z_n$ of distinct irreducible closed subsets of $X$.

Note that it is *false* in general that (i) for a non-empty open subset $U \subset X$ of an irreducible space $X$, we have $\dim X = \dim U$, and (ii) for an irreducible closed subset $Z \subset X$, we have $\dim Z + \operatorname{codim}_X Z = \dim X$. However these statements are true if $X$ is the topological space underlying an irreducible scheme of finite type over a field (see [H], II, 3.2.8 and Ex. 3.20). In this case, the dimension function has other good properties; for example, if $f : X \to Y$ is any morphism between irreducible schemes of finite type over a field, with dense image, then the image contains an open dense subset of $Y$, and $\dim f^{-1}(y) = \dim X - \dim Y$ for all $y$ in some (perhaps smaller) open dense subset of $Y$ (see [H], II, Ex. 3.22). Hence if $\dim X < \dim Y$, a morphism $f : X \to Y$ cannot have dense image, i.e., the closure of its image is a proper subvariety of $Y$. This is a very useful fact in practice when making general position arguments (see lemma (D.69) below for an illustration).

If $x \in X$ is a point, its codimension is defined to be the Krull dimension of $\mathcal{O}_{X,x}$. If $X$ is a scheme, we will use sometimes use $X^p$ to denote the set of points of $X$ of codimension $p$. This convention is used in Chapter 5 of the main text; see (5.20).

(D.31) The category of schemes has *fibered products*. Thus if $f : X \to S$, $g : Y \to S$ are morphisms of schemes, there exists a scheme $X \times_S Y$ together with morphisms $p : X \times_S Y \to X$, $q : X \times_S Y \to Y$ such that the diagram

$$\begin{array}{ccc} X \times_S Y & \stackrel{q}{\to} & Y \\ p \downarrow & & \downarrow g \\ X & \stackrel{f}{\to} & S \end{array}$$

commutes, and such that for any scheme $Z$ which fits into a commutative diagram

$$\begin{array}{ccc} Z & \stackrel{q'}{\to} & Y \\ p' \downarrow & & \downarrow g \\ X & \stackrel{f}{\to} & S \end{array}$$

there is a unique morphism of schemes $h : Z \to X \times_S Y$ such that $p' = p \circ h$, $q' = q \circ h$. This is proved in [H] (II, 3.3), by reducing to the case when $X = \operatorname{Spec} A$, $Y = \operatorname{Spec} B$, $S = \operatorname{Spec} C$ are all affine; in this case $X \times_S Y \cong \operatorname{Spec}(A \otimes_C B)$.

This leads to two important notions. First, if $f : X \to Y$ is a morphism of schemes, the *fiber* $X_y$ (also called the *scheme theoretic fiber*) of $f$ over a point $y \in Y$ is defined to be the fiber product $X \times_Y \operatorname{Spec} k(y)$. Here $k(y)$ is the residue field of $y$, and the morphism $f : \operatorname{Spec} k(y) \to (Y, \mathcal{O}_Y)$ is the inclusion of $\{y\}$ in $Y$, together with the map of sheaves $f^{-1}\mathcal{O}_Y = \mathcal{O}_{Y,y} \twoheadrightarrow k(y)$ (where we identify sheaves on $\{y\}$ with Abelian groups). The topological space underlying $X_y$ is the fiber of the continuous map $f : X \to Y$ over $y$ with its subspace topology; the scheme structure on $X_y$ includes more information (for example, even if $X$, $Y$ are integral, $X_y$ may not be reduced, corresponding to the geometric ideas of 'multiple fibers' or 'singularities' of the morphism $f$).

Another important notion is that of *base change*. If $f : X \to S$ is a morphism, we regard $X$ as a *scheme over* $S$; if $f : X \to S$, $g : Y \to S$ are schemes over $S$, an $S$-morphism from $X$ to $Y$ is a morphism $h : X \to Y$ such that $g \circ h = f$. For example, one may consider the category of schemes over a field $k$ (i.e., over $\operatorname{Spec} k$). One writes '$S$-scheme' instead 'scheme over $S$'; in the context of $S$-schemes, the morphism to $S$ is referred to as the *structure morphism*. Given any morphism $S' \to S$, we obtain a scheme $X' = S' \times_S X$ over $S'$. This gives a functor from schemes over $S$ to schemes over $S'$, called the *base change* corresponding to $S' \to S$. For example, one may take $k$ to be a field, $S = \operatorname{Spec} k$, $K$ an extension field, $S' = \operatorname{Spec} K$. A morphism $f : X \to S$ is *closed* if for any closed subset $Z \subset X$, $f(Z) \subset S$ is closed; $f$ is called *universally closed* if for any base change $S' \to S$, the corresponding morphism $f' : X' = S' \times_S X \to S'$ is closed. In a similar way, if $\mathcal{P}$ is a property of morphisms of schemes, one says a morphism is 'universally $\mathcal{P}$' if $\mathcal{P}$ holds for every morphism obtained by an arbitrary base change.

(D.32)   The notion of fiber products allows us to define the 'correct' analogues for schemes of the notions of Hausdorff spaces and proper maps of topological spaces; the definitions involve the *diagonal*. If $f : X \to S$ is a morphism, there is a unique induced morphism $\Delta$ from $X$ to the fiber product $X \times_S X$, called the diagonal morphism, such that the diagram

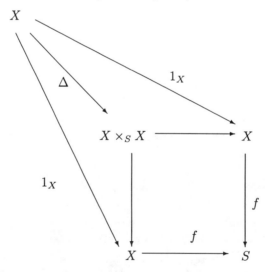

commutes. Here $1_X$ denotes the identity morphism on $X$. In particular, one sees at once that $\Delta$ is an immersion.

(D.33)   A morphism $f : X \to S$ is defined to be *separated* if $\Delta$ is a *closed immersion*. A scheme $X$ is separated if $X \to \operatorname{Spec} \mathbb{Z}$ is separated.

Since $\Delta$ is anyway an immersion, we see that $f$ is separated if and only if the image of $\Delta$ is a closed subset of $X \times_S X$. If $X$ is a scheme of finite type over $\mathbb{C}$, the field of complex numbers, then the set of closed points of $X$ is naturally identified with the set of points of a complex analytic space (this boils down to the assertion that a polynomial with complex coefficients is an analytic function). In particular, we can associate a topological space to this set of closed points; this space is Hausdorff precisely when $X \to \operatorname{Spec} \mathbb{C}$ is separated.

If $f : \operatorname{Spec} B \to \operatorname{Spec} A$ is any morphism of affine schemes, then $f$ is separated. In fact $\Delta$ is the closed immersion $\operatorname{Spec} B \to \operatorname{Spec}(B \otimes_A B)$ which corresponds to the surjection of rings $B \otimes_A B \to B$ given by $b_1 \otimes b_2 \mapsto b_1 b_2$.

One good property of domains of separated morphisms is the following: if $f : X \to \operatorname{Spec} A$ is separated, for some ring $A$, then for any pair of affine open subsets $U, V \subset X$, their intersection $U \cap V$ is also affine. Indeed, $U \cap V \cong (U \times_{\operatorname{Spec} A} V) \cap \Delta_X$, where the intersection on the right is as subschemes of $X \times_{\operatorname{Spec} A} X$, and $\Delta_X$ is the diagonal subscheme. Now $U, V$ are affine, so $U \times_{\operatorname{Spec} A} V$ is affine; if $f$ is separated, $\Delta_X$ is a closed subscheme, so that $U \cap V$ is isomorphic to a closed subscheme of the affine scheme $U \times_{\operatorname{Spec} A} V$; hence $U \cap V$ is affine.

(D.34) A morphism $f : X \to S$ is *proper* if $f$ is separated, of finite type, and universally closed.

(D.35) It may not be so easy to check directly, using the definition, that a morphism is proper, or even separated. However, it follows easily from the definitions that separated and proper morphisms are each stable under composition and arbitrary base change; further, an open immersion is separated, and a closed immersion is proper. Also, separatedness and properness are *local on the base*, i.e., to check either condition for $f : X \to S$, it suffices to check it for each of the morphisms $f^{-1}(U_i) \to U_i$ for some open cover $U_i$ of $S$. If $X$ is Noetherian, and $f : X \to Y$, $g : Y \to Z$ are morphisms, such that $g \circ f : X \to Z$ is either separated or proper, then $f$ has the same property. Finally, we have shown (see (D.27)) that a finite morphism is closed; since it is affine, it is separated; since any base change of a finite morphism is finite, any finite morphism is proper.

One can show that (i) a proper morphism between affine schemes is finite; (ii) a proper morphism with finite fibers is finite (see [H], III, Ex. 11.2).

(D.36) For morphisms $f : X \to Y$ with $X$ *Noetherian*, one has the *valuative criteria* for separatedness and properness. These are as follows.

Let $R$ be a valuation ring with quotient field $K$. Thus $R$ is a local integral domain with quotient field $K$ such that for any non-zero $x \in K$, either $x \in R$ or $x^{-1} \in R$. There is then a homomorphism (called a *valuation*) $v : K^* \to G$ to a totally ordered Abelian group $G$ such that $R - \{0\}$ consists of the elements $x \in K^*$ such that $v(x) \geq 0_G$, where $0_G \in G$ is the identity. Further, one has that $v(x+y) \geq \min(v(x), v(y))$ for all $x, y \in K^*$

with $x + y \neq 0$. Familiar examples of valuation rings are discrete valuation rings; another type of valuation ring, contained in the quotient field of a ring of formal power series, is described in the course of the proof of the Mercurjev-Suslin theorem in Chapter 8 (see the remark after lemma (8.20)).

The criterion for separatedness is as follows. Let $f : X \to S$ be a morphism with $X$ Noetherian. Then $f$ is separated $\Leftrightarrow$ for any valuation ring $R$, with a morphism $h : \operatorname{Spec} R \to S$, and for any morphism $g : \operatorname{Spec} K \to X$ such that the diagram

$$\begin{array}{ccc} \operatorname{Spec} K & \xrightarrow{g} & X \\ \downarrow & & \downarrow f \\ \operatorname{Spec} R & \xrightarrow{h} & S \end{array}$$

commutes, there exists *at most one* extension of $g$ to a morphism $\tilde{g} : \operatorname{Spec} R \to X$ such that $\tilde{g}$ lifts $g$.

Similarly, if $X$ is Noetherian and $f$ is of finite type, then $f$ is proper $\Leftrightarrow$ in the above situation, there exists a unique lift $\tilde{g}$ of $g$ as above.

The proofs of the above two valuative criteria are given in [H], II, Theorem 4.3 and Theorem 4.8.

(D.37)   We use the valuative criterion to show that $f : \mathbb{P}^n_A \to \operatorname{Spec} A$ is proper, for any ring $A$. If this is true for $A = \mathbb{Z}$, the ring of integers, then it is true for any $A$, since properness is preserved under base change. So we may assume $f : \mathbb{P}^n_\mathbb{Z} \to \operatorname{Spec} \mathbb{Z}$. Since $f$ is clearly of finite type, $\mathbb{P}^n_\mathbb{Z}$ is Noetherian, and it suffices to show that if $R$ is a valuation ring with quotient field $K$, then any morphism $g : \operatorname{Spec} K \to \mathbb{P}^n_\mathbb{Z}$ extends uniquely to a morphism $\tilde{g} : \operatorname{Spec} R \to \mathbb{P}^n_\mathbb{Z}$. (Note that there are unique morphisms $\operatorname{Spec} K \to \operatorname{Spec} \mathbb{Z}$ and $\operatorname{Spec} R \to \operatorname{Spec} \mathbb{Z}$.)

Let $x \in \mathbb{P}^n_\mathbb{Z}$ be the image of the unique point of $\operatorname{Spec} K$. Then the morphism $g$ determines an injective homomorphism $k(x) \to K$, where $k(x)$ is the residue field at $x$. We may explicitly realize this as follows. $\mathbb{P}^n_\mathbb{Z} = \operatorname{Proj} \mathbb{Z}[X_0, \ldots, X_n]$ is covered by the affine open subschemes $U_i = D_+(X_i) = \operatorname{Spec} \mathbb{Z}[X_0/X_i, X_1/X_i, \ldots, X_n/X_i]$; if $x \in U_i$, then the morphism $\operatorname{Spec} K \to \mathbb{P}^n_\mathbb{Z}$ factors through a morphism $\operatorname{Spec} K \to U_i$, i.e., corresponds to a homomorphism $\rho : \mathbb{Z}[X_0/X_i, \ldots, X_n/X_i] \to K$. The point $x \in U_i$ is the prime ideal which is the kernel of this homomorphism.

Let $v : K^* \to G$ be a valuation corresponding to the valuation ring $R \subset K$, where $G$ is a totally ordered Abelian group. Extend $v$ to a mapping $v : K \to G \cup \{\infty\}$ where $G \cup \{\infty\}$ is a totally ordered set, such that $\infty$ is an element larger than any element of $G$; we define $v(0) = \infty$. Choose an index $j$ such that

$$v(\rho(X_j/X_i)) = \min\{v(\rho(X_0/X_i)), v(\rho(X_1/X_i)), \ldots, v(\rho(X_n/X_i))\}.$$

Appendix D: Results from Algebraic Geometry    315

Since $\rho(X_i/X_i) = 1$, this minimal value of $v$ is an element of $G$. Then $\rho(X_j/X_i) \neq 0$, and so $\rho$ extends to a homomorphism
$$\mathbb{Z}[X_0/X_i, \ldots, X_n/X_i, (X_j/X_i)^{-1}] \to K,$$
which we also denote by $\rho$. Now $X_l/X_j = (X_l/X_i)(X_j/X_i)^{-1}$, and
$$v(\rho(X_l/X_j)) = v(\rho(X_l/X_i)) - v(\rho(X_j/X_i)) \geq 0.$$
Hence $\rho(X_l/X_j) \in R$ for all $l$, and we have a commutative diagram of rings
$$\begin{array}{ccc} \mathbb{Z}[X_0/X_j, \ldots, X_n/X_j] & \stackrel{\rho}{\to} & R \\ \downarrow & & \downarrow \\ \mathbb{Z}[X_0/X_i, \ldots, X_n/X_i, (X_j/X_i)^{-1}] & \stackrel{\rho}{\to} & K \end{array}$$
which yields a commutative diagram of schemes and morphisms
$$\begin{array}{ccc} \operatorname{Spec} R & \to & U_j \\ \uparrow & & \uparrow \\ \operatorname{Spec} K & \to & U_i \cap U_j \end{array}$$
In particular, the morphism $\operatorname{Spec} K \to \mathbb{P}^n_{\mathbb{Z}}$ extends to a morphism $\operatorname{Spec} R \to \mathbb{P}^n_{\mathbb{Z}}$. One verifies easily that this extension is unique, and in particular, does not depend on the choice of the index $j$.

In particular, if $k$ is an algebraically closed field, $\mathbb{P}^n_k \to \operatorname{Spec} k$ is proper. Hence the projection $\pi : \mathbb{P}^n_k \times_k \mathbb{A}^m_k \to \mathbb{A}^m_k$ is a closed map. In concrete terms, this means the following. Let $X_0, \ldots, X_n$ be variables corresponding to homogeneous coordinates on $\mathbb{P}^n_k$ (i.e., $\mathbb{P}^n_k = \operatorname{Proj} k[X_0, \ldots, X_n]$) and let $Y_1, \ldots, Y_m$ be the variables corresponding to coordinates on $\mathbb{A}^m_k$ (i.e., $\mathbb{A}^m_k = \operatorname{Spec} k[Y_1, \ldots, Y_m]$). A closed set $W$ in $\mathbb{P}^n_k \times_k \mathbb{A}^m_k$ is the zero locus of a finite set of polynomials
$$F_1(X_0, \ldots, X_n; Y_1, \ldots, Y_m), \ldots, F_s(X_0, \ldots, X_n; Y_1, \ldots, Y_m),$$
where the $F_j$ are homogeneous in the $X_\nu$. To say that its image $\pi(W)$ in $\mathbb{A}^m_k$ is closed is to say that $\pi(W)$ consists of the zeroes of a set
$$G_1(Y_1, \ldots, Y_m), \ldots, G_t(Y_1, \ldots, Y_m)$$
of polynomials in the $Y_j$ alone; if $I \subset k[X_0, \ldots, X_n, Y_1, \ldots, Y_m]$ is the ideal generated by the $F_i$, we take $G_j$ to be generators for the intersection of $I$ with the subring $k[Y_1, \ldots, Y_m]$, i.e., the $G_j$ are obtained from the $F_i$ by 'eliminating the variables $X_\nu$'. The assertion, that the zero locus of these $G_j$ describes the image $\pi(W)$, is the content of 'elimination theory' in classical algebraic geometry, and may be proved directly without appealing to schemes, valuation rings, etc. (for example, for the case when

$k = \mathbb{C}$, see D. Mumford, *Algebraic Geometry I: Complex Projective Varieties*, Grundlehren Math. 221, Springer-Verlag (1976), Theorem (2.23) and (2.25)).

(D.38) Let $S$ be a scheme. Define *projective n-space over $S$*, denoted $\mathbb{P}_S^n$, to be the fiber product $\mathbb{P}_\mathbb{Z}^n \times_{\operatorname{Spec} \mathbb{Z}} S$.

For $S = \operatorname{Spec} A$, this is consistent with our earlier definition of $\mathbb{P}_A^n$. Since properness is preserved under base change, the natural map $p : \mathbb{P}_S^n \to S$ is proper.

A morphism $f : X \to S$ is called *projective* if there is a closed immersion (see (D.29)) $i : X \to \mathbb{P}_S^n$ such that $f = p \circ i$, where $p : \mathbb{P}_S^n \to S$ is the natural map. We say that $X$ is *projective over $S$* if there is a projective morphism $f : X \to S$. A scheme $X$ is said to be *quasi-projective over $S$* if it is isomorphic to an open subscheme of a projective $S$-scheme $\overline{X}$.

Since a closed immersion is proper, and a composition of proper morphism is proper, we see that any projective morphism is proper. The converse is false, even if $S = \operatorname{Spec} \mathbb{C}$, where $\mathbb{C}$ is the field of complex numbers; however, if $f : X \to S$ is proper, and $S$ is Noetherian, then there is a projective morphism $g : X' \to S$ and a morphism $h : X' \to X$ such that (i) $g = f \circ h$ (ii) there is a dense open subset $U \subset X$ such that $h^{-1}(U) \to U$ is an isomorphism. This assertion is known as Chow's lemma; in practice, it reduces the proofs of many assertions about proper morphisms to the special case of projective morphisms. (Note that the morphism $h : X' \to X$ is also projective, since the closed immersion $i : X' \to \mathbb{P}_S^n$ used to factorize $g$ induces a closed immersion $i' : X' \to \mathbb{P}_X^n \cong \mathbb{P}_S^n \times_S X$ which factorizes $h$.)

The structure morphism $X \to S$ of a quasi-projective $S$-scheme $X$ is separated and of finite type.

(D.39) We will use the term *variety over a field $k$* to denote a separated $k$-scheme of finite type. A *quasi-projective variety over $k$* is a separated scheme of finite type over $k$ which is quasi-projective over $k$. This includes the notions of affine and projective varieties over $k$.

(D.40) A morphism $f : X \to Y$ is *flat* if for each $x \in X$, if $y = f(x)$, then the induced map of rings $\mathcal{O}_{Y,y} \to \mathcal{O}_{X,x}$ is flat (i.e., the functor $- \otimes_{\mathcal{O}_{Y,y}} \mathcal{O}_{X,x}$ is exact).

Flat morphisms of finite type between Noetherian schemes are *open*. This can be deduced using the following assertion from commutative algebra: if $f : R \to S$ is a flat homomorphism between Noetherian rings, then the *going down theorem* holds, i.e., given a pair of prime ideals $P \subset P'$ of $R$ and a prime $Q'$ of $S$ contracting to $P'$ (i.e., with $f^{-1}(Q') = P'$), there exists a prime ideal $Q \subset Q'$ of $S$ which contracts to $P$. To see this, by localising at $P'$ and $Q'$, we may assume $R, S$ are local with maximal ideals $P'$ and $Q'$ respectively, and $Q'$ contracts to $P'$; now $R/P \hookrightarrow R_P/PR_P$,

Appendix D: Results from Algebraic Geometry    317

so by flatness, $S/PS \hookrightarrow S \otimes_R (R_P/PR_P)$, and $S/PS \twoheadrightarrow S/Q' \neq 0$; hence $P(S \otimes_R R_P)$ is not the unit ideal. If $Q_1$ is any maximal ideal of $S \otimes_R R_P$ containing $P(S \otimes_R R_P)$, then under the homomorphism $R \to S \otimes_R R_P$, we see that $Q_1$ contracts to $P$ (the contraction contains $P$, but cannot be larger since $Q_1$ is not the unit ideal). Take $Q$ to be the contraction of $Q_1$ to $S$.

In particular, if $f : (R, \mathcal{M}_R) \to (S, \mathcal{M}_S)$ is a flat local homomorphism between Noetherian local rings, then $\dim S \geq \dim R$. If $\sqrt{\mathcal{M}_R S} = \mathcal{M}_S$, then $\dim R = \dim S$ (from dimension theory in commutative algebra, $\dim R = d \Leftrightarrow$ there exist $x_1, \ldots, x_d \in \mathcal{M}_R$ with $\sqrt{\sum R f_i} = \mathcal{M}_R$, and $d$ is the smallest such number; since also $\sqrt{\sum S f_i} = \mathcal{M}_S$, we have $\dim S \leq \dim R$).

This means that if $f : X \to Y$ is a flat morphism, and $y \in Y^p$ is a point of codimension $p$, and $x \in X$ is the generic point of an irreducible component of the fiber $X_y$, then $x \in X^p$. This means that $f^{-1}$ *preserves codimension* of closed subschemes; this is important for $K$-theory.

We give a criterion for a morphism to be flat: if $f : X \to Y$ is a morphism between irreducible varieties over a field $k$ (i.e., $X$, $Y$ are irreducible separated $k$-schemes of finite type) such that $Y$ is regular, $X$ is Cohen-Macaulay (for example, regular) and all fibers of $f$ have dimension $\dim X - \dim Y$, then $f$ is flat (see [H], III, Ex. 10.9). For example, any finite surjective morphism between integral regular schemes is flat.

(D.41)    If $(f, f^\#) : (X, \mathcal{O}_X) \to (Y, \mathcal{O}_Y)$ is a morphism of ringed spaces, and $\mathcal{F}$ is any sheaf of $\mathcal{O}_X$-modules, then the natural homomorphism $\mathcal{O}_Y \to f_*\mathcal{O}_X$ (induced by $f^\#$) makes the direct image $f_*\mathcal{F}$ into an $\mathcal{O}_Y$-module. Then $f_*$, considered as a functor from $\mathcal{O}_X$-modules to $\mathcal{O}_Y$-modules, has a left adjoint $f^*$, given by $f^*\mathcal{G} = \mathcal{O}_X \otimes_{f^{-1}\mathcal{O}_Y} f^{-1}\mathcal{G}$. We call $f^*\mathcal{G}$ the *module theoretic inverse image* of $\mathcal{G}$. Note that if $f$ is a flat morphism of schemes, then $f^*$ is an *exact* functor. For any $\mathcal{O}_X$-module $\mathcal{F}$ and any locally free $\mathcal{O}_Y$-module $\mathcal{E}$, we have a natural isomorphism $f_*(\mathcal{F} \otimes_{\mathcal{O}_X} f^*\mathcal{E}) \cong f_*(\mathcal{F}) \otimes_{\mathcal{O}_Y} \mathcal{E}$, called the *projection formula*.

## Coherent and Quasi-coherent Sheaves

(D.42)    For any ringed space $(X, \mathcal{O}_X)$, we have defined the notion of a sheaf of $\mathcal{O}_X$-modules. When $X = \operatorname{Spec} R$, there are certain distinguished $\mathcal{O}_X$-modules which correspond naturally to $R$-modules, in the sense that (i) the category of these modules is a full Abelian subcategory of the category of all $\mathcal{O}_X$-modules, and (ii) this full subcategory is naturally equivalent to the category of $R$-modules, in a fashion analogous to the equivalence of the category of affine schemes with the opposite of the category of rings.

First, to any $R$-module $M$, we associate an $\mathcal{O}_X$-module $\widetilde{M}$ as follows. For any $P \in X$, let $M_P$ denote the localization of $M$ at the multiplicative set $R - P$, so that $M_P \cong M \otimes_R R_P$. For any open set $D(I) \subset X$, let

$$\widetilde{M}(D(I)) = \{s \in \prod_{P \in D(I)} M_P \mid \text{locally on } D(I), s \text{ equals a quotient}$$

$$m/a \text{ with } m \in M \text{ and } a \in R\}.$$

Here, we mean that for each $P \in D(I)$, there is

(i) a neighborhood $D(f)$ of $P$ in $D(I)$ (i.e., $f \in I$, $f \notin P$), and

(ii) an element $m/f^n \in M_f$,

such that for each $Q \in D(f)$, $s$ and $m/f^n$ have the same image in $M_Q$. Because of the 'local' nature of the condition defining the sections $s$, the above definition yields a sheaf of $\mathcal{O}_X$-modules on $X = \operatorname{Spec} R$, such that (see [H], II, (5.1))

(i) the stalk of $\widetilde{M}$ at $P$ is $M_P$

(ii) there is a natural isomorphism $\widetilde{M}(D(f)) \cong M_f$; in particular, $\widetilde{M}(\operatorname{Spec} R) \cong M$.

In (ii) above, one can show that in fact there is an isomorphism of sheaves of $\mathcal{O}_{\operatorname{Spec} R_f}$-modules $\widetilde{M}|_{D(f)} \cong \widetilde{M_f}$. It is clear from the construction that $M \mapsto \widetilde{M}$ is functorial in $M$. We may also use the notation $M^\sim$ instead of $\widetilde{M}$.

It is shown in [H] (II, (5.2)) that:

(i) for any ring $R$, the functor $M \mapsto \widetilde{M}$ is exact and fully faithful

(ii) if $f : \operatorname{Spec} S \to \operatorname{Spec} R$ is a morphism, corresponding to a homomorphism $\varphi :\to S$, then the direct image functor $f_*$ and module theoretic inverse image functor $f^*$ have the following descriptions: for any $R$-module $M$, we have a natural isomorphism $f^*\widetilde{M} \cong (S \otimes_R M)^\sim$, and for any $S$-module $N$, the direct image $f_*\widetilde{N}$ is $\widetilde{N_R}$, where $N_R$ is $N$ regarded as an $R$-module via $\varphi$

(iii) for any family of $R$-modules $M_i$, we have $\widetilde{\oplus M_i} \cong \oplus_i \widetilde{M_i}$

(iv) for any pair of $R$-modules $M$ and $N$, we have $\widetilde{M} \otimes_{\mathcal{O}_{\operatorname{Spec} R}} \widetilde{N} \cong (M \otimes_R N)^\sim$.

It is easy to see from the definitions that the functor $M \mapsto \widetilde{M}$ from $R$-modules to $\mathcal{O}_{\operatorname{Spec} R}$-modules is left adjoint to the global section functor $\mathcal{F} \mapsto \Gamma(\operatorname{Spec} R, \mathcal{F})$.

Appendix D: Results from Algebraic Geometry    319

(D.43)  A sheaf $\mathcal{F}$ of $\mathcal{O}_X$-modules on a scheme $X$ is called *quasi-coherent* if there exists an affine open cover $U_i = \operatorname{Spec} R_i$ of $X$ and $R_i$-modules $M_i$ such that there are isomorphisms of $\mathcal{O}_{U_i}$-modules $\mathcal{F}|_{U_i} \cong \widetilde{M_i}$.

Note that $\mathcal{O}_X$ is quasi-coherent; more generally, any locally free $\mathcal{O}_X$-module is quasi-coherent. If $j : U \to X$ is the inclusion of an open subscheme of $X$, then in general $j_!\mathcal{O}_U$ is an $\mathcal{O}_X$-module (in fact a sheaf of ideals) which is *not* quasi-coherent.

One can show that quasi-coherent sheaves have the following properties (see the discussion below, and [H], II, (5.4), (5.7), (5.8), III, (8.5)).

(i) Let $X$ be a scheme. An $\mathcal{O}_X$-module $\mathcal{F}$ is quasi-coherent if and only if for each affine open subscheme $U = \operatorname{Spec} R$ of $X$, we have an isomorphism of $\mathcal{O}_U$-modules $\mathcal{F}|_U \cong (\mathcal{F}(U))\tilde{\;}$.

(ii) The kernel, cokernel and image of any $\mathcal{O}_X$-linear map between quasi-coherent $\mathcal{O}_X$-modules is quasi-coherent; any extension of quasi-coherent $\mathcal{O}_X$-modules is quasi-coherent. Any direct sum of quasi-coherent $\mathcal{O}_X$-modules is quasi-coherent; if $\mathcal{F}, \mathcal{G}$ are quasi-coherent, then $\mathcal{F} \otimes_{\mathcal{O}_X} \mathcal{G}$ is quasi-coherent.

(iii) For any morphism $f : X \to Y$, if $\mathcal{G}$ is a quasi-coherent $\mathcal{O}_Y$-module, then $f^*\mathcal{G}$ is a quasi-coherent $\mathcal{O}_X$-module. If $X$ is Noetherian, or $f$ is quasi-compact (inverse image of any affine open set is quasi-compact) and separated, then for any quasi-coherent $\mathcal{O}_X$-module $\mathcal{F}$, the direct image $f_*\mathcal{F}$ is quasi-coherent.

(iv) If $X$ is Noetherian, then for any morphism $f : X \to Y$ and any quasi-coherent $\mathcal{O}_X$-module $\mathcal{F}$, the higher direct images $R^i f_*\mathcal{F}$ are quasi-coherent, and for any affine open subscheme $U$ of $Y$, we have $R^i f_*\mathcal{F}|_U \cong H^i(f^{-1}(U), \mathcal{F}|_{f^{-1}(U)})\tilde{\;}$. For any quasi-coherent $\mathcal{O}_X$-module $\mathcal{F}$ and any locally free $\mathcal{O}_Y$-module $\mathcal{E}$, we have the *projection formula*
$$R^i f_*(\mathcal{F} \otimes_{\mathcal{O}_X} f^*\mathcal{E}) \xrightarrow{\cong} R^i f_*(\mathcal{F}) \otimes_{\mathcal{O}_Y} \mathcal{E}.$$

The assertion (ii) follows from the case when $X$ is affine; that an extension of quasi-coherent sheaves is quasi-coherent, reduces to the assertion that if $0 \to \mathcal{F}' \to \mathcal{F} \to \mathcal{F}'' \to 0$ is exact with $\mathcal{F}', \mathcal{F}''$ quasi-coherent, and $X$ is affine, then $0 \to \mathcal{F}'(X) \to \mathcal{F}(X) \to \mathcal{F}''(X) \to 0$ is exact ([H], II, (5.6)). In (iii), the point is that the inverse image of any affine open subset of $Y$ is a finite union of affine open subsets of $X$, such that the pairwise intersections of any two of these open subsets of $X$ is again a finite union of open affines. Hence there is an exact sequence of the form
$$0 \to (f_*\mathcal{F})|_U \to \oplus_i f_*(\mathcal{F}|_{U_i}) \to \oplus_{i,j,k} f_*(\mathcal{F}|_{U_{ijk}})$$

where $\{U_{ijk}\}_k$ is an affine open cover of $U_i \cap U_j$. The second and third term are quasi-coherent from the case when both $X$ and $Y$ are affine, since $\sim$ commutes with direct sums; by (ii), $f_*\mathcal{F}$ is also quasi-coherent. (iv) follows from (iii), that is the case $i = 0$, by induction on $i$, using the fact that any quasi-coherent $\mathcal{O}_X$-module on an affine scheme $X = \operatorname{Spec} R$ is a submodule of a flasque quasi-coherent $\mathcal{O}_X$-module (namely one of the form $\widetilde{I}$ where $I$ is an injective $R$-module; see [H], III, (3.4)); now one uses the fact that $R^i f_*$ and $H^i$ both vanish for a flasque sheaf, and the long exact sequences for $R^i f_*$ and for $H^i$. For the projection formula, we use the fact that tensoring with a locally free sheaf is exact, and the projection formula for the case $i = 0$.

In particular, if $i : Y \to X$ is the inclusion of a closed subscheme of $X$ (see (D.29)), let $\mathcal{I}_Y = \ker(\mathcal{O}_X \twoheadrightarrow i_*\mathcal{O}_Y)$ be the sheaf of ideals of $Y$ in $X$. Now $i_*\mathcal{O}_Y$ is quasi-coherent on $X$, since $i$ is quasi-compact and separated, and hence $\mathcal{I}_Y$ is quasi-coherent. Conversely, one sees ([H], II, (5.9)) that any quasi-coherent sheaf of ideals in $\mathcal{O}_X$ determines a subscheme of $X$. In particular, the closed subschemes of an affine scheme $\operatorname{Spec} R$ are precisely the schemes $\operatorname{Spec} R/I$ for all ideals $I$ of $R$.

Note that it is *not* true that for arbitrary quasi-coherent sheaves $\mathcal{F}, \mathcal{G}$, the $\mathcal{O}_X$-module $\mathcal{H}om_{\mathcal{O}_X}(\mathcal{F}, \mathcal{G})$ is quasi-coherent, since for modules $M, N$ over a ring $R$ and a multiplicative set $S$, it is in general not true that $S^{-1}\operatorname{Hom}_R(M, N) \cong \operatorname{Hom}_{S^{-1}R}(S^{-1}M, S^{-1}N)$. However this is true if $M$ is a finitely presented $R$-module. This partly motivates the following notion.

(D.44)  A sheaf $\mathcal{F}$ of $\mathcal{O}_X$-modules is *coherent* if

(i) for any $x \in X$, there exists an open neighborhood $U$ of $x$ and a surjection $\mathcal{O}_U^{\oplus r} \to \mathcal{F}|_U$, for some positive integer $r$ (i.e., $\mathcal{F}$ is *locally finitely generated*), and

(ii) for any open $U \subset X$ and any homomorphism $u : \mathcal{O}_U^{\oplus r} \to \mathcal{F}|_U$, the kernel of $u$ is locally finitely generated.

Note that a coherent sheaf is locally a cokernel of a map $\mathcal{O}_U^{\oplus s} \to \mathcal{O}_U^{\oplus r}$, and is hence quasi-coherent. This implies that if $\mathcal{F}$ is coherent and $\mathcal{G}$ is quasi-coherent, then $\mathcal{H}om_{\mathcal{O}_X}(\mathcal{F}, \mathcal{G})$ is quasi-coherent, and it is coherent if $\mathcal{G}$ is coherent. If $X$ is Noetherian, then $\mathcal{F}$ is a coherent $\mathcal{O}_X$-module if and only if for each affine open set $U = \operatorname{Spec} R$ of $X$, we have $\mathcal{F}|_U \cong \widetilde{\mathcal{F}(U)}$, where $\mathcal{F}(U)$ is a *finitely generated* $R$-module (see [H], (5.4)); it is enough to check this condition for all $U$ belonging to some affine open cover of $X$. One sees that if $0 \to \mathcal{F}' \to \mathcal{F} \to \mathcal{F}'' \to 0$ is exact, and any two of the three $\mathcal{O}_X$-modules is coherent, so is the third. If $X$ is Noetherian, the kernel, cokernel and image of any morphism between coherent sheaves is coherent. If $f : X \to Y$ is a morphism between Noetherian schemes, and $\mathcal{G}$ is coherent on $Y$, then $f^*\mathcal{G}$ is coherent on $X$. It is *not* true that the higher direct images (or even the direct image itself) of a coherent sheaf under a morphism between Noetherian schemes are coherent; however, this is true for *proper* morphisms (Grothendieck, EGA III, 3.2.1). For projective morphisms, this is shown in [H], III, (8.8); we discuss this further below (see Theorem (D.62)).

(D.45)  An important notion for $K$-theory is that of the *support* of a coherent $\mathcal{O}_X$-module $\mathcal{F}$ on a Noetherian scheme $X$. If $X = \operatorname{Spec} A$ is affine, and $\mathcal{F} = \widetilde{M}$, then $M$ is a finite $A$-module. The support of $\widetilde{M}$ is $\operatorname{Spec} A/\operatorname{ann} M$, where $\operatorname{ann} M = \{a \in A \mid aM = 0\}$ is the annihilator of $M$. Depending on the context, we may regard the support as a subscheme or a subset of $X$. In general, the support of $\mathcal{F}$ is defined to be the subscheme $Z$ of $X$ such that for any affine open subscheme $U = \operatorname{Spec} A$ of $X$, the scheme $U \cap Z$ is the support of $\mathcal{F}|_U = \widetilde{\mathcal{F}(U)}$. Equivalently, we can define the sheaf of annihilators of $\mathcal{F}$, which is a coherent sheaf of ideals in $\mathcal{O}_X$; then $Z$ is the corresponding subscheme. In some contexts, one regards the support of $\mathcal{F}$ as the closed subset of $X$ underlying the scheme $Z$.

We can also define the notion of an *associated point* of a coherent sheaf $\mathcal{F}$ on a Noetherian scheme $X$; this is a point $x \in X$ such that the maximal ideal of $\mathcal{O}_{X,x}$ is an associated prime (in the sense of commutative algebra) of $\mathcal{F}_x$. If $U = \operatorname{Spec} A$ is an affine open subscheme containing $x$, and $\mathcal{F}|_U = \widetilde{M}$, and $x$ corresponds to a prime ideal $P$, then $x$ is an associated point of $\mathcal{F}$ if and only if there is an $m \in M$ whose annihilator in $A$ is $P$, i.e., $P$ is an associated prime of $M$ in the sense of commutative algebra.

We remark that if $\mathcal{F}$ is a coherent $\mathcal{O}_X$-module on a Noetherian scheme $X$, then for any $r \geq 0$, the set

$$\{x \in X \mid \dim_{k(x)} \mathcal{F}_x \otimes_{\mathcal{O}_{X,x}} k(x) \geq r\}$$

is *closed*. This follows easily from the case when $X$ is affine. In particular, if $X$ is integral, with generic point $\eta$, the *rank* of $\mathcal{F}$ is the dimension of $\mathcal{F}_\eta$, which is a finite dimensional vector space over the function field of $X$; if $\mathcal{F}$ is of rank $r$, then

$$\{x \in X \mid \dim_{k(x)} \mathcal{F}_x \otimes_{\mathcal{O}_{X,x}} k(x) = r\}$$

is *open*, and is the largest subset of $X$ on which $\mathcal{F}$ restricts to a locally free sheaf of rank $r$ (we make use of the easy lemma that a finitely generated module of rank $r$ over a local integral domain is free $\Leftrightarrow$ tensored with the residue field, its dimension is $r$; this follows at once from Nakayama's lemma).

(D.46)   We now describe the correspondence between graded modules over a finitely generated graded ring $R$ and quasi-coherent and coherent sheaves on $\operatorname{Proj} R$. If $M$ is a graded $R$-module, we associate to it the sheaf $\widetilde{M}$ on $\operatorname{Proj} R$, defined as follows. For any homogeneous prime $P \in \operatorname{Proj} R$, recall that $S_{(P)}$ is the multiplicative set of homogeneous elements of $R - P$; let $M_{(P)}$ be the subgroup of $S_{(P)}^{-1} M$ consisting of elements $(1/b)m$ where $m \in M$, $b \in S_{(P)}$ are homogeneous of the same degree (so that $(1/b)m$ has degree 0). This is a module over the local ring $R_{(P)}$. Now define

$$\widetilde{M}(D_+(I)) = \{s \in \prod_{P \in D_+(I)} M_{(P)} \mid s \text{ is locally equal to}$$

an element $(1/b)m$ where $m \in M$, $b \in R$ have the same degree$\}$.

The condition on $s$ means that for each $P \in D_+(I)$, there exists a homogeneous $f \in I - P$, say of degree $d$, such that for some homogeneous element $m \in M_{nd}$ for some $n$, the elements $s$ and $(1/f^n)m$ have the same image in $M_{(Q)}$ for each $Q \in D_+(f)$.

One checks (see [H], II, (5.11)) that

(i) the stalk of $\widetilde{M}$ at $P$ is $M_{(P)}$

(ii) for any homogeneous element $f \in R$, if $M_{(f)}$ is the $R_{(f)}$-submodule of elements of degree 0 in the localization $M_f$, then $\widetilde{M} \mid_{D_+(f)} \cong \widetilde{M_{(f)}}$ (the sheaf on the right is the quasi-coherent sheaf on $D_+(f) = \operatorname{Spec} R_{(f)}$ associated to the $R_{(f)}$-module $M_{(f)}$)

(iii) $\widetilde{M}$ is a quasi-coherent $\mathcal{O}_{\operatorname{Proj} R}$-module; if $S$ is Noetherian and $M$ is finitely generated, then $\widetilde{M}$ is coherent.

(iv) there is a natural homomorphism $M_0 \to \Gamma(\operatorname{Proj} R, \widetilde{M})$.

Here $M_0$ is the group of elements of degree 0 in $M$.

For any graded $R$-module $M$, we can define a new graded $R$-module $M(n)$ by setting $M(n)_d = M_{n+d}$, with the evident $R$-module structure. In particular, let $R(n)$ be the graded $R$-module given by shifting the grading on $R$ by $n$, i.e., $R(n)_d = R_{n+d}$. Let $\mathcal{O}_{\operatorname{Proj} R}(n) = \widetilde{R(n)}$ for each $n$. For any $\mathcal{O}_{\operatorname{Proj} R}$-module $\mathcal{F}$, define $\mathcal{F}(n) = \mathcal{F} \otimes_{\mathcal{O}_{\operatorname{Proj} R}} \mathcal{O}_{\operatorname{Proj} R}(n)$. One can show that if $R$ is generated over $R_0$ by $R_1$, then

(i) $\mathcal{O}_{\operatorname{Proj} R}(1)$ is an invertible $\mathcal{O}_{\operatorname{Proj} R}$-module (locally free sheaf of rank 1), and $\mathcal{O}_{\operatorname{Proj} R}(n) = \mathcal{O}_{\operatorname{Proj} R}(1)^{\otimes n}$ (if $n < 0$, the right side of this isomorphism is defined to be $(\mathcal{O}_{\operatorname{Proj} R}(1)^{\otimes -n})^*$);

(ii) for any graded $R$-module $M$, we have $\widetilde{M(n)} = \widetilde{M}(n)$;

(iii) there is a natural homomorphism $M_n \to \Gamma(\operatorname{Proj} R, \widetilde{M(n)})$.

The operation $M \mapsto M(n)$ on graded modules (or $\mathcal{F} \mapsto \mathcal{F}(n)$ on sheaves) is called *twisting*. It allows one to associate a graded module to any quasi-coherent $\mathcal{O}_{\operatorname{Proj} R}$-module $\mathcal{F}$, by the formula $\Gamma_*(\mathcal{F}) = \oplus_{n \in \mathbb{Z}} \Gamma(\operatorname{Proj} R, \mathcal{F}(n))$. This is a graded $R$-module in a natural way, since elements of $R_n$ give global sections of $\mathcal{O}_{\operatorname{Proj} R}(n)$. In particular, we have a graded $R$-algebra $\Gamma_*(R)$. One has a natural isomorphism $R \to \Gamma_*(R)$, which is an isomorphism (see [H], II, (5.13)) if $R = A[X_0, \ldots, X_n]$ is a polynomial ring (with the usual grading), i.e., if $\operatorname{Proj} R = \mathbb{P}^n_A$.

One can show ([H], II, (5.15)) that under suitable assumptions, there is a natural isomorphism $\Gamma_*(\mathcal{F})\tilde{} \cong \mathcal{F}$ for any quasi-coherent sheaf on $\operatorname{Proj} R$. This follows from lemma (D.48) below, which also has other applications.

(D.47)   Let $X$ be a scheme, and $\mathcal{L}$ an invertible $\mathcal{O}_X$-module. Let $f \in \Gamma(X, \mathcal{L})$, and let $X_f$ be the set of points in $X$ where the image $f_x$ of $f$ in the stalk $\mathcal{L}_x$ does not lie in $\mathcal{M}_x \mathcal{L}_x$, where $\mathcal{M}_x \subset \mathcal{O}_{X,x}$ is the maximal ideal. We may express the condition that $f_x \notin \mathcal{M}_x \mathcal{L}_x$ by saying that "$f$ does not vanish at $x$", since the image of $f_x$ in $\mathcal{L}_x/\mathcal{M}_x \mathcal{L}_x = \mathcal{L}_x \otimes k(x)$ is the 'value' of the section $f$ in the fiber at $x$.

Now $f$ determines an $\mathcal{O}_X$-linear map $\mathcal{L}^{-1} \to \mathcal{O}_X$, whose image is a quasi-coherent sheaf of ideals on $X$, hence defines a closed subscheme; $X_f$ is the complement of this closed subscheme.

**Lemma D.48** *Let $X$, $\mathcal{L}$, $f \in \Gamma(X, \mathcal{L})$ and $X_f$ be as above. Let $\mathcal{F}$ be a quasi-coherent sheaf on $X$.*

(i) *Assume $X$ is quasi-compact, and $s \in \Gamma(X, \mathcal{F})$ such that $s \mapsto 0 \in \mathcal{F}(X_f)$. Then for some $n > 0$, we have $s \otimes f^n = 0 \in \Gamma(X, \mathcal{F} \otimes \mathcal{L}^{\otimes n})$.*

(ii) *Assume further that $X$ has a finite open cover by open affines $U_i$ such that $\mathcal{L}|_{U_i}$ is free, and $U_i \cap U_j$ is quasi-compact for each $i, j$. Given a section $t \in \mathcal{F}(X_f)$, there exists $n > 0$ such that $t \otimes f^n$ is the restriction to $X_f$ of a global section of $\mathcal{F} \otimes \mathcal{L}^{\otimes n}$.*

**Proof.**   (i) First choose a finite cover $\{U_i\}$ of $X$ by affine open sets, such that $\mathcal{L}|_{U_i} \cong \mathcal{O}_{U_i}$ for each $i$. Write $\mathcal{F}|_{U_i} = \widetilde{M_i}$ for some $A_i = \mathcal{O}_X(U_i)$-module $M_i$; then the restrictions of $s$ yield elements $s_i \in M_i$. Since $\mathcal{L}|_{U_i}$ is free for each $i$, the section $f$ yields an element $f_i \in A_i$, and $X_f \cap U_i = \operatorname{Spec}(A_i)_{f_i}$. We also have $\mathcal{F} \otimes \mathcal{L}^{\otimes n}(U_i) \cong M_i$, and $\mathcal{F} \otimes \mathcal{L}^{\otimes n}(U_i \cap X_f) = (M_i)_{f_i}$. Now $s$ vanishes on $X_f$; hence $s_i \in M_i$ vanishes in the localization $(M_i)_{f_i}$, for each $i$. Hence $f_i^{n_i} s_i = 0$ in $M_i$ for each $i$. Hence if $n = \max n_i$, then $s \otimes f^n$ restricts to 0 in each $M_i$, hence is 0, as a section of $\mathcal{F} \otimes \mathcal{L}^{\otimes n}$.

(ii) Let $\mathcal{F}|_{U_i} = \widetilde{M_i}$, and let $f|_{U_i} = f_i \in \Gamma(U_i, \mathcal{O}_{U_i})$. Let $t_i$ be the restriction of $t$ to $X_f \cap U_i$. Then $t_i \in (M_i)_{f_i}$ for each $i$. Hence there is an $n_1 > 0$ such that $t_i = t'_i f_i^{-n_1}$ for each $i$, for some $t'_i \in M_i$. Then $t'_i$ can be regarded as a section of $\mathcal{F} \otimes \mathcal{L}^{\otimes n_1}$ on $U_i$, for each $i$. Now $t'_i - t'_j$ is a section of this sheaf on $U_i \cap U_j$ which vanishes on $X_f \cap U_i \cap U_j$. Hence by (i), there exists $n_2 > 0$ such that $(t'_i - t'_j) \otimes f^{n_2} = 0$ in $\mathcal{F} \otimes \mathcal{L}^{\otimes(n_1+n_2)}(U_i \cap U_j)$, for all $i, j$. Hence the sections $t'_i \otimes f_i^{n_2}$ of $\mathcal{F} \otimes \mathcal{L}^{\otimes(n_1+n_2)}$ agree on the intersections $U_i \cap U_j$, and patch up to a global section of $\mathcal{F} \otimes \mathcal{L}^{\otimes n}$, with $n = n_1 + n_2$.

Note that the quasi compactness and finiteness assumptions of the lemma are valid if $X$ is Noetherian. If $X = \operatorname{Proj} R$, where $R$ is generated by a finite subset of $R_1$ as an algebra over $R_0$, the lemma easily implies that for any homogeneous element $f$ of degree 1 in $R$, which we may regard as a section of $\mathcal{L} = \mathcal{O}_{\operatorname{Proj} R}(1)$, we have a natural isomorphism $\widetilde{\Gamma_*(\mathcal{F})}(D_+(f)) \cong \mathcal{F}(D_+(f))$, for any quasi-coherent sheaf $\mathcal{F}$ (the finite generation assumption on $R$ implies that the hypotheses of the lemma are valid; note that for any homogeneous element $f$ of degree 1 in $R$, we have $\mathcal{L}|_{D_+(f)} \cong \mathcal{O}_{D_+(f)}$).

In particular, for any ring $A$, any closed subscheme $X \subset \mathbb{P}_A^n$ is defined by a graded ideal $I \subset A[X_0, \ldots, X_n]$. Indeed, if $\mathcal{I}$ is the ideal sheaf of $X$, and $I = \Gamma_*(\mathcal{I})$, then $I \subset \Gamma_*(\mathcal{O}_{\operatorname{Proj} R}) = R$ defines the subscheme of $\operatorname{Proj} R$ with ideal sheaf $\tilde{I} = \mathcal{I}$. Thus $X$ is projective over a ring $A$ if and only if $X = \operatorname{Proj} R$ where $R$ is a graded $A$-algebra generated over $R_0 = A$ by a finite set of elements of degree 1.

**Theorem D.49** (Serre) *Let $X$ be a projective scheme over a Noetherian ring $A$, and let $\mathcal{F}$ be a coherent $\mathcal{O}_X$-module. Then there exists an integer $n_0$ such that $\mathcal{F}(n)$ is generated by global sections for all $n \geq n_0$.*

**Proof.** Let $i : X \hookrightarrow \mathbb{P}_A^n$ be an embedding as a closed subscheme. Then $i_*\mathcal{F}$ is coherent on $\mathbb{P}_A^n$. So we reduce to the case when $X = \mathbb{P}_A^n$. Let $\mathbb{P}_A^n = \operatorname{Proj} R$, where $R = A[X_0, \ldots, X_n]$ is the homogeneous coordinate ring of $\mathbb{P}_A^n$. Let $U_i = D_+(X_i)$, so that the $U_i$ are a finite covering of $\mathbb{P}_A^n$ by affine open sets. Let $\mathcal{F}(U_i) = M_i$, so that $M_i$ is a finite $R_{(X_i)}$-module for each $i$. Choose finite sets of generators $m_{ij}$ of $M_i$. Then there exists an $n_0$ such that for each $n \geq n_0$, we have (by the above lemma) that $m_{ij} \otimes X_i^n$ extends to a global section of $\mathcal{F}(n)$ for each $i, j$; now one sees at once that these sections globally generate $\mathcal{F}$.

**Corollary D.50** *Let $X$, $\mathcal{F}$ be as above. Then $\mathcal{F}$ is a quotient of $\mathcal{O}_X(-n)^{\oplus m}$ for any sufficiently large $n > 0$, and a suitable $m > 0$ (depending only on $\mathcal{F}$).*

The applicability of the above theorem and corollary are extended by the following lemma, which is also useful elsewhere.

**Lemma D.51** *Let $X$ be a Noetherian scheme, $U \subset X$ an open subscheme, and let $\mathcal{F}$ be a coherent $\mathcal{O}_U$-module. Then there exists a coherent $\mathcal{O}_X$-module $\overline{\mathcal{F}}$ such that $\overline{\mathcal{F}}|_U \cong \mathcal{F}$.*

**Proof.** First suppose $X = \operatorname{Spec} A$ is affine. If $j : U \to X$ is the inclusion, then $j_*\mathcal{F}$ is quasi-coherent, hence of the form $\widetilde{M}$, and $\widetilde{M}|_U = \mathcal{F}$. If $f_1, \ldots, f_r \in A$ such that $\cup_i D(f_i) = U$, we are given that $M_{f_i} = \mathcal{F}(D(f_i))$ is a finite $A_{f_i}$-module for each $i$. Choose a finite subset $S$ of $M$ containing generators for each of the modules $M_{f_i}$, and let $N$ be the $A$-submodule of $M$ generated by $S$. Then $\widetilde{N}$ is coherent, and $\widetilde{N} \hookrightarrow \widetilde{M}$, so that $\widetilde{N}|_V \hookrightarrow \widetilde{M}|_V$ for any open set $V \subset X$. In particular, $\widetilde{N}|_{D(f_i)} = \widetilde{N_{f_i}} \hookrightarrow \widetilde{M_{f_i}}$, while $N_{f_i} \twoheadrightarrow M_{f_i}$. Hence $\widetilde{N_{f_i}} = \widetilde{M_{f_i}}$, and so $\widetilde{N}|_U = \mathcal{F}$.

In general, work by induction on the number of affine open subsets of $X$ needed to cover $X - U$. If there exists an affine open subset $V$ with $U \cup V = X$, then $\mathcal{F}|_{U \cap V}$ extends to a coherent sheaf $\mathcal{G}$ on $V$; now $\mathcal{F}$ and $\mathcal{G}$ patch up to give a coherent sheaf on $X$ (see (D.9)). In the inductive step, suppose $U \cup U_1 \cup \cdots \cup U_n = X$; by induction, $\mathcal{F}$ has a coherent extension $\mathcal{G}$ to $U \cup U_1 \cup \cdots \cup U_{n-1}$; by the first case, $\mathcal{G}$ now has a coherent extension to $X$.

**Corollary D.52** *Suppose $X$ is a quasi-projective scheme over a Noetherian ring $A$, and $\mathcal{F}$ is a coherent sheaf on $X$. Let $f : X \to \mathbb{P}^n_A$ be an immersion onto a locally closed subset of $\mathbb{P}^n_A$, and let $\mathcal{L} = f^*\mathcal{O}_{\mathbb{P}^n}(1)$. Then there exists $n_0$ such that $\mathcal{F} \otimes_{\mathcal{O}_X} \mathcal{L}^{\otimes n}$ is generated by global sections for all $n \geq n_0$. In particular, $\mathcal{F}$ is a quotient of a locally free $\mathcal{O}_X$-module of finite rank.*

(D.53) There is another situation in which any coherent sheaf is a quotient of a direct sum of invertible sheaves. Recall (see (D.25)) that a Noetherian scheme $X$ is called *regular* if each of the local rings $\mathcal{O}_{X,x}$ are regular local rings (i.e., $\mathcal{M}_x/\mathcal{M}_x^2$ is a $k(x)$-vector space of dimension equal to the Krull dimension of $\mathcal{O}_{X,x}$). If $x \in X^1$ is any point of codimension 1 in a regular scheme $X$, then $\mathcal{O}_{X,x}$ is a regular local ring of dimension 1, i.e., a discrete valuation ring. If $x \in X^1$, its closure $D \subset X$ (with the reduced structure) is a subscheme of $X$, such that for any $y \in D$, the stalk of the ideal sheaf $\mathcal{I}_{D,y} \subset \mathcal{O}_{X,y}$ is a prime ideal of height 1 (since $\mathcal{O}_{X,y}/\mathcal{I}_{D,y} \cong \mathcal{O}_{D,y}$ which is an integral domain, $\mathcal{I}_{D,y}$ is a prime ideal; further the 1-dimensional local ring $\mathcal{O}_{X,x}$ is the localization of $\mathcal{O}_{X,y}$ at $\mathcal{I}_{D,y}$). Since $\mathcal{O}_{X,y}$ is regular, hence is a unique factorization domain, any prime ideal of height 1 is a principal ideal. Hence $\mathcal{I}_D$ is a coherent $\mathcal{O}_X$-module, all of whose stalks are free modules of rank 1, i.e., $\mathcal{I}_D$ is an invertible $\mathcal{O}_X$-module. Define $\mathcal{O}_X(D) = \mathcal{H}om_{\mathcal{O}_X}(\mathcal{I}_D, \mathcal{O}_X)$ for any irreducible $D$. The inclusion $\mathcal{I}_D \subset \mathcal{O}_X$ gives a global section $s$ of $\mathcal{O}_X(D)$, hence an $\mathcal{O}_X$-linear map $\mathcal{O}_X \to \mathcal{O}_X(D)$ which is an isomorphism on $X - D$, and restricts to 0 on $D$.

Let $X$ be a Noetherian regular scheme. A *divisor* on $X$ is an element of the free Abelian group on $X^1$. A divisor is called *effective* if it is a positive linear combination of elements of $X^1$. If $D = \sum_{i=1}^{r} n_i D_i$ is any effective divisor, we may associate to it the invertible sheaf $\mathcal{O}_X(D) = \mathcal{O}_X(D_1)^{\otimes n_1} \otimes \cdots \otimes \mathcal{O}_X(D_r)^{\otimes n_r}$. If $s_i$ is the canonical global section of $\mathcal{O}_X(D_i)$ described above, for each $i$, then $s_1^{\otimes n_1} \otimes \cdots \otimes s_r^{\otimes n_r}$ is a canonically defined global section of $\mathcal{O}_X(D)$, which gives an isomorphism $\mathcal{O}_X \to \mathcal{O}_X(D)$ on $X - \cup_i D_i$, and restricts to 0 on $\cup_i D_i$. This inclusion $\mathcal{O}_X \to \mathcal{O}_X(D)$ also identifies the coherent sheaf $\mathcal{H}om_{\mathcal{O}_X}(\mathcal{O}_X(D), \mathcal{O}_X)$ with a sheaf of ideals $\mathcal{I}_D \subset \mathcal{O}_X$, which defines a canonical structure on $D$ of a closed subscheme of $X$.

Any irreducible closed proper subset of $X$ is contained in an irreducible divisor. We deduce that if $X$ is a Noetherian regular scheme, then there is a basis of open sets in $X$ consisting of sets $X_s = \{x \mid s : \mathcal{O}_X \to \mathcal{L}$ restricts to 0 on precisely the complement of $X_s\}$, for some section $s$ of an invertible $\mathcal{O}_X$-module $\mathcal{L}$.

The following is a particular case of a result of Kleiman.

**Theorem D.54** *Let $X$ be a Noetherian regular scheme. Then any coherent $\mathcal{O}_X$-module $\mathcal{F}$ is a quotient of a direct sum of invertible $\mathcal{O}_X$-modules.*

**Proof.** Suppose given an invertible $\mathcal{O}_X$-module $\mathcal{L}$ and a global section $s$. Let $X_s$ be the corresponding open subset of $X$. Let $a \in \mathcal{F}(X_s)$. Then by lemma (D.48), we see that $s^{\otimes n} \otimes a$ is the restriction to $X_s$ of a global section of $\mathcal{F} \otimes \mathcal{L}^{\otimes n}$. This yields a morphism $(\mathcal{L}^*)^{\otimes n} \to \mathcal{F}$, such that the image of the restriction to $X_s$ contains $a$. Since $X$ is Noetherian, hence quasi-compact, and $\mathcal{F}$ is coherent, we can find a finite number of such open sets $X_{s_i}$ (corresponding to invertible sheaves $\mathcal{L}_i$ and $s_i \in \mathcal{L}_i(X)$) and sections $a_i \in \mathcal{F}(X_{s_i})$, such that these sections generate the stalks of $\mathcal{F}$ at any point of $X$; now the corresponding map $\oplus_i (\mathcal{L}_i^*)^{\otimes n_i} \to \mathcal{F}$ is surjective.

**Corollary D.55** *Let $X$ be a Noetherian regular scheme of Krull dimension $\leq n$. Then for any coherent $\mathcal{O}_X$-module $\mathcal{F}$, there exists an $\mathcal{O}_X$-linear resolution*

$$0 \to \mathcal{E}_n \to \mathcal{E}_{n-1} \to \cdots \to \mathcal{E}_0 \to \mathcal{F} \to 0$$

*where the $\mathcal{E}_i$ are locally free of finite rank.*

**Proof.** From the above lemma, there exists such a resolution where the $\mathcal{E}_i$ are locally free for $i < n$, and $\mathcal{E}_n$ is coherent. To show that $\mathcal{E}_n$ is locally free, it suffices to show that for each $x \in X$, the stalk $(\mathcal{E}_n)_x$ is a free $\mathcal{O}_{X,x}$-module; here the stalks $(\mathcal{E}_i)_x$ are free for $i < n$. Since $\mathcal{O}_{X,x}$ is a regular local ring of Krull dimension $\leq n$, the finite $\mathcal{O}_{X,x}$-module $\mathcal{F}_x$ has projective dimension $\leq n$. This means that for any resolution

$$0 \to M \to F_{n-1} \to \cdots \to F_0 \to \mathcal{F}_x \to 0$$

where the $F_i$ are free $\mathcal{O}_{X,x}$-modules, the module $M$ is also free.

(D.56)   On any Noetherian scheme $X$, we say that a coherent $\mathcal{O}_X$-module $\mathcal{F}$ is of *finite homological dimension* if it has a finite resolution by locally free $\mathcal{O}_X$-modules of finite rank. If $\mathcal{F}$ has finite homological dimension, then for each $x \in X$, the stalk $\mathcal{F}_x$ has finite projective dimension over $\mathcal{O}_{X,x}$. If $X$ is a scheme on which any coherent sheaf is a quotient of a locally free sheaf of finite rank (for example, a scheme which is quasi-projective over a Noetherian ring, or a locally closed subscheme of a regular scheme), then $\mathcal{F}$ has finite homological dimension precisely when its stalks have finite projective dimension.

(D.57)   An invertible sheaf $\mathcal{L}$ on a Noetherian scheme $X$ is *ample* if for any coherent sheaf $\mathcal{F}$, there exists an integer $n_0$ such that $\mathcal{F} \otimes \mathcal{L}^{\otimes n}$ is generated by its global sections, for any $n \geq n_0$.

One example of an ample invertible sheaf is the following: if $X$ is quasi-projective over a Noetherian ring $A$, and $f : X \to \mathbb{P}_A^n$ is a locally closed immersion in $\mathbb{P}_A^n$, then $f^*\mathcal{O}_{\mathbb{P}^n}(1)$ is an ample invertible sheaf on $X$, by corollary (D.52). If $X$ is of finite type over a Noetherian ring $A$, then for any ample invertible sheaf $\mathcal{L}$ on $X$, some positive tensor power $\mathcal{L}^{\otimes n}$ is of this form ([H], II, Theorem 7.6). Clearly, if there is an ample invertible sheaf on $X$, then any coherent $\mathcal{O}_X$-module is a quotient of a locally free $\mathcal{O}_X$-module of finite rank.

(D.58)   The notions of quasi-coherent and coherent sheaves allow one to 'relativize' the notions of the spectrum of a ring, and of Proj of a graded ring, as follows. If $f : Z \to X$ is an affine morphism (i.e., $f^{-1}(U)$ is affine for any affine open set $U$) then $\mathcal{S} = f_*\mathcal{O}_Z$ is a quasi-coherent sheaf of $\mathcal{O}_X$-algebras, such that for any affine open set $U \subset X$, we have $f^{-1}(U) \cong \operatorname{Spec} \mathcal{S}(U)$. Conversely, given any quasi-coherent $\mathcal{O}_X$-algebra $\mathcal{S}$, we may define an $X$-scheme $Z = \mathbf{Spec}\,\mathcal{S}$ by taking patching together the affine schemes $Z_U = \operatorname{Spec} \mathcal{S}(U)$ for various affine open subsets $U$ of $X$ (see [H], II, Ex. 5.17). If $f : Z = \mathbf{Spec}\,\mathcal{S} \to X$ is the natural map, then $f_*$ induces an equivalence of categories from the category of quasi-coherent $\mathcal{O}_Z$-modules to the category of $\mathcal{S}$-modules which are quasi-coherent as $\mathcal{O}_X$-modules.

Similarly, suppose $\mathcal{S} = \oplus_{n \geq 0} \mathcal{S}_n$ is a quasi-coherent $\mathcal{O}_X$-algebra, where (i) $\mathcal{S}_0 = \mathcal{O}_X$ (ii) $\mathcal{S}_d$ is a coherent $\mathcal{O}_X$-module for each $d$ (iii) $\mathcal{S}$ is generated by $\mathcal{S}_1$ as an $\mathcal{O}_X$-algebra. Then we may define (see [H], II, §7) an $X$-scheme $\mathbf{Proj}\,\mathcal{S}$, with a structure morphism $f : \mathbf{Proj}\,\mathcal{S} \to X$, as follows: if $U \subset X$

is affine, then we may form the $\mathcal{O}_X(U)$-scheme $\operatorname{Proj} \mathcal{S}(U)$; these schemes glue together in a natural way to give the scheme **Proj** $\mathcal{S}$. The invertible sheaves $\mathcal{O}(1)$ on the various local schemes $\operatorname{Proj} \mathcal{S}(U)$ also glue to give an invertible sheaf $\mathcal{O}_{\mathbf{Proj}\,\mathcal{S}}(1)$ on **Proj** $\mathcal{S}$.

(D.59) Two applications of the above relativization of Spec and Proj are the following. A *geometric vector bundle* (in the terminology of Hartshorne's book; see II, Ex. 5.18) of rank $n$ on a scheme $X$ is a scheme $V$ with a morphism $\pi : V \to X$, such that for some open covering $\{U_i\}$ of $X$, we are given isomorphisms of $U_i$-schemes $\varphi_i : \pi^{-1}(U_i) \xrightarrow{\cong} \mathbb{A}^n_{U_i}$, such that the induced 'glueing' automorphisms $\varphi_j \circ \varphi_i^{-1}$ of $\mathbb{A}^n_{U_i \cap U_j}$ are linear along the fibers, i.e., are given by an element of $\operatorname{GL}_n(\mathcal{O}_X(U_i \cap U_j))$. Two such 'local triviality' data $\{U_i, \varphi_i\}$ and $\{V_j, \psi_i\}$ define the same structure of a vector bundle on $V$ if their union defines a structure of a vector bundle. A morphism of vector bundles on $X$ may be defined to be a morphism of $X$-schemes which is linear on the fibers, in the obvious sense defined using the local trivializations.

If $\pi : V \to X$ is a vector bundle, then the sheaf of sections of $\pi$, which is *a priori* a sheaf of sets, has a natural structure as a locally free $\mathcal{O}_X$-module of rank $n$; locally, this corresponds to the natural identification of the sheaf of sections of $\mathbb{A}^n_U \to U$ with $\mathcal{O}_U^{\oplus n}$. Conversely, given a locally free $\mathcal{O}_X$-module $\mathcal{E}$ of rank $n$, we may associate to it a vector bundle $\mathbb{V}(\mathcal{E}) = \mathbf{Spec}(\operatorname{Sym}(\mathcal{E}))$, where $\operatorname{Sym}(\mathcal{E})$ is the symmetric algebra of $\mathcal{E}$ over $\mathcal{O}_X$. Locally, if we choose an isomorphism $\mathcal{E} \mid_U \cong \mathcal{O}_U^{\oplus n}$, the symmetric algebra is identified with a polynomial algebra in $n$-variables, and we get an isomorphism $\mathbb{V}(\mathcal{E}) \mid_U \cong \mathbb{A}^n_U$; one checks easily that these local isomorphisms determine a structure of a vector bundle on $\mathbb{V}(\mathcal{E})$. Finally, one can show that the sheaf of sections of $\mathbb{V}(\mathcal{E})$ is naturally isomorphic to $\mathcal{E}^*$, the dual locally free sheaf. These construction provide an anti-equivalence between the categories of locally free sheaves (of constant rank) on $X$ and vector bundles on $X$. Thus one very often uses the term 'vector bundle' to mean 'locally free sheaf'. This convention is followed in the main text; the only place where 'vector bundle' is used to mean a geometric vector bundle is in Proposition (5.17).

In a similar fashion, if $\mathcal{E}$ is a locally free sheaf of rank $r$ on $X$, we may define the *projective bundle* associated to $\mathcal{E}$ to be $\mathbb{P}_X(\mathcal{E}) = \mathbb{P}(E) = \mathbf{Proj}\,(\operatorname{Sym}(\mathcal{E}))$. If $\pi : \mathbb{P}(\mathcal{E}) \to X$ is the natural map (the structure morphism) then there are natural isomorphisms $\operatorname{Sym}^n(\mathcal{E}) \cong \pi_* \mathcal{O}_{\mathbb{P}(\mathcal{E})}(n)$ for $n \geq 0$, which are obtained by applying $\pi_*$ to natural surjections $\pi^* \operatorname{Sym}^n(\mathcal{E}) \to \mathcal{O}_{\mathbb{P}(\mathcal{E})}(n)$; and $\pi_* \mathcal{O}_{\mathbb{P}(\mathcal{E})}(n) = 0$ for $n < 0$. The $X$-scheme $\mathbb{P}(\mathcal{E})$ has the following universal property: for any $X$-scheme $Y$, with structure morphism $f : Y \to X$, the $X$-morphisms $g : Y \to \mathbb{P}(\mathcal{E})$ are in bijection with isomorphism classes of surjections $f^*\mathcal{E} \to \mathcal{L}$ for invertible sheaves $\mathcal{L}$ (two such surjections are isomorphic if they have the same kernel).

## Cohomology and Direct Images of Quasi-coherent and Coherent Sheaves

(D.60)   As observed earlier (see (D.43)(iv)), if $f : X \to Y$ is a morphism between Noetherian schemes, then the higher direct images $R^i f_* \mathcal{F}$ of a quasi-coherent $\mathcal{O}_X$-module $\mathcal{F}$ are quasi-coherent, such that for any affine open subscheme $U = \operatorname{Spec} A$ of $Y$, we have $R^i f_* \mathcal{F}\,|_U = H^i(f^{-1}(U), \mathcal{F}\,|_{f^{-1}(U)})^{\sim}$.

On a Noetherian affine scheme $X = \operatorname{Spec} A$, one has $H^i(X, \mathcal{F}) = 0$ for any quasi-coherent sheaf $\mathcal{F}$. This is proved as follows (see [H], III, §3). First we claim that if $I$ is an injective $A$-module, then $\widetilde{I}$ is flasque. One uses the Artin-Rees lemma to see that for $f \in A$, the natural localization map $I \to I_f$ is surjective (i.e., $I$ is $f$-divisible); using this and Noetherian induction, one deduces that $\widetilde{I}$ is flasque. Now for any quasi-coherent sheaf $\mathcal{F}$, writing $\mathcal{F} = \widetilde{M}$, let

$$0 \to M \to I_0 \to I_1 \to \cdots$$

be a resolution by injective $A$-modules. Then the corresponding resolution of $\mathcal{O}_X$-modules

$$0 \to \widetilde{M} \to \widetilde{I_0} \to \widetilde{I_1} \to \cdots$$

is a flasque resolution, so that it may be used to compute cohomology; taking global sections, we recover the given injective resolution, so that $H^i(X, \mathcal{F}) = 0$ as claimed.

(D.61)   From Leray's theorem, we deduce that if $X$ is separated over a Noetherian ring $A$, then for any quasi-coherent sheaf $\mathcal{F}$, there is a natural isomorphism of $H^i(X, \mathcal{F})$ with the Čech cohomology group $\check{H}^i(\mathcal{U}, \mathcal{F})$ where $\mathcal{U} = \{U_i\}$ is a covering of $X$ by affine open subschemes. This is because the separatedness ensures that all finite intersections of the $U_i$ are affine schemes, on which quasi-coherent sheaves have vanishing $H^i$ (for $i > 0$). A direct proof of this isomorphism which does not appeal to Leray's theorem is given in [H] (III, 4.5).

The computation using Čech cohomology has the following consequence. Let $f : X \to Y$ be a separated morphism of finite type between Noetherian schemes, such that $d$ is the maximum of the dimensions of the fibers of $f$; then for any quasi-coherent $\mathcal{O}_X$-module $\mathcal{F}$, we have $R^i f_* \mathcal{F} = 0$ for $i > d$. This is local on the base; it suffices to show that for a suitable affine open covering $\{U_i\}$ of $Y$, the scheme $f^{-1}(U_i)$ has an affine open covering such that all $d+2$-fold intersections are empty. The existence of such coverings (for 'sufficiently small $U_i$') can be deduced from the assumption on dimensions of the fibers.

A basic theorem of Grothendieck is the following.

**Theorem D.62** *Let $f : X \to Y$ be a proper morphism between Noetherian schemes, and let $\mathcal{F}$ be a coherent $\mathcal{O}_X$-module. Then the $\mathcal{O}_Y$-modules $R^i f_* \mathcal{F}$ are coherent $\mathcal{O}_Y$-modules.*

We will outline a proof of this theorem in the case when $f$ is a projective morphism, i.e., when there is a factorization $f = \pi \circ i$ for a closed immersion $i : X \to \mathbb{P}^n_Y$, where $\pi : \mathbb{P}^n_Y \to Y$ is the structure morphism (see EGA III, 3.2.1 for the general case). We reduce immediately to the case when $X = \mathbb{P}^n_Y$, since $i_*$ is exact and preserves coherence and 'flasqueness'. Further, the theorem is local on the base, so we reduce to the case when $Y = \operatorname{Spec} A$ for a Noetherian ring $A$. In this case, the theorem amounts to the statement that $H^i(\mathbb{P}^n_A, \mathcal{F})$ is a finite $A$-module for each $n$.

We first state a result giving a computation of the cohomology groups of the invertible sheaves $\mathcal{O}_{\mathbb{P}^N_A}(n)$ (see [H], III, 5.1). Recall that the tensor product makes $\oplus_{n \geq 0} \mathcal{O}_{\mathbb{P}^N_A}(n)$ into a sheaf of graded $\mathcal{O}_{\mathbb{P}^N_A}$-algebras, making $\oplus_{n \geq 0} H^0(\mathbb{P}^N_A, \mathcal{O}_{\mathbb{P}^N_A}(n))$ into a graded $A$-algebra.

**Theorem D.63** *(a) The graded $A$-algebra $\oplus_{n \geq 0} H^0(\mathbb{P}^N_A, \mathcal{O}_{\mathbb{P}^N_A}(n))$ is naturally isomorphic to a polynomial algebra $A[X_0, \ldots, X_N]$; thus*

$$H^0(\mathbb{P}^N_A, \mathcal{O}_{\mathbb{P}^N_A}(n)) \cong A^{\oplus \binom{N+n}{n}}.$$

*We also have $H^0(\mathbb{P}^N_A, \mathcal{O}_{\mathbb{P}^N_A}(n)) = 0$ for $n < 0$.*

*(b) $H^i(\mathbb{P}^N_A, \mathcal{O}_{\mathbb{P}^N_A}(n)) = 0$ for $0 < i < N$ for all $n \in \mathbb{Z}$.*

*(c) $H^N(\mathbb{P}^N_A, \mathcal{O}_{\mathbb{P}^N_A}(-N-1))$ is a free $A$-module of rank 1, and the natural pairings*

$$H^0(\mathbb{P}^N_A, \mathcal{O}_{\mathbb{P}^N_A}(n)) \otimes_A H^N(\mathbb{P}^N_A, \mathcal{O}_{\mathbb{P}^N_A}(-N-1-n)) \to H^N(\mathbb{P}^N_A, \mathcal{O}_{\mathbb{P}^N_A}(-N-1)) \cong A$$

*are perfect pairings between free $A$-modules.*

The idea of the proof is to compute cohomology as Čech cohomology of the graded quasi-coherent sheaf $\mathcal{S} = \oplus_{n \in \mathbb{Z}} \mathcal{O}_{\mathbb{P}^N_A}(n)$ using the standard affine open covering of $\mathbb{P}^N_A$ by $N+1$ affine open subsets $D_+(X_i)$. The Čech complex of $\mathcal{S}$ has an explicit description in terms of localized modules over the polynomial ring $S = A[X_0, \ldots, X_N]$,

$$\check{C}^p(\mathcal{U}, \mathcal{S}) = \oplus_{0 \leq i_0 < \ldots < i_p \leq N} S_{X_{i_0} X_{i_1} \cdots X_{i_p}},$$

with differentials given by alternating sums of natural localization maps. From this explicit description, one may compute the cohomology directly by calculations with polynomials, or more elegantly as in [H].

(D.64)   Now we prove that $H^i(\mathbb{P}^N_A, \mathcal{F})$ is a finitely generated $A$-module, for any coherent sheaf $\mathcal{F}$ on $\mathbb{P}^N_A$. As seen earlier (see (D.50)), we may write $\mathcal{F}$ as a quotient of $\mathcal{O}_{\mathbb{P}^N_A}(-m)^{\oplus r}$ for some sufficiently large $m$, $r$, giving an exact sequence of coherent $\mathcal{O}_{\mathbb{P}^N_A}$-modules

$$0 \to \mathcal{G} \to \mathcal{O}_{\mathbb{P}^N_A}(-m)^{\oplus r} \to \mathcal{F} \to 0$$

From the induced long exact sequence of cohomology groups, we see that $H^i(\mathbb{P}^N_A, \mathcal{F})$ is finitely generated $\Leftrightarrow H^{i+1}(\mathbb{P}^n_A, \mathcal{G})$ is finitely generated. Since $H^i(\mathbb{P}^N_A, \mathcal{F}) = 0$ for any coherent $\mathcal{F}$ for $i > N$ (using the Čech complex for the standard open covering, for example), we see by descending induction on $i$ that $H^i(\mathbb{P}^N_A, \mathcal{F})$ is finitely generated for all coherent sheaves $\mathcal{F}$.

(D.65)   A similar argument shows that for any coherent sheaf $\mathcal{F}$, there is an integer $n_0$ such that for $n \geq n_0$, we have $H^i(\mathbb{P}^N_A, \mathcal{F}(n)) = 0$ for all $n \geq n_0$, for all $i > 0$. Indeed, the above exact sequence shows that for $n \geq m - N$, and any $i > 0$, there is an isomorphism $H^i(\mathbb{P}^N_A, \mathcal{F}(n)) \cong H^{i+1}(\mathbb{P}^N_A, \mathcal{G}(n))$. Again the result follows by descending induction.

(D.66)   We deduce from the above results that for any locally free sheaf $\mathcal{E}$ of rank $N+1$ on a Noetherian scheme $X$, if $\pi : \mathbb{P}(\mathcal{E}) \to X$ is the structure morphism, then

(i)  there are natural isomorphisms $\mathrm{Sym}^n(\mathcal{E}) \cong \pi_*\mathcal{O}_{\mathbb{P}(\mathcal{E})}(n)$, for $n \geq 0$, and $\pi_*\mathcal{O}_{\mathbb{P}(\mathcal{E})}(n) = 0$ for $n < 0$;

(ii) $R^i\pi_*\mathcal{O}_{\mathbb{P}(\mathcal{E})}(n) = 0$ for all $n \in \mathbb{Z}$ and $0 < i < N$;

(iii) there is a natural isomorphism $R^N\pi_*\mathcal{O}_{\mathbb{P}(\mathcal{E})}(-N-1) \to \mathcal{O}_X$, and the pairings

$$\pi_*\mathcal{O}_{\mathbb{P}(\mathcal{E})}(n) \otimes_{\mathcal{O}_X} R^N\pi_*\mathcal{O}_{\mathbb{P}(\mathcal{E})}(-N-1-n) \to R^N\pi_*\mathcal{O}_{\mathbb{P}(\mathcal{E})}(-N-1) \cong \mathcal{O}_X$$

are perfect pairings between locally free $\mathcal{O}_X$-modules;

(iv) for any coherent $\mathcal{O}_{\mathbb{P}(\mathcal{E})}$-module $\mathcal{F}$, the sheaves $R^i\pi_*\mathcal{F}(n)$ are coherent for any $i \geq 0$ and any $n \in \mathbb{Z}$, and for some $n_0 > 0$, they vanish for all $i > 0$ and $n \geq n_0$.

## Some Miscellaneous Topics

In this section we prove two lemmas about varieties, i.e., schemes of finite type over a field, which are needed in the text. We also discuss the construction of quasi-coherent sheaves using faithfully flat descent.

(D.67)   Let $B$ be an $A$-algebra. If $I = \ker(B \otimes_A B \xrightarrow{\mu} B)$, where $\mu$ is the multiplication map, then we define $\Omega_{B/A} = I/I^2$ to be the module of *Kähler differentials* of $B$ relative to $A$. There is an $A$-linear derivation $d : B \to \Omega_{B/A}$ (i.e., a homomorphism of $A$-modules satisfying $d(b_1 b_2) = b_1\, db_2 + b_2\, db_1$), given by the formula $db = b \otimes 1 - 1 \otimes b$. One verifies that this is universal for $A$-linear derivations from $B$ to $B$-modules.

If $L/K$ is a finitely generated extension of fields, then $\Omega_{L/K}$ is a finite dimensional $L$-vector space of dimension at least equal to the transcendence degree of $L$ over $K$; equality holds $\Leftrightarrow L$ is separably generated over $K$. The formation of $\Omega_{B/A}$ commutes with localization; hence if $A \to B$ is an inclusion of integral domains which are finitely generated $k$-algebras, and if $\dim B - \dim A = n$, then $\Omega_{B/A}$ is of rank $\geq n$, with equality if and only if the quotient field of $B$ is separably generated over that of $A$. Thus for any prime ideal $\mathcal{P}$ of $B$ with residue field $k(\mathcal{P})$, we have that $\dim \Omega_{B_\mathcal{P}/A} \otimes k(\mathcal{P}) \geq n$, and the set of primes for which equality holds is an open subset (perhaps empty) of $\operatorname{Spec} B$.

There are two standard exact sequences which are useful in working with Kähler differentials (see Matsumura, *Commutative Algebra*, Benjamin (1970), Th. 57 and Th. 58). First, if $B$ is an $A$-algebra, and $C$ is a $B$-algebra (and hence also an $A$-algebra), we have an exact sequence

$$C \otimes_B \Omega_{B/A} \to \Omega_{C/A} \to \Omega_{C/B} \to 0$$

where the maps are induced by the universal property of Kähler differentials. Next, if $B$ is an $A$-algebra, and $I \subset B$ is an ideal, then there is an exact sequence

$$I/I^2 \xrightarrow{\delta} \Omega_{B/A} \otimes_B B/I \to \Omega_{(B/I)/A} \to 0$$

where $\delta(f \pmod{I^2}) = df \otimes 1 \in \Omega_{B/A} \otimes B/I$.

Now we discuss the notion of smoothness. If $k$ is a field, and $A$ is an integral domain which is a finitely generated $k$-algebra of dimension $d$, we define $A$ to be *smooth* over $k$ if its module of Kähler differentials $\Omega_{A/k}$ is a projective $A$-module of rank $d$. From the second exact sequence above, one can show that this is equivalent to the following: if $A = k[X_1, \ldots, X_n]/(f_1, \ldots, f_r)$, then the minors of size $n - d + 1$ of the Jacobian matrix $[\frac{\partial f_i}{\partial X_j}]$ vanish in $A$ (i.e., lie in the ideal $(f_1, \ldots, f_r)$), and the minors of size $n - d$ generate the unit ideal in $A$. If $A$ is smooth over $k$, then $A$ is regular (see Matsumura, *Commutative Algebra*, Sec. 29).

More generally, if $S$ is a $k$-scheme of finite type, and $f : X \to S$ is of finite type, we say that $f$ is *smooth of relative dimension $d$* if (i) $f$ is flat (see (D.40)) (ii) if $X' \subset X$, $S' \subset S$ are irreducible components, with

Appendix D: Results from Algebraic Geometry 333

$f(X') \subset S'$, then $\dim X' - \dim S' = d$ (iii) for each $x \in X$, we have

$$\dim_{k(x)} (\Omega^1_{X/S})_x \otimes_{\mathcal{O}_{X,x}} k(x) = d.$$

Here $\Omega^1_{X/S}$ is the sheaf of relative Kähler differentials (see [H], II, §8); it is a quasi-coherent sheaf such that if $U = \operatorname{Spec} B \subset X$ and $V = \operatorname{Spec} A \subset S$ are affine open subsets such that $f(U) \subset V$, then $\Omega^1_{X/S}\,|_U \cong \widetilde{\Omega_{B/A}}$.

If $f : X \to S$ is a morphism between integral smooth $k$-schemes of finite type, and $d = \dim X - \dim Y$, then $f$ is smooth of relative dimension $d$ if and only if $\Omega^1_{X/Y}$ is locally free of rank $d$ on $X$ (see [H], III, 10.4 and II, Ex. 8.1). For any morphism $f : X \to S$, there is a largest open subscheme $U \subset X$ (possibly empty) such that $f\,|_U$ is smooth, namely that where $\Omega^1_{X/Y}$ restricts to a locally free sheaf of rank $d$. We will use this to prove a special case of Bertini's theorem.

**Lemma D.68** *Let $A$ be a smooth finitely generated $k$-algebra of dimension $n$, and let $T \subset \operatorname{Spec} A$ be a finite subset. Then there exist $x_1, \ldots, x_{n-1} \in A$ and an element $f \in A$ such that $f \notin \cap_{\mathcal{P} \in T} \mathcal{P}$, and the morphism $\operatorname{Spec} A_f \to \mathbb{A}^{n-1}_k$ is smooth of relative dimension 1. If $k$ is infinite, and $A$ is generated by elements $y_1, \ldots, y_m$, then the $x_i$ may be chosen to be 'general' $k$-linear combinations of the $y_j$.*

(Here 'general' means 'for all elements in a nonempty Zariski open subset of $k^{mn}$'.)

**Proof.** Since the subset of $X$ where a morphism to $\mathbb{A}^{n-1}$ is smooth is open, it suffices to consider the case when $T$ consists of closed points. Since $A$ is smooth over $k$, for any $x \in T$ we have $\dim_{k(x)}(\Omega^1_{X/k})_x \otimes k(x) = n$. Let $R$ be the semilocal ring of $A$ at the finite set of points of $T$ (the intersection of the local rings of $A$ at the points). Let $J$ be the Jacobson radical of $R$, and let $\overline{R} = R/J^2$. Then $\Omega_{R/k} \otimes R/J \cong \Omega_{\overline{R}/k} \otimes R/J$. From the Chinese remainder theorem, $\overline{R} \cong \prod_{x \in T} \overline{R}_x$, and $\Omega_{\overline{R}/k} \cong \prod_x \Omega_{\overline{R}_x/k}$. The elements $\{dr_x \mid x_r \in R\}$ generate $\Omega_{\overline{R}_x/k}$; so we can find $(r_1)_x, \ldots, (r_n)_x$ in $\overline{R}_x$ which map to a basis of $\Omega_{\overline{R}_x/k} \otimes k(x) \cong k(x)^{\oplus n}$. By the Chinese Remainder theorem, we can find $r_1, \ldots, r_n \in \overline{R}$ lifting the various $(r_i)_x$, and can then lift the $r_j$ to elements $x_j \in A$ (since $R/J^2 \cong A/(J^2 \cap A)$). Let $X \to \mathbb{A}^{n-1}_k$ be the morphism determined by $x_1, \ldots, x_{n-1}$. We check easily that for $x \in T$, $(\Omega^1_{X/\mathbb{A}^{n-1}})_x \otimes k(x)$ has dimension 1, since (by the second exact sequence for Kähler differentials) it is the quotient of $(\Omega^1_{X/k})_x \otimes k(x)$ by the subspace spanned by the images of $dx_i$. Now $X \to \mathbb{A}^{n-1}_k$ is smooth in a non-empty open neighborhood of $T$, hence on $\operatorname{Spec} A_f$ for some $f$ which is a unit at points of $T$.

If $k$ is infinite, and the $y_j$ generate $A$, then the images of the $dy_j$ generate $\Omega_{A/k}$. Hence in the above argument, we may choose the $r_i$ to be $k$-linear combinations $\overline{x}_i = \sum_{ij} a_{ij} \overline{y}_j$ of the images of the $y_j$; the desired properties of the $r_i$ are equivalent to the non-vanishing of certain determinants which are non-zero polynomials in the $a_{ij}$.

We also prove a special case of the Noether normalization lemma.

**Lemma D.69** *Let $A$ be a $k$-algebra of dimension $n$ generated by $y_1, \ldots, y_m$, where $k$ is an infinite field. Then for 'general' $k$-linear combinations $x_1, \ldots, x_n$ of the $y_j$, the morphism $f : X = \operatorname{Spec} A \to \mathbb{A}_k^n$ with coordinates $x_i$ is finite.*

**Proof.** Let $X = \operatorname{Spec} A$. The generators $y_j$ determine a closed embedding $X \hookrightarrow \mathbb{A}_k^m$.

We may regard $\mathbb{A}_k^m$ as the open subscheme $D_+(Y_0)$ of $\mathbb{P}_k^m$, which has homogeneous coordinates $Y_0, \ldots, Y_m$, and $y_i = Y_i/Y_0$. Then $H = \mathbb{P}_k^m - \mathbb{A}_k^m \cong \mathbb{P}_k^{m-1}$ is the hyperplane defined by $Y_0 = 0$. Let $\overline{X} \subset \mathbb{P}_k^m$ be the closure of $X$ in $\mathbb{P}_k^m$. Let $Y = \overline{X} \cap H$. Now any proper subset of an irreducible Noetherian space has strictly smaller dimension. Hence $Y \subset H \cong \mathbb{P}^{m-1}$ is a projective subscheme of dimension $\leq n - 1$, since each irreducible component $X_i$ of $X$ has dimension $\leq n$, and $Y = \cup_i (\overline{X}_i - X_i)$ where $\dim \overline{X}_i - X_i < \dim \overline{X}_i = \dim X_i \leq n$.

If $x_1, \ldots, x_n$ are homogenous linear polynomials in the $y_j$, let $X_1, \ldots, X_n$ be the corresponding homogeneous linear polynomials in the $Y_j$ (thus $X_j = x_j Y_0$). Let $L \subset H$ be the closed subscheme (isomorphic to $\mathbb{P}^{m-n-1}$) defined by $Y_0 = X_1 = \cdots = X_n = 0$. The inclusion of graded rings $k[Y_0, X_1, \ldots, X_n] \hookrightarrow k[Y_0, \ldots, Y_m]$ induces a morphism

$$(\operatorname{Proj} k[Y_0, \ldots, Y_m]) - V((Y_0, X_1, \ldots, X_n)) \to \operatorname{Proj} k[Y_0, X_1, \ldots, X_n],$$

that is to say a morphism $p_L : \mathbb{P}_k^m - L \to \mathbb{P}_k^n$ (this is called the *projection from the linear subspace $L$*). Further, if $\mathbb{A}_k^n = \mathbb{P}_k^n - \{Y_0 = 0\}$, then $p_L^{-1}(\mathbb{A}_k^n) = \mathbb{A}_k^m$. The induced morphism $\overline{X} - L \to \mathbb{P}_k^n$ restricts to the morphism $f : X \to \mathbb{A}_k^n$ determined by the $x_i$.

If $L \cap \overline{X} = \phi$, then we get an induced morphism $p_L : \overline{X} \to \mathbb{P}_k^n$, which is hence a projective morphism. Hence $\overline{X} \cap p_L^{-1}(\mathbb{A}_k^n) \to \mathbb{A}_k^n$ is projective, i.e., $f : X \to \mathbb{A}_k^n$ is projective. In particular $f_* \mathcal{O}_X$ is a coherent $\mathcal{O}_{\mathbb{A}_k^n}$-module, i.e., $A$ is a finite module over the polynomial ring $k[x_1, \ldots, x_n]$.

So we are reduced to proving that for 'general' linear combinations $x_i$ of the $y_j$, the resulting linear space $L$ is disjoint from $\overline{X}$, that is to say, from $Y$.

This is done by a dimension count. We may regard the linear spaces $L$ as parametrized by the $k$-rational points of $Z = (\mathbb{A}_k^m - \{0\})^n$; consider the incidence scheme $\Gamma \subset H \times_k Z \cong \mathbb{P}_Z^{m-1}$ defined by the vanishing of the polynomials $\sum_{j=1}^m t_{ij} Y_j$, $1 \leq i \leq n$, where the $t_{ij}$, $1 \leq j \leq m$ are the coordinate functions on the $i^{\text{th}}$ factor of $\mathbb{A}^m - \{0\}$ in $Z$. One finds easily that under the projection $\Gamma \to H$, the fiber of $\Gamma$ over any point $x$ of $H$ is isomorphic to $(\mathbb{A}^{m-1}_{k(x)} - \{0\})^n$, since the condition that a hyperplane pass through a point is a linear relation between the coefficients of its defining equation. Hence the inverse image $\Gamma_Y$ of $Y$ in $\Gamma$ has dimension $\dim Y + n(m-1) \leq n - 1 + n(m-1) < mn = \dim Z$. Hence (see (D.30)) the closure of the image of $\Gamma_Y$ in $Z$ is a proper subscheme; for any $k$-rational point in the complement (i.e., for a non-empty Zariski open set in $k^{mn}$) the corresponding linear space $L$ is disjoint from $Y$.

(D.70) We end this section with a brief discussion of the theory of faithfully flat descent, used in the main text in the discussion of Severi-Brauer schemes. A morphism $f : U \to X$ is faithfully flat if it is flat and surjective; this is equivalent to requiring that any diagram of $\mathcal{O}_Y$-modules $\mathcal{F}' \to \mathcal{F} \to \mathcal{F}''$ is an exact sequence if and only if $f^*\mathcal{F}' \to f^*\mathcal{F} \to f^*\mathcal{F}''$ is exact. One example of a faithfully flat morphism is as follows: if $\{U_i\}$ is a covering of $X$ by open subschemes, and $U = \coprod U_i$ is the disjoint union of the $U_i$ with the natural scheme structure, then the natural morphism $U \to X$ is faithfully flat. The reader has seen several constructions (including the definition of a scheme) where local objects were defined together with patching data of some kind. For example, we have seen (see (D.9)) that one may define a sheaf on $X$ by giving sheaves $\mathcal{F}_i$ on $U_i$, and isomorphisms

$$\varphi_{ij} : \mathcal{F}_i \mid_{U_i \cap U_j} \to \mathcal{F}_j \mid_{U_i \cap U_j},$$

such that $\varphi_{jk} \circ \varphi_{ij} = \varphi_{ik}$ in

$$\operatorname{Hom}(\mathcal{F}_i \mid_{U_i \cap U_j \cap U_k}, \mathcal{F}_k \mid_{U_i \cap U_j \cap U_k}).$$

The $\mathcal{F}_i$ determine a sheaf $\widetilde{\mathcal{F}}$ on $U$. If $p_1, p_2$ are the two projections $U \times_X U \to U$, the $\varphi_{ij}$ amount to an isomorphism $\varphi : p_1^* \widetilde{F} \to p_2^* \mathcal{F}$, satisfying a 'cocycle condition' $p_{23}^* \varphi \circ p_{12}^* \varphi = p_{13}^* \varphi$ on $U \times_X U \times_X U$.

More generally, suppose $f : U \to X$ is a faithfully flat morphism. One aspect of the theory of descent gives an equivalence of categories between quasi-coherent $\mathcal{O}_X$-modules, and quasi-coherent $\mathcal{O}_U$-modules with *descent data*, i.e., $\mathcal{O}_U$-modules $\widetilde{\mathcal{F}}$ with isomorphisms $\varphi : p_1^* \widetilde{\mathcal{F}} \to p_2^* \widetilde{\mathcal{F}}$ satisfying the cocycle condition.

To prove this, one reduces at once to the case when $X$ is affine; replacing $U$ by an affine open cover, we reduce further to the case when $U$ is affine. Now we must show that if $B$ is a faithfully flat $A$-algebra, then the category of $A$-modules is equivalent to the category of $B$-modules $M$ with additional data, consisting of an isomorphism of $B \otimes_A B$-modules $\varphi : M \otimes_A B \to B \otimes_A M$ such that

$$(1 \otimes \varphi) \circ (\varphi \otimes 1) : M \otimes_A B \otimes_A B \to B \otimes_A B \otimes_A M$$

equals $\tilde{\varphi}$, given by tensoring $\varphi$ with $B \otimes_A B \otimes_A B$, regarded as a $B \otimes_A B$-algebra via the first and last factors.

It is easy to see that if $N$ is an $A$-module, and $M = B \otimes_A N$, then $\varphi_N : (B \otimes_A N) \otimes_A B \to B \otimes_A (B \otimes_A N)$ given by $\varphi((b_1 \otimes n) \otimes b_2) = b_1 \otimes (b_2 \otimes n)$ satisfies the cocycle condition. We claim that $N = \{x \in B \otimes_A N \mid \varphi_N(x \otimes 1) = 1 \otimes x\}$, which amounts to saying that the sequence of $A$-modules

$$0 \to N \xrightarrow{\alpha} B \otimes_A N \xrightarrow{\beta} B \otimes_A B \otimes_A N \quad \cdots (*)$$

is exact, where $\alpha(n) = 1 \otimes n$, and $\beta(x) = 1 \otimes x - \varphi_N(x \otimes 1)$, i.e.,

$$\beta(b \otimes n) = 1 \otimes b \otimes n - b \otimes 1 \otimes n.$$

Since $B$ is faithfully flat over $A$, it suffices to check the exactness after tensoring (over $A$) with $B$. The tensored sequence is

$$0 \to B \otimes_A N \xrightarrow{1 \otimes \alpha} B \otimes_A B \otimes_A N \xrightarrow{1 \otimes \beta} B \otimes_A B \otimes_A B \otimes_A N$$

But $1 \otimes \alpha$ is split by the map $\alpha' : b \otimes b' \otimes n \mapsto bb' \otimes n$. Similarly there is a map

$$\beta' : B \otimes_A B \otimes_A B \otimes_A N \to B \otimes_A B \otimes_A N,$$

$$b_1 \otimes b_2 \otimes b_3 \otimes n \mapsto b_1 b_2 \otimes b_3 \otimes n.$$

One sees at once that $\beta \circ \beta' + \alpha' \circ \alpha$ is the identity on $B \otimes_A B \otimes_A N$, so that the sequence has no homology at this term.

In fact one can show by a similar argument that the above 3 term sequence extends to a resolution of $A$ whose terms are the $A$-algebras $B^{\otimes n}$.

Now suppose given a $B$-module $M$ with descent data, i.e., a suitable $\varphi$. Let $N = \{x \in M \mid x \otimes 1 = \varphi(1 \otimes x)\}$. Then $N$ is an $A$-module, and there is a canonical $B$-linear map $f : B \otimes_A N \to M$. We claim $f$ is an isomorphism, and the map $\varphi$ corresponds to $\varphi_N$. By construction, the sequence

$$0 \to N \to M \xrightarrow{\gamma} B \otimes_A M$$

is exact, where $\gamma(x) = 1 \otimes x - \varphi(x \otimes 1)$. Since $B$ is faithfully flat over $A$, the tensored sequence

$$0 \to N \otimes_A B \to M \otimes_A B \xrightarrow{\gamma \otimes 1} B \otimes_A M \otimes_A B$$

is also exact. We see that the composite

$$N \otimes_A B \to M \otimes_A B \xrightarrow{\varphi} B \otimes_A M$$

is given by $n \otimes b \mapsto 1 \otimes bn$.

We claim that there is a commutative diagram

$$\begin{array}{ccc} M \otimes_A B & \xrightarrow{\gamma \otimes 1} & B \otimes_A M \otimes_A B \\ \varphi \downarrow & & \downarrow (1 \otimes \varphi) \\ B \otimes_A M & \xrightarrow{\mu} & B \otimes_A B \otimes_A M \end{array}$$

where $\mu(b \otimes m) = 1 \otimes b \otimes m - b \otimes 1 \otimes m$. Indeed,

$$(1 \otimes \varphi) \circ (\gamma \otimes 1)(m \otimes b) = (1 \otimes \varphi)(1 \otimes m \otimes b) - (1 \otimes \varphi)(\varphi(m \otimes 1) \otimes b)$$
$$= 1 \otimes \varphi(m \otimes b) - \widetilde{\varphi}(m \otimes b \otimes 1) = \mu \circ \varphi(m \otimes b),$$

where the second equality is because $\varphi$ satisfies the cocycle condition. Since the vertical arrows in the diagrams are isomorphisms, $\varphi$ induces an isomorphism $\ker(\gamma \otimes 1) \to \ker(\mu)$. But the sequence

$$0 \to M \to B \otimes_A M \xrightarrow{\mu} B \otimes_A B \otimes_A M$$

is exact, since $B \otimes_A B$ is a faithfully flat $B$-algebra; this is just the sequence corresponding to (∗) in this situation. Hence $\varphi$ induces an isomorphism $N \otimes_A B \to M$, which we computed earlier to be the map given by $n \otimes b \mapsto bn$. This means the canonical map $f : B \otimes_A N \to M$ is an isomorphism.

This is applied in the main text as follows. We say that $f : X \to S$ is a *Severi-Brauer* scheme over $S$ if there is a faithfully flat morphism $\pi : S' \to S$ such that $X' = X \times_S S' \to S'$ is isomorphic to $\mathbb{P}^r_{S'}$, for some $r \geq 0$. For example, we could be considering varieties over a field which become isomorphic to projective space over the algebraic closure of the field. Then after replacing $S'$ by a further faithfully flat extension, if needed, we may assume that the natural automorphism of $X' \times_{S'} X'$ obtained by switching the factors is induced by an element $\varphi$ of $\mathrm{GL}_{r+1}(\Gamma(\mathcal{O}'_S))$. Hence any $\mathrm{GL}_{r+1}$-equivariant quasi-coherent sheaf $\mathcal{F}_0$ on $\mathbb{P}^r_{\mathbb{Z}}$ will determine a quasi-coherent sheaf $\mathcal{F}'$ on $X'$, together with an isomorphism $\varphi^* : p_1^* \mathcal{F}' \to p_2^* \mathcal{F}'$. This isomorphism will in general only satisfy the cocycle condition up to composing with a scalar automorphism; however if $\mathcal{F}_0$ is a $\mathrm{PGL}_{r+1}$-equivariant sheaf, then $\mathcal{F}'$ satisfies the cocycle condition, and yields a quasi-coherent sheaf $\mathcal{F}$ on $X$; we say $\mathcal{F}$ is the sheaf on $X$ determined by $\mathcal{F}_0$ via descent theory.

As another illustration of these ideas, suppose $k$ is a field, $L$ a finite Galois extension, $G = \mathrm{Gal}(L/k)$. Then $B = L$ is a faithfully flat $A = k$-algebra, and by the normal basis theorem, $B \otimes_A B$ is naturally isomorphic to a direct product of copies $L_g$ of $L$ indexed by elements $g \in G$. A coherent sheaf on $\mathrm{Spec}\, B$ is just a finite dimensional $L$-vector space $V$; if $\dim V = n$ and we choose a basis for $V$, an isomorphism $\varphi : V \otimes_A B \to B \otimes_A V$ of $B \otimes_A B$-modules amounts to a function $G \to \mathrm{GL}_n(L)$, and the cocycle condition means that this is a cocycle in the sense of group cohomology, for the natural action of $G$ on $\mathrm{GL}_n(L)$ (see Serre, *Local Fields*, for example). The assertion that $V$ 'descends' to a $k$-vector space $V_0$ such that $B \otimes_A V_0 = L \otimes_k V_0 \cong V$, under which the 'obvious' isomorphism $\varphi_0 : B \otimes_A (B \otimes_A V_0) \to (B \otimes_A V_0) \otimes_A B$ corresponds to $\varphi$, just means that the above cocycle for group cohomology is a coboundary. Hence in this special situation, descent theory reduces to Hilbert's Theorem 90, and its generalization to $\mathrm{GL}_n$, i.e., that $\check{H}^1(L/k, \mathrm{GL}_n(L)) = 0$.

For a further discussion of descent theory, see Milne, *Étale Cohomology*, I, §2 and the more detailed references given there.

# Bibliography

H. Bass: *Algebraic K-Theory*, Benjamin, New York (1968).

A. A. Beilinson: Coherent sheaves on $\mathbb{P}^n$ and problems of linear algebra, *J. Funct. Anal. Appl.* 12 (1978) 214–215 (English translation).

S. Bloch, A. Ogus: Gersten's Conjecture and the Homology of Schemes, *Ann. Sci. E.N.S.* 7 (1974) 181–201.

S. Bloch: *Lectures on Algebraic Cycles*, Duke Univ. Math. Ser. IV, Durham (1976).

A. Borel: Stable real cohomology of arithmetic groups, *Ann. Sci. E.N.S.* 7 (1974) 235–272.

H. Cartan, S. Eilenberg: *Homological Algebra*, Princeton Math. Ser. 19, Princeton (1956).

A. Dold, R. Thom: Quasifaserungen und unendliche symmetrische Produkte, *Ann. Math.* 67 (1958) 239–281.

S. Dutta, M. Hochster, J. E. McLaughlin: Modules of finite projective dimension with negative intersection multiplicity, *Invent. Math.* 79 (1985) 253–291.

W. Fulton: *Intersection Theory*, Ergeb. Math. 3 (2), Springer-Verlag, New York (1984).

P. Gabriel: Des catégories abéliennes, *Bull. Math. Soc. France* 90 (1962) 323–448.

P. Gabriel, M. Zisman: *Calculus of Fractions and Homotopy Theory*, Ergeb. Math. 35, Springer-Verlag, New York (1967).

H. Gillet: Riemann–Roch Theorems for Higher Algebraic K-Theory, *Adv. Math. 40 (1981) 203–289.*

D. Grayson: *Higher Algebraic K-Theory II (after Daniel Quillen)*, Lect. Notes in Math. 551, Springer-Verlag, New York (1976).

D. Grayson: Localization for flat modules in algebraic K-theory, *J. Alg.* 61 (1979) 463–496.

P. A. Griffiths, J. W. Morgan: *Rational Homotopy Theory and Differential Forms*, Prog. Math. 16, Birkhäuser, Boston (1981).

R. Hartshorne: *Algebraic Geometry*, Grad. Texts in Math. 52, Springer-Verlag, New York (1977).

M. Levine: Modules of finite length and K-groups of surface singularities, *Compos. Math.* 59 (1986) 21–40.

M. Levine: Localization on Singular Varieties, *Invent. Math.* 91 (1988) 423–464.

J.-L. Loday: K-théorie algébrique et représentations de groupes, *Ann. Sci. E.N.S.* 9 (1976) 309–377.

J.-L. Loday: *Cyclic Homology*, Grundlehren Math. 301, Springer-Verlag, New York (1992).

S. Mac Lane: *Categories for the Working Mathematician*, Grad. Texts in Math. 5, Springer-Verlag, New Yrok (1972).

J. P. May: *Simplicial Objects in Algebraic Topology*, Midway Reprints, Univ. Chicago Press (1982).

J. P. May (with contributions by F. Quinn, N. Ray and J. Tornehave): $E_\infty$ ring spaces and $E_\infty$ ring spectra, Lect. Notes in Math. 533, Springer-Verlag, New York (1976).

A. S. Merkurjev, A. A. Suslin: $\mathcal{K}$-cohomology for Severi–Brauer varieties and the norm residue homomorphism, *Math. USSR Izv.* 21 (1983) 307–340 (English translation).

A. S. Merkurjev: $K_2$ of fields and the Brauer group, in *Applications of Algebraic K-Theory to Algebraic Geometry and Number Theory, Part II*, Contemp. Math. 55, A.M.S. (1986).

J. S. Milne: *Étale Cohomology*, Princeton Math. Ser. 33, Princeton (1980).

J. Milnor: The geometric realization of a semi-simplicial complex, *Ann. Math.* 65 (1957) 357–362.

J. Milnor: On spaces having the homotopy type of a CW complex, *Trans. A.M.S.* 90 (1959) 272–280.

J. Milnor: *Introduction to Algebraic K-Theory*, Ann. Math. Studies 72, Princeton University Press, Princeton (1971).

J. Milnor, J. C. Moore: On the structure of Hopf algebras, *Ann. Math.* 81 (1965) 211–264.

P. Olum: Obstructions to extensions and homotopies, *Ann. Math.* 52 (2950) 1–50.

A. I. Panin: On Algebraic K-theory of Generalized Flag Fibre Bundles and their Twisted Forms, in *Algebraic K-Theory*, ed. A. A. Suslin, Advances in Soviet Math. Vol. 4, A.M.S. (1991) 21–46. Amer. Math. Soc. (1991).

S. Priddy: On $\Omega^\infty S^\infty$ and the infinite symmetric group, *Proc. Symp. Pure Math.* 22, A.M.S. (1971) 217–220.

D. Quillen: On the cohomology and K-theory for the general linear group over a finite field, *Ann. Math.* 96 (1972) 552–586.

D. Quillen: *Higher Algebraic K-Theory I*, Lect. Notes in Math. 341, Springer-Verlag, New York (1973).

A. A. Rojtman: The torsion of the group of 0-cycles modulo rational equivalence, *Ann. Math.* 111 (1980) 553–569.

J. Rosenberg: *Algebraic K-Theory and Its Applications*, Grad. Texts in Math. 147, Springer-Verlag, New York (1994).

G. Segal: Classifying Spaces and Spectral Sequences, *Publ. Math. I.H.E.S.* 34 (1968) 105–112.

J.-P. Serre: *Local Fields*, Grad. Texts. in Math. 67, Springer-Verlag, New York (1979).

C. Soulé: $K_2$ et le groupe de Brauer (d'après A. S. Merkurjev et A. A. Suslin), Sém. Bourbaki 601, Nov. 1982, Astérique 105–106, Soc. Math. France (1983).

# Bibliography

V. Srinivas: Zero Cycles on a Singular Surface I, II, *J. Reine Ang. Math.* 359 (1985) 90–105 and 362 (1985) 4–27.

N. Steenrod: A convenient categroy of topological spaces, *Mich. Math. J.* 14 (1967) 133–152.

M. Stein: Surjective stability in dimension 0 for $K_2$ and related functors, *Trans. A.M.S.* 178 (1973) 165–191.

A. A. Suslin: Algebraic K-theory and the norm residue homomorphism, *J. Soviet Math.* 30 (1985) 2556–2611.

A. A. Suslin: Torsion in $K_2$ of fields, *K-Theory* 1 (1987) 5–29.

R. Swan: K-theory of quadric hypersurfaces, *Ann. Math.* 122 (1985) 113–153.

J. Tate: Relations between $K_2$ and Galois cohomology, *Invent. Math.* 36 (1976) 257–274.

R. W. Thomason: Beware the phony multiplication on Quillen's $\mathcal{A}^{-1}\mathcal{A}$, *Proc. Amer. Math. Soc.* 80 (1980) 569–573.

R. W. Thomason, T. Trobaugh: Higher Algebraic K-Theory of Schemes and of Derived Categories, in *The Grothendieck Festschrift, Volume III*, Prog. in Math. 88, Birkhäuser, Boston (1990).

W. van der Kallen: The $K_2$ of rings with many units, *Ann. Sci. E.N.S* 10 (1977) 473–515.

Sh. Wang: On the commutator subgroup of a simple algebra, *Amer. J. Math.* 72 (1950) 323–334.

A. Weil: *Basic Number Theory*, Grundlehren Math. 144, Springer-Verlag, New York (1974).

G. W. Whitehead: *Elements of Homotopy Theory*, Grad. Texts in Math. 61, Springer-Verlag, New York (1978).